PRACTICAL IMAGE AND VIDEO PROCESSING USING MATLAB®

PRACTICAL IMAGE AND VIDEO PROCESSING USING MATLAB®

OGE MARQUES
Florida Atlantic University

A JOHN WILEY & SONS, INC., PUBLICATION

About the Cover (by Roger Dalal)
The elegant Nautilus, with its progressive chambers and near-perfect logarithmic spiral, demonstrates the beauty of mathematics and the power of digital image processing. Created exclusively for *Practical Image and Video Processing Using MATLAB®*, this composition features multiple layers and processing techniques. The primary image is doubly sharpened with an 8 pixel radius, and enhanced with posterizing and edge detection algorithms. The outer, secondary image is indexed to 20 colors, pixelized at two percent resolution of the center image, and partially hidden by a fading, offset radial mask.

Published by John Wiley & Sons, Inc., Hoboken, New Jersey.
Published simultaneously in Canada.

For general information on our other products and services please contact our Customer Care Department with the U.S. at 877-762-2974, outside the U.S. at 317-572-3993 or fax 317-572-4002.

Wiley also publishes its books in a variety of electronic formats. Some content that appears in print, however, may not be available in electronic format.

Library of Congress Cataloging-in-Publication Data

Marques, Oge.
Practical image and video processing using MATLAB® / Oge Marques.
p. cm.
Includes bibliographical references and index.
ISBN 978-0-470-04815-3 (hardback)
1. Image processing–Mathematics. 2. Digital video–Mathematics. 3. Image processing–Digital techniques. 4. MATLAB®. I. Title.
TA1637.M3375 2011
502.85'66–dc22

2011008249

oBook ISBN: 978111093467
ePDF ISBN: 9781118093481
ePub ISBN: 9781118093474

Printed in The United States of America.

To my Son Nicholas, whose Precious Existence has Provided the Greatest Motivation to Pursue this Project.

And in Loving Memory of my Father, Ogê Aby Marques.

CONTENTS

PART II VIDEO PROCESSING

LIST OF FIGURES

LIST OF TABLES

FOREWORD

In packing for an office move earlier this year, I was struck by how many of my books (in many cases books sent to me free by their publishers for evaluation and potential use) were knockoffs: books that repackaged material that had been thoroughly consolidated in good textbooks years or even decades ago. Oge Marques' textbook is not a knockoff, even though much its subject matter has been around for years. It is a thoughtful and original compilation of material that is highly relevant today for students of imaging science, imaging technology, image understanding, and, foremost, image processing.

Imaging is my principal field of expertise. My interest in this book is great because imaging and image processing have grown together in recent years. Some forms of imaging—computational imaging is the buzzword that comes first to mind—presume that there will be (must be, in fact) postdetection processing of the raw information collected by the sensor system. Indeed, the sensor may output something that makes little or no sense to the observer in the absence of critical postprocessing operations.

Ultimately, in commercial mass-produced systems, image processing is implemented by specialized hardware. In the research and development stages of an imaging system, however, the processing is almost certain to be implemented using MATLAB®. Marques' book addresses this fact by linking directly to MATLAB® the many processing operations that are described.

There are of course numerous texts that describe digital image processing operations and algorithms. None, however, emphasizes as this one does the human vision system and the interaction and intercomparison between that system and machine vision systems.

The book contains a wealth of practical material and an invaluable up-to-date list of references, including journals, periodicals, and web sites. I would hope that the book sees subsequent editions, with correspondingly updated lists. Also invaluable to the teacher is the inclusion, now characteristic of many contemporary textbooks, of concise chapter summaries that address the question "What have we learned?" Tutorials punctuate the text, taking the students through important material in an active-learning process.

I write this foreword not just because I think this book addresses its intended image processing audience well but also because I see it influencing the thinking of my own students, students interested in imaging systems from a physics and technological standpoint but who must understand the relationship between imaging systems and image processing systems.

WILLIAM T. RHODES

William T. Rhodes is Emeritus Professor of Electrical & Computer Engineering at Georgia Institute of Technology. In 2005 he joined the Electrical Engineering faculty at Florida Atlantic University and became Associate Director of that university's Imaging Technology Center. A Fellow of the Optical Society of America and of the SPIE, he is editor-in-chief of the Springer Series in Optical Sciences and editor-in-chief of the online journal *SPIE Reviews*.

PREFACE

The prospect of using computers to emulate some of the attributes of the human visual system has attracted the interest of scientists, engineers, and mathematicians for more than 30 years, making the field of image processing one of the fastest-growing branches of applied computer science research. During the past 15 years, the fields of image and video processing have experienced tremendous growth and become more popular and accessible. This growth has been driven by several factors: widely available and relatively inexpensive hardware; a variety of software tools for image and video editing, manipulation, and processing; popularization of the Web and its strong emphasis on visual information, a true revolution in photography that has rendered film-based cameras all but obsolete; advances in the movie industry; and groundbreaking changes on the way we watch, record, and share TV programs and video clips.

APPROACH

This book provides a practical introduction to the most important topics in image and video processing, using MATLAB® (and its Image Processing Toolbox) as a tool to demonstrate relevant techniques and algorithms. The word *Practical* in its title is not meant to suggest a coverage of all the latest consumer electronics products in these fields; this knowledge would be superficial at best and would be obsolete shortly after (or even before!) the publication of the book. The word *Practical* should rather be interpreted in the sense of "enabling the reader/student to develop practical projects, that is, working prototypes, using the knowledge gained from the book." It also has other implications, such as the adoption of a "just enough math" philosophy, which favors the computational, algorithmic, and conceptual aspects of the techniques described along the book, over excessive mathematical formalism.

As a result, the book should appeal not only to its original target audience when used as a textbook (namely, upper-level undergraduate and early graduate students in Computer Science, Computer Engineering, Electrical Engineering, and related courses) but also to researchers and practitioners who have access to MATLAB®, solid computing/programming skills, and want to teach themselves the fundamentals of image and video processing.

KEY FEATURES

- This is the first book to combine image processing, video processing, and a practical, MATLAB®-oriented approach to experimenting with image and video algorithms and techniques.
- Complete, up-to-date, technically accurate, and practical coverage of essential topics in image and video processing techniques.
- 37 MATLAB® tutorials, which can be used either as step-by-step guides to exploring image and video processing techniques using MATLAB® on your own or as lab assignments by instructors adopting the textbook.
- More than 330 figures and 30 tables illustrating and summarizing the main techniques and concepts described in the text.
- This book adopts a "just enough math" philosophy. Many students are intimidated by image and video processing books with heavy emphasis on the mathematical aspects of the techniques. This book addresses this issue by offering the minimal mathematical treatment necessary to fully understand a technique without sacrificing the integrity of its explanation.
- The book emphasizes and encourages practical experimentation. After presenting a topic, it invites the readers to play on their own, reinforcing and expanding what they have just learned and venturing into new avenues along the same theme.
- The book has been designed to answer the most basic questions a student/reader is likely to have when first presented with a topic. It builds on my experience teaching image and video processing courses for 20 years, and the insights acquired along the way.
- The book includes many extra features to reinforce the understanding of its topics and allow the reader to learn more about them, such as exercises and programming projects, useful Web sites, and an extensive list of bibliographical references at the end of the chapters.

A TOUR OF THE BOOK

This book has been organized into two parts: *Image Processing* and *Video Processing*.

Part I (Image Processing) starts with an introduction and overview of the field (Chapter 1) that should motivate students to devote time and effort to the material in the remaining chapters. Chapter 2 introduces the fundamental concepts, notation, and terminology associated with image representation and basic image processing operations. Chapters 3 and 4 are devoted to MATLAB® and its Image Processing Toolbox, respectively, and establish the beginning of a series of chapters with hands-on activities, presented in the form of step-by-step tutorials at the end of each chapter from this point onward (except Chapter 5). Chapter 5 discusses the factors involved in image acquisition and digitization. Chapter 6 presents arithmetic and logic operations and introduces region of interest (ROI) processing. Chapter 7 covers geometric operations, such as resizing, rotation, cropping, and warping. Chapters 8–10 are devoted to point-based (Chapter 8), histogram-based (Chapter 9), and neighborhood-based (Chapter 10) image enhancement techniques. Chapter 11 extends the reach of image processing operations to the frequency domain and presents the Fourier transform and relevant frequency-domain image filtering techniques. Solutions to the problem of image restoration—particularly in cases of noise and blurring—are discussed in Chapter 12. Chapter 13 presents a detailed coverage of mathematical morphology and its use in image processing. Chapter 14 is devoted to edge detection techniques. Chapter 15 covers image segmentation. Chapter 16 transitions from grayscale to color images and presents representative color image processing techniques and algorithms. Image compression and coding, including the most recent and relevant standards, are the subject of Chapter 17. Chapter 18 looks at the problem of feature extraction and representation and leads naturally to Chapter 19 where the resulting feature vectors could be used for classification and recognition purposes.

Part II (Video Processing) starts by presenting the main concepts and terminology associated with analog video signals and systems and digital video formats and standards (Chapter 20). It then proceeds to describe the technically involved problem of standards conversion (Chapter 21). Chapter 22 discusses motion estimation and compensation techniques, shows how video sequences can be filtered, and concludes with an example of a simple solution to the problem of object detection and tracking in video sequences using MATLAB®.

The book contains two appendices. Appendix A presents selected aspects of the human visual system that bear implications in the design of image and video processing systems. Appendix B provides a tutorial on how to develop graphical user interfaces (GUIs) in MATLAB®.

NOTES TO INSTRUCTORS

This book can be used for upper-level undergraduate or introductory graduate courses in image and video processing, for one or two semesters. Most of the material included in this book has been extensively tested in many such courses during the past 20 years. The following is a summary of recommendations for instructors adopting this textbook.

Part I is organized around a typical machine vision system, from image acquisition to pattern classification. All chapters (except Chapters 16 and 17) in Part I follow a natural logic sequence, which covers all the steps involved in acquiring images, preprocessing them to remove imperfections or improve their properties, segmenting them into objects of interest, extracting objects' features, and classifying the objects into categories. The goal of Chapter 1 is to provide breadth, perspective, early examples of what can be achieved with image processing algorithms, and a systemic view of what constitutes a machine vision system. Some instructors may want to combine this information with the material from Chapter 2 as they introduce the topic early in their courses.

The material from Chapters 3 and 4 has been carefully selected to make the book self-contained, providing students all the MATLAB® and Image Processing Toolbox information they might need for the corresponding tutorials. Readers will likely keep these two chapters for future reference should they ever require MATLAB®-related help later in the course. Instructors with limited lecture time may choose to cover both chapters briefly, assign the corresponding tutorials, and monitor students' progress as they work on the tutorials and answer the associated questions.

Chapter 5 briefly introduces the topic of image sensing and acquisition. Its main goal is to equip the reader with information on the steps needed to convert a three-dimensional (3D) real-world scene into a two-dimensional (2D) digitized version of it. Instructors teaching courses with a strong emphasis on image capture and acquisition hardware may want to supplement this material with detailed references, for example, on sensors that operate outside the visible spectrum, stereo-mounted cameras, camera calibration, and many other topics.

Chapters 6–10 are straightforward and cover essential topics in any image processing class. They also provide room for many interesting discussions, lab assignments, and small projects.

Chapter 11 may be a bit challenging to some students, due to the mathematical formalism associated with the Fourier transform. Instructors may find the interactive MATLAB® frequency-domain demo (`fddemo`) introduced in that chapter a valuable tool to develop students' confidence on their understanding of the basic concepts of frequency-domain filtering techniques. Chapter 12 builds on the knowledge from Chapters 10 and 11, with focus on noise reduction and deblurring techniques. Some instructors may prefer to tone down the discussion of noise models (Sections 12.1 and 12.2) and present the techniques described in Sections 12.3–12.5 earlier on as applications of spatial-domain and frequency-domain filtering techniques.

Chapter 13 is self-contained, which gives instructors the flexibility to adjust their level of coverage—from skipping it altogether, to covering it in detail—without major impact on the other topics in their courses.

Chapters 14 and 15 provide introductory coverage of two essential topics in any image processing course. Instructors who want to present some of these contents earlier in the course or in a different sequence should be able to easily extract the associated sections and move them to a different point in time.

Chapter 16 comprises information on color image processing and is somehow related to earlier chapters (particularly, those on enhancement, segmentation, and edge extraction). We have made a conscious decision of keeping color in a separate chapter rather than spreading color image processing throughout the text. We believe that by the time readers reach Chapter 16, they will be able to easily navigate through its contents focusing on the differences between what they learned earlier for grayscale images and later for their color equivalent. Instructors who do not agree with this decision can easily bring sections of Chapter 16 to an earlier point in their courses.

Chapter 17 deals with image compression and coding, very extensive and technically complex topics upon which entire books have been written. Since the focus of the book is on building *practical* image processing and machine vision solutions using MATLAB®, we decided to approach the topic of image coding and compression from a broad perspective (standards in use today, categories of compression techniques and their chief characteristics, etc.) instead of attempting to embed a deeper discussion of these topics that could be potentially distracting and would most likely add little value. From a pragmatic viewpoint, since the reader's goal is to process images using MATLAB® and its rich capabilities for reading and writing images from/to a wide variety of formats (most of which use some type of compression), we focused on how to use these capabilities in a meaningful way. Instructors may want to proceed in different ways, depending on their goals, ranging from expanding the material in Chapter 17 with additional references (if image coding and compression is an important part of their course syllabus) to skipping the chapter altogether (if the course's main goal is to build a machine vision solution to a practical problem, which probably would not require that type of knowledge).

Chapters 18 and 19 are tightly interrelated. They provide the information needed to design and implement two of the most critical stages of image processing and machine vision solutions: feature extraction and pattern classification. Chapter 18 offers a wide array of choices for feature extraction and representation techniques, depending on the type of image and the specific needs of the solution being designed. Instructors may appreciate the fact that Chapter 19 provides all the basic concepts that students may need from the associated fields of pattern recognition, data mining, and information retrieval, without requiring additional references. This is particularly important if the course does not enforce prerequisites in any of these areas. The tutorial at the end of Chapter 19 was created to put the selection, design, and fine-tuning of the algorithms presented in Chapters 18 and 19 under perspective. It is my hope that at this point in the book, students will not only be fluent in MATLAB® and image processing but will also have acquired the ability to look back and reflect critically on what works, what does not, and why.

Part II is organized in three chapters, which can be used in the later part of a one- or two-semester course that combines image and video processing or at the early stages of course devoted exclusively to video processing. In the latter case, the instructor may want to supplement the material in Part II with additional references (e.g., scholarly papers in video processing and related topics in the case of graduate-level courses).

Chapter 20 covers a very broad range of topics, from basic analog video concepts to digital video standards and codecs. It offers room for expansion in multiple directions, from a deeper study of TV broadcasting systems to a more detailed analysis of contemporary video compression schemes and standards. Chapter 21 covers the topic of standards conversion and discusses the most popular techniques used to accomplish it. Chapter 22 expands the discussion to include motion estimation and compensation, as well as (interframe and intraframe) video filtering techniques. It concludes with a practical project implemented in MATLAB® by one of my former students: an object detection and tracking system in video sequences with fixed camera and moving, complex, background. The goal of including this case study is to conclude the discussion of Part II (and the book) reminding the reader that at this point they should be knowledgeable enough to attempt similar projects (which instructors may assign as end-of-course projects).

The material in Appendix A is very relevant to image and video processing systems because it explains the relationship between properties of the human visual system and their impact on design decisions involved in building such systems. Some instructors may choose to present (part of) it earlier in their courses.

Appendix B is a practical guide to the development of GUIs for MATLAB® applications. It should empower students to develop visually attractive, interactive, and functional interfaces to their MATLAB® projects.

A note about MATLAB® and the tutorials at the end of chapters. Having used MATLAB® (and its Image Processing Toolbox) for more than a decade, I wholeheartedly agree with Rudra Pratap [Pra02] who wrote, "MATLAB®'s ease of use is its main feature." MATLAB® has a shallow learning curve, which allows the user to engage in an interactive learning style that accommodates the right degree of challenge needed to raise the user's skills by a certain amount, and so on, in a staircase-like progression. The MATLAB® tutorials included in this book have been conceived under this philosophy.

Web site

The book's companion web site (http://www.ogemarques.com) contains many supplementary materials for students and instructors: MATLAB® code for all tutorials in the book, MATLAB® code for selected figures, test images and video sequences, supplementary problems, tutorials, and projects (that could not make it to the printed version), and an ever-growing and frequently maintained list of useful web sites—including (links to) image processing conferences, software, hardware, research groups, test image databases, and much more.

OGE MARQUES

Boca Raton, FL

ACKNOWLEDGMENTS

I am deeply grateful to many people who have collaborated—directly or indirectly—on this project. This book would not have been possible without their help.

I want to thank many inspiring professors, supervisors, and colleagues who have guided my steps in the fields of image and video processing and related areas, particularly Maria G. Te Vaarwerk, Wim Hoeks, Bart de Greef, Eric Persoon, John Bernsen, Borko Furht, and Bob Cooper.

A very special thank you to my friend and colleague Hugo Vieira Neto, who has been a great supporter of this project from its early planning stages to its completion.

I am deeply indebted to Gustavo Benvenutti Borba for his excellent work in the creation of most of the figures in this book, his insightful reviews and comments, and continuous encouragement, investing many hours of his time in return for not much more than these few lines of thanks.

Many thanks to Liam M. Mayron for his encouragement, support, and expert help throughout all the steps of this project.

This book could not have been produced without the invaluable contributions of Jeremy Jacob, who wrote, revised, and documented most of the MATLAB® code associated with the book and its tutorials and also contributed the contents for Appendix B.

Special thanks to the MathWorks Book Program (Courtney Esposito, Naomi Fernandes, and Meg Vulliez) for their support over the years.

Several friends and colleagues reviewed draft versions of selected portions of the text: Liam M. Mayron, Hugo Vieira Neto, Mathias Lux, Gustavo Benvenutti Borba, Pierre Baillargeon, Humberto Remigio Gamba, Vladimir Nedovic, Pavani Chilamakuri, and Joel Gibson. I would like to thank them for their careful reviews and insightful comments and suggestions. I have done my best to correct the mistakes they pointed out and improve the contents of the book according to their suggestions. If any error remains, it is entirely my responsibility, not theirs. If you should find any errors, please e-mail me at omarques@ieee.org, and I will correct them in future printings of this book.

My biggest thanks to my publisher George J. Telecki and his wonderful staff at John Wiley & Sons, Inc. who have patiently worked with me throughout the lifetime of this project: Lucy Hitz, Rachel Witmer, and Melissa Valentine. Their kindness and professionalism have made the creation of this book a very enjoyable process.

Thanks to Amy Hendrickson (TeXnology Inc.) for her expert help with LaTeX issues.

I am also indebted to Roger Dalal who designed the inspiring cover art for the book.

Last but certainly not least, I want to thank my family for their unfailing love, patience, and understanding.

Oge Marques

PART I

IMAGE PROCESSING

CHAPTER 1

INTRODUCTION AND OVERVIEW

WHAT WILL WE LEARN?

- What is image processing?
- What are the main applications of image processing?
- What is an image?
- What is a digital image?
- What are the goals of image processing algorithms?
- What are the most common image processing operations?
- Which hardware and software components are typically needed to build an image processing system?
- What is a machine vision system (MVS) and what are its main components?
- Why is it so hard to emulate the performance of the human visual system (HVS) using cameras and computers?

1.1 MOTIVATION

Humans have historically relied on their vision for tasks ranging from basic instinctive survival skills to detailed and elaborate analysis of works of art. Our ability to guide our actions and engage our cognitive abilities based on visual input is a remarkable trait of the human species, and much of how exactly we do what we do—and seem to do it so well—remains to be discovered.

Practical Image and Video Processing Using MATLAB®. By Oge Marques.

The need to extract information from images and interpret their contents has been one of the driving factors in the development of image processing[1] and computer vision during the past decades.

Image processing applications cover a wide range of human activities, such as the following:

- *Medical Applications*: Diagnostic imaging modalities such as digital radiography, PET (positron emission tomography), CAT (computerized axial tomography), MRI (magnetic resonance imaging), and fMRI (functional magnetic resonance imaging), among others, have been adopted by the medical community on a large scale.
- *Industrial Applications*: Image processing systems have been successfully used in manufacturing systems for many tasks, such as safety systems, quality control, and control of automated guided vehicles (AGVs).
- *Military Applications*: Some of the most challenging and performance-critical scenarios for image processing solutions have been developed for military needs, ranging from detection of soldiers or vehicles to missile guidance and object recognition and reconnaissance tasks using unmanned aerial vehicles (UAVs). In addition, military applications often require the use of different imaging sensors, such as range cameras and thermographic forward-looking infrared (FLIR) cameras.
- *Law Enforcement and Security*: Surveillance applications have become one of the most intensely researched areas within the video processing community. Biometric techniques (e.g., fingerprint, face, iris, and hand recognition), which have been the subject of image processing research for more than a decade, have recently become commercially available.
- *Consumer Electronics*: Digital cameras and camcorders, with sophisticated built-in processing capabilities, have rendered film and analog tape technologies obsolete. Software packages to enhance, edit, organize, and publish images and videos have grown in sophistication while keeping a user-friendly interface. High-definition TVs, monitors, DVD players, and personal video recorders (PVRs) are becoming increasingly popular and affordable. Image and video have also successfully made the leap to other devices, such as personal digital assistants (PDAs), cell phones, and portable music (MP3) players.
- *The Internet, Particularly the World Wide Web*: There is a huge amount of visual information available on the Web. Collaborative image and video uploading, sharing, and annotation (tagging) have become increasingly popular. Finding and retrieving images and videos on the Web based on their contents remains an open research challenge.

[1]From this point on, the use of the phrase *image processing* should be interpreted as *digital image processing*. We shall only use the *digital* qualifier when it becomes relevant (e.g., after an analog image has been converted to a digital representation).

1.2 BASIC CONCEPTS AND TERMINOLOGY

In this section, we define the most frequently used terms in Part I of this book. Although there is no universal agreement on the terminology used in this field, the definitions presented here are consistently used throughout the book. This section is structured in a question-and-answer format.

What Is an Image?

An *image* is a visual representation of an object, a person, or a scene produced by an optical device such as a mirror, a lens, or a camera. This representation is two dimensional (2D), although it corresponds to one of the infinitely many projections of a real-world, three-dimensional (3D) object or scene.

What Is a Digital Image?

A digital image is a representation of a two-dimensional image using a finite number of points, usually referred to as *picture elements*, *pels*, or *pixels*. Each pixel is represented by one or more numerical values: for monochrome (grayscale) images, a single value representing the intensity of the pixel (usually in a [0, 255] range) is enough; for color images, three values (e.g., representing the amount of red (R), green (G), and blue (B)) are usually required. Alternative ways of representing color images, such as the *indexed color image* representation, are described in Chapter 2.

What Is Digital Image Processing?

Digital image processing can be defined as the science of modifying digital images by means of a digital computer. Since both the images and the computers that process them are digital in nature, we will focus exclusively on *digital* image processing in this book. The changes that take place in the images are usually performed *automatically* and rely on carefully designed algorithms. This is in clear contrast with another scenario, such as touching up a photo using an airbrush tool in a photo editing software, in which images are processed *manually* and the success of the task depends on human ability and dexterity. We refer to the latter as *image manipulation* to make this distinction more explicit.

What Is the Scope of Image Processing?

In this book, we adopt the terminology used in [GW08] (among others) and employ the term *image processing* to refer to all the techniques and applications described in Part I of this book, whether the output is a modified (i.e., processed) version of the input image, an encoded version of its main attributes, or a nonpictorial description of its contents.

Moreover, we distinguish among three levels of image processing operations [GW08]:

- *Low Level*: Primitive operations (e.g., noise reduction, contrast enhancement, etc.) where both the input and the output are images.
- *Mid Level*: Extraction of attributes (e.g., edges, contours, regions, etc.) from images.
- *High Level*: Analysis and interpretation of the contents of a scene.

This book does not cover the area of computer graphics or *image synthesis*, the process by which a 2D or 3D image is rendered from numerical data. In fact, we are often interested in the opposite process (sometimes referred to as *image analysis*), by which textual and numerical data can be extracted from an array of pixels.

Image processing is a multidisciplinary field, with contributions from different branches of science (particularly mathematics, physics, and computer science) and computer, optical, and electrical engineering. Moreover, it overlaps other areas such as pattern recognition, machine learning, artificial intelligence, and human vision research. This combination of cross-disciplinary research and intersecting fields can be seen in the list of magazines and journals presented in Section 1.6.

1.3 EXAMPLES OF TYPICAL IMAGE PROCESSING OPERATIONS

Image processing covers a wide and diverse array of techniques and algorithms, which will be described in detail in the remainder of Part I of this book. In this section, we provide a preview of the most representative image processing operations that you will learn about in forthcoming chapters.

1. *Sharpening (Figure 1.1)*: A technique by which the edges and fine details of an image are enhanced for human viewing. Chapters 8–10 will discuss how this is done in the spatial domain, whereas Chapter 11 will extend the discussion to frequency-domain techniques.
2. *Noise Removal (Figure 1.2)*: Image processing filters can be used to reduce the amount of noise in an image before processing it any further. Depending on the type of noise, different noise removal techniques are used, as we will learn in Chapter 12.
3. *Deblurring (Figure 1.3)*: An image may appear blurred for many reasons, ranging from improper focusing of the lens to an insufficient shutter speed for a fast-moving object. In Chapter 12, we will look at image deblurring algorithms.
4. *Edge Extraction (Figure 1.4)*: Extracting edges from an image is a fundamental preprocessing step used to separate objects from one another before identifying their contents. Edge detection algorithms and techniques are discussed in Chapter 14.

(a) (b)

FIGURE 1.1 Image sharpening: (a) original image; (b) after sharpening.

5. *Binarization (Figure 1.5)*: In many image analysis applications, it is often necessary to reduce the number of gray levels in a monochrome image to simplify and speed up its interpretation. Reducing a grayscale image to only two levels of gray (black and white) is usually referred to as *binarization*, a process that will be discussed in more detail in Chapter 15.

(a) (b)

FIGURE 1.2 Noise removal: (a) original (noisy) image; (b) after removing noise.

(a) (b)

FIGURE 1.3 Deblurring: (a) original (blurry) image; (b) after removing the (motion) blur. Original image: courtesy of MathWorks.

6. *Blurring (Figure 1.6)*: It is sometimes necessary to blur an image in order to minimize the importance of texture and fine detail in a scene, for instance, in cases where objects can be better recognized by their shape. Blurring techniques in spatial and frequency domain will be discussed in Chapters 10 and 11, respectively.
7. *Contrast Enhancement (Figure 1.7)*: In order to improve an image for human viewing as well as make other image processing tasks (e.g., edge extraction)

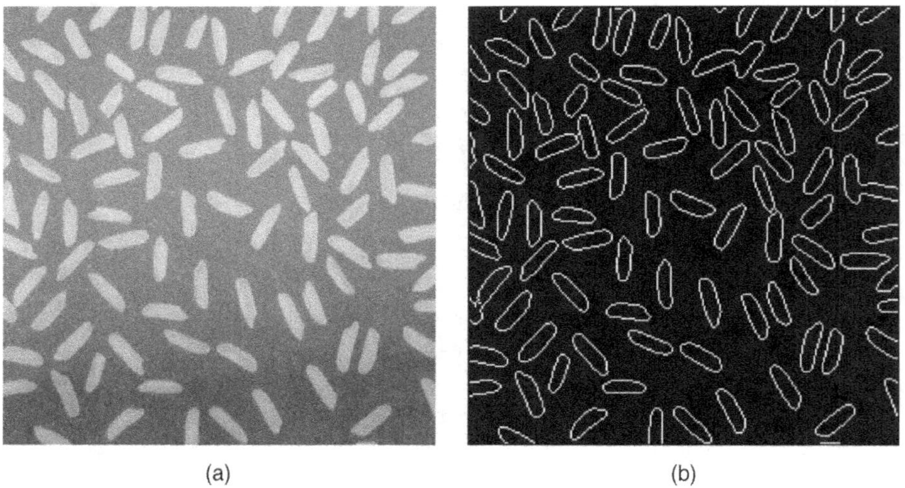

(a) (b)

FIGURE 1.4 Edge extraction: (a) original image; (b) after extracting its most relevant edges. Original image: courtesy of MathWorks.

FIGURE 1.5 Binarization: (a) original grayscale image; (b) after conversion to a black-and-white version. Original image: courtesy of MathWorks.

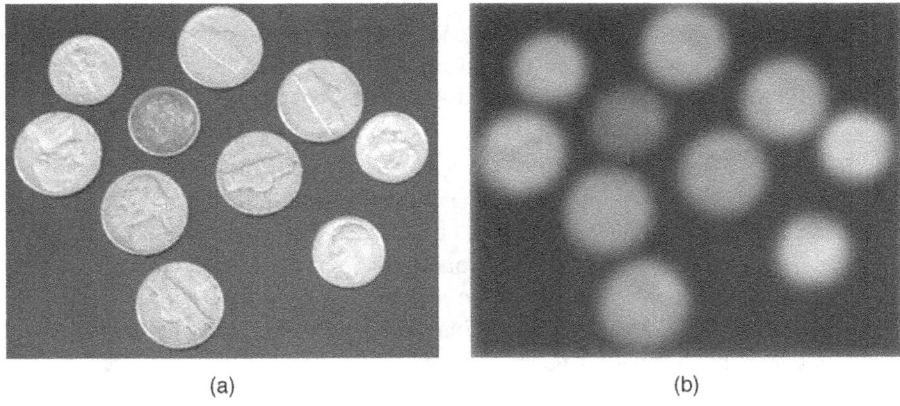

FIGURE 1.6 Blurring: (a) original image; (b) after blurring to remove unnecessary details. Original image: courtesy of MathWorks.

FIGURE 1.7 Contrast enhancement: (a) original image; (b) after histogram equalization to improve contrast.

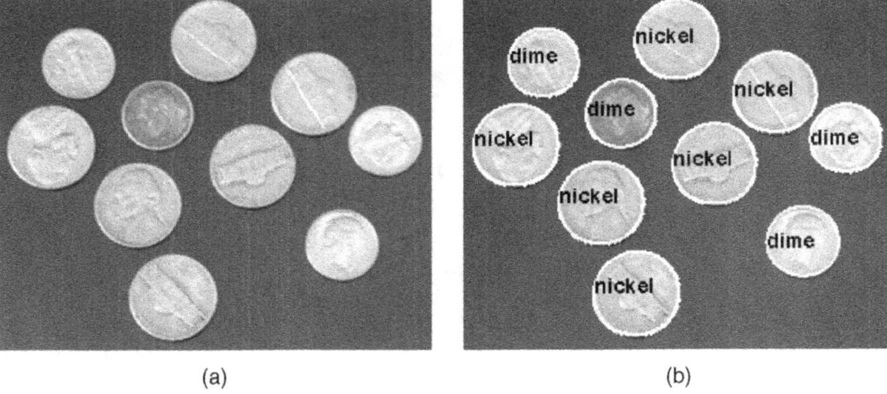

(a) (b)

FIGURE 1.8 Object segmentation and labeling: (a) original image; (b) after segmenting and labeling individual objects. Original image: courtesy of MathWorks.

easier, it is often necessary to enhance the contrast of an image. Contrast enhancement techniques using transformation functions and histogram processing will be discussed in Chapters 8 and 9, respectively.

8. *Object Segmentation and Labeling (Figure 1.8)*: The task of segmenting and labeling objects within a scene is a prerequisite for most object recognition and classification systems. Once the relevant objects have been segmented and labeled, their relevant features can be extracted and used to classify, compare, cluster, or recognize the objects in question. Segmentation and labeling of connected components from an image will be discussed in Chapters 13 and 15. Feature extraction and representation, and pattern recognition will be covered in Chapters 18 and 19, respectively.

1.4 COMPONENTS OF A DIGITAL IMAGE PROCESSING SYSTEM

In this section, we present a generic digital image processing system and discuss its main building blocks (Figure 1.9). The system is built around a computer in which most image processing tasks are carried out, but also includes hardware and software for image acquisition, storage, and display. The actual hardware associated with each block in Figure 1.9 changes as technology evolves. In fact, even contemporary digital still cameras can be modeled according to that diagram: the CCD sensor corresponds to the *Acquisition* block, flash memory is used for *storage*, a small LCD monitor for *display*, and the digital signal processor (DSP) chip becomes the '*Computer*', where certain image processing operations (e.g., conversion from RAW format to JPEG[2]) take place.

[2]See Section 2.2 for information on image file formats.

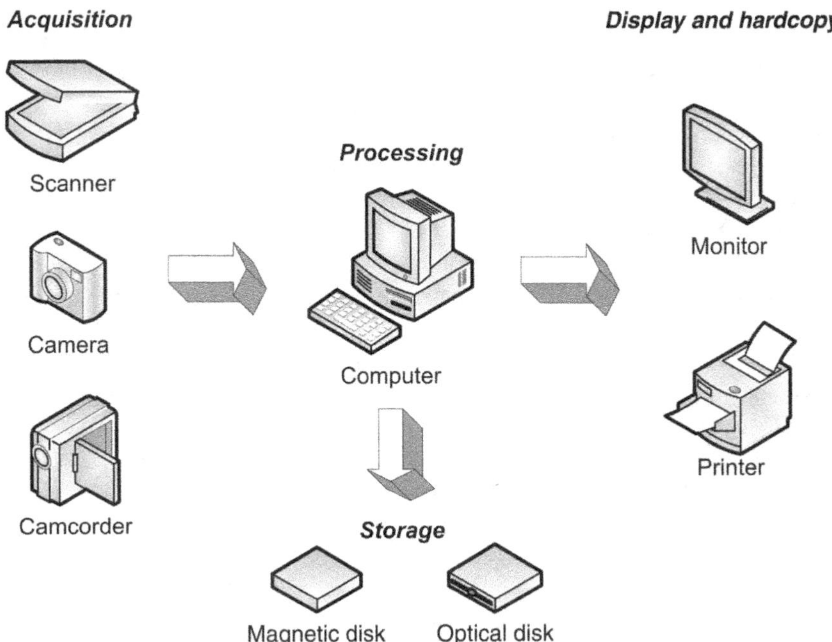

FIGURE 1.9 Components of a digital image processing system. Adapted and redrawn from [Umb05].

Hardware

The hardware components of a digital image processing system typically include the following:

- *Acquisition Devices*: Responsible for capturing and digitizing images or video sequences. Examples of general-purpose acquisition devices include scanners, cameras, and camcorders. Acquisition devices can be interfaced with the main computer in a number of ways, for example, USB, FireWire, Camera Link, or Ethernet. In cases where the cameras produce analog video output, an image digitizer—usually known as *frame grabber*—is used to convert it to digital format.
- *Processing Equipment*: The main computer itself, in whatever size, shape, or configuration. Responsible for running software that allows the processing and analysis of acquired images.
- *Display and Hardcopy Devices*: Responsible for showing the image contents for human viewing. Examples include color monitors and printers.
- *Storage Devices*: Magnetic or optical disks responsible for long-term storage of the images.

Software

The software portion of a digital image processing system usually consists of modules that perform specialized tasks. The development and fine-tuning of software for image processing solutions is iterative in nature. Consequently, image processing researchers and practitioners rely on programming languages and development environments that support modular, agile, and iterative software development.

In this book, the software of choice is MATLAB® (MATrix LABoratory), a multiplatform, data analysis, prototyping, and visualization tool with built-in support for matrices and matrix operations, rich graphics capabilities, and a friendly programming language and development environment. MATLAB offers programmers the ability to edit and interact with the main functions and their parameters, which leads to valuable time savings in the software development cycle.

MATLAB has become very popular with engineers, scientists, and researchers in both industry and academia, due to many factors, such as the availability of *toolboxes* containing specialized functions for many application areas, ranging from data acquisition to image processing (which is the main focus of our interest and will be discussed in Chapter 4).

1.5 MACHINE VISION SYSTEMS

In this section, we introduce the main components of a machine vision system (Figure 1.10) using a practical example application: recognizing license plates at a highway toll booth. Image processing is not a one-step process: most solutions follow a sequential processing scheme whose main steps are described next.

The *problem domain*, in this case, is the automatic recognition of license plates. The goal is to be able to extract the alphanumeric contents of the license plate of a vehicle passing through the toll booth in an automated and unsupervised way, that is, without need for human intervention. Additional requirements could include 24/7 operation (under artificial lighting), all-weather operation, minimal acceptable success rate, and minimum and maximum vehicle speed.

The *acquisition* block is in charge of acquiring one or more images containing a front or rear view of the vehicle that includes its license plate. This can be implemented using a CCD camera and controlling the lighting conditions so as to ensure that the image will be suitable for further processing. The output of this block is a digital image that contains a (partial) view of the vehicle. Several factors should be considered in the design of this block and will likely impact the quality of the resulting image as well as the performance of the whole system, such as the maximum speed allowed for the vehicle without risk of blurring the picture, illumination aspects (e.g., number, type, and positioning of light sources), choice of lenses, and the specification (resolution and speed) of the image digitizer hardware.

The goal of the *preprocessing* stage is to improve the quality of the acquired image. Possible algorithms to be employed during this stage include contrast improvement, brightness correction, and noise removal.

The *segmentation* block is responsible for partitioning an image into its main components: relevant foreground objects and background. It produces at its output a number of labeled regions or "subimages." It is possible that in this particular case segmentation will be performed at two levels: (1) extracting the license plate from the rest of the original image; and (2) segmenting characters within the plate area. Automatic image segmentation is one of the most challenging tasks in a machine vision system.

The *feature extraction* block (also known as *representation and description*) consists of algorithms responsible for encoding the image contents in a concise and descriptive way. Typical features include measures of color (or intensity) distribution, texture, and shape of the most relevant (previously segmented) objects within the image. These features are usually grouped into a *feature vector* that can then be used as a numerical indicator of the image (object) contents for the subsequent stage, where such contents will be recognized (classified).

Once the most relevant features of the image (or its relevant objects, in this case individual characters) have been extracted and encoded into a feature vector, the next step is to use this K-dimensional numerical representation as an input to the *pattern classification* (also known as *recognition and interpretation*) stage. At this point, image processing meets classical pattern recognition and benefits from many of its tried-and-true techniques, such as minimum distance classifiers, probabilistic classifiers, neural networks, and many more. The ultimate goal of this block is to classify (i.e., assign a label to) each individual character, producing a string (or ASCII file) at the output, containing the license plate contents.

In Figure 1.10, all modules are connected to a large block called *knowledge base*. These connections—inspired by a similar figure in [GW08]—are meant to indicate that the successful solution to the license plate recognition problem will depend on how much knowledge about the problem domain has been encoded and stored in the MVS. The role of such knowledge base in the last stages is quite evident (e.g., the

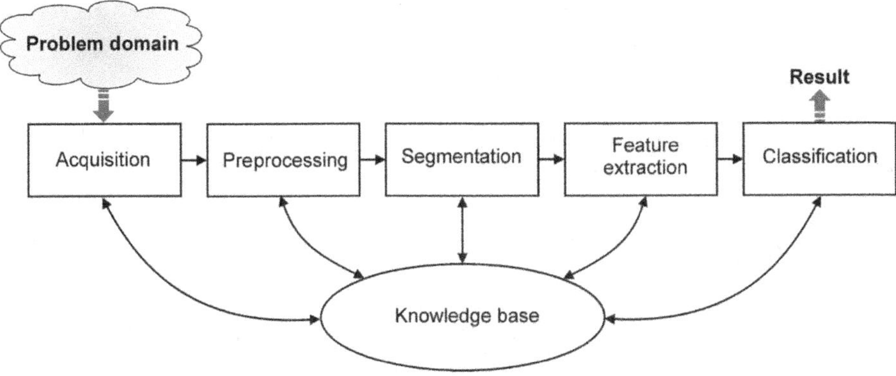

FIGURE 1.10 Diagram of a machine vision system. Adapted and redrawn from [GW08].

knowledge that the first character must be a digit may help disambiguate between a "0" and an "O" in the *pattern classification* stage). In a less obvious way, the knowledge base should (ideally) help with all tasks within the MVS. For example, the *segmentation* block could benefit from rules specifying known facts about the license plates, such as shape and aspect ratio, most likely location within the original image, number of characters expected to appear within the plate, size and position information about the characters, and relevant background patterns that may appear in the plate area.

The human visual system and a machine vision system have different strengths and limitations and the designer of an MVS must be aware of them. A careful analysis of these differences provides insight into why it is so hard to emulate the performance of the human visual system using cameras and computers. Three of the biggest challenges stand out:

- The HVS can rely on a very large database of images and associated concepts that have been captured, processed, and recorded during a lifetime. Although the storage of the images themselves is no longer an expensive task, mapping them to high-level semantic concepts and putting them all in context is a very hard task for an MVS, for which there is no solution available.
- The very high speed at which the HVS makes decisions based on visual input. Although several image processing and machine vision tasks can be implemented at increasingly higher speeds (often using dedicated hardware or fast supercomputers), many implementations of useful algorithms still cannot match the speed of their human counterpart and cannot, therefore, meet the demands of real-time systems.
- The remarkable ability of the HVS to work under a wide range of conditions, from deficient lighting to less-than-ideal perspectives for viewing a 3D object. This is perhaps the biggest obstacle in the design of machine vision systems, widely acknowledged by everyone in the field. In order to circumvent this limitation, most MVS must impose numerous constraints on the operating conditions of the scene, from carefully controlled lighting to removing irrelevant distractors that may mislead the system to careful placing of objects in order to minimize the problems of shades and occlusion.

Appendix A explores selected aspects of the HVS in more detail.

1.6 RESOURCES

In this section, we have compiled a list of useful resources for readers who want to deepen their knowledge and tackle more advanced concepts, algorithms, and mathematical constructs. A similar section on video processing resources appears in Chapter 20.

Books

The following is a list of selected books on image processing and related fields:

- Burger, W. and Burge, M. J., *Digital Image Processing: An Algorithmic Introduction Using Java*, New York: Springer, 2008.
- Gonzalez, R. C. and Woods, R. E., *Digital Image Processing*, 3rd ed., Upper Saddle River, NJ: Prentice Hall, 2008.
- Sonka, M, Hlavac, V., and Boyle, R., *Image Processing, Analysis, and Computer Vision*, 3rd ed., Toronto, Ontario: Thomson Learning, 2008.
- Pratt, W. K., *Digital Image Processing*, 4th ed., New York: Wiley–Interscience, 2007.
- Jahne, B., *Digital Image Processing: Concepts, Algorithms, and Scientific Applications*, 6th ed., Berlin: Springer, 2005.
- Davies, E. R., *Machine Vision: Theory, Algorithms, Practicalities*, 3rd ed., San Francisco, CA: Morgan Kaufmann, 2005.
- Umbaugh, S. E., *Computer Imaging: Digital Image Analysis and Processing*, Boca Raton, FL: CRC Press, 2005.
- Gonzalez, R. C., Woods, R. E., and Eddins, S. L., *Digital Image Processing Using MATLAB*, Upper Saddle River, NJ: Prentice Hall, 2004.
- McAndrew, A., *Introduction to Digital Image Processing with MATLAB*, Boston, MA: Thomson, 2004.
- Forsyth, D. F. and Ponce, J., *Computer Vision: A Modern Approach*, Upper Saddle River, NJ: Prentice Hall, 2002.
- Duda, R. O., Hart, P. E., and Stork, D. G., *Pattern Classification*, 2nd ed., New York: Wiley, 2001.
- Shapiro, L. G. and Stockman, G. C., *Computer Vision*, Upper Saddle River, NJ: Prentice Hall, 2001.
- Bovik, A. (Ed.), *Handbook of Image and Video Processing*, San Diego, CA: Academic Press, 2000.
- Efford, N., *Digital Image Processing: A Practical Introduction Using Java*, Harlow, UK: Addison-Wesley, 2000.
- Seul, M., O'Gorman, L., and Sammon, M. J., *Practical Algorithms for Image Analysis*, Cambridge, UK: Cambridge University Press, 2000.
- Schalkoff, R. J., *Digital Image Processing and Computer Vision*, New York: Wiley, 1989.

Magazines and Journals

The following are some of the magazines and journals that publish research results in image and video processing and related areas (in alphabetical order): *Artificial Intelligence*, *Computer Vision and Image Understanding*, *EURASIP Journal on Advances in Signal Processing*, *EURASIP Journal on Image and Video Processing*,

IEEE Multimedia Magazine, *IEEE Transactions on Circuits and Systems for Video Technology*, *IEEE Transactions on Image Processing*, *IEEE Transactions on Medical Imaging*, *IEEE Transactions on Multimedia*, *IEEE Transactions on Pattern Analysis and Machine Intelligence*, *Image and Vision Computing*, *International Journal of Computer Vision*, *Journal of Electronic Imaging (SPIE and IS&T)*, *Journal of Mathematical Imaging and Vision*, *Journal of Visual Communication and Image Representation*, *Machine Vision and Applications*, *Pattern Recognition*, and *Signal Processing: Image Communication*.

In addition, there are a number of useful trade magazines, such as *Advanced Imaging Magazine*, *Imaging and Machine Vision Europe Magazine*, *Studio Monthly*, and *Vision Systems Design Magazine*.

Web Sites

This section contains a selected number of useful web sites. Some of them are portals to image processing and computer vision information (e.g., conferences, research groups, and publicly available databases) and contain many links to other relevant sites.

Since web pages often move and change URLs (and some disappear altogether), the current URL for the web sites suggested in the book will be available (and maintained up-to-date) in the book's companion web site (http://www.ogemarques.com). The book's companion web site will also add relevant sites that were not available at the time of this writing.

- *CVonline*: This ever-growing Compendium of Computer Vision, edited and maintained by Professor Robert B. Fisher, School of Informatics, University of Edinburgh, UK, includes a collection of hypertext summaries on more than 1400 topics in computer vision and related subjects.
  http://homepages.inf.ed.ac.uk/rbf/CVonline
- *Computer Vision Home Page*: Arguably the best-known portal for image processing research. Excellent starting point when searching for publications, test images, research groups, and much more.
  http://www.cs.cmu.edu/ cil/vision.html
- *USC Annotated Computer Vision Bibliography*: An extensive and structured compilation of relevant bibliography in the fields of computer vision and image processing.
  http://iris.usc.edu/Vision-Notes/bibliography/contents.html
- *USC Computer Vision Conference Listing*: A structured list of computer vision and image processing conferences and pertinent details.
  http://iris.usc.edu/Information/Iris-Conferences.html
- *Vision-Related Books*: A list of more than 500 books—including online books and book support sites—edited and maintained by Bob Fisher.
  http://homepages.inf.ed.ac.uk/rbf/CVonline/books.htm

- *Mathtools.net*: A technical computing portal containing links to thousands of relevant code for MATLAB (among others).
http://www.mathtools.net/
- *The MathWorks Central File Exchange*: User-contributed MATLAB code for image processing and many other fields.
http://www.mathworks.com/matlabcentral/fileexchange/

WHAT HAVE WE LEARNED?

- Digital image processing is the science of modifying digital images using a digital computer.
- Digital image processing is closely related to other areas such as *computer vision* and *pattern recognition*.
- Digital image processing algorithms, techniques, and applications usually take an image as input and produce one of the following outputs: a modified (i.e., processed) image, an encoded version of the main attributes present in the input image, or a nonpictorial description of the input image's contents.
- Digital image processing has found applications in almost every area of modern life, from medical imaging devices to quality control in manufacturing systems, and from consumer electronics to law enforcement and security.
- An *image* is a visual representation of an object, a person, or a scene produced by an optical device such as a mirror, a lens, or a camera. This representation is two dimensional, although it corresponds to one of the infinitely many projections of a real-world, three-dimensional object or scene.
- A digital image is a representation of a two-dimensional image using a finite number of pixels, each of which indicates the gray level or color contents of the image at that point.
- Image manipulation techniques consist of manually modifying the contents of an image using preexisting tools (e.g., airbrush).
- Representative image processing operations include image sharpening, noise removal, edge extraction, contrast enhancement, and object segmentation and labeling.
- A digital image processing system is usually built around a general-purpose computer equipped with hardware for image and video acquisition, storage, and display. The software portion of the system usually consists of modules that perform specialized tasks. In this book, we shall use MATLAB (and its Image Processing Toolbox) as the software of choice.
- A machine vision system is a combination of hardware and software designed to solve problems involving the analysis of visual scenes using intelligent algorithms. Its main components are acquisition, preprocessing, segmentation, feature extraction, and classification.

- It is extremely difficult to emulate the performance of the human visual system—in terms of processing speed, previously acquired knowledge, and the ability to resolve visual scenes under a wide range of conditions—using machine vision systems.

LEARN MORE ABOUT IT

- Chapter 1 of [GW08] contains a detailed account of the history of image processing and its most representative applications.
- Another good overview of image processing applications is provided by Baxes [Bax94].
- Chapter 16 of [SHB08] contains insightful information on design decisions involved in the development of a few selected case studies in machine vision.

1.7 PROBLEMS

1.1 Use the block diagram from Figure 1.10 as a starting point to design a machine vision system to read the label of the main integrated circuit (IC) on a printed circuit board (PCB) (see Figure 1.11 for an example). Explain what each block will do, their input and output, what are the most challenging requirements, and how they will be met by the designed solution.

FIGURE 1.11 Test image for the design of a machine vision system to read the label of the main integrated circuit on a printed circuit board.

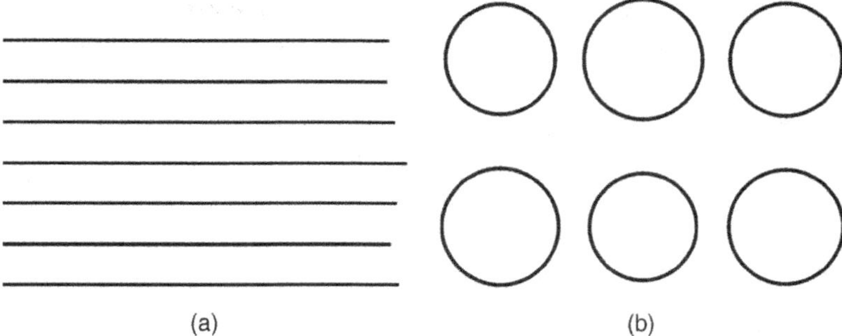

FIGURE 1.12 (a) Test image for distance estimation: parallel lines with up to 5% difference in length. (b) Test image for area estimation: circles with up to 10% difference in radius. Both images are adapted and redrawn from [Jah05].

1.2 In our discussion on machine vision systems, we indicated that the following are the three biggest difficulties in emulating the human visual system: its huge database (images and concepts captured, processed, and recorded during a lifetime), its high speed for processing visual data and making decisions upon them, and the ability to perform under a wide range of work conditions. Explain each of these challenges in your own words, and comment on which ones are more likely to be minimized, thanks to advances in image processing hardware and software.

1.3 Who do you think would perform better at the following tasks: man (HVS) or computer (MVS)? Please explain why.

(a) Determining which line is the shortest in Figure 1.12a.

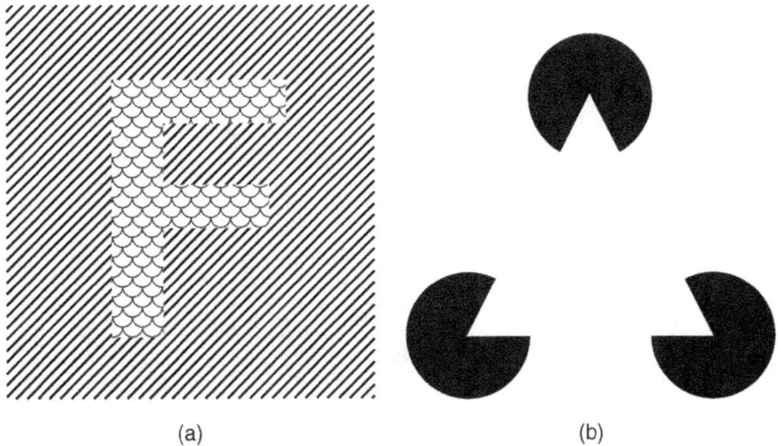

FIGURE 1.13 (a) Test image for texture-based object segmentation. (b) Test image for object segmentation based on "interpolation" of object boundaries. Both images are adapted and redrawn from [Jah05].

(b) Determining which circle is the smallest in Figure 1.12b.

(c) Segmenting the image containing the letter "F" from the background in Figure 1.13a.

(d) Segmenting the white triangle (this triangle—known as "Kanizsa's triangle"—is a well-known optical illusion) in Figure 1.13b.

CHAPTER 2

IMAGE PROCESSING BASICS

WHAT WILL WE LEARN?

- How is a digital image represented and stored in memory?
- What are the main types of digital image representation?
- What are the most popular image file formats?
- What are the most common types of image processing operations and how do they affect pixel values?

2.1 DIGITAL IMAGE REPRESENTATION

A digital image—whether it was obtained as a result of sampling and quantization of an analog image or created already in digital form—can be represented as a two-dimensional (2D) matrix of real numbers. In this book, we adopt the convention $f(x, y)$ to refer to monochrome images of size $M \times N$, where x denotes the row number (from 0 to $M - 1$) and y represents the column number (between 0 and $N - 1$) (Figure 2.1):

$$f(x, y) = \begin{bmatrix} f(0,0) & f(0,1) & \cdots & f(0,N-1) \\ f(1,0) & f(1,1) & \cdots & f(1,N-1) \\ \vdots & \vdots & & \vdots \\ f(M-1,0) & f(M-1,1) & \cdots & f(M-1,N-1) \end{bmatrix} \tag{2.1}$$

Practical Image and Video Processing Using MATLAB®. By Oge Marques.

FIGURE 2.1 A monochrome image and the convention used to represent rows (x) and columns (y) adopted in this book.

The value of the two-dimensional function $f(x, y)$ at any given pixel of coordinates (x_0, y_0), denoted by $f(x_0, y_0)$, is called the *intensity* or *gray level* of the image at that pixel. The maximum and minimum values that a pixel intensity can assume will vary depending on the data type and convention used. Common ranges are as follows: 0.0 (black) to 1.0 (white) for `double` data type and 0 (black) to 255 (white) for `uint8` (unsigned integer, 8 bits) representation.

The convention expressed by equation (2.1) and Figure 2.1 is consistent with programming languages that use 0-based array notation (e.g., Java, C, C++) and several other textbooks, but *not* with MATLAB and its Image Processing Toolbox (IPT), which use 1-based array notation. Whenever the situation calls for explicit disambiguation between the two conflicting conventions, we shall use the notation `f(p,q)` to refer to the MATLAB representation of $f(x, y)$ (where `p` denotes row and `q` denotes column):

$$\texttt{f(p,q)} = \begin{bmatrix} \texttt{f(1,1)} & \texttt{f(1,2)} & \cdots & \texttt{f(1,N)} \\ \texttt{f(2,1)} & \texttt{f(2,2)} & \cdots & \texttt{f(2,N)} \\ \vdots & \vdots & & \vdots \\ \texttt{f(M,1)} & \texttt{f(M,2)} & \cdots & \texttt{f(M,N)} \end{bmatrix} \tag{2.2}$$

Monochrome images are essentially 2D *matrices* (or *arrays*). Since MATLAB treats matrices as a built-in data type, it is easier to manipulate them without resorting to common programming language constructs (such as double `for` loops to access each individual element within the array), as we shall see in Chapter 3.

Images are represented in digital format in a variety of ways. At the most basic level, there are two different ways of encoding the contents of a 2D image in digital format: *raster* (also known as *bitmap*) and *vector*. Bitmap representations use one or more two-dimensional arrays of pixels, whereas vector representations use a series of drawing commands to represent an image. Each encoding method has its pros and cons: the greatest advantages of bitmap graphics are their quality and display speed; their main disadvantages include larger memory storage requirements and size dependence (e.g., enlarging a bitmap image may lead to noticeable artifacts). Vector representations require less memory and allow resizing and geometric manipulations without introducing artifacts, but need to be rasterized for most presentation devices.

In either case, there is no such a thing as a perfect digital representation of an image. Artifacts due to finite resolution, color mapping, and many others will always be present. The key to selecting an adequate representation is to find a suitable compromise between size (in bytes), subjective quality, and interoperability of the adopted format or standard. For the remaining of this book, we shall focus exclusively on bitmap images.

2.1.1 Binary (1-Bit) Images

Binary images are encoded as a 2D array, typically using 1 bit per pixel, where a 0 usually means "black" and a 1 means "white" (although there is no universal agreement on that). The main advantage of this representation—usually suitable for images containing simple graphics, text, or line art—is its small size. Figure 2.2 shows a binary image (the result of an edge detection algorithm) and a 6×6 detailed region, where pixels with a value of 1 correspond to edges and pixels with a value of 0 correspond to the background.

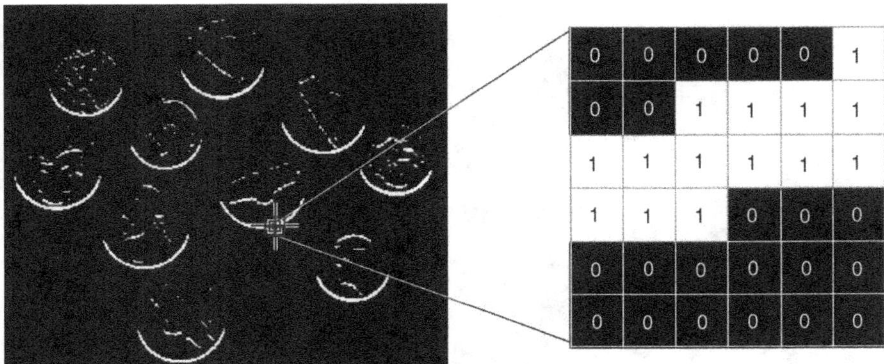

FIGURE 2.2 A binary image and the pixel values in a 6×6 neighborhood. Original image: courtesy of MathWorks.

In MATLAB

Binary images are represented in MATLAB using a *logical* array of 0's and 1's. Although we could also use an array of `uint8` and restrict the array values to be only 0 or 1, that would not, technically, be considered a binary image in MATLAB [GWE04]. Conversion from numerical to logical arrays can be accomplished using function `logical`.

Logical arrays can also be created and processed using relational and logical operators (see Chapter 6).

2.1.2 Gray-Level (8-Bit) Images

Gray-level (also referred to as *monochrome*) images are also encoded as a 2D array of pixels, usually with 8 bits per pixel, where a pixel value of 0 corresponds to "black," a pixel value of 255 means "white," and intermediate values indicate varying shades of gray. The total number of gray levels is larger than the human visual system requirements (which, in most cases, cannot appreciate any improvements beyond 64 gray levels), making this format a good compromise between subjective visual quality and relatively compact representation and storage.

Figure 2.3 shows a grayscale image and a 6 × 6 detailed region, where brighter pixels correspond to larger values.

In MATLAB

Intensity images can be represented in MATLAB using different data types (or *classes*). For monochrome images with elements of integer classes `uint8` and `uint16`, each pixel has a value in the [0, 255] and the [0, 65,535] range, respectively. Monochrome images of class `double` have pixel values in the [0.0, 1.0] range.

FIGURE 2.3 A grayscale image and the pixel values in a 6 × 6 neighborhood.

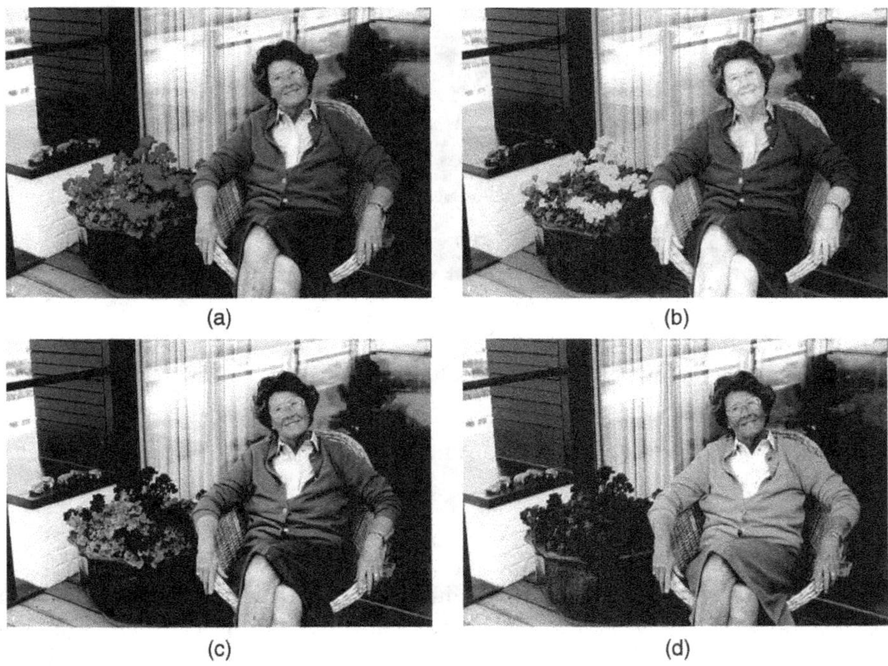

(a) (b) (c) (d)

FIGURE 2.4 Color image (a) and its R (b), G (c), and B (d) components.

2.1.3 Color Images

Representation of color images is more complex and varied. The two most common ways of storing color image contents are *RGB* representation—in which each pixel is usually represented by a 24-bit number containing the amount of its red (R), green (G), and blue (B) components—and *indexed* representation—where a 2D array contains indices to a color palette (or *lookup table* - (*LUT*)).

24-Bit (RGB) Color Images Color images can be represented using three 2D arrays of same size, one for each color channel: red (R), green (G), and blue (B) (Figure 2.4).[1] Each array element contains an 8-bit value, indicating the amount of red, green, or blue at that point in a [0, 255] scale. The combination of the three 8-bit values into a 24-bit number allows 2^{24} (16,777,216, usually referred to as 16 million or 16 M) color combinations. An alternative representation uses 32 bits per pixel and includes a fourth channel, called the *alpha channel*, that provides a measure of transparency for each pixel and is widely used in image editing effects.

[1]Color images can also be represented using alternative color models (or *color spaces*), as we shall see in Chapter 16.

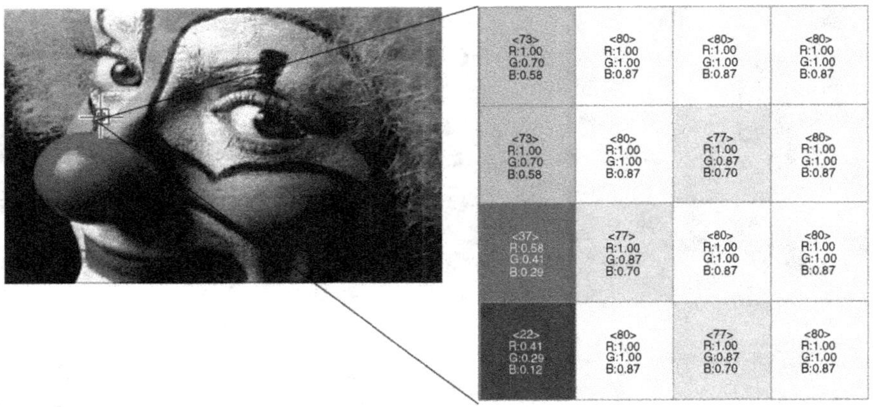

FIGURE 2.5 An indexed color image and the indices in a 4×4 neighborhood. Original image: courtesy of MathWorks.

Indexed Color Images A problem with 24-bit color representations is backward compatibility with older hardware that may not be able to display the 16 million colors simultaneously. A solution—devised before 24-bit color displays and video cards were widely available—consisted of an indexed representation, in which a 2D array of the same size as the image contains indices (pointers) to a *color palette* (or *color map*) of fixed maximum size (usually 256 colors). The color map is simply a list of colors used in that image.

Figure 2.5 shows an indexed color image and a 4×4 detailed region, where each pixel shows the index and the values of R, G, and B at the color palette entry that the index points to.

2.1.4 Compression

Since raw image representations usually require a large amount of storage space (and proportionally long transmission times in the case of file uploads/downloads), most image file formats employ some type of compression. Compression methods can be *lossy*—when a tolerable degree of deterioration in the visual quality of the resulting image is acceptable—or *lossless*—when the image is encoded in its full quality. The overall results of the compression process in terms of both storage savings—usually expressed in terms of compression ratio or bits per pixel (bpp)—and resulting quality loss (for the case of lossy techniques) may vary depending on the technique, format, options (such as the quality setting for JPEG), and actual image contents. As a general guideline, lossy compression should be used for general-purpose photographic images, whereas lossless compression should be preferred when dealing with line art, drawings, facsimiles, or images in which no loss of detail may be tolerable (most notably, space images and medical images). The topic of image compression will be discussed in detail in Chapter 17.

2.2 IMAGE FILE FORMATS

Most of the image file formats used to represent bitmap images consist of a *file header* followed by (often compressed) *pixel data*. The image file header stores information about the image, such as image height and width, number of bands, number of bits per pixel, and some signature bytes indicating the file type. In more complex file formats, the header may also contain information about the type of compression used and other parameters that are necessary to decode (i.e., decompress) the image.

The simplest file formats are the BIN and PPM formats. The BIN format simply consists of the raw pixel data, without any header. Consequently, the user of a BIN file must know the relevant image parameters (such as height and width) beforehand in order to use the image. The PPM format and its variants (PBM for binary images, PGM for grayscale images, PPM for color images, and PNM for any of them) are widely used in image processing research and many free tools for format conversion include them. The headers for these image formats include a 2-byte signature, or "magic number," that identifies the file type, the image width and height, the number of bands, and the maximum intensity value (which determines the number of bpp per band).

The Microsoft Windows bitmap (BMP) format is another widely used and fairly simple format, consisting of a header followed by raw pixel data.

The JPEG format is the most popular file format for photographic quality image representation. It is capable of high degrees of compression with minimal perceptual loss of quality. The technical details of the JPEG compression algorithm (and its presumed successor, the JPEG 2000 standard) will be discussed in Chapter 17.

Two other image file formats are very widely used in image processing tasks: GIF (Graphics Interchange Format) and TIFF (Tagged Image File Format). GIF uses an indexed representation for color images (with a palette of a maximum of 256 colors), the LZW (Lempel–Ziv–Welch) compression algorithm, and a 13-byte header. TIFF is a more sophisticated format with many options and capabilities, including the ability to represent truecolor (24 bpp) and support for five different compression schemes.

Portable Network Graphics (PNG) is an increasingly popular file format that supports both indexed and truecolor images. Moreover, it provides a patent-free replacement for the GIF format.

Some image processing packages adopt their own (sometimes proprietary) formats. Examples include XCF (the native image format of the GIMP image editing program) and RAW (which is a family of formats mostly adopted by camera manufacturers).

Since this book focuses on using MATLAB and its IPT, which provide built-in functionality for reading from and writing to most common image file formats,[2] we will not discuss in detail the specifics of the most popular image file formats any further. If you are interested in knowing more about the internal representation and details of these formats, refer to "Learn More About It" section at the end of the chapter.

[2]In the words of Professor Alasdair McAndrew [McA04], "you can use MATLAB for image processing very happily without ever really knowing the difference between GIF, TIFF, PNG, and all the other formats."

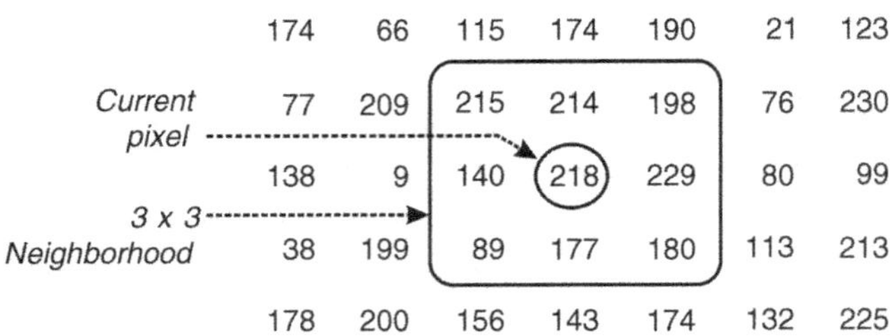

FIGURE 2.6 Pixels within a neighborhood.

2.3 BASIC TERMINOLOGY

This section introduces important concepts and terminology used to understand and express the properties of an image.

Image Topology It involves the investigation of fundamental image properties—usually done on binary images and with the help of morphological operators (see Chapter 13)—such as number of occurrences of a particular object, number of separate (not connected) regions, and number of holes in an object, to mention but a few.

Neighborhood The pixels surrounding a given pixel constitute its *neighborhood*, which can be interpreted as a smaller matrix containing (and usually centered around) the reference pixel. Most neighborhoods used in image processing algorithms are small square arrays with an odd number of pixels, for example, the 3×3 neighborhood shown in Figure 2.6.

In the context of image topology, neighborhood takes a slightly different meaning. It is common to refer to the *4-neighborhood* of a pixel as the set of pixels situated above, below, to the right, and to the left of the reference pixel (p), whereas the set of all of p's immediate neighbors is referred to as its *8-neighborhood.* The pixels that belong to the 8-neighborhood, but not to the 4-neighborhood, make up the *diagonal neighborhood* of p (Figure 2.7).

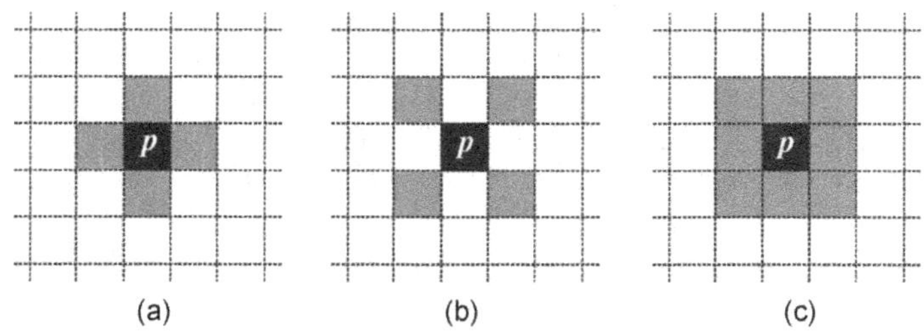

FIGURE 2.7 Concept of neighborhood of pixel p (from an image topology perspective): (a) 4-neighborhood; (b) diagonal neighborhood; (c) 8-neighborhood.

Adjacency In the context of image topology, two pixels p and q are *4-adjacent* if they are 4-neighbors of each other and *8-adjacent* if they are 8-neighbors of one another. A third type of adjacency—known as *mixed adjacency* (or simply *m-adjacency*)—is sometimes used to eliminate ambiguities (i.e., redundant paths) that may arise when 8-adjacency is used.

Paths In the context of image topology, a *4-path* between two pixels p and q is a sequence of pixels starting with p and ending with q such that each pixel in the sequence is 4-adjacent to its predecessor in the sequence. Similarly, an *8-path* indicates that each pixel in the sequence is 8-adjacent to its predecessor.

Connectivity If there is a 4-path between pixels p and q, they are said to be *4-connected*. Similarly, the existence of an 8-path between them means that they are *8-connected*.

Components A set of pixels that are connected to each other is called a *component*. If the pixels are 4-connected, the expression *4-component* is used; if the pixels are 8-connected, the set is called an *8-component*. Components are often labeled (and optionally pseudocolored) in a unique way, resulting in a *labeled image*, $L(x, y)$, whose pixel values are symbols of a chosen alphabet. The symbol value of a pixel typically denotes the outcome of a decision made for that pixel—in this case, the unique number of the component to which it belongs.

In MATLAB

MATLAB's IPT contains a function `bwlabel` for labeling connected components in binary images. An associated function, `label2rgb`, helps visualize the results by painting each region with a different color.

Figure 2.8 shows an example of using `bwlabel` and `label2rgb` and highlights the fact that the number of connected components will vary from 2 (when 8-connectivity is used, Figure 2.8b) to 3 (when 4-connectivity is used, Figure 2.8c).

Distances Between Pixels There are many image processing applications that require measuring distances between pixels. The most common distance measures

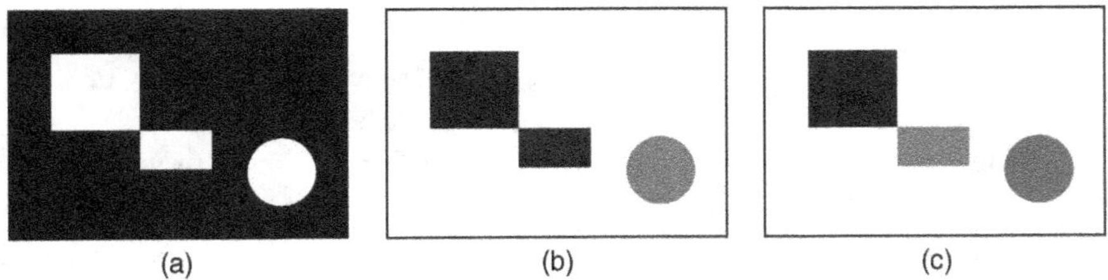

FIGURE 2.8 Connected components: (a) original (binary) image; (b) results for 8-connectivity; (c) results for 4-connectivity.

between two pixels p and q, of coordinates (x_0, y_0) and (x_1, y_1), respectively, are as follows:

- Euclidean distance:

$$D_e(p, q) = \sqrt{(x_1 - x_0)^2 + (y_1 - y_0)^2} \tag{2.3}$$

- D_4 (also known as *Manhattan* or *city block*) distance:

$$D_4(p, q) = |x_1 - x_0| + |y_1 - y_0| \tag{2.4}$$

- D_8 (also known as *chessboard*) distance:

$$D_8(p, q) = \max(|x_1 - x_0|, |y_1 - y_0|) \tag{2.5}$$

It is important to note that the distance between two pixels depends only on their coordinates, not their values. The only exception is the D_m distance, defined as "the shortest m-path between two m-connected pixels."

2.4 OVERVIEW OF IMAGE PROCESSING OPERATIONS

In this section, we take a preliminary look at the main categories of image processing operations. Although there is no universal agreement on a taxonomy for the field, we will organize them as follows:

- *Operations in the Spatial Domain*: Here, arithmetic calculations and/or logical operations are performed on the original pixel values. They can be further divided into three types:
  - *Global Operations*: Also known as *point operations*, in which the entire image is treated in a uniform manner and the resulting value for a processed pixel is a function of its original value, regardless of its location within the image. *Example:* contrast adjustment (Chapters 8 and 9).
  - *Neighborhood-Oriented Operations*: Also known as *local* or *area operations*, in which the input image is treated on a pixel-by-pixel basis and the resulting value for a processed pixel is a function of its original value and the values of its neighbors. *Example:* spatial-domain filters (Chapter 10).
  - *Operations Combining Multiple Images*: Here, two or more images are used as an input and the result is obtained by applying a (series of) arithmetic or logical operator(s) to them. *Example:* subtracting one image from another for detecting differences between them (Chapter 6).
- *Operations in a Transform Domain*: Here, the image undergoes a mathematical transformation—such as Fourier transform (FT) or discrete cosine transform (DCT)—and the image processing algorithm works in the transform domain. *Example:* frequency-domain filtering techniques (Chapter 12).

2.4.1 Global (Point) Operations

Point operations apply the same mathematical function, often called *transformation function*, to all pixels, regardless of their location in the image or the values of their neighbors. Transformation functions in the spatial domain can be expressed as

$$g(x, y) = T\left[f(x, y)\right] \tag{2.6}$$

where $g(x, y)$ is the processed image, $f(x, y)$ is the original image, and T is an operator on $f(x, y)$.

Since the actual coordinates do not play any role in the way the transformation function processes the original image, a shorthand notation can be used:

$$s = T\left[r\right] \tag{2.7}$$

where r is the original gray level and s is the resulting gray level after processing.

Figure 2.9 shows an example of a transformation function used to reduce the overall intensity of an image by half: $s = r/2$. Chapter 8 will discuss point operations and transformation functions in more detail.

2.4.2 Neighborhood-Oriented Operations

Neighborhood-oriented (also known as *local* or *area*) operations consist of determining the resulting pixel value at coordinates (x, y) as a function of its original value and the value of (some of) its neighbors, typically using a *convolution* operation. The

(a)

(b)

FIGURE 2.9 Example of intensity reduction using a transformation function: (a) original image; (b) output image.

W_1	W_2	W_3
W_4	W_5	W_6
W_7	W_8	W_9

FIGURE 2.10 A 3 × 3 convolution mask, whose generic weights are W_1, ..., W_9.

convolution of a source image with a small 2D array (known as *window*, *template*, *mask*, or *kernel*) produces a destination image in which each pixel value depends on its original value and the value of (some of) its neighbors. The convolution mask determines which neighbors are used as well as the relative weight of their original values. Masks are normally 3 × 3, such as the one shown in Figure 2.10. Each mask coefficient ($W_1, \ldots, W_9$) can be interpreted as a *weight*. The mask can be thought of as a small window that is overlaid on the image to perform the calculation on one pixel at a time. As each pixel is processed, the window moves to the next pixel in the source image and the process is repeated until the last pixel has been processed.

Convolution operations are widely used in image processing. Depending on the choice of kernel, the same basic operation can be used to blur an image, enhance it, find its edges, or remove noise from it. Chapter 10 will explain convolution in detail and discuss image enhancement using neighborhood operations.

2.4.3 Operations Combining Multiple Images

There are many image processing applications that combine two images, pixel by pixel, using an arithmetic or logical operator, resulting in a third image, Z:

$$X \quad opn \quad Y = \quad Z \tag{2.8}$$

where X and Y may be images (arrays) or scalars, Z is necessarily an array, and opn is a binary mathematical (+, −, ×, /) or logical (AND, OR, XOR) operator. Figure 2.11 shows schematically how pixel-by-pixel operations work. Chapter 6 will discuss arithmetic and logic pixel-by-pixel operations in detail.

2.4.4 Operations in a Transform Domain

A *transform* is a mathematical tool that allows the conversion of a set of values to another set of values, creating, therefore, a new way of representing the same information. In the field of image processing, the original domain is referred to as *spatial domain*, whereas the results are said to lie in the *transform domain*. The

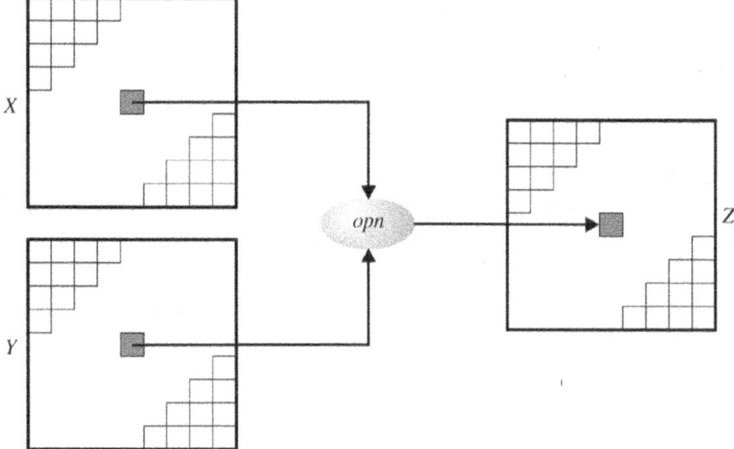

FIGURE 2.11 Pixel-by-pixel arithmetic and logic operations.

motivation for using mathematical transforms in image processing stems from the fact that some tasks are best performed by transforming the input images, applying selected algorithms in the transform domain, and eventually applying the inverse transformation to the result (Figure 2.12). This is what happens when we filter an image in the 2D frequency domain using the FT and its inverse, as we shall see in Chapter 11.

WHAT HAVE WE LEARNED?

- Images are represented in digital format in a variety of ways. *Bitmap* (also known as *raster*) representations use one or more two-dimensional arrays of pixels (picture elements), whereas *vector* representations use a series of drawing commands to represent an image.
- Binary images are encoded as a 2D array, using 1 bit per pixel, where usually—but not always—a 0 means "black" and a 1 means "white."

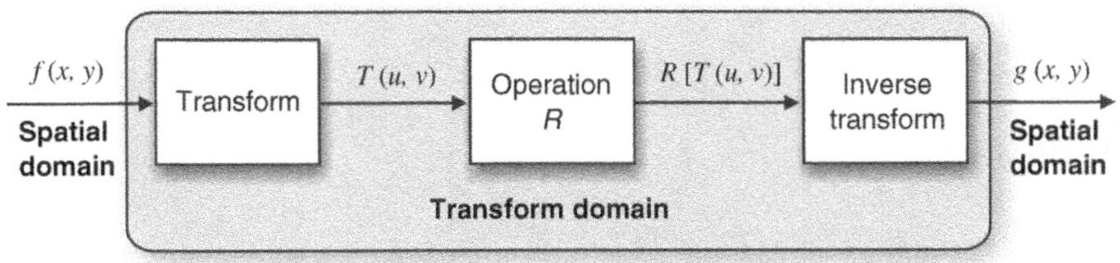

FIGURE 2.12 Operations in a transform domain.

- Gray-level (monochrome) images are encoded as a 2D array of pixels, using 8 bits per pixel, where a pixel value of 0 usually means "black" and a pixel value of 255 means "white," with intermediate values corresponding to varying shades of gray.
- The two most common ways of storing color image contents are *RGB* representation—in which each pixel is usually represented by a 24-bit number containing the amount of its red (R), green (G), and blue (B) components—and *indexed* representation—where a 2D array contains indices to a color palette (or *look up table*).
- Some of the most popular image file formats in use today are BMP, GIF, JPEG, TIFF, and PNG.
- MATLAB's built-in functions for reading images from files and writing images to files (`imread` and `imwrite`, respectively) support most file formats and their variants and options.
- Image topology is the field of image processing concerned with investigation of fundamental image properties (e.g., number of connected components and number of holes in an object) using concepts such as adjacency and connectivity.
- Image processing operations can be divided into two big groups: *spatial domain* and *transform domain*. Spatial-domain techniques can be further divided into pixel-by-pixel (*point*) or neighborhood-oriented (*area*) operations.

LEARN MORE ABOUT IT

- The book by Miano [Mia99] is a useful guide to graphic file formats.

CHAPTER 3

MATLAB BASICS

WHAT WILL WE LEARN?

- What is MATLAB and why has it been selected to be the tool of choice for this book?
- What programming environment does MATLAB offer?
- What are M-files?
- What is the difference between MATLAB scripts and functions?
- How can I get started with MATLAB?

3.1 INTRODUCTION TO MATLAB

MATLAB (MATrix LABoratory) is a data analysis, prototyping, and visualization tool with built-in support for matrices and matrix operations, excellent graphics capabilities, and a high-level programming language and development environment.

MATLAB has become very popular with engineers, scientists, and researchers in both industry and academia, due to many factors, among them, the availability of rich sets of specialized functions—encapsulated in *toolboxes*—for many important areas,

Practical Image and Video Processing Using MATLAB®. By Oge Marques.

from neural networks to finances to image processing (which is the main focus of our interest and will be discussed in Chapter 4).[1]

MATLAB has an extensive built-in documentation. It contains descriptions of MATLAB's main functions, sample code, relevant demos, and general help pages. MATLAB documentation can be accessed in a number of different ways, from command-line text-only help to hyperlinked HTML pages (which have their online counterpart in MathWorks's web site: http://www.mathworks.com).

MATLAB's basic data type is the *matrix* (or *array*). MATLAB does not require dimensioning, that is, memory allocation prior to actual usage. All data are considered to be matrices of some sort. Single values are considered by MATLAB to be 1×1 matrices.

The MATLAB algorithm development environment provides a command-line interface, an interpreter for the MATLAB programming language, an extensive set of numerical and string manipulation functions, 2D and 3D plotting functions, and the ability to build graphical user interfaces (GUIs). The MATLAB programming language interprets commands, which shortens programming time by eliminating the need for compilation.

If you want to get started with MATLAB right away, this is a good time to try *Tutorial 3.1: MATLAB—A Guided Tour* on page 44.

3.2 BASIC ELEMENTS OF MATLAB

In this section, we present the basic elements of MATLAB, its environment, data types, and command-line operations.

3.2.1 Working Environment

The MATLAB working environment consists of the following:

- *MATLAB Desktop*: This typically contains five subwindows: the *Command Window*, the *Workspace Browser*, the *Current Directory Window*, the *Command History Window*, and one or more *Figure Windows*, which are visible when the user displays a graphic, such as a plot or image.
- *MATLAB Editor*: This is used to create and edit M-files. It includes a number of useful functions for saving, viewing, and debugging M-files.
- *Help System*: This includes particularly the *Help Browser*, which displays HTML documents and contains a number of search and display options.

In *Tutorial 3.1: MATLAB—A Guided Tour* (page 44), you will have a hands-on introduction to the MATLAB environment.

[1] In 2003, MathWorks released the *Image Acquisition Toolbox*, which contains a collection of functions for image acquisition. This toolbox is briefly described in Section 5.3.2 and will not be explored in more detail in this book.

TABLE 3.1 MATLAB Data Classes

Data Class	Description
`uint8`	8-Bit unsigned integers (1 byte per element)
`uint16`	16-Bit unsigned integers (2 bytes per element)
`uint32`	32-Bit unsigned integers (4 bytes per element)
`int8`	8-Bit signed integers (1 byte per element)
`int16`	16-Bit signed integers (2 bytes per element)
`int32`	32-Bit signed integers (4 bytes per element)
`single`	Single-precision floating numbers (4 bytes per element)
`double`	Double-precision floating numbers (8 bytes per element)
`logical`	Values are 0 (false) or 1 (true) (1 byte per element)
`char`	Characters (2 bytes per element)

3.2.2 Data Types

MATLAB supports the data types (or data classes) listed in Table 3.1.

The first eight data types listed in the table are known as *numeric* classes. A numeric value in MATLAB is of class `double` unless specified otherwise. Conversion between data classes (also known as *typecasting*) is possible (and often necessary). A *string* in MATLAB is simply a $1 \times n$ array of characters.

3.2.3 Array and Matrix Indexing in MATLAB

MATLAB arrays (vectors and matrices) are indexed using a 1-based convention. Therefore, `a(1)` is the syntax to refer to the first element of a one-dimensional array `a` and `f(1,1)` is the syntax to refer to the first element of a two-dimensional array, such as the top left pixel in a monochrome image `f`. The colon (`:`) operator provides powerful indexing capabilities, as you will see in Tutorial 3.2 (page 46).

MATLAB does not restrict the number of dimensions of an array, but has an upper limit on the maximum number of elements allowed in an array. If you type

```
[c, maxsize] = computer
```

variable `maxsize` will contain the maximum number of elements allowed in a matrix on the computer and version of MATLAB that you are using.

3.2.4 Standard Arrays

MATLAB has a number of useful, built-in standard arrays:

- `zeros(m,n)` creates an `m` × `n` matrix of zeros.
- `ones(m,n)` creates an `m` × `n` matrix of ones.
- `true(m,n)` creates an `m` × `n` matrix of logical ones.

- `false(m,n)` creates an `m` × `n` matrix of logical zeros.
- `eye(n)` returns an $n \times n$ *identity matrix*.
- `magic(m)` returns a *magic square*[2] of order `m`.
- `rand(m,n)` creates an `m` × `n` matrix whose entries are pseudorandom numbers uniformly distributed in the interval [0, 1].
- `randn(m,n)` creates an `m`×`n` matrix whose entries are pseudorandom numbers that follow a normal (i.e., Gaussian) distribution with mean 0 and variance 1.

3.2.5 Command-Line Operations

Much of the functionality available in MATLAB can be accessed from the command line (as can be seen in the first steps of *Tutorial 3.2: MATLAB Data Structures* on page 46). A command-line operation is equivalent to executing one line of code, usually calling a built-in function by typing its name and parameters at the prompt (>>). If the line does not include a variable to which the result should be assigned, it is assigned to a built-in variable `ans` (as in `ans`wer). If the line does not end with a semicolon, the result is also echoed back to the command window.

Considering the large number of available built-in functions (and the parameters that they may take) in MATLAB, the command-line interface is a very effective way to access these functions without having to resort to a complex series of menus and dialog boxes. The time spent learning the names, syntax, and parameters of useful functions is also well spent, because whatever works on a command-line (one command at a time) fashion will also work as part of a larger program or user-defined function. Moreover, MATLAB provides additional features to make the command-line interaction more effective, such as the ability to easily access (with arrow keys) previous operations (and save typing time). Successful command-line operations can also be easily selected from the *Command History Window* and combined into a script.

3.3 PROGRAMMING TOOLS: SCRIPTS AND FUNCTIONS

The MATLAB development environment interprets commands written in the MATLAB programming language, which is extremely convenient when our major goal is to rapidly prototype an algorithm. Once an algorithm is stable, it can be compiled with the MATLAB compiler (available as an add-on to the main product) for faster execution, which is particularly important for large data sets. The MATLAB compiler MATCOM converts MATLAB native code into C++ code, compiles the C++ code, and links it with the MATLAB libraries. Compiled code may be up to 10 times as fast as their interpreted equivalent, but the speedup factor depends on how vectorized the original code was; highly optimized vectorized code may not experience any speedup at all.

[2] A *magic square* is a square array in which the sum along any row, column, or main diagonal results in the same value.

For faster computation, programmers may dynamically link C routines as MATLAB functions through the MEX utility.

3.3.1 M-Files

M-files in MATLAB can be *scripts* that simply execute a series of MATLAB *commands* (or *statements*) or can be *functions* that accept *arguments* (*parameters*) and produce one or more *output values*. User-created functions extend the capabilities of MATLAB and its toolboxes to address specific needs or applications.

An M-file containing a *script* consists of a sequence of commands to be interpreted and executed. In addition to calls to built-in functions, scripts may also contain variable declarations, calls to user-created functions (which may be stored in separate files), decision statements, and repetition loops. Scripts are usually created using a text editor (e.g., the built-in MATLAB Editor) and stored with a meaningful name and the `.m` extension. Once a script has been created and saved, it can be invoked from the command line by simply typing its name.

An M-file containing a *function* has the following components:

- *Function Definition Line*: It has the form
  `function [outputs] = function_name(inputs)`
  The keyword `function` is required. The output arguments are enclosed within square brackets, whereas the input arguments are enclosed within parentheses. If the function does not return any value, only the word `function` is used and there is no need for the brackets or the equal sign. Function names must begin with a letter, must not contain spaces, and be limited to 63 characters in length.
- *H1 Line*: It is a single comment line that follows the function definition line. There can be no blank lines or leading spaces between the H1 line and the function definition line. The H1 line is the first text that appears when the user types
  `>> help function_name`[3]
  in the Command Window. Since this line provides important summary information about the contents and purpose of the M-file, it should be as descriptive as possible.
- *Help Text*: It is a block of text that follows the H1 line, without any blank lines between the two. The help text is displayed after the H1 line when a user types `help function_name` at the prompt.
- *Function Body*: This contains all the MATLAB code that performs computation and assigns values to output parameters.[4]
- *Comments*: In MATLAB, these are preceded by the `%` symbol.

[3]Note that the prompt is different in the educational version of MATLAB (EDU≫).

[4]Even though the MATLAB programming language includes a "return" command, it does not take any arguments. This is in contrast with other programming languages that may use "return" followed by a parameter as a standard way of returning a single value or a pointer to where multiple values reside in memory.

Here is an example of a simple function (`raise_to_power`) that will be used in Tutorial 3.3 on page 53:

```
function z = raise_to_power(val,exp)
%RAISE_TO_POWER Calculate power of a value
% z = raise_to_power(val,exp) raise val to a power with value of exp
% and store it in z.

z = val ^ exp;
```

3.3.2 Operators

MATLAB operators can be grouped into three main categories:

- *Arithmetic Operators*: Perform numeric computations on matrices.
- *Relational Operators*: Compare operands.
- *Logical Operators*: Perform standard logical functions (e.g., AND, NOT, and OR)

Since MATLAB considers a matrix and its standard built-in data type, the number of array and matrix operators available in MATLAB far exceeds the traditional operators found in a conventional programming language. Table 3.2 contains a summary of them. All operands can be real or complex.

Table 3.3 shows a list of some of the most useful specialized matrix operations that can also be easily performed in MATLAB.

TABLE 3.2 MATLAB Array and Matrix Arithmetic Operators

Operator	Name	MATLAB Function
+	Array and matrix addition	`plus(a,b)`
-	Array and matrix subtraction	`minus(a,b)`
.*	Element-by-element array multiplication	`times(a,b)`
*	Matrix multiplication	`mtimes(a,b)`
./	Array right division	`rdivide(a,b)`
.\	Array left division	`ldivide(a,b)`
/	Matrix right division	`mrdivide(a,b)`
\	Matrix left division	`mldivide(a,b)`
.^	Array power	`power(a,b)`
.^	Matrix power	`mpower(a,b)`
.'	Vector and matrix transpose	`transpose(a)`
'	Vector and matrix complex conjugate transpose	`ctranspose(a)`
+	Unary plus	`uplus(a)`
—	Unary minus	`uminus(a)`
:	Colon	`colon(a,b)` or `colon(a,b,c)`

TABLE 3.3 Examples of MATLAB Specialized Matrix Operations

Name	MATLAB Operator or Function
Matrix transpose	Apostrophe (') operator
Inversion	inv function
Matrix determinant	det function
Flip up and down	flipud function
Flip left and right	fliplr function
Matrix rotation	rot90 function
Matrix reshape	reshape function
Sum of the diagonal elements	trace function

Since monochrome images are essentially 2D arrays, that is, matrices, all operands in Tables 3.2 and 3.3 are applicable to images. However, the MATLAB Image Processing Toolbox (IPT) also contains specialized versions of the main arithmetic operations involving images, whose main advantage is the support of integer data classes (MATLAB math operators require inputs of class double). These functions are listed in Table 3.4 and will be discussed in more detail in Chapter 6.

The relational operators available in MATLAB parallel the ones you would expect in any programming language. They are listed in Table 3.5.

TABLE 3.4 Specialized Arithmetic Functions Supported by the IPT

Function	Description
imadd	Adds two images or adds a constant to an image
imsubtract	Subtracts two images or subtracts a constant from an image
immultiply	Multiplies two images (element-by-element) or multiplies a constant times an image
imdivide	Divides two images (element-by-element) or divides an image by a constant
imabsdiff	Computes the absolute difference between two images
imcomplement	Complements an image
imlincomb	Computes a linear combination of two or more images

TABLE 3.5 Relational Operators

Operator	Name
<	Less than
<=	Less than or equal to
>	Greater than
>=	Greater than or equal to
==	Equal to
~=	Not equal to

TABLE 3.6 Logical Operators

Operator	Name
&	AND
|	OR
~	NOT

TABLE 3.7 Logical Functions

Function	Description
`xor`	Performs the exclusive-or (XOR) between two operands
`all`	Returns a 1 if all the elements in a vector are nonzero or a 0 otherwise. Operates columnwise on matrices
`any`	Returns a 1 if any of the elements in a vector are nonzero or a 0 otherwise. Operates columnwise on matrices

MATLAB includes a set of standard logical operators. They are listed in Table 3.6. MATLAB also supports the logical functions listed in Table 3.7 and a vast number of functions that return a logical 1 (true) or logical 0 (false) depending on whether the value or condition in their arguments is `true` or `false`, for example, `isempty(a)`, `isequal(a,b)`, `isnumeric(a)`, and many others (refer to the MATLAB online documentation).

3.3.3 Important Variables and Constants

MATLAB has a number of built-in variables and constants, some of which are listed in Table 3.8.

3.3.4 Number Representation

MATLAB can represent numbers in conventional decimal notation (with optional decimal point and leading plus or minus sign) as well as in scientific notation (using the letter `e` to specify a power-of-10 exponent). Complex numbers are represented using either `i` or `j` as a suffix for the imaginary part.

TABLE 3.8 Selected Built-In Variables and Constants

Name	Value Returned
`ans`	Most recent answer
`eps`	Floating-point relative accuracy
`i` (or `j`)	Imaginary unit ($\sqrt{-1}$)
`NaN` (or `nan`)	Not-a-number (e.g., the result of 0/0)
`Inf`	Infinity (e.g., the result of a division by 0)

All numbers are stored internally using the IEEE floating-point standard, resulting in a range of approximately 10^{-308}–10^{+308}.

3.3.5 Flow Control

The MATLAB programming language supports the usual flow control statements found in most other contemporary high-level programming languages: `if` (with optional `else` and `elseif`) and `switch` decision statements, `for` and `while` loops and the associated statements (`break` and `continue`), and the `try...catch` block for error handling. Refer to the online documentation for specific syntax details.

3.3.6 Code Optimization

As a result of the matrix-oriented nature of MATLAB, the MATLAB programming language is a *vectorized language*, which means that it can perform many operations on numbers grouped as vectors or matrices without explicit loop statements. Vectorized code is more compact, more efficient, and parallelizable.[5] In addition to using vectorized loops, MATLAB programmers are encouraged to employ other optimization tricks, such as preallocating the memory used by arrays. These ideas—and their impact on the execution speed of MATLAB code—are discussed in Tutorial 3.3 (page 53).

3.3.7 Input and Output

Basic input and output functionality can be achieved with functions `input` (to request user input and read data from the keyboard) and `disp` (to display a text or array on the screen). MATLAB also contains many support functions to read from and write to files.

3.4 GRAPHICS AND VISUALIZATION

MATLAB has a rich set of primitives for plotting 2D and 3D graphics. Tutorial 9.1 (page 188) will explore some of the 2D plotting capabilities in connection with the plotting of image histograms and transformation functions, whereas Tutorials in Chapter 11 will show examples of 3D plots in connection with the design of image processing filters in the frequency domain.

MATLAB also includes a number of built-in functions to display (and inspect the contents of) images, some of which will be extensively used throughout the book (and discussed in more detail in Tutorial 4.2 on page 74).

[5]Getting used to writing code in a vectorized way—as opposed to the equivalent nested for loops in a conventional programming language—is not a trivial process and it may take time to master it.

3.5 TUTORIAL 3.1: MATLAB—A GUIDED TOUR

Goal

The goal of this tutorial is to give a brief overview of the MATLAB environment.

Objectives

- Become familiar with the working environment in MATLAB.
- Learn how to use the working directory and set paths.
- Become familiar with the MATLAB help system.
- Explore functions that reveal system information and other useful built-in functions.

Procedure

The environment in MATLAB has a simple layout, consisting of several key areas (Figure 3.1).[6] A description of each is given here:

- A: This pane consists of two tabbed areas: one that displays all files in your current *working directory*, and another that displays your *workspace*. The workspace lists all the variables you are currently using.
- B: This pane shows your *history* of commands.
- C: This is where you can modify your current *working directory*. To change the current directory, you can either type it directly in the text box or click on the button to select the directory. You can also change the working directory using the `path` command. See help documents for more information.
- D: This is the *command window*. Here you control MATLAB by typing in commands.

In order for you to use files such as M-files and images, MATLAB must know where these files are located. There are two ways this can be done: by setting the *current directory* to a specific location, or by adding the location to a list of set paths known to MATLAB. The *current directory* should be used for temporary locations or when you need to access a directory only once. This directory is reset every time MATLAB is restarted. To change the current directory, see description C above. Setting a path is a permanent way of telling MATLAB where files are located—the location will remain in the settings when MATLAB is closed. The following steps will illustrate how to set a path in MATLAB:

1. From the **File** menu, select **Set Path...**

[6]The exact way your MATLAB window will look may differ from Figure 3.1, depending on the operating system, MATLAB version, and which windows and working areas have been selected—under the *Desktop* menu—to be displayed.

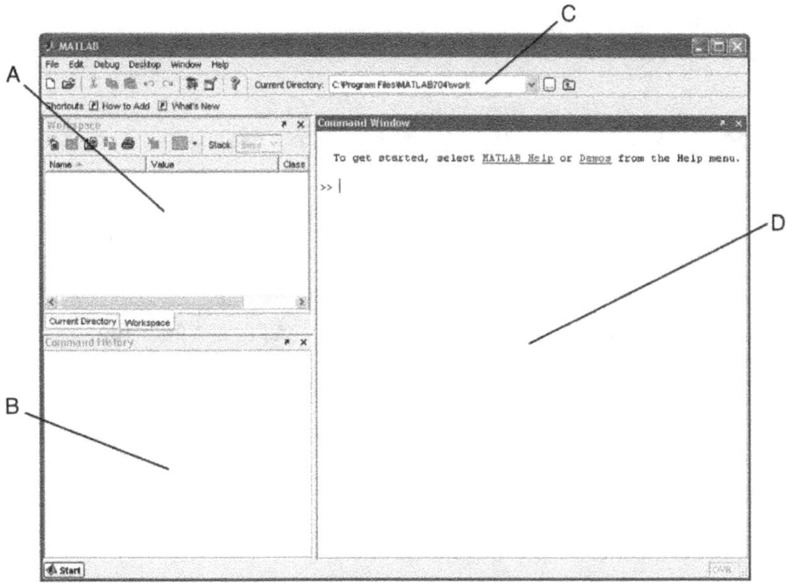

FIGURE 3.1 MATLAB environment.

The paths are presented in the list box in decreasing precedence order.

2. If the directory you wish to add has subfolders within it, then use the **Add with Subfolders...** button. If your directory only contains files, you can use the **Add Folder...** button. To change the precedence of your directory, use the **Move** buttons.
3. **Save** your changes and close the **Set Path** window.

The help system in MATLAB is very useful. It provides information about functions, such as syntax, descriptions, required and optional input parameters, and output parameters. It also includes a number of demos. The following steps will show you how to access help documents and navigate the help window.

4. To access the help system, type `doc` in the command window. Open the help system now.

The left pane of the help window displays a tree of help information. The right pane will show the help document that you choose. If you know exactly what function you need help for, you can use the `doc`, `helpwin`, or `help` commands to directly access the help information.

5. In the command window in MATLAB, execute the following commands to view the help document for the function `computer`.

```
help computer
helpwin computer
doc computer
```

Question 1 What is the difference between the commands `help`, `helpwin`, and `doc`?

There are several commands that can be used to gain information about the system. Explore them in this step.

6. Execute the following commands, one at a time, to see their function.

```
realmin
realmax
bitmax
computer
ver
version
hostid
license
```

Question 2 What is the difference between `ver` and `version`?

7. For some MATLAB humor, repeatedly type the command `why`.

3.6 TUTORIAL 3.2: MATLAB DATA STRUCTURES

Goal

The goal of this tutorial is to learn how to create, initialize, and access some of the most useful data structures available in MATLAB.

Objectives

- Learn how to use MATLAB for basic calculations involving variables, arrays, matrices, and vectors.
- Explore multidimensional and cell arrays.
- Review matrix operations.
- Learn how to use MATLAB structures.
- Explore useful functions that can be used with MATLAB data structures.

Procedure

1. Execute the following lines of code one at a time in the *Command Window* to see how MATLAB can be used as a calculator.

```
2 + 3
2*3 + 4*5 + 6*7;
```

Question 1 What is the variable `ans` used for?

Question 2 What is the purpose of using a semicolon (`;`) at the end of a statement?

2. Perform calculations using variables.

```
fruit_per_box = 20; num_of_boxes = 5;
total_num_of_fruit = fruit_per_box * num_of_boxes
```

Question 3 Experiment with creating your own variables. Are variables case sensitive?

Question 4 What is the value/purpose of these variables: `pi`, `eps`, `inf`, `i`? Is it possible to overwrite any of these variables? If so, how can it be undone?

3. Execute the commands `who` and `whos`, one at a time, to see their function and the difference between them.

There are several commands that will keep the MATLAB environment clean. Use them whenever you feel your command window or workspace is cluttered with statements and variables.

4. Clear a variable in the workspace. After execution, note how the variable disappears from the workspace.

```
clear fruit_per_box
```

5. Clear the command window and all variables with the following lines of code (one at a time to see their effects individually).

```
clc
clear all
```

6. Create a 3×3 matrix by executing the following line of code.

```
A = [1 2 3;4 5 6;7 8 9]
```

Question 5 What is the use of the semicolon in this statement?

The Colon Operator

7. A very useful operator in MATLAB is the colon (`:`). It can be used to create a vector of numbers.

```
1:5
```

8. A third parameter determines how to count between the starting and ending numbers. It is specified between the start and end values.

```
1:1:5
5:-1:1
1:2:9
9:-2:1
```

Question 6 Write a statement that will generate a vector of values ranging from 0 to π in increments of $\pi/4$.

9. The colon operator can also be used to return an entire row or column of a matrix.

```
A = [1 2 3;4 5 6;7 8 9]
A(:,1)
A(1,:)
```

Question 7 Write a line of code that would generate the same 3×3 matrix as in the variable `A` above, but using the colon operator to generate the sequence of numbers in each row instead of explicitly writing them.

10. The colon operator can be replaced with the function `colon`, which performs the same operation.

```
colon(1,5)
```

As seen in the steps above, creating a vector of evenly spaced numbers is easily done with the colon (`:`) operator when we know where the vector starts, ends, and how large the space in between each value is. In certain cases, we may wish to create a vector of numbers that range between two numbers, but we only know the quantity of values needed (for example, create a vector that consists of 4 values between $\pi/4$ and π) . To do this, we use the `linspace` function.

11. Execute this command to see how the function `linspace` operates.

```
linspace(pi/4,pi,4)
```

12. Compare the result from the previous step with these values.

```
pi/4
pi/2
3*pi/4
pi
```

Special Built-In Matrices

MATLAB has several built-in functions that will generate frequently used matrices automatically.

13. Execute the following lines of code one at a time.

```
zeros(3,4)
ones(3,4)
ones(3,4) * 10
rand(3,4)
randn(3,4)
```

Question 8 What is the difference between the functions `rand(M,N)` and `randn(M,N)`?

Matrix Concatenation

Concatenation of matrices is done with brackets ([ ]) or using the `cat` function. Take, for example, the statement

```
A = [1 2 3;4 5 6;7 8 9]
```

The brackets are combining three rows. Instead of explicitly defining each row all at once, they can be defined individually as vectors and then combined into a matrix using brackets.

14. Combine the three individual vectors into a 3 × 3 matrix.

```
X = [1 2 3]; Y = [4 5 6]; Z = [7 8 9];
A = [X;Y;Z]
B = cat(1,X,Y,Z)
```

Similarly, the brackets can be used to delete a row of a matrix.

15. Delete the last row (row 3) of the matrix `A`. Note how the colon operator is used to specify the entire row.

```
A(3,:) = []
```

A vector with N elements is an array with one row and N columns. An element of a vector can be accessed easily by addressing the number of the element, as in `X(5)`, which would access the fifth element of vector `X`. An element of a two-dimensional matrix is accessed by first specifying the row, then the column, as in `X(2,5)`, which would return the element at row 2, column 5. Matrices of dimensions higher than 2 can be accessed in a similar fashion. It is relevant to note that arrays in MATLAB are 1-based—the first element of an array is assigned or accessed using 1, as opposed to 0, which is the standard in many programming languages.

16. Use the `ones` and `rand` functions to create multidimensional arrays.

```
A = ones(4,3,2);
B = rand(5,2,3);
size(A)
size(B)
disp(A)
disp(B)
```

Question 9 What does the `size` function do?

Question 10 What does the `disp` function do?

Operations Involving Matrices

Performing arithmetic operations on matrices can be achieved with the operators `+ - * /`. The default for the multiply (`*`) and divide (`/`) operators is matrix multiplication and matrix division. To perform arithmetic operations on individual elements of a matrix, precede the operator with a dot (`.`).

17. Perform matrix multiplication on two matrices.

```
X = [1 2 -2; 0 -3 4; 7 3 0]
Y = [1 0 -1; 2 3 -5; 1 3 5]
X * Y
```

18. Perform element-by-element multiplication.

```
X .* Y
```

19. Perform another matrix multiplication on two matrices.

```
X = eye(3,4)
Y = rand(4,2)
X * Y
Y * X
```

Question 11 Why did the last operation fail?

20. Use the `diag` and `trace` functions to perform operations on the diagonal of a matrix.

```
Y = rand(3,3)*4
Y_diag = diag(Y)
Y_trace = trace(Y)
```

Question 12 What does the `diag` function do?

Question 13 What does the `trace` function do?

Question 14 Write an alternative statement that would produce the same results as the `trace` function.

21. Calculate the transpose of a matrix.

```
Y
Y_t = Y'
```

22. Calculate the inverse of a matrix and show that $YY^{-1} = Y^{-1}Y = I$, where I is the identity matrix.

```
Y_inv = inv(Y)
Y * Y_inv
Y_inv * Y
```

23. Calculate the determinant of a matrix.

```
Y_det = det(Y)
```

Cell Array

As demonstrated earlier, a matrix is the fundamental data type in MATLAB. It resembles the classical definition of an array as a homogeneous data structure, that is, one in which all its components are of the same type. *Cell arrays*, on the other hand, are another type of array where each cell can be any data type allowed by MATLAB. Each cell is independent of another and, therefore, can contain any data type that MATLAB supports. When using cell arrays, one must be careful when accessing or assigning a value to a cell; instead of parentheses, curly braces ({ }) must be used.

24. Execute the following lines of code one at a time to see how cell arrays are handled in MATLAB.

```
X{1} = [1 2 3;4 5 6;7 8 9];    %Cell 1 is a matrix
X{2} = 2+3i;                   %Cell 2 is complex
X{3} = 'String';               %Cell 3 is a string
X{4} = 1:2:9;                  %Cell 4 is a vector
X
celldisp(X)
X(1)
X{1}
```

Question 15 What does the `celldisp` function do?

Question 16 What does the percent (`%`) character do?

Question 17 What is the difference between the last two lines in the code above (`X(1)` as opposed to `X{1}`)?

There is another way to assign values to a cell array that is syntactically different, but yields the same results. Note in the next step how the cell index is enclosed within normal parentheses (( )), but the data that will be saved to the cell is encapsulated by curly braces ({ }).

25. Execute this line to see another way of assigning cell array values.

```
X(1) = {[1 2 3;4 5 6;7 8 9]};
```

26. The next few lines of code will demonstrate proper and improper ways of cell array assignment when dealing with strings.

```
X(3) = 'This produces an error'
X(3) = {'This is okay'}
X{3} = 'This is okay too'
```

Structures

Structures are yet another way of storing data in MATLAB. The syntax for structures is similar to that of other programming languages. We use the dot (`.`) operator to refer to different fields in a structure. Structures with identical layout (number of fields, their names, size, and meaning) can be combined in an array (of structures).

27. Create an array of two structures that represents two images and their sizes.

```
my_images(1).imagename = 'Image 1';
my_images(1).width = 256;
my_images(1).height = 256;
```

```
my_images(2).imagename = 'Image 2';
my_images(2).width = 128;
my_images(2).height = 128;
```

28. View details about the structure and display the contents of a field.

```
my_images(1)
my_images(2).imagename
```

29. Display information about the structure.

```
num_of_images = prod(size(my_images))
fieldnames(my_images)
class(my_images)
isstruct(my_images)
isstruct(num_of_images)
```

Question 18 What does the `fieldnames` function do?

Question 19 What does it mean when the result from the function `isstruct` is 1? What does it mean when it is 0?

Question 20 Use the help system to determine what function can be used to delete a field from a structure.

3.7 TUTORIAL 3.3: PROGRAMMING IN MATLAB

Goal

The goal of this tutorial is to explore programming concepts using scripts and functions.

Objectives

- Learn how to write and execute scripts.
- Explore the MATLAB Editor.
- Explore functions and how to write them.
- Learn the basics of loop vectorization.

What You Will Need

- `script3_3_1.m`
- `script3_3_2.m`
- `raise_to_power.m`

Procedure

Although the command window is simple and convenient to use, it does not provide a way of saving or editing your code. The commands stored in the *Command History* window can, however, be easily made into a script file. A script is a text file with the extension **.m** and is used to store MATLAB code. Scripts can be created and modified using the built-in editor.

1. To start a new script, navigate to **File > New > M-File**.

The MATLAB Editor may open in a separate window or it may be docked within the MATLAB environment, depending on how your environment was previously set up.

2. Open the file named `script3_3_1.m`. If the M-file is located in the current directory, you may also open it from the *Current Directory* listing by double-clicking the file name.

M-files are syntax color coded to aid in reading them. As you may have noticed, comments can be added to any script using two methods: percent (`%`) signs, or wrapping the comments with `%{` and `%}`. The second method is used for the header information in the script.

Question 1 Based on the script in file `script3_3_1.m`, what is the difference between using a percent (`%`) sign and using `%{` and `%}`?

To execute one or several lines of code in a script, highlight the code and press F9.[7]

3. Highlight all the code in `script3_3_1.m` and press F9 to execute the script.

The Cell Mode

A new editing mode (available in MATLAB 7.0 and above) called *cell mode* allows you to make minor adjustments to variables, and re-execute a block of code easily.

[7]Shortcut keys may vary for different versions of MATLAB for different operating systems. The shortcut keys listed in this book work for the PC version of MATLAB.

4. Open file `script3_3_2.m`.
5. Enable cell mode by navigating to **Cell > Enable Cell Mode** in the MATLAB environment.

With cell mode activated, you will note that the comments that are preceded with two percent signs (`%%`) are slightly bolder than the other comments. This corresponds to a *block title*. To create a cell block, all you need to do is give it a block title. Cell blocks allow you to modify and execute one block of code at a time. More than one cell block can exist in a single script.

6. Execute all the code in the script.

This script is an example of stretching the histogram of a monochrome image to the full dynamic range of allowed values, achieved by the `imadjust` function. This function also allows us to make gamma adjustments to the image. At this time, the concepts of histogram stretching and gamma correction are not important; they will be discussed in detail in Chapter 10. Instead, let us focus on how cell mode may be useful to us later on. By making small changes to the value of gamma, we can see the effects of gamma on the adjusted image. Cell mode makes this task easy.

7. Locate the line of code where the variable gamma is assigned a value of 1. Highlight *only* the value 1.

When cell mode is active, a new menu bar appears in the Editor. Among the icons, there are two text boxes that allow us to make adjustments to the value we just selected in the code. The first text box allows us to subtract and add, and the other allows us to divide and multiply the value we highlighted in the code.

8. In the add/subtract text box, type a value of 0.1 and press the plus (+) sign to the right.

Question 2 What happened to the value of the variable `gamma` in the code?

Question 3 In addition to the modification of code, what else happened when the plus button was pressed? What happens when it is pressed again?

Question 4 Other than the ones described in this tutorial, what features are available when cell mode is active?

Functions

Functions are also **.m** files that contain MATLAB code, but with the added ability of being able to be called from another script or function as well as receiving parameters. Open the `raise_to_power` function to see how the file is constructed.

9. Open `raise_to_power.m` in the Editor.

The M-file of a function contains all necessary information to call the function, execute it, and display its help documentation. The first line in the file defines the function with its input and output parameters. In the case of the `raise_to_power` function, it takes two parameters and returns one. Any comments immediately after the function definition are considered help documentation.

10. Ensure that the `raise_to_power.m` file is located in a directory that is part of either a set path or the current directory.
11. View the help information for the function by typing the following statement into the command window:

```
help raise_to_power
```

Question 5 Compare the help documentation with the comments in the M-file. How does MATLAB determine which comments are to be displayed for help and which are just comments in the function code?

Because we are using MATLAB for image manipulation, it is important to ensure that we are taking advantage of CPU and memory-efficient coding practices—particularly *loop vectorization*. Take, for example, the following block of pseudocode that could easily be implemented in any programming language. *NB*: Depending on your computer's speed, you may want to change the value of `MAX_CNT` in the code snippets below.

```
MAX_CNT = 10000
for i = 1 to MAX_CNT
   do x(i) = i ^ 2
```

Here, we are populating an array in which each element is the index of that element squared. The following MATLAB code implements the pseudocode above using typical programming techniques. In these examples, the `tic` and `toc` commands are used to calculate the execution time.

12. Implement the pseudocode above with the following statements.

```
tic
MAX_CNT = 10000
for i = 1:MAX_CNT
   x(i) = i ^ 2;
end
toc
```

We could greatly influence the speed of this loop by simply preallocating the memory used by the array. This is effective because every time we add data to the matrix, new memory must be allocated to hold the larger variable if there is no room for it. If the correct amount of memory is already allocated, then MATLAB only needs to alter the data in each cell.

13. Implement the previous loop, but with preallocation.

```
tic
MAX_CNT = 10000
x = zeros(1,MAX_CNT);
for i = 1:MAX_CNT
   x(i) = i ^ 2;
end
toc
```

Question 6 By what factor did preallocating the array x increase performance?

In MATLAB, this is still considered poor programming technique. MATLAB acts as an interpreter between the code you write and the hardware of your computer. This means that each time a statement is encountered, it is interpreted to a lower level language that is understood by the hardware. Because of this, loops, such as the one above, cause MATLAB to interpret the statement in the loop, however, many times the loop executes—in the case above, 10,000 times. Interpretation is slow compared to the speed of computer hardware. Since MATLAB operates on matrices natively, we can perform the same operation using *loop vectorization*, thus getting rid of the loop.

14. Implement a more efficient version of the loop.

```
tic
MAX_CNT = 10000
i = 1:MAX_CNT;
x = i .^ 2;
toc
```

Question 7 In the code above, why do we use `.^` instead of just `^`?

Question 8 How much faster is loop vectorization than our previous two implementations?

Here we do not need to explicitly tell MATLAB to perform the operation on each element because that is the nature of the MATLAB environment. Loop vectorization also takes care of the preallocation problem we saw in the first implementation.

Question 9 Consider the following pseudocode. Write a vectorized equivalent.

```
i = 0
for t = 0 to 2*pi in steps of pi/4
   do i = i + 1
      x(i) = sin(t)
```

Earlier it was mentioned that a script can be made from the *Command History* window. Now that we have entered some commands, we can see how this can be done.

15. To create a script from the *Command History* window, locate the last four statements entered. To select all the four statements, hold down the **ctrl** key and select each statement, and then right-click the highlighted statements, and select **Create M-File**.

WHAT HAVE WE LEARNED?

- MATLAB (MATrix LABoratory) is a data analysis, prototyping, and visualization tool with built-in support for matrices and matrix operations, excellent graphics capabilities, and a high-level programming language and development environment. It has been selected as the tool of choice for this book because of its ease of use and extensive built-in support for image processing operations, encapsulated in the Image Processing Toolbox.
- The MATLAB working environment consists of the *MATLAB Desktop*, the *MATLAB Editor*, and the *Help System*.
- M-files in MATLAB can be *scripts* that simply execute a series of MATLAB *commands* (or *statements*) or can be *functions* that can accept *arguments* (*parameters*) and produce one or more *output values*. User-created functions extend the capabilities of MATLAB and its toolboxes to address specific needs or applications.
- An M-file containing a *script* consists of a sequence of commands to be interpreted and executed much like a batch file. An M-file containing a *function*, on the other hand, contains a piece of code that performs a specific task, has a meaningful name (by which other functions will call it), and (optionally) receives parameters and returns results.
- The best ways to get started with MATLAB are to explore its excellent built-in product documentation and the online resources at MathWorks web site.

LEARN MORE ABOUT IT

There are numerous books on MATLAB, such as the following:

- Hanselman, D. and Littlefield, B., *Mastering MATLAB 7*, Upper Saddle River, NJ: Prentice Hall, 2005. Excellent reference book.
- Pratap, R., *Getting Started with MATLAB*, New York, NY: Oxford University Press, 2002. Great book for beginners.

ON THE WEB

There are many free MATLAB tutorials available online. The book's companion web site (http: //www.ogemarques.com/) contains links to several of them.

3.8 PROBLEMS

3.1 Using MATLAB as a calculator, perform the following calculations:

(a) 24.4×365

(b) $\cos(\pi/4)$

(c) $\sqrt[3]{45}$

(d) $e^{-0.6}$

(e) 4.6^5

(f) $y = \sum_{i=1}^{6} (i^2 - 3)$

3.2 Use the `format` function to answer the following questions:

(a) What is the precision used by MATLAB for floating-point calculations?

(b) What is the default number of decimal places used to display the result of a floating-point calculation in the command window?

(c) How can you change it to a different number of decimal places?

3.3 Initialize the following variables: $x = 345.88$; $y = \log_{10}(45.8)$; and $z = \sin(3\pi/4)$. Note that the MATLAB function `log` calculates the natural logarithm; you need to use `log10` to calculate the common (base-10) logarithm of a number.

Use them to calculate the following:

(a) $e^{(z^3 - y^{1.2})}$

(b) $x^2 - \sqrt{(4y + z)}$

(c) $\frac{x-y}{z+5y}$

3.4 Initialize the following matrices:

$$X = \begin{bmatrix} 4 & 5 & 1 \\ 0 & 2 & 4 \\ 3 & 4 & 1 \end{bmatrix}$$

$$Y = \begin{bmatrix} -7 & 6 & -1 \\ 3 & -2 & 0 \\ 13 & -4 & 1 \end{bmatrix}$$

$$Z = \begin{bmatrix} 0 & 7 & 1 \\ 0 & 2 & -5 \\ 3 & 3 & -1 \end{bmatrix}$$

Use them to calculate the following:

(a) $3X + 2Y - Z$

(b) $X^2 - Y^3$

(c) $X^T Y^T$

(d) XY^{-1}

3.5 What is the difference between the following MATLAB functions:

(a) `log` and `log10`

(b) `rand` and `randn`

(c) `power` and `mpower`

(d) `uminus` and `minus`

3.6 What is the purpose of function `lookfor`? Provide an example in which it can be useful.

3.7 What is the purpose of function `which`? Provide an example in which it can be useful.

CHAPTER 4

THE IMAGE PROCESSING TOOLBOX AT A GLANCE

WHAT WILL WE LEARN?

- How do I read an image from a file using MATLAB?
- What are the main data classes used for image representation and how can I convert from one to another?
- Why is it important to understand the data class and range of pixel values of images stored in memory?
- How do I display an image using MATLAB?
- How can I explore the pixel contents of an image?
- How do I save an image to a file using MATLAB?

4.1 THE IMAGE PROCESSING TOOLBOX: AN OVERVIEW

The Image Processing Toolbox (IPT) is a collection of functions that extend the basic capability of the MATLAB environment to enable specialized signal and image processing operations, such as the following:

- Spatial transformations (Chapter 7)
- Image analysis and enhancement (Chapters 8–12, 14–15, and 18)

Practical Image and Video Processing Using MATLAB®. By Oge Marques.

- Neighborhood and block operations (Chapter 10)
- Linear filtering and filter design (Chapters 10 and 11)
- Mathematical transforms (Chapters 11 and 17)
- Deblurring (Chapter 12)
- Morphological operations (Chapter 13)
- Color image processing (Chapter 16)

Most of the toolbox functions are MATLAB M-files, whose source code can be inspected by opening the file using the MATLAB editor or simply typing `type function_name` at the prompt. The directory in which the M-file has been saved can be found by typing `which function_name` at the prompt. To determine which version of the IPT is installed on your system, type `ver` at the prompt.

You can extend the capabilities of the IPT by writing your own M-files, modifying and expanding existing M-files, writing wrapping functions around existing ones, or using the IPT in combination with other toolboxes.

4.2 ESSENTIAL FUNCTIONS AND FEATURES

This section presents the IPT functions used to perform essential image operations, such as reading the image contents from a file, converting between different data classes used to store the pixel values, displaying an image on the screen, and saving an image to a file.

4.2.1 Displaying Information About an Image File

MATLAB's IPT has a built-in function to display information about image files (without opening them and storing their contents in the workspace) `imfinfo`.

■ EXAMPLE 4.1

The MATLAB code below reads information about a built-in image file `pout.tif`.

```
imfinfo('pout.tif');
```

The resulting structure (stored in variable `ans`) will contain the following information:[1]

```
       Filename: '/.../pout.tif'
    FileModDate: '04-Dec-2000 13:57:50'
       FileSize: 69004
         Format: 'tif'
```

[1] All empty fields have been omitted for space reasons.

```
                     Width: 240
                    Height: 291
                  BitDepth: 8
                 ColorType: 'grayscale'
           FormatSignature: [73 73 42 0]
                 ByteOrder: 'little-endian'
            NewSubFileType: 0
             BitsPerSample: 8
               Compression: 'PackBits'
 PhotometricInterpretation: 'BlackIsZero'
              StripOffsets: [9x1 double]
           SamplesPerPixel: 1
              RowsPerStrip: 34
           StripByteCounts: [9x1 double]
               XResolution: 72
               YResolution: 72
            ResolutionUnit: 'Inch'
       PlanarConfiguration: 'Chunky'
               Orientation: 1
                 FillOrder: 1
          GrayResponseUnit: 0.0100
           MaxSampleValue: 255
            MinSampleValue: 0
              Thresholding: 1
```

Many of these fields are too technical and some are file format dependent. Nonetheless, you should still be able to locate information about the image size (240×291 pixels), the file size (69,004 bytes), the type of image (grayscale), the number of bits per pixel (8), and its minimum and maximum values (0 and 255).

Repeating the procedure for the built-in image `coins.png` will produce the following results:[2]

```
       Filename: '/.../coins.png'
    FileModDate: '16-Apr-2003 02:05:30'
       FileSize: 37906
         Format: 'png'
          Width: 300
         Height: 246
       BitDepth: 8
      ColorType: 'grayscale'
FormatSignature: [137 80 78 71 13 10 26 10]
  InterlaceType: 'none'
```

[2]All empty fields have been omitted for space reasons.

```
   Transparency: 'none'
   ImageModTime: '19 Sep 2002 20:31:12 +0000'
      Copyright: 'Copyright The MathWorks, Inc.'
```

Once again, you should be able to find basic information about the image, regardless of the several (potentially obscure) fields that are file format dependent. You may also have noticed that several fields have changed between the first and second files as a consequence of the specific file formats (TIFF and PNG, respectively).

Finally, let us repeat this command for a truecolor (24 bits per pixel) image `peppers.png`. The result will be[3]

```
       Filename: '/.../peppers.png'
    FileModDate: '16-Dec-2002 06:10:58'
       FileSize: 287677
         Format: 'png'
          Width: 512
         Height: 384
       BitDepth: 24
      ColorType: 'truecolor'
FormatSignature: [137 80 78 71 13 10 26 10]
  InterlaceType: 'none'
   Transparency: 'none'
   ImageModTime: '16 Jul 2002 16:46:41 +0000'
    Description: 'Zesty peppers'
      Copyright: 'Copyright The MathWorks, Inc.'
```

4.2.2 Reading an Image File

MATLAB's IPT has a built-in function to open and read the contents of image files in most popular formats (e.g., TIFF, JPEG, BMP, GIF, and PNG), `imread`.

The `imread` function allows you to read image files of almost any type, in virtually any format, located anywhere. This saves you from a (potentially large) number of problems associated with file headers, memory allocation, file format conventions, and so on, and allows to focus on what you want to do to the image once it has been read and stored into a variable in the MATLAB workspace.

The IPT also contains specialized functions for reading DICOM (Digital Imaging and Communications in Medicine) files (`dicomread`), NITF (National Imagery Transmission Format) files (`nitfread`), and HDR (high dynamic range) files (`hdrread`).

[3] All empty fields have been omitted for space reasons.

TABLE 4.1 IPT Functions to Perform Image Data Class Conversion

Name	Converts an Image to Data Class
`im2single`	`single`
`im2double`	`double`
`im2uint8`	`uint8`
`im2uint16`	`uint16`
`im2int16`	`int16`

4.2.3 Data Classes and Data Conversions

The IPT ability to read images of any type, store their contents into arrays, and make them available for further processing and display does not preclude the need to understand how the image contents are represented in memory. This section follows up on our discussion in Section 3.2.2 and explains the issue of data classes, data conversions, and when they may be needed.

The most common data classes for images are as follows:

- `uint8`: 1 byte per pixel, in the [0, 255] range
- `double`: 8 bytes per pixel, usually in the [0.0, 1.0] range
- `logical`: 1 byte per pixel, representing its value as *true* (1 or white) or *false* (0 or black)

Once the contents of an image have been read and stored into one or more variables, you are encouraged to inspect the data class of these variables and their range of values to understand *how* the pixel contents are represented and what is their allowed range of values. You should ensure that the data class of the variable is compatible with the input data class expected by the IPT functions that will be applied to that variable. If you are writing your own functions and scripts, you must also ensure data class compatibility, or perform the necessary conversions.[4]

MATLAB allows data class conversion (*typecasting*) to be done in a straightforward way, but this type of conversion does not handle the range problem and is usually *not* what you want to do. To convert an image (or an arbitrary array for that matter) to a data class and range suitable for image processing, you are encouraged to use one of the specialized functions listed in Table 4.1. The input data class for any of those functions can be `logical`, `uint8`, `uint16`, `int16`, `single`, or `double`.

[4]Some of the problems that may arise by overlooking these issues will only be fully appreciated once you acquire some hands-on experience with the IPT. They include (but are not limited to) the following: unwillingly truncating intermediate calculation results (e.g., by using an unsigned data class that does not allow negative values), setting wrong thresholds for pixel values (e.g., by assuming that the pixels range from 0 to 255, when in fact they range from 0.0 to 1.0), and masking out some of the problems by displaying the image contents using the *scale for display purposes* option. These issues will be referred to occasionally throughout the book.

The IPT also contains other functions (not listed in Table 4.1) to convert the following:

- An image (of data class `uint8`, `uint16`, `single`, `int16`, or `double`) to a binary image (of data class `logical`) based on threshold: `im2bw`. Note that this is not simply a type (data class) conversion, but instead an implementation of a global thresholding algorithm, which will be presented in Chapter 15.
- An image to an instance of the Java image class `java.awt.Image`: `im2java`.
- An image to an instance of the Java image class `java.awt.image.BufferedImage`: `im2java2d`.
- An image to a movie frame, `im2frame`, which will be discussed—together with related functions such as `frame2im` and `movie`—in Part II of the book.
- A matrix to a grayscale image, `mat2gray`, where the input can be `logical` or any numeric class, and the output is an array of data class `double` (with values within the [0.0, 1.0] range).

■ EXAMPLE 4.2

It is possible to convert a generic 2 × 2 array of class `double` (A) into its `uint8` equivalent (B) by simply typecasting it.

```
A =

   -8.0000    4.0000
         0    0.5000

>> B = uint8(A)

B =

    0    4
    0    1
```

As you may have noticed, the conversion consisted of truncation (all negative values became zero) and rounding off (0.5 became 1).

Using `im2uint8` will lead to results within a normalized range ([0, 255])

```
>> C = im2uint8(A)

C =

    0   255
    0   128
```

from which we can infer that it treated the original values as if they were in the [0, 1] range for data class `double`, that is, 0.5 became 128 (midrange point), anything less than or equal to 0 was truncated to 0, and anything greater than or equal to 1 was truncated to 255.

Let us now apply `mat2gray` to A and inspect the results:

```
>> D = mat2gray(A)

D =

         0    1.0000
    0.6667    0.7083
```

This result illustrates an important point: since A was already of data class `double`, there was no data class convention *per se*, but simply a range conversion—the smallest value (−8.0) became 0.0, the largest value (4.0) became 1.0, and all intermediate values were scaled within the new range.

Finally, let us use `im2bw` to convert our images to binary equivalents.

```
>> E = im2bw(D, 0.4)

E =

     0     1
     1     1
```

The results work as expected: any value greater than 0.4 becomes 1 (true), otherwise it becomes 0 (false). Note that since `im2bw` expects a threshold luminance level as a parameter, and requires it to be a nonnegative number between 0 and 1, it *implicitly* assumes that the input variable (*D*, in this case) is a normalized grayscale image of class `double`. In other words, if you try to use A as an input image (`E = im2bw(A, 0.4)`), the function will work without any error or warning, but the result may not make sense.

The IPT also includes functions to convert between RGB (truecolor), indexed image, and grayscale image, which are listed in Table 4.2. In Tutorial 4.2 (page 74), you will have a chance to experiment with some of these functions.

TABLE 4.2 IPT Functions to Perform Image Data Class Conversion

Name	Description	
	Converts	Into
`ind2gray`	An indexed image	Its grayscale equivalent
`gray2ind`	A grayscale image	An indexed representation
`rgb2gray`	An RGB (truecolor) image	Its grayscale equivalent
`rgb2ind`	An RGB (truecolor) image	An indexed representation
`ind2rgb`	An indexed color image	Its RGB (truecolor) equivalent

4.2.4 Displaying the Contents of an Image

MATLAB has several functions for displaying images:

- `image`: displays an image using the current color map.[5]
- `imagesc`: scales image data to the full range of the current color map and displays the image.
- `imshow`: displays an image and contains a number of optimizations and optional parameters for property settings associated with image display.
- `imtool`: displays an image and contains a number of associated tools that can be used to explore the image contents.

■ EXAMPLE 4.3

The following MATLAB code is used to open an image file and display it using different `imshow` options:

```
I = imread('pout.tif');
imshow(I)
figure, imshow(I,[])
figure, imshow(I,[100 160])
```

The first call to `imshow` displays the image in its original state. The following line opens a new figure and displays a scaled (for display purposes) version of the same image.[6] The last line specifies a range of gray levels, such that all values lower than 100 will be displayed as black and all values greater than 160 will be displayed as white. The three results are shown side by side in Figure 4.1.

[5]The color map array is an M × 3 matrix of class *double*, where each element is a floating-point value in the range [0, 1]. Each row in the color map represents the R (red), G (green), and B (blue) values for that particular row.

[6]The fact that the result is a much improved version of the original suggests that the image could be improved by image processing techniques, such as *histogram equalization* or *histogram stretching*, which will be described in Chapter 9.

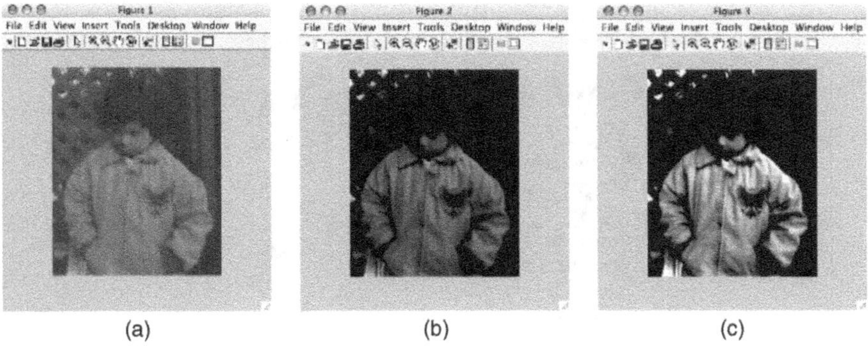

FIGURE 4.1 Displaying an image: (a) without scaling; (b) scaling for display purposes; (c) selecting only pixels within a specified range. Original image: courtesy of MathWorks.

4.2.5 Exploring the Contents of an Image

Image processing researchers and practitioners often need to inspect the contents of an image more closely. In MATLAB, this is usually done using the `imtool` function, which provides all the image display capabilities of `imshow` as well as access to other tools for navigating and exploring images, such as the *Pixel Region* tool (Figure 4.2), *Image Information* tool (Figure 4.3), and *Adjust Contrast* tool (Figure 4.4). These tools can also be directly accessed using their library functions `impixelinfo`, `imageinfo`, and `imcontrast`, respectively.

Before `imtool` was introduced, the syntax `imshow(I), pixval on` was used to enable the display of coordinates and values of a moused-over pixel.

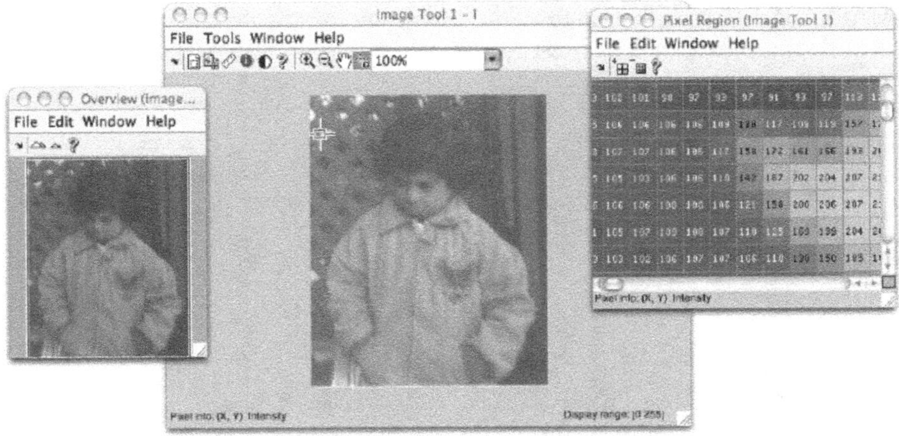

FIGURE 4.2 Displaying an image and exploring its contents with the *Pixel Region* tool. Original image: courtesy of MathWorks.

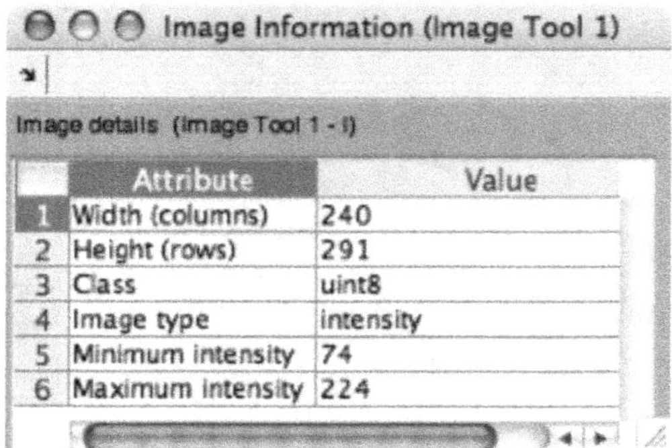

FIGURE 4.3 The *Image Information* tool.

Since `pixval` is now obsolete, the alternative way to do it is by using `imshow(I), impixelinfo`.

Two other relevant IPT functions for inspecting image contents and pixel values are

- `impixel`: returns the red, green, and blue color values of image pixels specified with the mouse.
- `imdistline`: creates a *Distance* tool—a draggable, resizable line that measures the distance between its endpoints. You can use the mouse to move and resize the line to measure the distance between any two points within an image (Figure 4.5).

4.2.6 Writing the Resulting Image onto a File

MATLAB's IPT has a built-in function, `imwrite`, to write the contents of an image in one of the most popular graphic file formats.

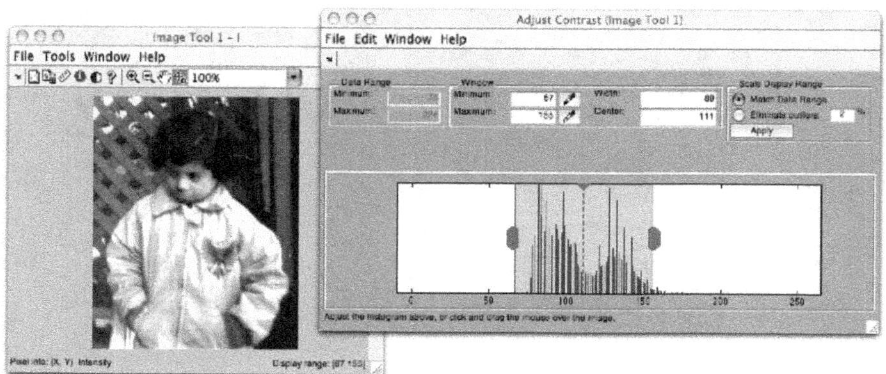

FIGURE 4.4 The *Adjust Contrast* tool. Original image: courtesy of MathWorks.

FIGURE 4.5 The *Distance* tool. Original image: courtesy of MathWorks.

If the output file format uses lossy compression[7] (e.g., JPEG), `imwrite` allows the specification of a *quality* parameter, used as a trade-off between the resulting image's subjective quality and the file size.

■ EXAMPLE 4.4

In this example, we will read an image from a PNG file and save it to a JPG file using three different quality parameters: 75 (default), 5 (poor quality, small size), and 95 (better quality, larger size).

```
I = imread('peppers.png');
imwrite(I, 'pep75.jpg');
imwrite(I, 'pep05.jpg', 'quality', 5);
imwrite(I, 'pep95.jpg', 'quality', 95);
```

The results are displayed in Figure 4.6. Figure 4.6c is clearly of lower visual quality than Figure 4.6b and d. On the other hand, the differences between Figure 4.6b and d are not too salient.

[7]You will learn about compression in Chapter 17.

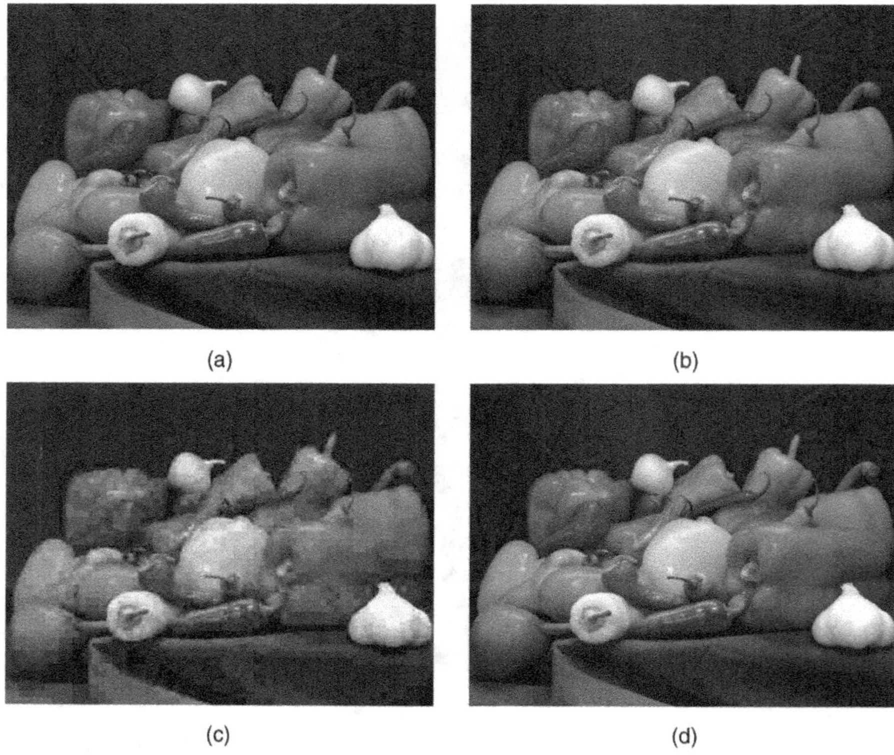

FIGURE 4.6 Reading and writing images: (a) Original image (PNG); (b) compressed image (JPG, $q = 75$, file size = 24 kB); (c) compressed image (JPG, $q = 5$, file size = 8 kB); (d) compressed image (JPG, $q = 95$, file size = 60 kB). Original image: courtesy of MathWorks.

4.3 TUTORIAL 4.1: MATLAB IMAGE PROCESSING TOOLBOX—A GUIDED TOUR

Goal

The goal of this tutorial is to introduce the capabilities of the Image Processing Toolbox and demonstrate how functions can be combined to solve a particular problem.

Objectives

- Learn how to use the help documentation for the IPT.
- Explore the *Identifying Round Objects Demo.*

Procedure

The MATLAB IPT offers rich functionality that may be used as building blocks for an imaging system. Let us begin by first exploring the help documentation that comes with the toolbox.

1. Open the help documentation for the IPT by navigating from the start menu located in the lower left corner of the MATLAB environment: **Start > Toolboxes > Image Processing > Help**.

The home page of the help document for the IPT provides links to several key areas of the document, such as categorical or alphabetical listings of functions, user guides, and demos. The user guides are broken up even further, including categories such as *Spatial Transformations* and *Image Segmentation*.

An easy and quick way to get an idea of what the IPT is capable of is to look at some of the demos provided with the toolbox. An example of an IPT demo is the *Identifying Round Objects Demo*, a sequence of steps used to demonstrate how IPT functions can be used and combined to develop algorithms for a particular application—in this case, identifying round objects. Most of the syntax may be unfamiliar to you at this point, but following the explanations associated with the code can give insight into its functionality. Any block of code in these demos can be evaluated by highlighting the code and selecting the **Evaluate Selection** option from the context-sensitive menu. Alternatively, you can open the `ipexroundness.m` file in the MATLAB Editor and evaluate one cell at a time. You can also evaluate each of the (six) main blocks of code by clicking on the **Run in the Command Window** hyperlink at the top right of the demo HTML page.[8]

2. Open the *Identifying Round Objects Demo* by navigating to the IPT demos: **Start > Toolboxes > Image Processing > Demos**. It will be under the **Measuring Image Features** category.
3. Run the first block of code.

As expected, the image is displayed (using the `imshow` function).

4. Execute the remaining code in the demo one block at a time. Focus on understanding *what* each step (block) does, rather than *how* it is done.

Question 1 Run (at least) two other demos from different categories within the IPT and comment on how each demo illustrates the capabilities of the IPT.

[8]You can run the entire demo by typing ipexroundness at the command prompt, but you would only see the final result and miss all the intermediate steps and results, which is the point of the demo.

4.4 TUTORIAL 4.2: BASIC IMAGE MANIPULATION

Goal

The goal of this tutorial is to explore basic image manipulation techniques using MATLAB and IPT.

Objectives

- Explore the different image types supported by MATLAB and IPT.
- Learn how to read images into MATLAB.
- Explore image conversion.
- Learn how to display images.
- Learn how to write images to disk.

Procedure

The IPT supports images of type binary, indexed, intensity, and truecolor. Before an image can be processed in MATLAB, it must first be loaded into memory. To read in an image, we use the `imread` function.

1. Load the image `coins.png` by executing the following statement:

```
I = imread('coins.png');
```

Question 1 What type of image is `coins.png`?

Question 2 Why do we use the semicolon (`;`) operator after the `imread` statement? What happens if we omit it?

Binary, intensity, and truecolor images can all be read with the `imread` function as demonstrated above. When reading in an indexed image, we must specify variables for both the image and its color map. This is illustrated in the following step:

2. Load the image `trees.tif`.

```
[X,map] = imread('trees.tif');
```

Some operations may require you to convert an image from one type to another. For example, performing image adjustments on an indexed image may not give the results you are looking to achieve because the calculations are performed on the index values and not the representative RGB values. To make this an easier task, we can convert the indexed image to an RGB image using `ind2rgb`.

3. Convert the indexed image `X` with color map `map` to an RGB image, `X_rgb`.

```
X_rgb = ind2rgb(X,map);
```

Question 3 How many dimensions does the variable `X_rgb` have and what are their sizes?

4. Convert the indexed image `X` with color map `map` to an intensity image.

```
X_gray = ind2gray(X,map);
```

Question 4 What class type is `X_gray`?

5. We can verify that the new intensity image consists of pixel values in the range [0, 255].

```
max(X_gray(:))
min(X_gray(:))
```

Question 5 Why are we required to use the colon operator (`:`) when specifying the `X_gray` variable? What happens if we omit it?

It was demonstrated in the previous step that the `X_gray` image contained values in the range [0, 255] (in this particular image, they happened to be exactly 0 and 255, which is just a coincidence). Let us see what happens when we convert the image to class `double`.

6. Convert the variable `X_gray` to class `double`.

```
X_gray_dbl = im2double(X_gray);
```

Question 6 What is the range of values for the new variable `X_gray_dbl`?

Similarly, you can convert to other class types by using `im2uint8` and `im2uint16`, for example. When converting a `uint16` image to `uint8`, you must be careful because the conversion quantizes the 65,536 possible values to 256 possible values.

MATLAB comes with built-in image displaying functions. The `image` function can be used to display image data, and the `imagesc` function will perform the same operation but in addition will scale the image data to the full range of values. The IPT provides an enhanced image displaying function that optimizes settings on the image axes to provide a better display of image data: `imshow`.

7. Use the `imshow` function (with the `impixelinfo` option) to display the `coins.png` image that is currently loaded in the variable `I`.

```
imshow(I), impixelinfo
```

Binary, intensity, and truecolor images can all be displayed as demonstrated previously. To display indexed images, we must specify the color map along with the image data.

8. Display the indexed image `trees.tif`. The image data are stored in variable `X` and the color map in `map`. Note that the `impixelinfo` option provides a clear hint that this is an indexed color image.

```
imshow(X,map), impixelinfo
```

Question 7 Consider an image where the range of possible values for each pixel is not [0, 255], but a nonstandard range such as [0, 99]. How would we display the image so that a value of 99 represents white and a value of 0 represents black?

The `impixel` function allows the inspection of the contents of selected pixels of interest within the image.

9. Use the `impixel` function to explore interactively the pixel contents of selected points in the image. Use the mouse to click on the points of interest: normal button clicks are used to select pixels, pressing **Backspace** or **Delete** removes the previously selected pixel, a double-click adds a final pixel and ends the selection, and pressing **Return** finishes the selection without adding a final pixel.

```
RGB = imread('peppers.png');
[c,r,p] = impixel(RGB);
```

Question 8 What is the meaning of the values stored in variables `r`, `c`, and `p`?

The `improfile` function can be used to compute and plot the intensity values along a line or a multiline path in an image.

10. Use the `improfile` function to explore the contents of a line in the `coins.png` image that is currently loaded in the variable `I`.

```
r1 = 17; c1 = 18; r2 = 201; c2 = 286;
imshow(I)
line([c1, c2], [r1, r2], 'Color', 'g', 'LineWidth', 2);
figure
improfile(I, [c1, c2], [r1, r2]);
ylabel('Gray level');
```

The `imtool` function is the latest and richest IPT function for displaying images. It provides all the image display capabilities of `imshow` as well as access to other

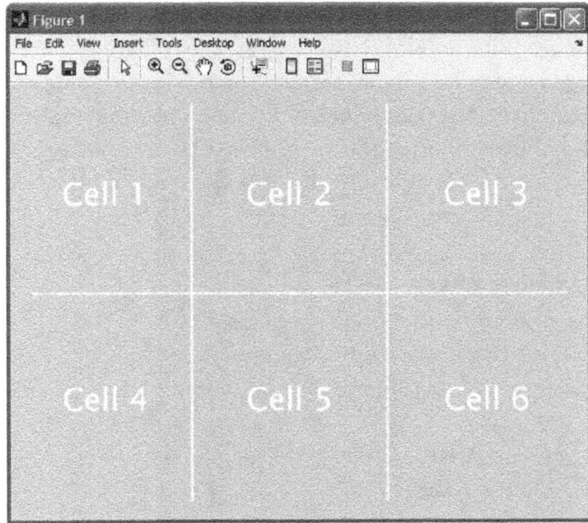

FIGURE 4.7 Division of a figure using subplot.

tools for navigating and exploring images, such as the *Pixel Region* tool, the *Image Information* tool, and the *Adjust Contrast* tool.

11. Use the `imtool` function to display the image currently loaded in the variable `X_rgb`. Note that a secondary window (Overview) will open as well. Explore the additional functionality, including the possibility of measuring distances between two points within the image.

```
imtool(X_rgb)
```

We can display multiple images within one figure using the `subplot` function. When using this function, the first two parameters specify the number of rows and columns to divide the figure. The third parameter specifies which subdivision to use. In the case of `subplot(2,3,3)`, we are telling MATLAB to divide the figure into two rows and three columns and set the third cell as active. This division is demonstrated in Figure 4.7.

12. Close any open figures (`close all`).
13. Execute the following statements to create a subplot with two images:

```
A = imread('pout.tif');
B = imread('cameraman.tif');
figure
subplot(1,2,1), imshow(A)
subplot(1,2,2), imshow(B)
```

Question 9 What is the range of values for image A and image B?

In the previous step, we displayed two images, both of which were intensity images. Even though there is no color map associated with intensity images, MATLAB uses a grayscale color map to display an intensity image (this happens in the background and is usually invisible to the user). Let us consider the case where an intensity image and an indexed image are both displayed in one figure, using the subplot function as before.

14. Close any open figures.
15. Display the coins.png (loaded in variable I) and the trees.tif (loaded in variable X and its color map in variable map) images in a subplot. Execute each statement at a time to see the effect on the images as they are displayed.

```
figure
subplot(1,2,1), imshow(I)
subplot(1,2,2), imshow(X,map)
```

Question 10 What happened to the coins image just after the trees image was displayed? Explain your answer.

To properly display images with different color maps, we must use the subimage function.

16. Use the subimage function to display multiple images with different color maps.

```
figure
subplot(1,2,1), subimage(I), axis off
subplot(1,2,2), subimage(X,map), axis off
```

The subimage function converts the image to an equivalent RGB image and then displays that image. We can easily do this ourselves, but there is no direct conversion from intensity to RGB, so we must first convert from intensity to indexed, and then from indexed to RGB.

17. Manually convert the intensity image coins (loaded in the variable I) to an indexed image and then to RGB. Note that the trees image (loaded in variable X with its color map in variable map) has already been converted to RGB in step 3 (saved in variable X_rgb).

```
[I_ind,I_map] = gray2ind(I,256);
I_rgb = ind2rgb(I_ind,I_map);
```

Question 11 What do the variables I_ind and I_map contain?

18. Display the truecolor images using the imshow function.

```
figure
subplot(1,2,1), imshow(I_rgb)
subplot(1,2,2), imshow(X_rgb)
```

19. Use `imwrite` to save two of the modified images in this tutorial to files for further use. Use the JPEG format for one of them and the PNG extension for the other. For example,

```
imwrite(X_rgb, 'rgb_trees.jpg');
imwrite(X_gray, 'gray_trees.png');
```

WHAT HAVE WE LEARNED?

- MATLAB's IPT has a built-in function to open and read the contents of image files in most popular formats, `imread`.
- The most common data classes for images are `uint8` (1 byte per pixel, [0, 255]), `double` (8 bytes per pixel, [0.0, 1.0]), and `logical` (1 byte per pixel, *true*—white or *false*—black).
- Data class compatibility is a critical prerequisite for image processing algorithms to work properly, which occasionally requires data class conversions. In addition to standard data class conversion (*typecasting*), MATLAB has numerous specialized functions for image class conversions.
- A good understanding of the different data classes (and corresponding ranges of pixel values) of images stored in memory is essential to the success of image processing algorithms. Ignoring or overlooking these aspects may lead to unwillingly truncating of intermediate calculation results, setting of wrong thresholds for pixel values, and many other potential problems.
- MATLAB has several functions for displaying images, such as `image`, `imagesc`, and `imshow`.
- To inspect the pixel contents of an image more closely, MATLAB includes the `imtool` function that provides all the image display capabilities of `imshow` as well as access to other tools for navigating and exploring images.
- To save the results of your image processing to a file, use MATLAB's built-in function `imwrite`.

LEARN MORE ABOUT IT

The best ways to learn more about the IPT are

- IPT demos: the MATLAB package includes many demos, organized in categories, that provide an excellent opportunity to learn most of the IPT's capabilities.

- MATLAB webinars: short (1 h or less) technical presentations by MathWorks developers and engineers.

ON THE WEB

- MATLAB Image Processing Toolbox home page.
  http://www.mathworks.com/products/image/
- Steve Eddins's blog: Many useful hints about image processing concepts, algorithm implementations, and MATLAB.
  http://blogs.mathworks.com/steve/

4.5 PROBLEMS

4.1 Create a 2 × 2 array of type `double`

```
A = [1.5 -2 ; 0.5 0]
```

and use it to perform the following operations:

(a) Convert to `uint8` using typecasting and interpret the results.

(b) Convert to a normalized ([0, 255]) grayscale image of data class `uint8` using `im2uint8` and interpret the results.

(c) Convert to a normalized ([0.0, 1.0]) grayscale image of data class `double` using `mat2gray` and interpret the results.

(d) Convert the result from the previous step to a binary image using `im2bw` (with a threshold level of 0.5) and interpret the results.

4.2 Create a 2 × 2 array of type `double`

```
A = [1 4 ; 5 3]
```

and write MATLAB statements to perform the following operations:

(a) Convert to a normalized ([0.0, 1.0]) grayscale image of data class `double`.

(b) Convert to a binary image of data class `logical`, such that any values greater than 2 (in the original image) will be interpreted as 1 (true).

(c) Repeat the previous step, this time producing a result of data class `double`.

4.3 The IPT function `gray2ind` allows conversion from a grayscale image to its indexed equivalent. What interesting property will the resulting color map (palette) display?

4.4 How does the IPT function `rgb2ind` handle the possibility that the original RGB image contains many more colors than the maximum palette (color map) size (65,536 colors)?

4.5 If you type `help imdemos` in MATLAB, you will see (among other things) a list of all the sample images that come with its IPT. Select five of those images and collect the following information about each of them:

- File name
- File format (extension)
- Type (binary, grayscale, truecolor, or indexed color)
- Size (bytes)
- Width (pixels)
- Height (pixels)

4.6 Select five images available in MATLAB (they may be the same as from the previous problem, or not, you choose). Open each of them using `imread`, save it (using `imwrite`) to (at least three) different file formats, and compare the resulting file size (in bytes) for each output format.

CHAPTER 5

IMAGE SENSING AND ACQUISITION

WHAT WILL WE LEARN?

- What are the main parameters involved in the design of an image acquisition solution?
- How do contemporary image sensors work?
- What is image digitization and what are the main parameters that impact the digitization of an image or video clip?
- What is sampling?
- What is quantization?
- How can I use MATLAB to resample or requantize an image?

5.1 INTRODUCTION

In Chapter 1, we described an image as a two-dimensional (2D) representation of a real-world, three-dimensional (3D) object or scene and indicated that the existence of a light source illuminating the scene is a requirement for such an image to be produced. We also introduced the concept of a digital image as a representation of a two-dimensional image using a finite number of pixels.

In this chapter, we will expand upon those concepts. More specifically, we will look at relevant issues involved in acquiring and digitizing an image, such as the

Practical Image and Video Processing Using MATLAB®. By Oge Marques.

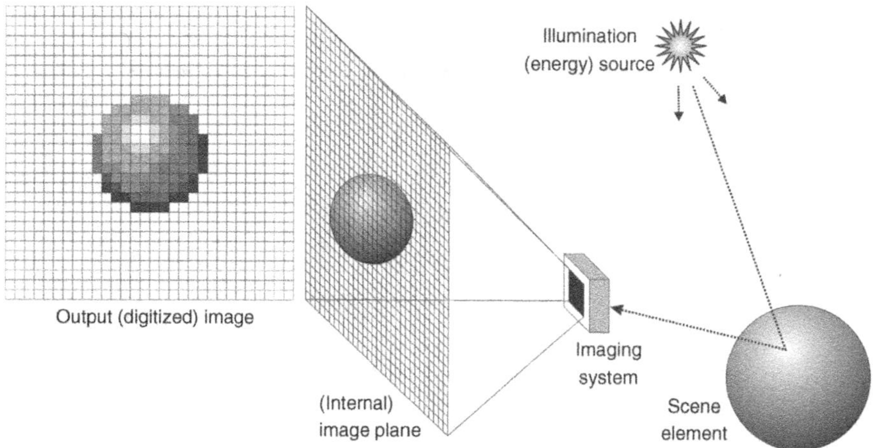

FIGURE 5.1 Image acquisition, formation, and digitization. Adapted and redrawn from [GW08].

principles of image formation as a result of reflection of light on an object or scene, the sensors typically used to capture the reflected energy, and the technical aspects involved in selecting the appropriate number of (horizontal and vertical) samples and quantization levels for the resulting image. In other words, we will present the information necessary to understand how we go from real-world scenes to 2D digital representations of those scenes. For the remaining chapters of Part I, we shall assume that such digital representations are available and will learn many ways of processing them, without looking back at how they were acquired.

Figure 5.1 shows a schematic view of the image acquisition, formation, and digitization process. The figure also highlights the need for an illumination (energy) source (see Section 5.2), the existence of an imaging sensor capable of converting optical information into its electrical equivalent (see Section 5.3), and the difference between the analog version of the image and its digitized equivalent, after having undergone sampling and quantization (see Section 5.4).

5.2 LIGHT, COLOR, AND ELECTROMAGNETIC SPECTRUM

The existence of light—or other forms of electromagnetic (EM) radiation—is an essential requirement for an image to be created, captured, and perceived. In this section, we will look at basic concepts related to light, the perception of color, and the electromagnetic spectrum.

5.2.1 Light and Electromagnetic Spectrum

Light can be described in terms of electromagnetic waves or particles, called *photons*. A photon is a tiny packet of vibrating electromagnetic energy that can be

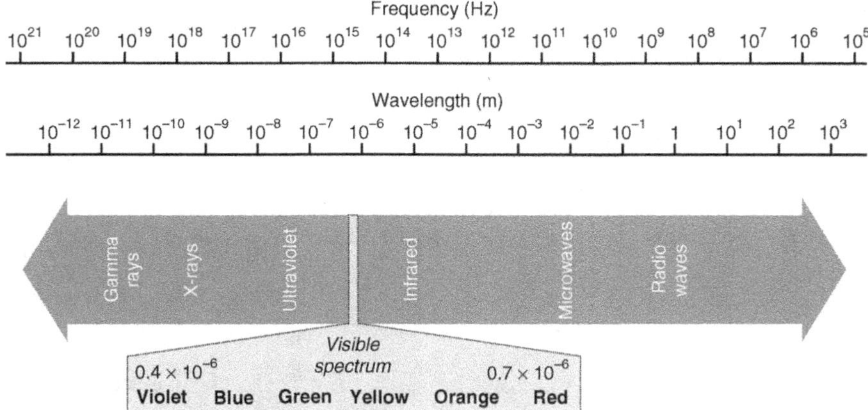

FIGURE 5.2 Electromagnetic spectrum.

characterized by its wavelength or frequency. Wavelength is usually measured in meters (and its multiples and submultiples). Frequency is measured in hertz (Hz) and its multiples. Wavelength (λ) and frequency (f) are related to each other by the following expression:

$$\lambda = \frac{v}{f} \tag{5.1}$$

where v is the velocity at which the wave travels, usually approximated to be equal to the speed of light (c): 2.998×10^8 m/s.

The human visual system (HVS) is sensitive to photons of wavelengths between 400 and 700 nm, where 1 nm $= 10^{-9}$ m. As shown in Figure 5.2, this is a fairly narrow slice within the EM spectrum, which ranges from *radio waves* (wavelengths of 1 m or longer) at one end to *gamma rays* (wavelengths of 0.01 nm or shorter) at the other end.

Even though much of the progress in image processing has been fostered by work on images outside the visible spectrum, captured with specialized sensors, this book will focus exclusively on images within the visible range of the EM spectrum. Light is the preferred energy source for most imaging tasks because it is safe, cheap, easy to control and process with optical hardware, easy to detect using relatively inexpensive sensors, and readily processed by signal processing hardware.

5.2.2 Types of Images

Images can be classified into three categories according to the type of interaction between the source of radiation, the properties of the objects involved, and the relative positioning of the image sensor (Figure 5.3) [Bov00c]:

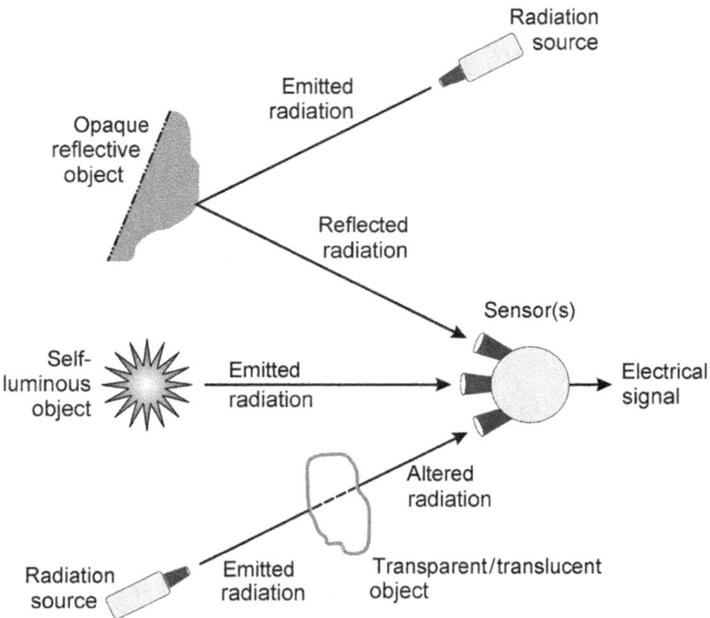

FIGURE 5.3 Recording the various types of interaction of radiation with objects and surfaces. Redrawn from [Bov00a].

- *Reflection Images*: These are the result of radiation that has been reflected from the surfaces of objects. The radiation may be *ambient* or *artificial*. Most of the images we perceive in our daily experiences are reflection images. The type of information that can be extracted from reflection images is primarily about surfaces of objects, for example, their shapes, colors, and textures.
- *Emission Images*: These are the result of objects that are self-luminous, such as stars and light bulbs (both within the visible light range), and—beyond visible light range—thermal and infrared images.
- *Absorption Images*: These are the result of radiation that passes through an object and results in an image that provides information about the object's internal structure. The most common example is X-ray image.

5.2.3 Light and Color Perception

Light is a particular type of EM radiation that can be sensed by the human eye. Colors perceived by humans are determined by the nature of the light reflected by the object, which is a function of the spectral properties of the light source as well as the absorption and reflectance properties of the object.

In 1666, Sir Isaac Newton discovered that a beam of sunlight passing through a prism undergoes decomposition into a continuous spectrum of components

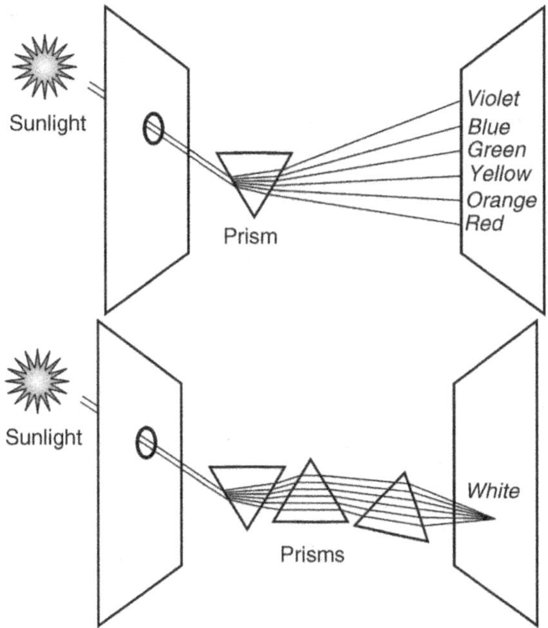

FIGURE 5.4 Newton's prism: many "colors" in the sunlight.

(Figure 5.4). Each of these components produces a different color experience, ranging from what we call *red* at one end to *violet* at the other. Newton's experiments and theories gave fundamental insights into the physical properties of light. They taught us that sunlight is actually composed of many different "colors" of light rather than just one. More important, the "colors" were not in the light itself but in the effect of the light on the visual system.

The radiance (physical power) of a light source is expressed in terms of its spectral power distribution (SPD). Figure 5.5 shows examples of SPDs of physical light sources commonly found in imaging systems: sunlight, tungsten lamp, light-emitting diode (LED), mercury arc lamp, and helium–neon laser. The human perception of each of these light sources will vary—from the yellowish nature of light produced by tungsten light bulbs to the extremely bright and pure red laser beam.

5.2.4 Color Encoding and Representation

Color can be encoded using three numerical components and appropriate spectral weighting functions. *Colorimetry* is the science that deals with the quantitative study of color perception. It is concerned with the representation of *tristimulus values*, from which the perception of color is derived. The simplest way to encode color in cameras and displays is by using the red (R), green (G), and blue (B) values of each pixel.

Human perception of light—and, consequently, color—is commonly described in terms of three parameters:

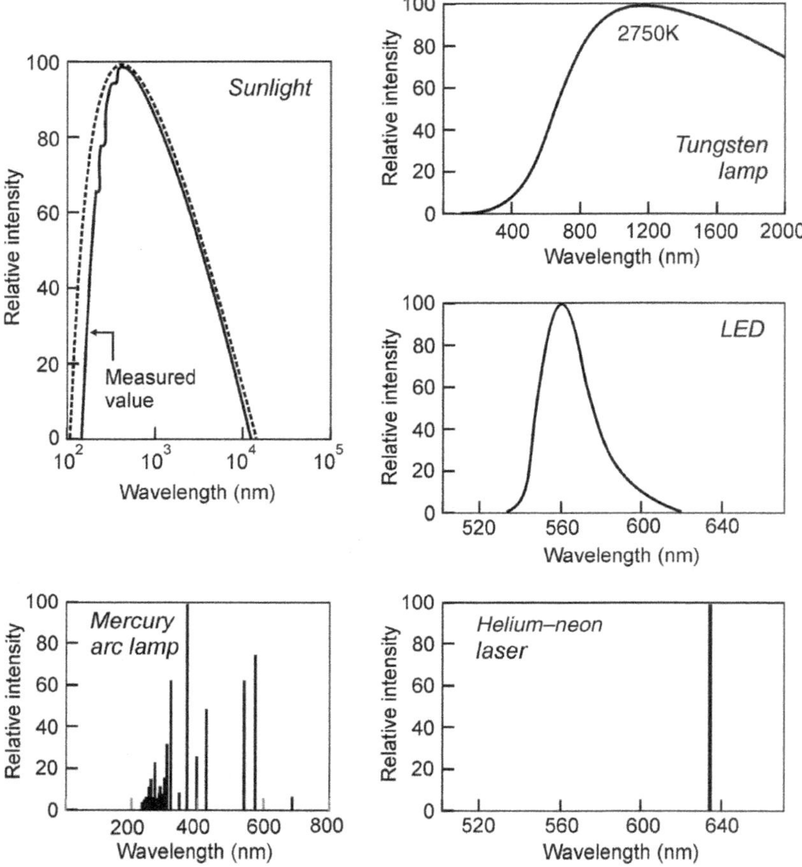

FIGURE 5.5 Spectral power distributions of common physical light sources. Redrawn from [Pra07].

- *Brightness*: The *subjective* perception of (achromatic) luminous intensity, or "the attribute of a visual sensation according to which an area appears to emit more or less light" [Poy03].
- *Hue*: "The attribute of a visual sensation according to which an area appears to be similar to one of the perceived colors, red, yellow, green and blue, or a combination of two of them" [Poy03]. From a spectral viewpoint, hue can be associated with the dominant wavelength of an SPD.
- *Saturation*: "The colorfulness of an area judged in proportion to its brightness" [Poy03], which usually translates into a description of the whiteness of the light source. From a spectral viewpoint, the more an SPD is concentrated at one wavelength, the more saturated will be the associated color. The addition of white light, that is, light that contains power at all wavelengths, causes color desaturation.

It is important to note that saturation and brightness are perceptual quantities that cannot be measured. The *Commission Internationale de L'Éclairage* (International Commission on Illumination, or simply CIE)—the international body responsible for standards in the field of color—has defined an objective quantity that is related to brightness: it is called *luminance*. Luminance can be computed as a weighted sum of red, green and blue components present in the image. We shall resume the discussion of color perception and representation in Chapter 16.

5.3 IMAGE ACQUISITION

In this section, we describe the basics of image acquisition and its two main building blocks: the image sensor and the optics associated with it.

5.3.1 Image Sensors

The main goal of an image sensor is to convert EM energy into electrical signals that can be processed, displayed, and interpreted as images. The way this is done varies significantly from one technology to another. The technology for image sensors has changed dramatically over the past 50 years, and sensors have shifted from vacuum tubes to solid-state devices, notably based on CCDs (charge-coupled devices) and CMOS (complementary metal oxide semiconductor) technologies.

Two of the most popular and relatively inexpensive devices used for image acquisition are the digital camera and the flatbed scanner. Cameras typically use 2D (area) CCD sensors, whereas scanners employ 1D (line) CCDs that move across the image as each row is scanned. CCDs have become the sensor of choice in many imaging applications because they do not suffer from geometric distortions and have a linear response to incident light [Eff00].

A CCD sensor is made up of an array of light-sensitive cells called *photosites*, manufactured in silicon, each of which produces a voltage proportional to the intensity of light falling on them. A photosite has a finite capacity of about 10^6 energy carriers, which imposes an upper bound on the brightness of the objects to be imaged. A saturated photosite can overflow, corrupting its neighbors and causing a defect known as *blooming*.

The *nominal resolution* of a CCD sensor is the size of the scene element that images to a single pixel on the image plane. For example, if a 20 cm $\times$ 20 cm square sheet of paper is imaged to form a 500 $\times$ 500 digital image, then the nominal resolution of the sensor is 0.04 cm.

The *field of view* (FOV) of an imaging sensor is a measure of how much of a scene it can see, for example, 10 cm $\times$ 10 cm. Since this may vary with depth, it is often more meaningful to refer to the *angular field of view*, for example, $55^\circ \times 40^\circ$.

A CCD camera sometimes plugs into a computer board, called *frame buffer*, which contains fast access memory (typically 0.1 ms per image) for the images captured by the camera. After being captured and temporarily stored in the frame buffer, images can be processed or copied to a long-term storage device, for example, a hard drive.

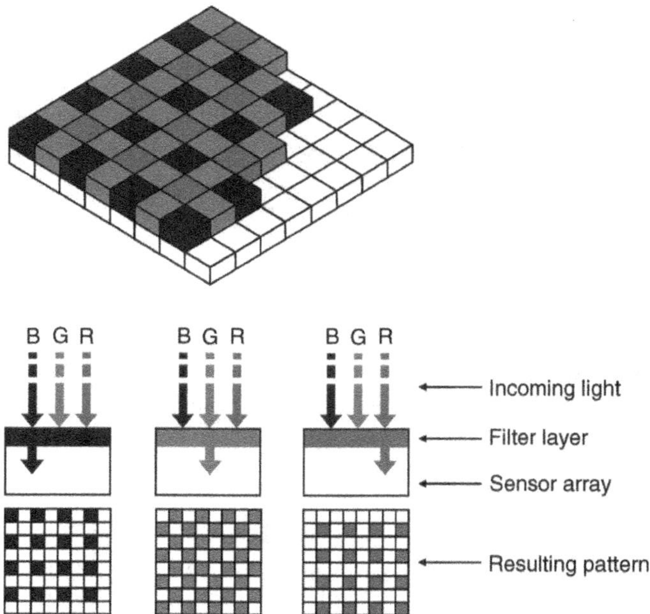

FIGURE 5.6 The Bayer pattern for single-CCD cameras.

Many contemporary cameras interface with the computer via fast standardized digital interfaces such as Firewire (IEEE 1394), Camera Link, or fast Ethernet. Such interfaces are already digital and therefore do not require a frame grabber. Moreover, most digital cameras and camcorders also include a generous amount of local storage media (e.g., from recordable CD or DVD to a myriad of specialized cards and memory sticks) that allow a significant amount of information to be locally stored in the device before being transferred to a computer.

In single-CCD cameras, colors are obtained by using a *tricolor imager* with different photosensors for each primary color of light (red, green, and blue), usually arranged in a Bayer pattern (Figure 5.6). In those cases, each pixel actually records only one of the three primary colors; to obtain a full-color image, a *demosaicing algorithm*—which can run inside the actual camera, before recording the image in JPEG format, or in a separate computer, working on the raw output from the camera—is used to interpolate a set of complete R, G, and B values for each pixel.

More expensive cameras use three CCDs, one for each color, and an *optical beam splitter* (Figure 5.7). Beam splitters have been around since the days of Plumbicon tubes. They are made of prisms with *dichroic* surfaces, that is, capable of reflecting light in one region of the spectrum and transmitting light that falls elsewhere.

An alternative technology to CCDs is CMOS. CMOS chips have the advantages of being cheaper to produce and requiring less power to operate than comparable CCD chips. Their main disadvantage is the increased susceptibility to noise, which limits their performance at low illumination levels. CMOS sensors were initially

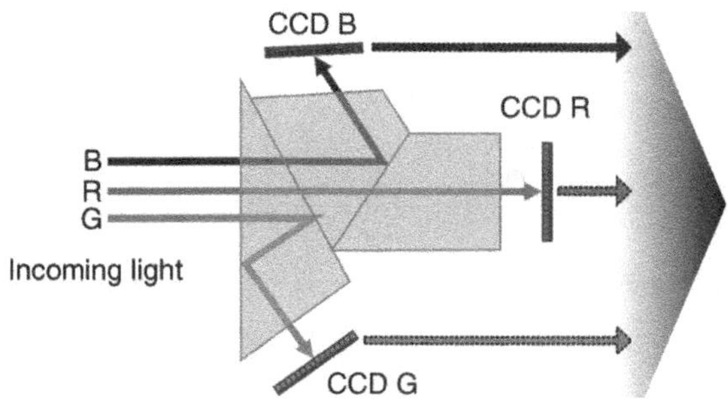

FIGURE 5.7 The beam splitter for three-CCD color cameras.

used in low-end cameras, such as webcams, but have recently been extended to much more sophisticated cameras, including the Panavision HDMAX 35 mm video camera.

A representative recent example of CMOS sensors is the Foveon X3 sensor (Figure 5.8), a CMOS image sensor for digital cameras, designed by Foveon, Inc. and manufactured by National Semiconductor. It has been designed as a layered sensor stack, in which each location in a grid has layered photosensors sensitive to all three primary colors, in contrast to the mosaic Bayer filter sensor design commonly used in digital camera sensors where each location is a single photosensor (pixel) sensitive to only one primary color. To perform its function, the Foveon sensor utilizes

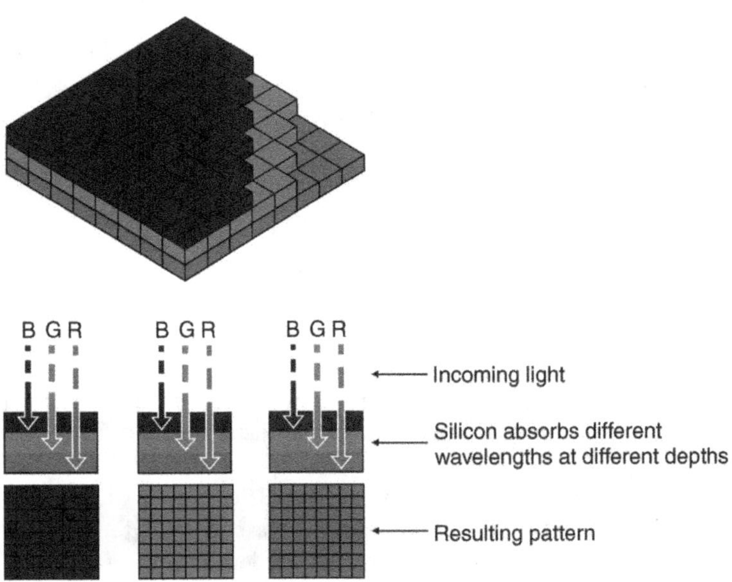

FIGURE 5.8 X3 color sensor.

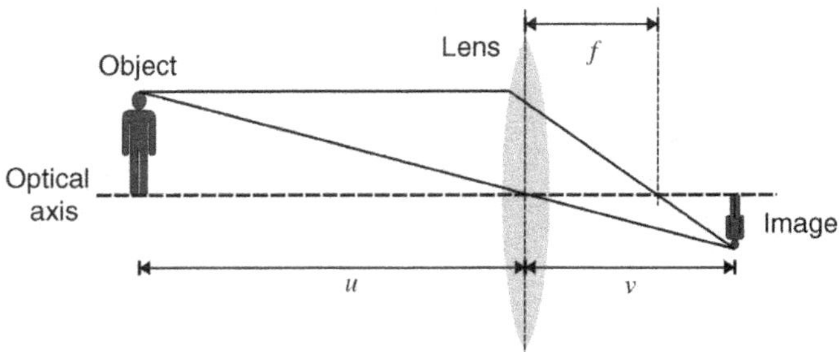

FIGURE 5.9 Image formation using a lens.

the physical property that different wavelengths of light penetrate silicon to different depths.

5.3.2 Camera Optics

A camera uses a lens to focus part of the scene onto the image sensor. Two of the most important parameters of a lens are its magnifying power and light gathering capacity. Magnifying power can be specified by a *magnification factor* (m), which is the ratio between image size and object size:

$$m = \frac{v}{u} \tag{5.2}$$

where u is the distance from an object to the lens and v is the distance from the lens to the image plane (Figure 5.9).

The magnifying power of a lens is usually expressed in terms of its *focal length*, f (in millimeters), the distance from the lens to the point at which parallel incident rays converge (Figure 5.9), given by the lens equation:

$$\frac{1}{f} = \frac{1}{u} + \frac{1}{v} \tag{5.3}$$

Combining equations (5.2) and (5.3), we can express f as a function of u and m:

$$f = \frac{um}{m+1} \tag{5.4}$$

Equation (5.4) provides a practical and convenient way to determine the focal length for a certain magnification factor and object distance and select the appropriate lens for the job.

The light gathering capacity of a camera lens is determined by its *aperture*, which is often expressed as an "f number"—a dimensionless value that represents the ratio between focal length and aperture diameter. Most lenses have a sequence of fixed apertures (e.g., $f2.8$, $f4$, $f5.6$, $f8$, $f11$) that progressively reduce the total amount of light reaching the sensor by half.

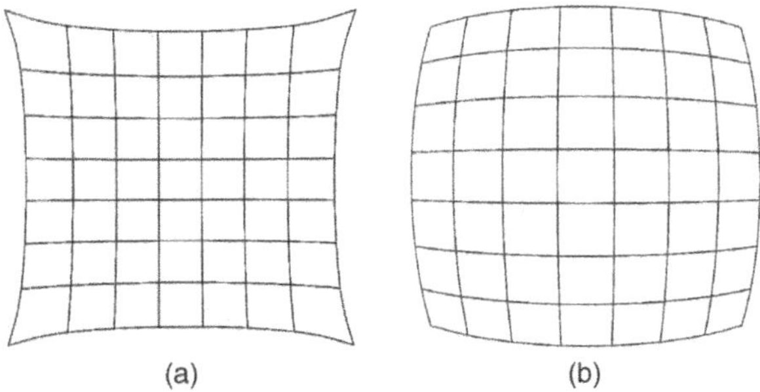

FIGURE 5.10 Examples of lens aberrations: (a) pincushion distortion; (b) barrel distortion.

Lenses may suffer from *aberrations*, which can affect image quality and generate undesired distortions on the resulting image. Examples of such aberrations include—among many others—the *pincushion distortion* and the *barrel distortion* (Figure 5.10). If such defects are not taken care of by the lens system itself, certain image processing techniques[1] can be used to correct them.

In MATLAB

The MATLAB Image Acquisition Toolbox (IAT) is a collection of functions that extend the capability of MATLAB, allowing image acquisition operations from a variety of image acquisition devices, from professional-grade frame grabbers to USB-based webcams. The IAT software uses components called *hardware device adaptors* to connect to devices through their drivers (Figure 5.11). At the time of this writing, the IAT supports a variety of devices and drivers, such as the IIDC 1394-based Digital Camera Specification (DCAM), and devices that provide Windows Driver Model (WDM) or Video for Windows (VFW) drivers, such as USB and IEEE 1394 (FireWire, i.LINK) Web cameras, digital video (DV) camcorders, and TV tuner cards. Since the release of Version 3.0, the functionality of the IAT software is available in a desktop application.

5.4 IMAGE DIGITIZATION

The image digitization stage bridges the gap between the analog natural world, from which scenes are acquired, and the digital format expected by computer algorithms in charge of processing, storing, or transmitting this image.

Digitization involves two processes (Figure 5.12): *sampling* (in time or space) and *quantization* (in amplitude). These operations may occur in any sequence, but usually sampling precedes quantization [Poy03]. Sampling involves selecting a finite number of points within an interval, whereas quantization implies assigning an amplitude

[1] Some of these techniques will be discussed in Chapter 7.

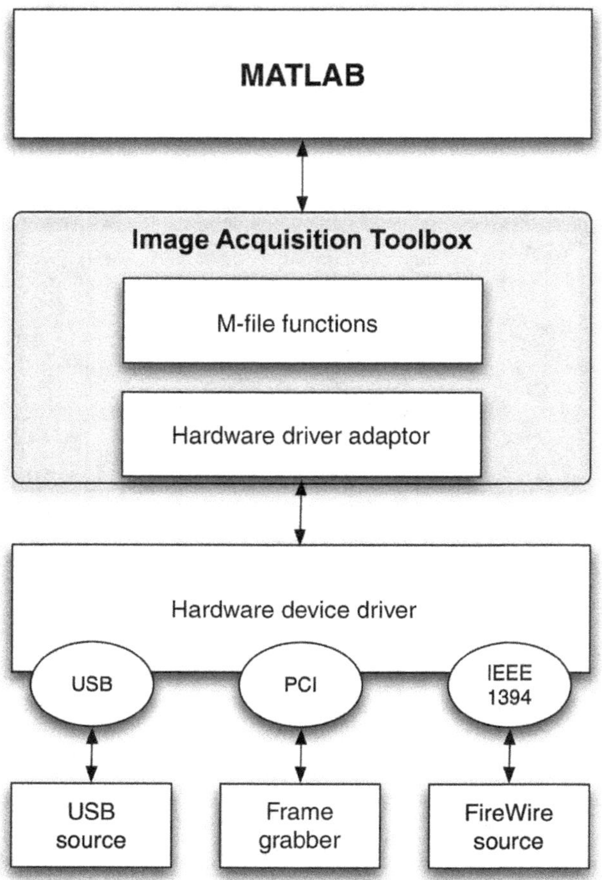

FIGURE 5.11 The main components of the MATLAB Image Acquisition Toolbox.

value (within a finite range of possible values) to each of those points. The result of the digitization process is a *pixel array*, which is a rectangular matrix of picture elements whose values correspond to their intensities (for monochrome images) or color components (for color images).

For consumer cameras and camcorders, it has become common to refer to the size of the pixel array by the product of the number of pixels along each dimension and express the result in megapixels (Mpx).[2] Figure 5.13 shows representative contemporary pixel arrays ranging from the QCIF videoconferencing standard (to be discussed in Chapter 20) to the 1920 × 1080 HDTV standard.

[2]It has also become common practice to associate image quality with the size of the camera's resulting pixel array, which is certainly one of the parameters to keep in mind when shopping for a new camera, but by no means the only one. Many other features—from the amount of optical zoom to the ability of working under low lighting—may turn out to be more relevant to the user.

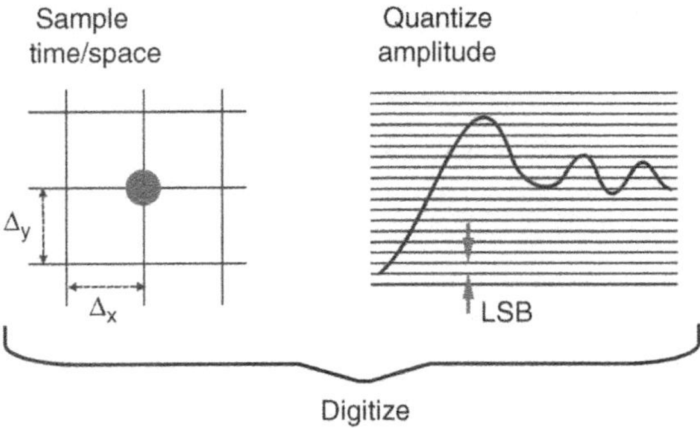

FIGURE 5.12 Digitization = sampling + quantization. Redrawn from [Poy03].

Digitization can take place in many different portions of a machine vision system (MVS) and has been moving progressively closer to the camera hardware during the past few years, making products such as "video capture cards" or "frame grabbers" more of a rarity.

5.4.1 Sampling

Sampling is the process of measuring the value of a 2D function at discrete intervals along the x and y dimensions. A system that has equal horizontal and vertical sampling densities is said to have *square sampling*. Several imaging and video systems use sampling lattices where the horizontal and the vertical sample pitch are unequal, that is, *nonsquare sampling*.

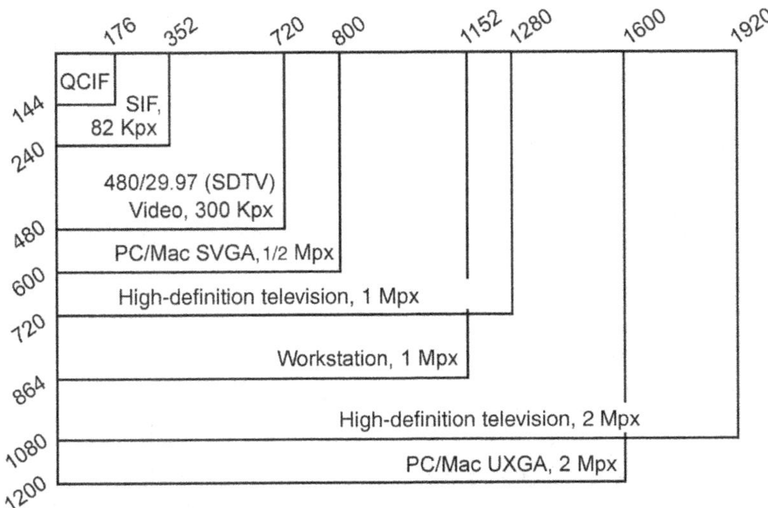

FIGURE 5.13 Pixel arrays of several imaging standards. Redrawn from [Poy03].

Two parameters must be taken into account when sampling images:

1. The *sampling rate*, that is, the number of samples across the height and width of the image. The choice of an appropriate sampling rate will impact image quality, as we shall see in Section 5.4.3. Inadequate values may lead to a phenomenon known as *aliasing*, which will be discussed later.
2. The *sampling pattern*, that is, the physical arrangement of the samples. A rectangular pattern, in which pixels are aligned horizontally and vertically into rows and columns, is by far the most common form, but other arrangements are possible, for example, the *hexagonal* and *log-polar* sampling patterns (see [SSVPB02] for an example).

If sampling takes place at a rate lower than twice the highest frequency component of the signal (the *Nyquist criterion*), there will not be enough points to ensure proper reconstruction of the original signal, which is referred to as *undersampling* or *aliasing*. Figure 5.14 illustrates the concept for 1D signals: part (a) shows the sampling process as a product of a train of sampling impulses and the analog signal being sampled, part (b) shows the result of reconstructing a signal from an appropriate number of samples, and part (c) shows that the same number of samples as in (b) would be insufficient to reconstruct a signal with higher frequency. The effect of aliasing on images is typically perceived in the form of Moiré patterns (which can be seen in the changes in the tablecloth pattern in Figure 5.16c and especially d).

In the case of temporal sampling (to be discussed again in Chapters 20 and 21), a very familiar example of the aliasing phenomenon is the *wagon wheel effect*, in which the spoked wheels of a wagon appear to be moving backward if the wagon moves at a speed that violates Nyquist theorem for the minimum number of temporal samples per second that would be required to display the proper motion.

5.4.2 Quantization

Quantization is the process of replacing a continuously varying function with a discrete set of quantization levels. In the case of images, the function is $f(x, y)$ and the

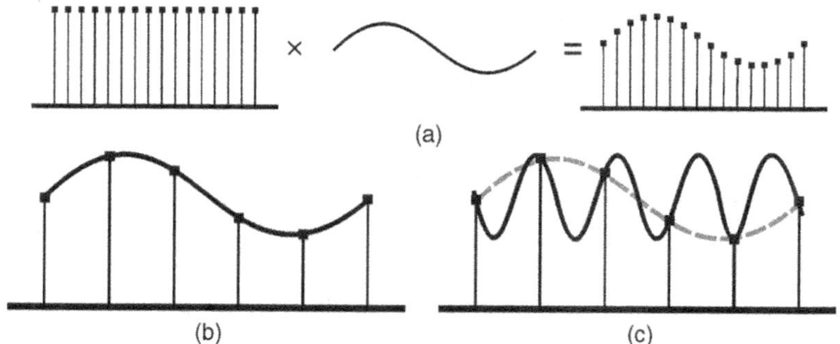

FIGURE 5.14 1D aliasing explanation. Redrawn from [Wat00].

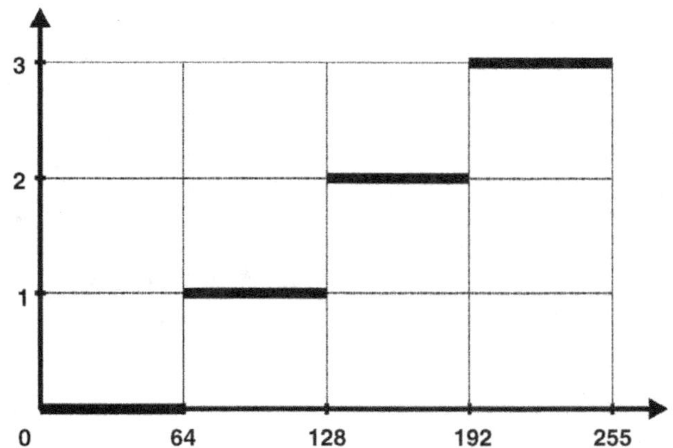

FIGURE 5.15 A mapping function for uniform quantization ($N = 4$).

quantization levels are also known as *gray levels*. It is common to adopt N quantization levels for image digitization, where N is usually an integral power of 2 that is, $N = 2^n$, where n is the number of bits needed to encode each pixel value. The case where $n = 2^8 = 256$ produces images where each pixel is represented by an unsigned byte, with values ranging from 0 (black) to 255 (white).

Image quantization can be described as a mapping process by which groups of data points (several pixels within a range of gray values) are mapped to a single point (i.e., a single gray level). This process is illustrated in Figure 5.15, which depicts the case where the number of gray levels is reduced from 256 to 4 by *uniform quantization*, meaning that the input gray level range is divided into N equal intervals of length 64.

5.4.3 Spatial and Gray-Level Resolution

Spatial resolution is a way of expressing the density of pixels in an image: the greater the spatial resolution, the more pixels are used to display the image within a certain fixed physical size. It is usually expressed quantitatively using units such as dots per inch (dpi).

■ EXAMPLE 5.1

Figure 5.16 shows the effects of reducing the spatial resolution of a 256-gray-level image. The original image (Figure 5.16a) is of size 1944 × 2592 and it is displayed at 1250 dpi. The three other images (Figure 5.16b through d) have reduced spatial resolution to 300, 150, and 72 dpi, respectively. They have been zoomed back to their original sizes in order to make meaningful comparisons. A close inspection of the results will show that the quality loss between the original image and its 300 dpi equivalent (Figure 5.16a and b) is not very noticeable, but the pixelation, jaggedness, loss of detail, and even the appearance of Moiré patterns present on the other two images (bottom row) are easy to notice.

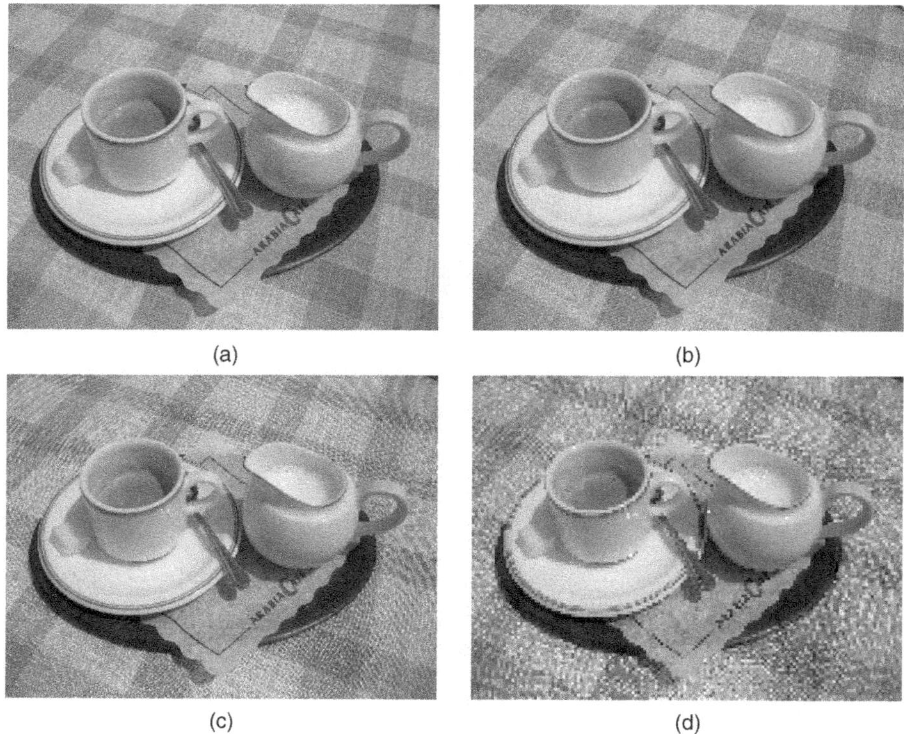

FIGURE 5.16 Effects of sampling resolution on image quality: (a) A 1944 × 2592 image, 256 gray levels, at a 1250 dpi resolution. The same image resampled at (b) 300 dpi; (c) 150 dpi; (d) 72 dpi.

Gray-level resolution refers to the smallest change in intensity level that the HVS can discern. The adoption of 8 bits per pixel for monochrome images is a good compromise between subjective quality and practical implementation (each pixel value is neatly aligned with a byte). Higher end imaging applications may require more than 8 bits per color channel and some image file formats support such need (e.g., 12-bit RAW and 16-bit TIFF files).

In MATLAB

(Re-)quantizing an image in MATLAB can be accomplished using the `grayslice` function, as illustrated in the following example.

■ EXAMPLE 5.2

Figure 5.17 shows the effects of quantization (gray) levels on image quality for an image with 480 × 640 pixels, starting with 256 gray levels and reducing it by a factor of 2 several times, until arriving at a binary version of the original image. A close inspection of the results will show that the quality loss between the original image and its 32 gray levels equivalent is not very noticeable (Figure 5.17a–d), but the quality

FIGURE 5.17 (a) A 480 × 640 image, 256 gray levels; (b–h) image requantized to 128, 64, 32, 16, 8, 4, and 2 gray levels.

of the last four images is unacceptable for most purposes due to the presence of *false contouring* and loss of relevant detail.

MATLAB Code

```
I1 = imread('ml_gray_640_by_480_256.png');
I2 = grayslice(I1,128); figure, imshow(I2,gray(128));
I3 = grayslice(I1,64); figure, imshow(I3,gray(64));
```

```
I4 = grayslice(I1,32); figure, imshow(I4,gray(32));
I5 = grayslice(I1,16); figure, imshow(I5,gray(16));
I6 = grayslice(I1,8); figure, imshow(I6,gray(8));
I7 = grayslice(I1,4); figure, imshow(I7,gray(4));
I8 = grayslice(I1,2); figure, imshow(I8,gray(2));
```

The choice of sampling and quantization parameters (total number of pixels and the number of gray levels per pixel) has also an impact on the resulting image file size. For example, a 1024×1024 image with 256 gray levels (8 bits per pixel) would require 1024^2 bytes, that is, 1 MB of disk space (neglecting any extra bytes needed for additional information, typically stored in the file header). Not too long ago, these image sizes were considered prohibitively large, which fostered the progress of image compression techniques and the standardization of file formats such as GIF and JPEG (see Chapter 17 for more details). In recent years, with significant increase in storage capacity at decreasing costs, the concern for disk space usage has been alleviated to some degree. Many digital image users would rather adopt more expansive file sizes than sacrifice the quality of their digital images. This has been seen in digital photography, with the creation and popularization of RAW file formats and increasing interest in images with more than 8 bits per pixel per color channel.

In summary, the choice of the number of samples per unit of distance (or area) (spatial resolution) and the number of colors or quantization (gray) levels used when digitizing an image should be guided by a trade-off between the impact on the storage size and the perceived resulting quality.

WHAT HAVE WE LEARNED?

- Images are formed as a result of the interaction between the source of radiation (e.g., visible light), the properties of the objects and surfaces involved, and the relative positioning and properties of the image sensor.
- Images can be classified into three categories according to the type of interaction between the source of radiation, the properties of the objects involved, and the relative positioning of the image sensor: reflection images, emission images, and absorption images.
- Contemporary image sensors (*imagers*) are usually built upon CCD or CMOS solid-state technologies. A CCD sensor is made up of an array of light-sensitive cells called "photosites," each of which produces a voltage proportional to the intensity of light falling on them. The cells are combined into a (1D or 2D) array that can be read sequentially by a computer input process.
- Image digitization is the process of sampling a continuous image (in space) and quantizing the resulting amplitude values so that they fall within a finite range.
- Some of the main parameters that impact the digitization of an image are the total number of pixels and the maximum number of colors (or gray levels) per pixel.

- Sampling is the process of measuring the value of a function at discrete intervals. (Re-)sampling an image in MATLAB can be accomplished using the `imresize` function.
- Quantization is the process of replacing a continuously varying function with a discrete set of quantization levels. (Re-)quantizing an image in MATLAB can be accomplished using the `grayslice` function.

LEARN MORE ABOUT IT

- For a much more detailed treatment of photometry and colorimetry, we recommend Chapter 3 of [Pra07] and the Appendix B of [Poy03].
- For a brief discussion and visual examples of images acquired in different ranges of the EM spectrum, we recommend Section 2.2.2 of [Umb05] and Section 1.3 of [GW08].
- Chapter 9 of [LI99], Section 6 of [WB00], and Chapter 3 of [GH99] provide a good overview of video cameras' components, principles, and technologies.
- The development of the X3 sensor by Foveon and the implications of that technological breakthrough have become the subject of a book by Gilder [Gil05].

ON THE WEB

- Charles Poynton's *Frequently Asked Questions about Color*
  http://poynton.com/ColorFAQ.html
- The MATLAB Image Acquisition Toolbox
  http://www.mathworks.com/products/imaq/
- Edmund Optics
  http://www.edmundoptics.com/
- The Imaging Source
  http://www.theimagingsource.com/en/products/

5.5 PROBLEMS

5.1 Experiment with a scanner's settings (in dpi) and scan the same material (e.g., a photo) several times using different settings but the same file format. Compare the resulting file size and quality for each resulting scanned image.

5.2 Repeat Problem 5.1, but this time keeping the settings the same and changing only the file format used to save the resulting image. Are there significant differences in file size? What about the subjective quality?

5.3 Assuming a monochrome image with 1024×1024 pixels and 256 gray levels,

(a) Calculate the total file size (in bytes), assuming a header (containing basic information, such as width and height of the image) of 32 bytes and no compression.

(b) Suppose the original image has been subsampled by a factor of 2 in both dimensions. Calculate the new file size (in bytes), assuming the header size has not changed.

(c) Suppose the original image has been requantized to allow encoding 2 pixels per byte. Calculate the new file size (in bytes), assuming the header size has not changed.

(d) How many gray levels will the image in part (c) have?

CHAPTER 6

ARITHMETIC AND LOGIC OPERATIONS

WHAT WILL WE LEARN?

- Which arithmetic and logic operations can be applied to digital images?
- How are they performed in MATLAB?
- What are they used for?

6.1 ARITHMETIC OPERATIONS: FUNDAMENTALS AND APPLICATIONS

Arithmetic operations involving images are typically performed on a pixel-by-pixel basis; that is, the operation is independently applied to each pixel in the image. Given a 2D array (X) and another 2D array of the same size or a scalar (Y), the resulting array, Z, is obtained by calculating

$$X \quad opn \quad Y = \quad Z \tag{6.1}$$

where *opn* is a binary arithmetic ($+$, $-$, $\times$, $/$) operator.

This section describes each arithmetic operation in more detail, focusing on how they can be performed and what are their typical applications.

Practical Image and Video Processing Using MATLAB®. By Oge Marques.

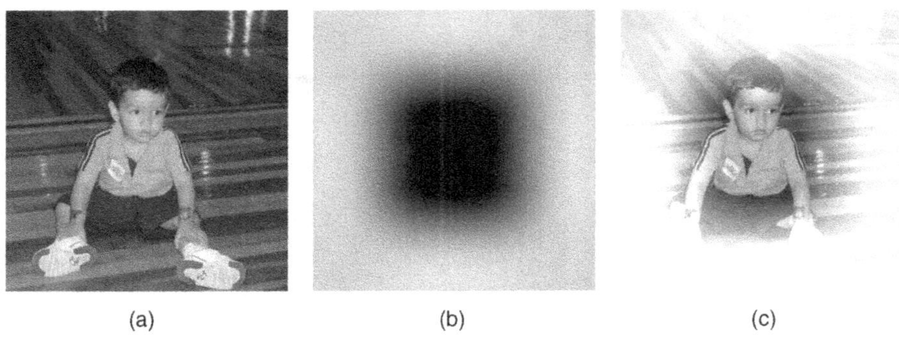

(a) (b) (c)

FIGURE 6.1 Adding two images: (a) first image (X); (b) second image (Y); (c) result ($Z = X + Y$).

6.1.1 Addition

Addition is used to blend the pixel contents from two images or add a constant value to pixel values of an image. Adding the contents of two monochrome images causes their contents to blend (Figure 6.1). Adding a constant value (scalar) to an image causes an increase (or decrease if the value is less than zero) in its overall brightness, a process sometimes referred to as *additive image offset* (Figure 6.2). Adding random amounts to each pixel value is a common way to simulate additive noise (Figure 6.3). The resulting (noisy) image is typically used as a test image for restoration algorithms such as those described in Chapter 12.

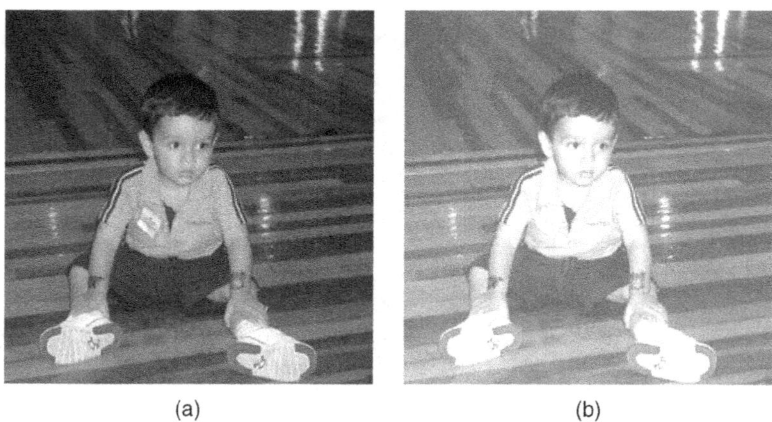

(a) (b)

FIGURE 6.2 Additive image offset: (a) original image (X); (b) brighter version ($Z = X + 75$).

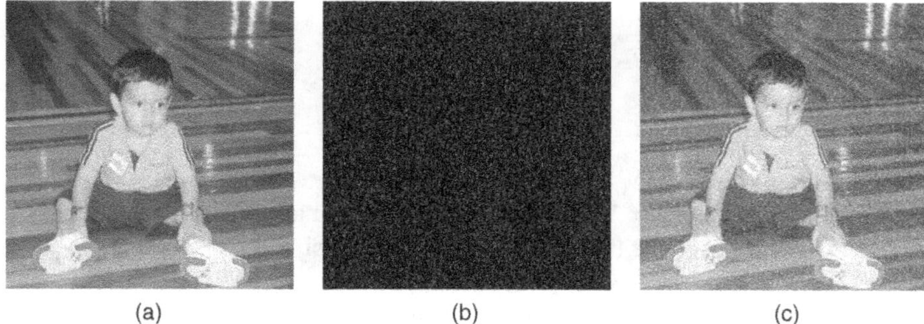

(a) (b) (c)

FIGURE 6.3 Adding noise to an image: (a) original image (X); (b) zero-mean Gaussian white noise (variance = 0.01) (N); (c) result ($Z = X + N$).

In MATLAB

MATLAB's Image Processing Toolbox (IPT) has a built-in function to add two images or add a constant (scalar) to an image: `imadd`. In Tutorial 6.1 (page 113), you will have a chance to experiment with this function.

When adding two images, you must be careful with values that exceed the maximum pixel value for the data type being used. There are two ways of dealing with this *overflow* issue: *normalization* and *truncation*. Normalization consists in storing the intermediate result in a temporary variable (W) and calculating each resulting pixel value in Z using equation (6.2).

$$g = \frac{L_{max}}{f_{max} - f_{min}}(f - f_{min}) \tag{6.2}$$

where f is the current pixel in W, L_{max} is the maximum possible intensity value (e.g., 255 for `uint8` or 1.0 for `double`), g is the corresponding pixel in Z, f_{max} is the maximum pixel value in W, and f_{min} is the minimum pixel value in W.

Truncation consists in simply limiting the results to the maximum positive number that can be represented with the adopted data type.

■ EXAMPLE 6.1

For the two 3×3 monochrome images below (X and Y), each of which represented as an array of unsigned integers, 8-bit (`uint8`), calculate $Z = X + Y$, using (a) normalization and (b) truncation.

$$X = \begin{bmatrix} 200 & 100 & 100 \\ 0 & 10 & 50 \\ 50 & 250 & 120 \end{bmatrix}$$

$$Y = \begin{bmatrix} 100 & 220 & 230 \\ 45 & 95 & 120 \\ 205 & 100 & 0 \end{bmatrix}$$

Solution

The intermediate array W (an array of unsigned integers, 16-bit, `uint16`) is obtained by simply adding the values of X and Y on a pixel-by-pixel basis:

$$W = \begin{bmatrix} 300 & 320 & 330 \\ 45 & 105 & 170 \\ 255 & 350 & 120 \end{bmatrix}$$

(a) Normalizing the [45, 350] range to the [0, 255] interval using equation (6.2), we obtain

$$Z_{\mathrm{a}} = \begin{bmatrix} 213 & 230 & 238 \\ 0 & 50 & 105 \\ 175 & 255 & 63 \end{bmatrix}$$

(b) Truncating all values above 255 in W, we obtain

$$Z_{\mathrm{b}} = \begin{bmatrix} 255 & 255 & 255 \\ 45 & 105 & 170 \\ 255 & 255 & 120 \end{bmatrix}$$

MATLAB code:

```
X = uint8([200 100 100; 0 10 50; 50 250 120])
Y = uint8([100 220 230; 45 95 120; 205 100 0])
W = uint16(X) + uint16(Y)
fmax = max(W(:))
fmin = min(W(:))
Za = uint8(255.0*double((W-fmin))/double((fmax-fmin)))
Zb = imadd(X,Y)
```

6.1.2 Subtraction

Subtraction is often used to detect differences between two images. Such differences may be due to several factors, such as artificial addition to or removal of relevant contents from the image (e.g., using an image manipulation program), relative object motion between two frames of a video sequence, and many others. Subtracting a

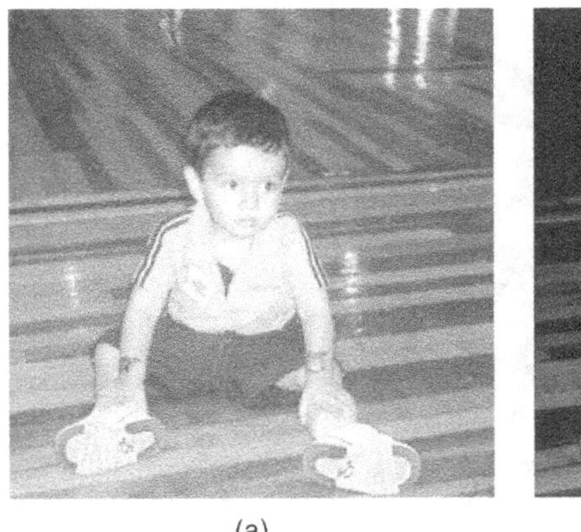

(a)

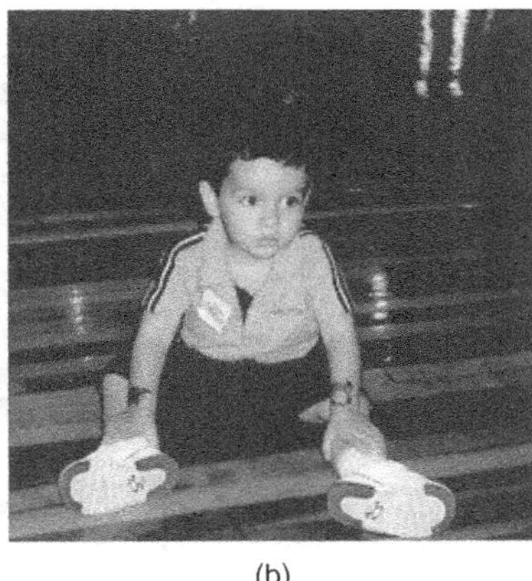

(b)

FIGURE 6.4 Subtractive image offset: (a) original image (X); (b) darker version ($Z = X - 75$).

constant value (scalar) from an image causes a decrease in its overall brightness, a process sometimes referred to as *subtractive image offset* (Figure 6.4).

When subtracting one image from another or a constant (scalar) from an image, you must be careful with the possibility of obtaining negative pixel values as a result. There are two ways of dealing with this *underflow* issue: treating subtraction as absolute difference (which will always result in positive values proportional to the difference between the two original images without indicating, however, which pixel was brighter or darker) and truncating the result, so that negative intermediate values become zero.

In MATLAB

The IPT has a built-in function to subtract one image from another, or subtract a constant from an image: `imsubtract`. The IPT also has a built-in function to calculate the absolute difference of two images: `imabsdiff`. The IPT also includes a function for calculating the negative (complement) of an image, `imcomplement`. In Tutorial 6.1 (page 113), you will have a chance to experiment with these functions.

■ EXAMPLE 6.2

For the two 3 × 3 monochrome images below (X and Y), each of which represented as an array of unsigned integers, 8-bit (`uint8`), calculate (a) $Z = X - Y$, (b) $Z = Y - X$, and (c) $Z = |Y - X|$. For parts (a) and (b), use truncation to deal with possible negative values.

$$X = \begin{bmatrix} 200 & 100 & 100 \\ 0 & 10 & 50 \\ 50 & 250 & 120 \end{bmatrix}$$

$$Y = \begin{bmatrix} 100 & 220 & 230 \\ 45 & 95 & 120 \\ 205 & 100 & 0 \end{bmatrix}$$

Solution

MATLAB's `imsubtract` will take care of parts (a) and (b), while `imabsdiff` will be used for part (c).

(a)

$$Z_a = \begin{bmatrix} 100 & 0 & 0 \\ 0 & 0 & 0 \\ 0 & 150 & 120 \end{bmatrix}$$

(b)

$$Z_b = \begin{bmatrix} 0 & 120 & 130 \\ 45 & 85 & 70 \\ 155 & 0 & 0 \end{bmatrix}$$

(c)

$$Z_c = \begin{bmatrix} 100 & 120 & 130 \\ 45 & 85 & 70 \\ 155 & 150 & 120 \end{bmatrix}$$

MATLAB code:

```
X = uint8([200 100 100; 0 10 50; 50 250 120])
Y = uint8([100 220 230; 45 95 120; 205 100 0])
Za = imsubtract(X,Y)
Zb = imsubtract(Y,X)
Zc = imabsdiff(Y,X)
```

Image subtraction can also be used to obtain the *negative* of an image (Figure 6.5):

$$g = -f + L_{max} \tag{6.3}$$

where L_{max} is the maximum possible intensity value (e.g., 255 for `uint8` or 1.0 for `double`), f is the pixel value in X, g is the corresponding pixel in Z.

(a) (b)

FIGURE 6.5 Example of an image negative: (a) original image; (b) negative image.

6.1.3 Multiplication and Division

Multiplication and division by a scalar are often used to perform brightness adjustments on an image. This process—sometimes referred to as *multiplicative image scaling*—makes each pixel value brighter (or darker) by multiplying its original value by a scalar factor: if the value of the scalar multiplication factor is greater than one, the result is a brighter image; if it is greater than zero and less than one, it results in a darker image (Figure 6.6). Multiplicative image scaling usually produces better subjective results than the additive image offset process described previously.

In MATLAB

The IPT has a built-in function to multiply two images or multiply an image by a constant: `immultiply`. The IPT also has a built-in function to divide one image

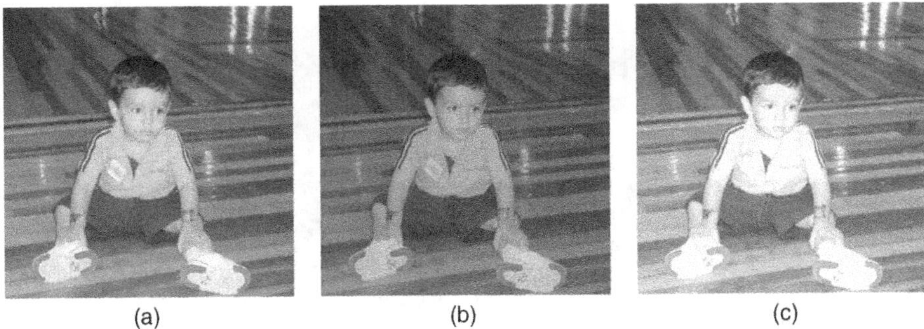

(a) (b) (c)

FIGURE 6.6 Multiplication and division by a constant: (a) original image (X); (b) multiplication result ($X \times 0.7$); (c) division result ($X/0.7$).

into another or divide an image by a constant: `imdivide`. In Tutorial 6.1 (page 113), you will have a chance to experiment with these functions.

6.1.4 Combining Several Arithmetic Operations

It is sometimes necessary to combine several arithmetic operations applied to one or more images, which may compound the problems of overflow and underflow discussed previously. To achieve more accurate results without having to explicitly handle truncations and round-offs, the IPT offers a built-in function to perform a linear combination of two or more images: `imlincomb`. This function computes each element of the output individually, in double-precision floating point. If the output is an integer array, `imlincomb` truncates elements that exceed the range of the integer type and rounds off fractional values.

■ EXAMPLE 6.3

Calculate the average of the three 3 × 3 monochrome images below (X, Y, and Z), each of which represented as an array of unsigned integers, 8-bit (`uint8`), using (a) `imadd` and `imdivide` without explicitly handling truncation and round-offs; (b) `imadd` and `imdivide`, but this time handling truncation and round-offs; and (c) `imlincomb`.

$$X = \begin{bmatrix} 200 & 100 & 100 \\ 0 & 10 & 50 \\ 50 & 250 & 120 \end{bmatrix}$$

$$Y = \begin{bmatrix} 100 & 220 & 230 \\ 45 & 95 & 120 \\ 205 & 100 & 0 \end{bmatrix}$$

$$Z = \begin{bmatrix} 200 & 160 & 130 \\ 145 & 195 & 120 \\ 105 & 240 & 150 \end{bmatrix}$$

Solution

(a)

$$S_{\text{a}} = \begin{bmatrix} 85 & 85 & 85 \\ 63 & 85 & 85 \\ 85 & 85 & 85 \end{bmatrix}$$

(b)

$$S_b = \begin{bmatrix} 167 & 160 & 153 \\ 63 & 100 & 97 \\ 120 & 197 & 90 \end{bmatrix}$$

(c)

$$S_c = \begin{bmatrix} 167 & 160 & 153 \\ 63 & 100 & 97 \\ 120 & 197 & 90 \end{bmatrix}$$

MATLAB code:

```
X = uint8([200 100 100; 0 10 50; 50 250 120])
Y = uint8([100 220 230; 45 95 120; 205 100 0])
Z = uint8([200 160 130; 145 195 120; 105  240 150])
Sa = imdivide(imadd(X,imadd(Y,Z)),3)
a = uint16(X) + uint16(Y)
b = a + uint16(Z)
Sb = uint8(b/3)
Sc = imlincomb(1/3,X,1/3,Y,1/3,Z,'uint8')
```

The result in (a) is incorrect due to truncation of intermediate results. Both (b) and (c) produce correct results, but the solution using `imlincomb` is much more elegant and concise.

6.2 LOGIC OPERATIONS: FUNDAMENTALS AND APPLICATIONS

Logic operations are performed in a bit-wise fashion on the binary contents of each pixel value. The AND, XOR, and OR operators require two or more arguments, whereas the NOT operator requires only one argument. Figure 6.7 shows the most common logic operations applied to binary images, using the following convention: 1 (true) for white pixels and 0 (false) for black pixels.

Figures 6.8–6.11 show examples of AND, OR, XOR, and NOT operations on monochrome images. The AND and OR operations can be used to combine images for special effects purposes. They are also used in masking operations, whose goal is to extract a region of interest (ROI) from an image (see Tutorial 6.2). The XOR operation is often used to highlight differences between two monochrome images. It is, therefore, equivalent to calculating the absolute difference between two images. The NOT operation extracts the binary complement of each pixel value, which is equivalent to applying the "negative" effect on an image.

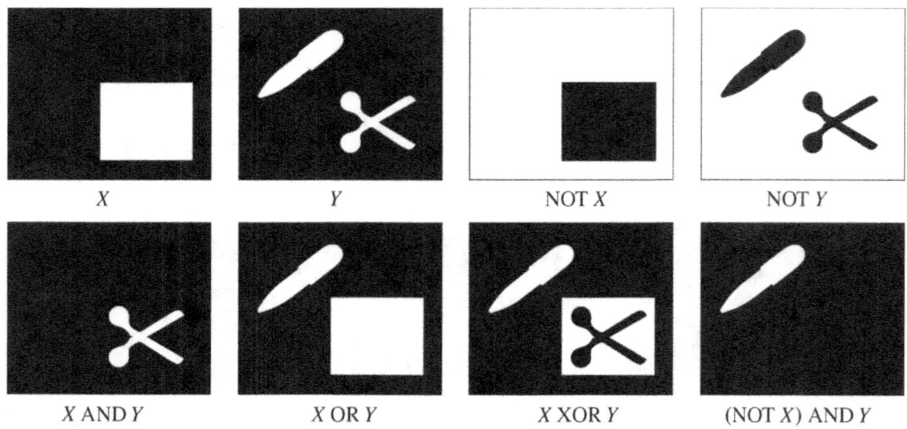

FIGURE 6.7 Logic operations on binary images.

FIGURE 6.8 The AND operation applied to monochrome images: (a) X; (b) Y; (c) X AND Y.

FIGURE 6.9 The OR operation applied to monochrome images: (a) X; (b) Y; (c) X OR Y.

In MATLAB

MATLAB has built-in functions to perform logic operations on arrays: `bitand`, `bitor`, `bitxor`, and `bitcmp`. In Tutorial 6.2 (page 118), you will have a chance to experiment with these functions.

(a) (b) (c)

FIGURE 6.10 The XOR operation applied to monochrome images: (a) X; (b) Y; (c) X XOR Y.

(a) (b)

FIGURE 6.11 The NOT operation applied to a monochrome image: (a) X; (b) NOT X.

6.3 TUTORIAL 6.1: ARITHMETIC OPERATIONS

Goal

The goal of this tutorial is to learn how to perform arithmetic operations on images.

Objectives

- Learn how to perform image addition using the `imadd` function.
- Explore image subtraction using the `imsubtract` function.
- Explore image multiplication using the `immultiply` function.
- Learn how to use the `imdivide` function for image division.

What You Will Need

- `cameraman2.tif`
- `earth1.tif`
- `earth2.tif`
- `gradient.tif`
- `gradient_with_text.tif`

Procedure

The IPT offers four functions to aid in image arithmetic: `imadd`, `imsubtract`, `immultiply`, and `imdivide`. You could use MATLAB's arithmetic functions (`+, −, *, /`) to perform image arithmetic, but it would probably require additional coding to ensure that the operations are performed in double precision, as well as setting cutoff values to be sure that the result is within grayscale range. The functions provided by the IPT do this for you automatically.

Image addition can be used to brighten (or darken) an image by adding (subtracting) a constant value to (from) each pixel value. It can also be used to blend two images into one.

1. Use the `imadd` function to brighten an image by adding a constant (scalar) value to all its pixel values.

```
I = imread('tire.tif');
I2 = imadd(I,75);
figure
subplot(1,2,1), imshow(I), title('Original Image');
subplot(1,2,2), imshow(I2), title('Brighter Image');
```

Question 1 What are the maximum and minimum values of the original and the adjusted image? Explain your results.

Question 2 How many pixels had a value of 255 in the original image and how many have a value of 255 in the resulting image?

2. Use the `imadd` function to blend two images.

```
Ia = imread('rice.png');
Ib = imread('cameraman.tif');
Ic = imadd(Ia,Ib);
figure
imshow(Ic);
```

Image subtraction is useful when determining whether two images are the same. By subtracting one image from another, we can highlight the differences between the two.

3. Close all open figures and clear all workspace variables.
4. Load two images and display them.

```
I = imread('cameraman.tif');
J = imread('cameraman2.tif');
```

```
figure
subplot(1,2,1), imshow(I), title('Original Image');
subplot(1,2,2), imshow(J), title('Altered Image');
```

While it may not be obvious at first how the altered image differs from the original image, we should be able to see where the difference is located after using the `imsubtract` function.

5. Subtract both images and display the result.

```
diffim = imsubtract(I,J);
figure
subplot(2,2,1), imshow(diffim), title('Subtracted Image');
```

6. Use the zoom tool to zoom into the right area of the difference image about halfway down the image. You will notice that a small region of pixels is faintly white.
7. To zoom back out, double-click anywhere on the image.

Now that you know where the difference is located, you can look at the original images to see the change. The difference image above does not quite seem to display all the details of the missing building. This is because when we performed image subtraction, some of the pixels resulted in negative values, but were then set to 0 by the `imsubtract` function (the function does this on purpose to keep the data within grayscale range). What we really want to do is calculate the absolute value of the difference between two images.

8. Calculate the absolute difference. Make sure Figure 2 is selected before executing this code.

```
diffim2 = imabsdiff(I,J);
subplot(2,2,2), imshow(diffim2), title('Abs Diff Image');
```

9. Use the zoom-in tool to inspect the new difference image.

Even though the new image may look the same as the previous one, it represents both positive and negative differences between the two images. To see this difference better, we will scale both difference images for display purposes, so their values occupy the full range of the gray scale.

10. Show scaled versions of both difference images.

```
subplot(2,2,3), imshow(diffim,[]), ...
     title('Subtracted Image Scaled');
```

```
subplot(2,2,4), imshow(diffim2,[]), ...
     title('Abs Diff Image Scaled');
```

11. Use the zoom tool to see the differences between all four difference images.

Question 3 How did we scale the image output?

Question 4 What happened when we scaled the difference images?

Question 5 Why does the last image show more detail than the others?

Multiplication is the process of multiplying the values of each pixel of same coordinates in two images. This can be used for a brightening process known as *dynamic scaling*, which results in a more naturally brighter image compared to directly adding a constant to each pixel.

12. Close all open figures and clear all workspace variables.
13. Use `immultiply` to dynamically scale the `moon` image.

```
I = imread('moon.tif');
I2 = imadd(I,50);
I3 = immultiply(I,1.2);
figure
subplot(1,3,1), imshow(I), title('Original Image');
subplot(1,3,2), imshow(I2), title('Normal Brightening');
subplot(1,3,3), imshow(I3), title('Dynamic Scaling');
```

Question 6 When dynamically scaling the `moon` image, why did the dark regions around the moon not become brighter as in the normally adjusted image?

Image multiplication can also be used for special effects such as an artificial 3D look. By multiplying a flat image with a gradient, we create the illusion of a 3D textured surface.

14. Close all open figures and clear all workspace variables.
15. Create an artificial 3D planet by using the `immultiply` function to multiply the `earth1` and `earth2` images.

```
I = im2double(imread('earth1.tif'));
J = im2double(imread('earth2.tif'));
K = immultiply(I,J);
figure
subplot(1,3,1), imshow(I), title('Planet Image');
subplot(1,3,2), imshow(J), title('Gradient');
subplot(1,3,3), imshow(K,[]), title('3D Planet');
```

Image division can be used as the inverse operation to dynamic scaling. Image division is accomplished with the `imdivide` function. When using image division for this purpose, we can achieve the same effect using the `immultiply` function.

16. Close all open figures and clear all workspace variables.
17. Use image division to dynamically darken the moon image.

```
I = imread('moon.tif');
I2 = imdivide(I,2);
figure
subplot(1,3,1), imshow(I), title('Original Image');
subplot(1,3,2), imshow(I2), title('Darker Image w/ Division')
```

18. Display the equivalent darker image using image multiplication.

```
I3 = immultiply(I,0.5);
subplot(1,3,3), imshow(I3), ...
    title('Darker Image w/ Multiplication');
```

Question 7 Why did the multiplication procedure produce the same result as division?

Question 8 Write a small script that will verify that the images produced from division and multiplication are equivalent.

Another use of the image division process is to extract the background from an image. This is usually done during a preprocessing stage of a larger, more complex operation.

19. Close all open figures and clear all workspace variables.
20. Load the images that will be used for background subtraction.

```
notext = imread('gradient.tif');
text = imread('gradient_with_text.tif');
figure, imshow(text), title('Original Image');
```

This image could represent a document that was scanned under inconsistent lighting conditions. Because of the background, the text in this image cannot be processed directly—we must preprocess the image before we can do anything with the text. If the background were homogeneous, we could use image thresholding to extract the text pixels from the background. Thresholding is a simple process of converting an image to its binary equivalent by defining a threshold to be used as a cutoff value: anything below the threshold will be discarded (set to 0) and anything above it will be kept (set to 1 or 255, depending on the data class we choose).

21. Show how thresholding fails in this case.

```
level = graythresh(text);
BW = im2bw(text,level);
figure, imshow(BW)
```

Although the specifics of the thresholding operation (using built-in functions `graythresh` and `im2bw`) are not important at this time, we can see that even though we attempted to segregate the image into dark and light pixels, it produced only part of the text we need (on the upper right portion of the image). If an image of the background with no text on it is available, we can use the `imdivide` function to extract the letters. To obtain such background image in a real scenario, such as scanning documents, a blank page that would show only the inconsistently lit background could be scanned.

22. Divide the background from the image to get rid of the background.

```
fixed = imdivide(text,notext);
figure
subplot(1,3,1), imshow(text), title('Original Image');
subplot(1,3,2), imshow(notext), title('Background Only');
subplot(1,3,3), imshow(fixed,[]), title('Divided Image')
```

Question 9 Would this technique still work if we were unable to obtain the background image?

6.4 TUTORIAL 6.2: LOGIC OPERATIONS AND REGION OF INTEREST PROCESSING

Goal

The goal of this tutorial is to learn how to perform logic operations on images.

Objectives

- Explore the `roipoly` function to generate image masks.
- Learn how to logically AND two images using the `bitand` function.
- Learn how to logically OR two images using the `bitor` function.
- Learn how to obtain the negative of an image using the `bitcmp` function.
- Learn how to logically XOR two images using the `bitxor` function.

What You Will Need

- `lindsay.tif`
- `cameraman2.tif`

Procedure

Logic operators are often used for image masking. We will use the `roipoly` function to create the image mask. Once we have a mask, we will use it to perform logic operations on the selected image.

1. Use the MATLAB help system to learn how to use the `roipoly` function when only an image is supplied as a parameter.

Question 1 How do we add points to the polygon?

Question 2 How do we delete points from the polygon?

Question 3 How do we end the process of creating a polygon?

2. Use the `roipoly` function to generate a mask for the `pout` image.

```
I = imread('pout.tif');
bw = roipoly(I);
```

Question 4 What class is the variable bw?

Question 5 What does the variable bw represent?

Logic functions operate at the bit level; that is, the bits of each image pixel are compared individually, and the new bit is calculated based on the operator we are using (AND, OR, or XOR). This means that we can compare only two images that have the same number of bits per pixel as well as equivalent dimensions. In order for us to use the bw image in any logical calculation, we must ensure that it consists of the same number of bits as the original image. Because the bw image already has the correct number of rows and columns, we need to convert only the image to `uint8`, so that each pixel is represented by 8 bits.

3. Convert the mask image to class `uint8`.

```
bw2 = uint8(bw);
```

Question 6 In the above conversion step, what would happen if we used the `im2uint8` function to convert the bw image as opposed to just using `uint8(bw)`? (*Hint*: after conversion, check what is the maximum value of the image bw2.)

4. Use the `bitand` function to compute the logic AND between the original image and the new mask image.

```
I2 = bitand(I,bw2);
imshow(I2);
```

Question 7 What happens when we logically AND the two images?

To see how to OR two images, we must first visit the `bitcmp` function, which is used for complementing image bits (NOT).

5. Use the `bitcmp` function to generate a complemented version of the `bw2` mask.

```
bw_cmp = bitcmp(bw2);
figure
subplot(1,2,1), imshow(bw2), title('Original Mask');
subplot(1,2,2), imshow(bw_cmp), title('Complemented Mask');
```

Question 8 What happened when we complemented the `bw2` image?

We can now use the complemented mask in conjunction with `bitor`.

6. Use `bitor` to compute the logic OR between the original image and the complemented mask.

```
I3 = bitor(I,bw_cmp);
figure, imshow(I3)
```

Question 9 Why did we need to complement the mask? What would have happened if we used the original mask to perform the OR operation?

The IPT also includes function `imcomplement`, which performs the same operation as the `bitcmp` function, complementing the image. The function `imcomplement` allows input images to be binary, grayscale, or RGB, whereas `bitcmp` requires that the image be an array of unsigned integers.

7. Complement an image using the `imcomplement` function.

```
bw_cmp2 = imcomplement(bw2);
```

Question 10 How can we check to see that the `bw_cmp2` image is the same as the `bw_cmp` image?

The XOR operation is commonly used for finding differences between two images.

8. Close all open figures and clear all workspace variables.
9. Use the `bitxor` function to find the difference between two images.

```
I = imread('cameraman.tif');
I2 = imread('cameraman2.tif');
I_xor = bitxor(I,I2);
figure
subplot(1,3,1), imshow(I), title('Image 1');
```

```
subplot(1,3,2), imshow(I2), title('Image 2');
subplot(1,3,3), imshow(I_xor,[]), title('XOR Image');
```

Logic operators are often combined to achieve a particular task. In next steps, we will use all the logic operators discussed previously to darken an image only within a region of interest.

10. Close all open figures and clear all workspace variables.
11. Read in image and calculate an adjusted image that is darker using the `imdivide` function.

```
I = imread('lindsay.tif');
I_adj = imdivide(I,1.5);
```

12. Generate a mask by creating a region of interest polygon.

```
bw = im2uint8(roipoly(I));
```

13. Use logic operators to show the darker image only within the region of interest, while displaying the original image elsewhere.

```
bw_cmp = bitcmp(bw);                %mask complement
roi = bitor(I_adj,bw_cmp);          %roi image
not_roi = bitor(I,bw);              %non_roi image
new_img = bitand(roi,not_roi);      %generate new image
imshow(new_img)                     %display new image
```

Question 11 How could we modify the above code to display the original image within the region of interest and the darker image elsewhere?

WHAT HAVE WE LEARNED?

- Arithmetic operations can be used to blend two images (addition), detect differences between two images or video frames (subtraction), increase an image's average brightness (multiplication/division by a constant), among other things.
- When performing any arithmetic image processing operation, pay special attention to the data types involved, their ranges, and the desired way to handle overflow and underflow situations.
- MATLAB's IPT has built-in functions for image addition (`imadd`), subtraction (`imsubtract` and `imabsdiff`), multiplication (`immultiply`), and division (`imdivide`). It also has a function (`imlincomb`) that can be used to perform several arithmetic operations without having to worry about underflow or overflow of intermediate results.

- Logic operations are performed on a bit-by-bit basis and are often used to mask out a portion of an image (the *region of interest*) for further processing.
- MATLAB's IPT has built-in functions for performing basic logic operations on digital images: AND (`bitand`), OR (`bitor`), NOT (`bitcmp`), and XOR (`bitxor`).

6.5 PROBLEMS

6.1 What would be the result of adding a positive constant (scalar) to a monochrome image?

6.2 What would be the result of subtracting a positive constant (scalar) from a monochrome image?

6.3 What would be the result of multiplying a monochrome image by a positive constant greater than 1.0?

6.4 What would be the result of multiplying a monochrome image by a positive constant less than 1.0?

6.5 Given the 3 × 3 images X and Y below, obtain (a) X AND Y; (b) X OR Y; (c) X XOR Y.

$$X = \begin{bmatrix} 200 & 100 & 100 \\ 0 & 10 & 50 \\ 50 & 250 & 120 \end{bmatrix}$$

$$Y = \begin{bmatrix} 100 & 220 & 230 \\ 45 & 95 & 120 \\ 205 & 100 & 0 \end{bmatrix}$$

6.6 What happens when you add a `uint8` [0, 255] monochrome image to itself?

6.7 What happens when you multiply a `uint8` [0, 255] monochrome image by itself?

6.8 What happens when you multiply a `double` [0, 1.0] monochrome image by itself?

6.9 What happens when you divide a `double` [0, 1.0] monochrome image by itself?

6.10 Would pixel-by-pixel division be a better way to find the differences between two monochrome images than subtraction, absolute difference, or XOR? Explain.

6.11 Write a MATLAB function to perform brightness correction on monochrome images. It should take as arguments a monochrome image, a number between 0 and 100 (amount of brightness correction, expressed in percentage terms), and a third parameter indicating whether the correction is intended to brighten or darken the image.

6.12 Write a MATLAB script that reads an image, performs brightness correction using the function written for Problem 6.11, and displays a window with the image and its histogram,[1] before and after the brightness correction operation. What do the histograms tell you? Would you be able to tell from the histograms alone what type of brightness correction was performed? Would you be able to estimate from the histogram information how much brightening or darkening the image experienced?

[1]Histograms will be introduced in Chapter 9, so you may try this problem after reading that chapter.

CHAPTER 7

GEOMETRIC OPERATIONS

WHAT WILL WE LEARN?

- What do geometric operations do to an image and what are they used for?
- What are the techniques used to enlarge/reduce a digital image?
- What are the main interpolation methods used in association with geometric operations?
- What are affine transformations and how can they be performed using MATLAB?
- How can I rotate, flip, crop, or resize images in MATLAB?
- What is image registration and where is it used?

7.1 INTRODUCTION

Geometric operations modify the geometry of an image by repositioning pixels in a constrained way. In other words, rather than changing the pixel values of an image (as most techniques studied in Part I of this book do), they modify the spatial relationships between groups of pixels representing features or objects of interest within the image. Figure 7.1 shows examples of typical geometric operations whose details will be presented later in this chapter.

Practical Image and Video Processing Using MATLAB®. By Oge Marques.

(a) (b) (c) (d)

FIGURE 7.1 Examples of typical geometric operations: (a) original image; (b) translation (shifting); (c) scaling (resizing); (d) rotation.

Geometric operations can be used to accomplish different goals, such as the following:

- Correcting geometric distortions introduced during the image acquisition process (e.g., due to the use of a fish-eye lens).
- Creating special effects on existing images, such as twirling, bulging, or squeezing a picture of someone's face.
- As part of *image registration*—the process of matching the common features of two or more images of the same scene, acquired from different viewpoints or using different equipment.

Most geometric operations consist of two basic components:

1. *Mapping Function*: This is typically specified using a set of *spatial transformation equations* (and a procedure to solve them) (Section 7.2).
2. *Interpolation Methods*: These are used to compute the new value of each pixel in the spatially transformed image (Section 7.3).

7.2 MAPPING AND AFFINE TRANSFORMATIONS

A geometric operation can be described mathematically as the process of transforming an input image $f(x, y)$ into a new image $g(x', y')$ by modifying the *coordinates* of image pixels:

$$f(x, y) \rightarrow g(x', y') \tag{7.1}$$

that is, the pixel value originally located at coordinates (x, y) will be relocated to coordinates (x', y') in the output image.

To model this process, a *mapping function* is needed. The mapping function specifies the new coordinates (in the output image) for each pixel in the input image:

$$(x', y') = T(x, y) \tag{7.2}$$

This mapping function is an arbitrary 2D function. It is often specified as two separate functions, one for each dimension:

$$x' = T_x(x, y) \tag{7.3}$$

and

$$y' = T_y(x, y) \tag{7.4}$$

where T_x and T_y are usually expressed as polynomials in x and y. The case where T_x and T_y are linear combinations of x and y is called *affine transformation* (or *affine mapping*):

$$x' = a_0 x + a_1 y + a_2 \tag{7.5}$$

$$y' = b_0 x + b_1 y + b_2 \tag{7.6}$$

Equations (7.5) and (7.6) can also be expressed in matrix form as follows:

$$\begin{bmatrix} x' \\ y' \\ 1 \end{bmatrix} = \begin{bmatrix} a_0 & a_1 & a_2 \\ b_0 & b_1 & b_2 \\ 0 & 0 & 1 \end{bmatrix} \begin{bmatrix} x \\ y \\ 1 \end{bmatrix} \tag{7.7}$$

Affine mapping transforms straight lines to straight lines, triangles to triangles, and rectangles to parallelograms. Parallel lines remain parallel and the distance ratio

TABLE 7.1 Summary of Transformation Coefficients for Selected Affine Transformations

Transformation	a_0	a_1	a_2	b_0	b_1	b_2
Translation by Δ_x, Δ_y	1	0	Δ_x	0	1	Δ_y
Scaling by a factor $[s_x, s_y]$	s_x	0	0	0	s_y	0
Counterclockwise rotation by angle θ	$\cos\theta$	$\sin\theta$	0	$-\sin\theta$	$\cos\theta$	0
Shear by a factor $[\text{sh}_x, \text{sh}_y]$	1	sh_y	0	sh_x	1	0

between points on a straight line does not change. Four of the most common geometric operations—translation, scaling, rotation, and shearing—are all special cases of equations (7.5) and (7.6), as summarized in Table 7.1.

The six parameters of the 2D affine mapping (equation (7.7)) are uniquely determined by three pairs of corresponding points. Given the coordinates of relevant points before and after the transformation, one can write n equations in x and y and solve them to find the n transformation coefficients. Figure 7.2 shows an example of a triangle (three vertices, $n = 6$), before and after the affine transformation.

In MATLAB

The IPT has two functions associated with affine transforms: `maketform` and `imtransform`. The `maketform` function is used to define the desired 2D spatial transformation. It creates a MATLAB structure (called a `TFORM`) that contains all the parameters required to perform the transformation. In addition to affine transformations, `maketform` also supports the creation of projective and custom transformations. After having defined the desired transformation, it can be applied to an input image using function `imtransform`. In Tutorial 7.2 (page 142), you will have a chance to learn more about these functions.

■ EXAMPLE 7.1

Generate the affine transformation matrix for each of the operations below: (a) rotation by 30°; (b) scaling by a factor 3.5 in both dimensions; (c) translation by [25, 15] pixels;

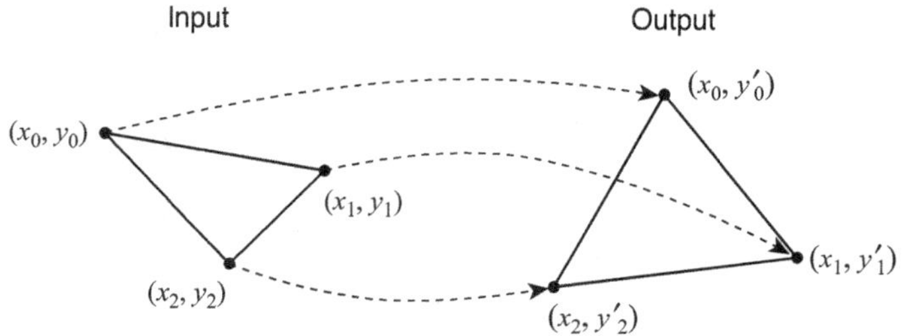

FIGURE 7.2 Mapping one triangle onto another by an affine transformation.

(d) shear by a factor [2, 3]. Use MATLAB to apply the resulting matrices to an input image of your choice.

Solution

Plugging the values into Table 7.1, we obtain the following:
(a) Since cos 30° = 0.866 and sin 30° = 0.500:

$$\begin{bmatrix} 0.866 & -0.500 & 0 \\ 0.500 & 0.866 & 0 \\ 0 & 0 & 1 \end{bmatrix}$$

(b)

$$\begin{bmatrix} 3.5 & 0 & 0 \\ 0 & 3.5 & 0 \\ 0 & 0 & 1 \end{bmatrix}$$

(c)

$$\begin{bmatrix} 1 & 0 & 0 \\ 0 & 1 & 0 \\ 25 & 15 & 1 \end{bmatrix}$$

(d)

$$\begin{bmatrix} 1 & 3 & 0 \\ 2 & 1 & 0 \\ 0 & 0 & 1 \end{bmatrix}$$

MATLAB code:

```
filename = 'any image of your choice'
I =  imread(filename);

% Rotation
Ta = maketform('affine', ...
     [cosd(30) -sind(30) 0; sind(30) cosd(30) 0; 0 0 1]');
Ia = imtransform(I,Ta);

%Scaling
Tb = maketform('affine',[3.5 0 0; 0 3.5 0; 0 0 1]');
Ib = imtransform(I,Tb);
```

```
% Translation
xform = [1 0 25; 0 1 15; 0 0 1]';
Tc = maketform('affine',xform);
Ic = imtransform(I,Tc, 'XData', ...
     [1 (size(I,2)+xform(3,1))], 'YData', ...
     [1 (size(I,1)+xform(3,2))],'FillValues', 128 );

% Shearing
Td = maketform('affine',[1 3 0; 2 1 0; 0 0 1]');
Id = imtransform(I,Td);
```

7.3 INTERPOLATION METHODS

7.3.1 The Need for Interpolation

After a geometric operation has been performed on the original image, the resulting value for each pixel can be computed in two different ways. The first one is called *forward mapping*—also known as *source-to-target mapping* (Figure 7.3)—and consists of iterating over every pixel of the input image, computing its new coordinates, and copying the value to the new location. This approach has a number of problems, such as the following:

- Many coordinates calculated by the transformation equations are not integers, and need to be rounded off to the closest integer to properly index a pixel in the output image.
- Many coordinates may lie out of bounds (e.g., negative values).

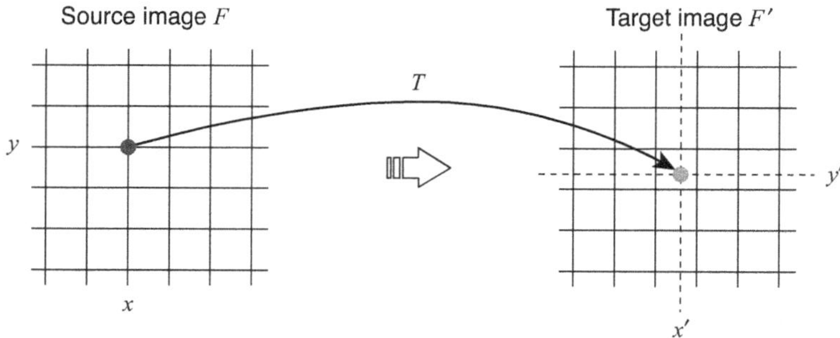

FIGURE 7.3 Forward mapping: for each pixel position in the input image, the corresponding (continuous) target position—resulting from applying a geometric transformation T—is found in the output image. In general, the target position (x', y') does not coincide with any discrete raster point, and the value of the pixel in the input image is copied to one of the adjacent target pixels. Redrawn from [BB08].

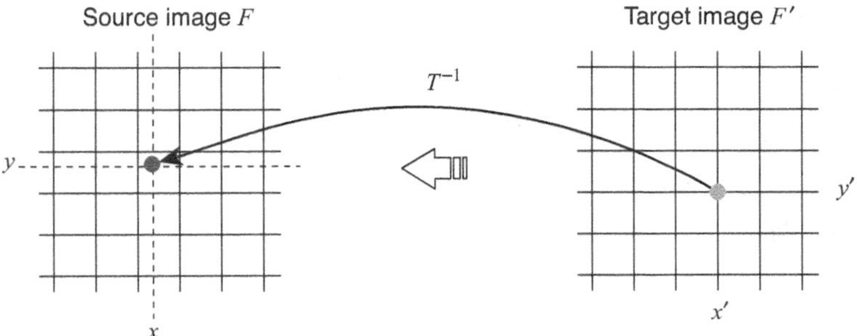

FIGURE 7.4 Backward mapping: for each discrete pixel position in the output image, the corresponding continuous position in the input image (x, y) is found by applying the inverse mapping function T^{-1}. The new pixel value is found by interpolation among the neighbors of (x, y) in the input image. Redrawn from [BB08].

- Many output pixels' coordinates are addressed several times during the calculations (which is wasteful) and some are not addressed at all (which leads to "holes" in the output image, meaning that no pixel value was computed for that coordinate pair).

The solution to the limitations listed above usually comes in the form of a *backward mapping*—also known as *target-to-source mapping* (Figure 7.4)—approach, which consists of visiting every pixel in the output image and applying the *inverse* transformation to determine the coordinates in the input image from which a pixel value must be sampled. Since this backward mapping process often results in coordinates outside the sampling grid in the original image, it usually requires some type of *interpolation* to compute the best value for that pixel.

7.3.2 A Simple Approach to Interpolation

If you were asked to write code to enlarge or reduce an image by a certain factor (e.g., a factor of 2 in both directions), you would probably deal with the problem of removing pixels (in the case of shrinking) by subsampling the original image by a factor of 2 in both dimensions, that is, skipping every other pixel along each row and column. Conversely, for the task of enlarging the image by a factor of 2 in both dimensions, you would probably opt for copying each original pixel to an $n \times n$ block in the output image. These simple interpolation schemes (*pixel removal* and *pixel duplication*, respectively) are fast and easy to understand, but suffer from several limitations, such as the following:

- The "blockiness" effect that may become noticeable when enlarging an image.
- The possibility of removing essential information in the process of shrinking an image.

- The difficulty in extending these approaches to arbitrary, noninteger, resizing factors.

Other simplistic methods—such as using the mean (or median) value of the original $n \times n$ block in the input image to determine the value of each output pixel in the shrunk image—also produce low-quality results and are bound to fail in some cases. These limitations call for improved interpolation methods, which will be briefly described next.

7.3.3 Zero-Order (Nearest-Neighbor) Interpolation

This baseline interpolation scheme rounds off the calculated coordinates (x', y') to their nearest integers. Zero-order (or *nearest-neighbor*) interpolation is simple and computationally fast, but produces low-quality results, with artifacts such as blockiness effects—which are more pronounced at large-scale factors—and jagged straight lines—particularly after rotations by angles that are not multiples of 90° (see Figure 7.5b).

7.3.4 First-Order (Bilinear) Interpolation

First-order (or bilinear) interpolation calculates the gray value of the interpolated pixel (at coordinates (x', y')) as a weighted function of the gray values of the four pixels surrounding the reference pixel in the input image. Bilinear interpolation produces visually better results than the nearest-neighbor interpolation at the expense of additional CPU time (see Figure 7.5c).

7.3.5 Higher Order Interpolations

Higher order interpolations are more sophisticated—and computationally expensive—methods for interpolating the gray value of a pixel. The third-order interpolation scheme implemented in several MATLAB functions is also known as *bicubic interpolation*. It takes into account the 4×4 neighborhood around the reference pixel and computes the resulting gray level of the interpolated pixel by performing the convolution of the 4×4 neighborhood with a cubic function.

Figure 7.5 shows the results of using different interpolation schemes to rotate an image by 35°. The jagged edge effect of the zero-order interpolation is visible (in part b), but there is little—if any—perceived difference between the bipolar (part c) and bicubic (part d) results.

7.4 GEOMETRIC OPERATIONS USING MATLAB

In this section, we will present a summary of typical geometric operations involving digital images that can easily be implemented using MATLAB and the IPT.

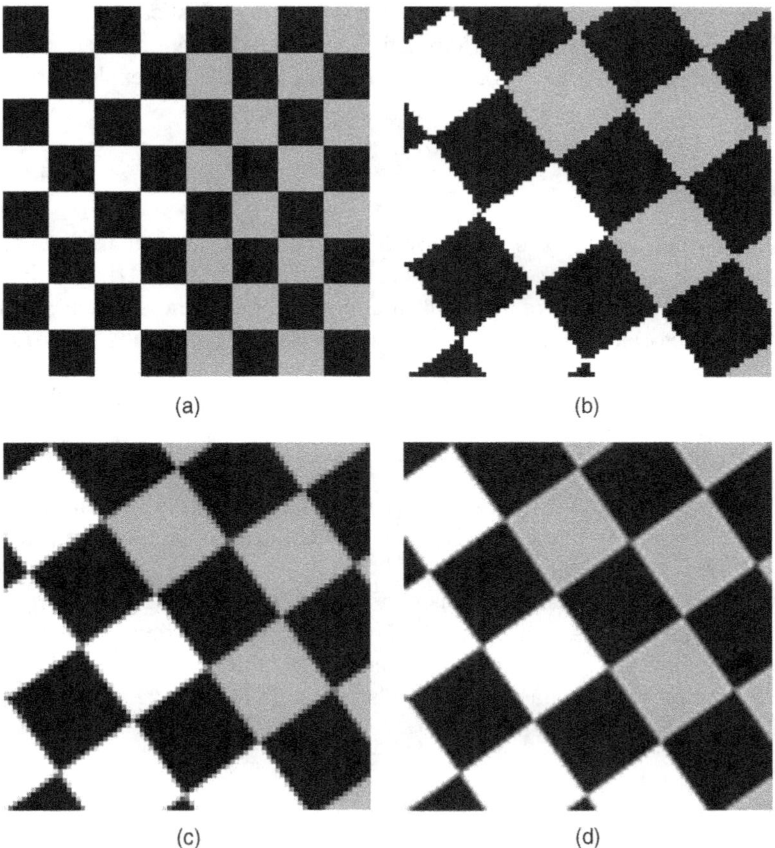

FIGURE 7.5 Effects of different interpolation techniques on rotated images: (a) original image; zoomed-in versions of rotated (35°) image using (b) zero-order (nearest-neighbor) interpolation; (c) first-order (bilinear) interpolation; (d) third-order (bicubic) interpolation.

7.4.1 Zooming, Shrinking, and Resizing

One of the most common geometric operations is the resize operation. In this book, we distinguish between *true image resizing*—where the resulting image size (in pixels) is changed—and *resizing an image for human viewing*—which we will refer to as *zooming (in)* and *shrinking* (or *zooming out*). They are both useful image processing operations and often rely on the same underlying algorithms. The main difference lies in the fact that zooming and shrinking are usually performed interactively (with a tool such as IPT's `imshow` or `imtool`) and their results last for a brief moment, whereas resizing is typically accomplished in a noninteractive way (e.g., as part of a MATLAB script) and its results are stored for longer term use.

The IPT has a function for resizing images, `imresize`. The `imresize` function allows the user to specify the interpolation method used (nearest-neighbor, bilinear, or bicubic interpolation—the default method). It is a rather sophisticated function, which also allows specification of an interpolation kernel and additional parameter/value pairs. In Tutorial 7.1 (page 138), you will use this function and explore some of its options.

7.4.2 Translation

Translation of an input image $f(x, y)$ with respect to its Cartesian origin to produce an output image $g(x', y')$ where each pixel is displaced by $[\Delta_x, \Delta_y]$ (i.e., $x' = x + \Delta_x$ and $y' = y + \Delta_y$) consists of a special case of affine transform (as discussed in Section 7.2). In Tutorial 7.2 (page 142), you will use `maketform` and `imtransform` to perform image translation.

7.4.3 Rotation

Rotation of an image constitutes another special case of affine transform (as discussed in Section 7.2). Consequently, image rotation can also be accomplished using `maketform` and `imtransform`.

The IPT also has a specialized function for rotating images, `imrotate`. Similar to `imresize`, `imrotate` allows the user to specify the interpolation method used: nearest-neighbor (the default method), bilinear, or bicubic. It also allows specification of the size of the output image. In Tutorials 7.1 (page 138), and 7.2 (page 142), you will explore the `imrotate` function.

7.4.4 Cropping

The IPT has a function for cropping images, `imcrop`, which crops an image to a specified rectangle. The crop rectangle can be specified interactively (with the mouse) or its coordinates be passed as parameters to the function. In Tutorial 7.1 (page 138), you will experiment with both options for using this function.

7.4.5 Flipping

The IPT has two functions for flipping matrices (which can also be used for raster images, of course): `flipud`—which flips a matrix up to down—and `fliplr`—which flips a matrix left to right. In Tutorial 7.1 (page 138), you will experiment with both functions.

7.5 OTHER GEOMETRIC OPERATIONS AND APPLICATIONS

7.5.1 Warping

Warping can be defined as the "transformation of an image by reparameterization of the 2D plane" [FDHF+05]. Warping techniques are sometimes referred to as *rubber sheet transformations*, because they resemble the process of applying an image to a sheet of rubber and stretching it according to a predefined set of rules.

The *quadratic warp* is a particular case of *polynomial warping*, where the transformed coordinates (x', y') for a pixel whose original coordinates are (x, y) are given by the following equations:

$$x' = a_0x^2 + a_1y^2 + a_2xy + a_3x + a_4y + a_5 \tag{7.8}$$

$$y' = b_0x^2 + b_1y^2 + b_2xy + b_3x + b_4y + b_5 \tag{7.9}$$

where the coefficients $a_0, ..., a_5, b_0, ..., b_5$ are typically chosen to introduce more complex distortions into an image, for example, turning straight lines into curves. One practical application of the quadratic warping method is to compensate for lens distortions, particularly the *barrel* and *pincushion* distortions discussed in Chapter 5. A variant of equations (7.8) and (7.9), using third-degree polynomials and 20 coefficients, is called a *cubic warp*.

Warping operations of order 2 or higher are usually specified using control points in the source image and mapping them to specified locations in the destination image. Control points are usually associated with key locations and features in the image, such as corners of objects. It is possible to specify more than the minimally required number of control points; in these cases, a least-square method is used to determine the coefficients that best match the desired displacements.

Piecewise warping is an alternative to polynomial warping. It allows the desired warping to be specified with the help of a control grid on top of the input image. The user specifies which control points should be moved by dragging the intersections of the gridlines to new locations using the mouse.

7.5.2 Nonlinear Image Transformations

Nonlinear image transformations usually involve a conversion from rectangular to polar coordinates followed by a deliberate distortion of the resulting points.

Twirling The twirl transformation causes an image to be rotated around an anchor point of coordinates (x_c, y_c) with a space-variant rotation angle: the angle has a value of α at the anchor point and decreases linearly with the radial distance from the center. The effect is limited to a region within the maximum radius r_{max}. All pixels outside this region remain unchanged.

Since this transformation uses backward mapping, we are interested in the equations for the *inverse* mapping function:

$$T_x^{-1} : x = \begin{cases} x_c + r\cos(\theta) & \text{for } r \leq r_{max} \\ x' & \text{for } r > r_{max} \end{cases} \tag{7.10}$$

and

$$T_y^{-1}: y = \begin{cases} y_c + r\sin(\theta) & \text{for } r \le r_{\max} \\ y' & \text{for } r > r_{\max} \end{cases} \tag{7.11}$$

where

$d_x = x' - x_c, \quad r = \sqrt{{d_x}^2 + {d_y}^2}$

$d_y = y' - y_c, \quad \theta = \arctan(d_x, d_y) + \alpha \cdot \left(\frac{r_{\max} - r}{r_{\max}}\right)$

Rippling The ripple transformation causes a local wave-like displacement of the image along both directions, x and y. The parameters for this mapping function are the (nonzero) period lengths L_x, L_y (in pixels) and the associated amplitude values A_x, A_y. The inverse transformation function is given by the following:

$$T_x^{-1}: x = x' + A_x \cdot \sin\left(\frac{2\pi \cdot y'}{L_x}\right) \tag{7.12}$$

$$T_y^{-1}: y = y' + A_y \cdot \sin\left(\frac{2\pi \cdot x'}{L_y}\right) \tag{7.13}$$

Twirling, rippling, and many other nonlinear transformations are often used to create an artistic (or humorous) effect on an image by causing controllable deformations (Figure 7.6).

FIGURE 7.6 Image deformation effects using *Photo Booth*.

7.5.3 Morphing

Morphing is a geometric transformation technique that gradually converts an image into another. The idea is to produce a visible *metamorphosis* effect as intermediate images are being displayed. Image morphing was quite popular in TV, movies, and advertisements in the 1980s and 1990s, but has lost impact since then.

Morphing can be seen as a modified version of piecewise warping, in which the user specifies control points in both the initial and final images. These control points are then used to generate two meshes (one from each image). Affine transformations relate the resulting meshes. An important aspect of morphing is that the warp is computed incrementally, one small step at a time, in combination with a dissolve effect from the initial image to the final one.

An alternative method for image morphing, *fields-based morphing*, originally proposed by Beier and Neely [BN92], does not use meshes. It relies on pairs of reference lines drawn on both images and computes the perpendicular distance between each pixel and each control line. It then uses distance and relative position to determine the correct position where a pixel should be placed in the final image. In this method, all control lines influence, to some extent, the outcome for a certain pixel: the closer the line, the stronger the influence.

7.5.4 Seam Carving

Seam carving [AS07] is a recently proposed—but already very popular—image operator for content-aware image resizing. The basic idea of the algorithm is to find *seams*[1] in the original image and use that information to either (i) reduce the image size by removing ("carving out") the seams that contribute the least to the image's contents or (ii) enlarge the image by inserting additional seams. By applying these operators in both directions, the image can be retargeted to a new size with very little loss of meaningful contents. Figure 7.7 shows an example of seam carving for content-aware image resizing.

7.5.5 Image Registration

Image registration is the process of aligning two or more images of the same scene. First, each input image is compared with a *reference image*—also known as *base image*. Next, a spatial transformation is applied to the input image in order to align it with the base image. The key step in image registration is determining the parameters of the spatial transformation required to bring the images into alignment. This is a complex and fascinating topic for which many books have been written (see "Learn More About It" section at the end of the chapter).

[1]A seam is an "optimal 8-connected path of pixels on a single image from top to bottom, or left to right, where optimality is defined by an image energy function" [AS07].

(a) (b)

FIGURE 7.7 Using seam carving for content-aware resizing: (a) original image (334 × 500 pixels); (b) cropped image (256 × 256 pixels). Original image from Flickr. Seam carving results were obtained using the publicly available implementation by Mathias Lux: http://code.google.com/p/java-imageseams/.

The IPT contains a *Control Point Selection* tool for interactive image registration (Figure 7.10). You will learn how to use that tool to perform basic image registration tasks in Tutorial 7.2 (page 142).

7.6 TUTORIAL 7.1: IMAGE CROPPING, RESIZING, FLIPPING, AND ROTATION

Goal

The goal of this tutorial is to learn how to crop, resize, and rotate digital images.

Objectives

- Learn how to crop an image using the `imcrop` function.
- Learn how to resize an image using the `imresize` function.
- Learn how to flip an image upside down and left–right using `flipud` and `fliplr`.
- Learn how to rotate an image using the `imrotate` function.
- Explore interpolation methods for resizing and rotating images.

Procedure

In the first part of this tutorial, you will learn how to crop an image. Cropping in MATLAB can be done interactively—using the *Crop Image* option in the Image Tool (`imtool`) toolbar—or programmatically—using the `imcrop` function.

1. Open the `cameraman` image and use the *Crop Image* option in the Image Tool (`imtool`) toolbar to crop it such that only the portion of the image containing the tallest building in the background is selected to become the cropped image. Pay attention to (and write down) the coordinates of the top left and bottom right corners as you select the rectangular area to be cropped. You will need this information for the next step.
2. Double-click inside the selected area to complete the cropping operation.
3. Save the resulting image using the **File > Save as...** option in the `imtool` menu. Call it `cropped_building.png`.

```
I = imread('cameraman.tif');
imtool(I)
```

Question 1 Which numbers did you record for the top left and bottom right coordinates and what do they mean? *Hint*: Pay attention to the convention used by the **Pixel info** status bar at the bottom of the `imtool` main window. The IPT occasionally uses a so-called *spatial coordinate system*, whereas *y* represents rows and *x* represents columns. This does *not* correspond to the image axis coordinate system defined in Chapter 2.

4. Open and display the cropped image.

```
I2 = imread('cropped_building.png');
imshow(I2)
```

5. We shall now use the coordinates recorded earlier to perform a similar cropping from a script.
6. The `imcrop` function expects the crop rectangle—a four-element vector `[xmin ymin width height]`—to be passed as a parameter.
7. Perform the steps below replacing my values for `x1`, `y1`, `x2`, and `y2` with the values you recorded earlier.

```
x1 = 186;  x2 = 211;  y1 = 105;  y2 = 159;
xmin = x1;  ymin = y1;  width = x2-x1;  height = y2-y1;
I3 = imcrop(I, [xmin ymin width height]);
imshow(I3)
```

Resizing an image consists of enlarging or shrinking it, using nearest-neighbor, bilinear, or bicubic interpolation. Both resizing procedures can be executed using the `imresize` function. Let us first explore enlarging an image.

8. Enlarge the `cameraman` image by a scale factor of 3. By default, the function uses bicubic interpolation.

```
I_big1 = imresize(I,3);
figure, imshow(I), title('Original Image');
figure, imshow(I_big1), ...
    title('Enlarged Image w/ bicubic interpolation');
```

As you have seen in Chapter 4, the IPT function `imtool` can be used to inspect the pixel values of an image.[2] The `imtool` function provides added functionality to visual inspection of images, such as zooming and pixel inspection.

9. Use the `imtool` function to inspect the resized image, `I_big1`.

```
imtool(I_big1)
```

10. Scale the image again using nearest-neighbor and bilinear interpolations.

```
I_big2 = imresize(I,3,'nearest');
I_big3 = imresize(I,3,'bilinear');
figure, imshow(I_big2), ...
    title('Resized w/ nearest-neighbor interpolation');
figure, imshow(I_big3), ...
    title('Resized w/ bilinear interpolation');
```

Question 2 Visually compare the three resized images. How do they differ?

One way to shrink an image is by simply deleting rows and columns of the image.

11. Close any open figures.
12. Reduce the size of the cameraman image by a factor of 0.5 in both dimensions.

```
I_rows = size(I,1);
I_cols = size(I,2);
I_sm1 = I(1:2:I_rows, 1:2:I_cols);
figure, imshow(I_sm1);
```

Question 3 How did we scale the image?

Question 4 What are the limitations of this technique?

Although the technique above is computationally efficient, its limitations may require us to use another method. Just as we used the `imresize` function for enlarging, we can just as well use it for shrinking. When using the `imresize` function, a scale factor larger than 1 will produce an image larger than the original, and a scale factor smaller than 1 will result in an image smaller than the original.

[2]Note that this function is available only in MATLAB Version 7 and above.

13. Shrink the image using the `imresize` function.

```
I_sm2 = imresize(I,0.5,'nearest');
I_sm3 = imresize(I,0.5,'bilinear');
I_sm4 = imresize(I,0.5,'bicubic');
figure, subplot(1,3,1), imshow(I_sm2), ...
    title('Nearest-neighbor Interpolation');
subplot(1,3,2), imshow(I_sm3), title('Bilinear Interpolation');
subplot(1,3,3), imshow(I_sm4), title('Bicubic Interpolation');
```

Note that in the case of shrinking using either bilinear or bicubic interpolation, the `imresize` function automatically applies a low-pass filter to the image (whose default size is 11×11), slightly blurring it before the image is interpolated. This helps to reduce the effects of aliasing during resampling (see Chapter 5).

Flipping an image upside down or left–right can be easily accomplished using the `flipud` and `fliplr` functions.

14. Close all open figures and clear all workspace variables.
15. Flip the `cameraman` image upside down.
16. Flip the `cameraman` image from left to right.

```
I =  imread('cameraman.tif');
J = flipud(I);
K = fliplr(I);
subplot(1,3,1), imshow(I), title('Original image')
subplot(1,3,2), imshow(J), title('Flipped upside-down')
subplot(1,3,3), imshow(K), title('Flipped left-right')
```

Rotating an image is achieved through the `imrotate` function.

17. Close all open figures and clear all workspace variables.
18. Rotate the `eight` image by an angle of 35°.

```
I = imread('eight.tif');
I_rot = imrotate(I,35);
imshow(I_rot);
```

Question 5 Inspect the size (number of rows and columns) of `I_rot` and compare it with the size of `I`. Why are they different?

Question 6 The previous step rotated the image counterclockwise. How would you rotate the image 35° clockwise?

We can also use different interpolation methods when rotating the image.

19. Rotate the same image using bilinear interpolation.

```
I_rot2 = imrotate(I,35,'bilinear');
figure, imshow(I_rot2)
```

Question 7 How did bilinear interpolation affect the output of the rotation? *Hint:* The difference is noticeable between the two images near the edges of the rotated image and around the coins.

20. Rotate the same image, but this time crop the output.

```
I_rot3 = imrotate(I,35,'bilinear','crop');
figure, imshow(I_rot3)
```

Question 8 How did the `crop` setting change the size of our output?

7.7 TUTORIAL 7.2: SPATIAL TRANSFORMATIONS AND IMAGE REGISTRATION

In this tutorial, we will explore the IPT's functionality for performing spatial transformations (using the `imtransform`, `maketform`, and other related functions). We will also show a simple example of selecting control points (using the IPT's *Control Point Selection* tool) and using spatial transformations in the context of image registration.

In the first part of this tutorial, you will use `imtransform` and `maketform` to implement affine transformations (see Table 7.1), apply them to a test image, and inspect the results.

1. Open the `cameraman` image.
2. Use `maketform` to make an affine transformation that resizes the image by a factor $[s_x, s_y]$. The `maketform` function can accept transformation matrices of various sizes for N-dimensional transformations. But since `imtransform` only performs 2D transformations, you can only specify 3×3 transformation matrices. For affine transformations, the first two columns of the 3×3 matrices will have the values $a_0, a_1, a_2, b_0, b_1, b_2$ from Table 7.1, whereas the last column must contain `0 0 1`.
3. Use `imtransform` to apply the affine transformation to the image.
4. Compare the resulting image with the one you had obtained using `imresize`.

```
I1 = imread('cameraman.tif');
sx = 2; sy = 2;
```

```
T = maketform('affine',[sx 0 0; 0 sy 0; 0 0 1]');
I2 = imtransform(I1,T);
imshow(I2), title('Using affine transformation')
I3 = imresize(I1, 2);
figure, imshow(I3), title('Using image resizing')
```

Question 1 Compare the two resulting images (`I2` and `I3`). Inspect size, gray-level range, and visual quality. How are they different? Why?

5. Use `maketform` to make an affine transformation that rotates an image by an angle θ.
6. Use `imtransform` to apply the affine transformation to the image.
7. Compare the resulting image with the one you had obtained using `imrotate`.

```
I1 = imread('cameraman.tif');
theta = 35*pi/180
xform = [cos(theta) sin(theta) 0; -sin(theta) cos(theta) 0; 0 0 1]'
T = maketform('affine',xform);
I4 = imtransform(I1, T);
imshow(I4), title('Using affine transformation')
I5 = imrotate(I1, 35);
figure, imshow(I5), title('Using image rotating')
```

Question 2 Compare the two resulting images (`I4` and `I5`). Inspect size, gray-level range, and visual quality. How are they different? Why?

8. Use `maketform` to make an affine transformation that translates an image by Δ_x, Δ_y.
9. Use `imtransform` to apply the affine transformation to the image and use a fill color (average gray in this case) to explicitly indicate the translation.
10. Display the resulting image.

```
I1 = imread('cameraman.tif');
delta_x = 50;
delta_y = 100;
xform = [1 0 delta_x; 0 1 delta_y; 0 0 1]'
tform_translate = maketform('affine',xform);
I6 = imtransform(I1, tform_translate,...
     'XData', [1 (size(I1,2)+xform(3,1))],...
     'YData', [1 (size(I1,1)+xform(3,2))],...
     'FillValues', 128 );
figure, imshow(I6)
```

Question 3 Compare the two images (`I1` and `I6`). Inspect size, gray-level range, and visual quality. How are they different? Why?

11. Use `maketform` to make an affine transformation that performs shearing by a factor [sh_x, sh_y] on an input image.
12. Use `imtransform` to apply the affine transformation to the image.
13. Display the resulting image.

```
I = imread('cameraman.tif');
sh_x = 2; sh_y = 1.5;
xform = [1 sh_y 0; sh_x 1 0; 0 0 1]'
T = maketform('affine',xform);
I7 = imtransform(I1, T);
imshow(I7)
```

Image Registration

In the last part of the tutorial, you will learn how to use spatial transformations in the context of image registration. The main steps are illustrated in a block diagram format in Figure 7.8.

14. Open the base image (Figure 7.9a) and the unregistered image (Figure 7.9b).

```
base = imread('klcc_a.png');
unregistered = imread('klcc_b.png');
```

15. Specify control points in both images using `cpselect` (Figure 7.10). This is an interactive process that is explained in detail in the IPT online documentation. For the purpose of this tutorial, we will perform the following:
    - Open the *Control Point Selection* tool.
    - Choose a zoom value that is appropriate and lock the ratio.
    - Select the *Control Point Selection* tool in the toolbar.
    - Select a total of 10 control points per image, making sure that after we select a point in one image with click on the corresponding point in the other image, thereby establishing a match for that point. See Figure 7.11 for the points I chose.
    - Save the resulting control points using the **File > Export Points to Workspace** option in the menu.

```
cpselect(unregistered, base);
```

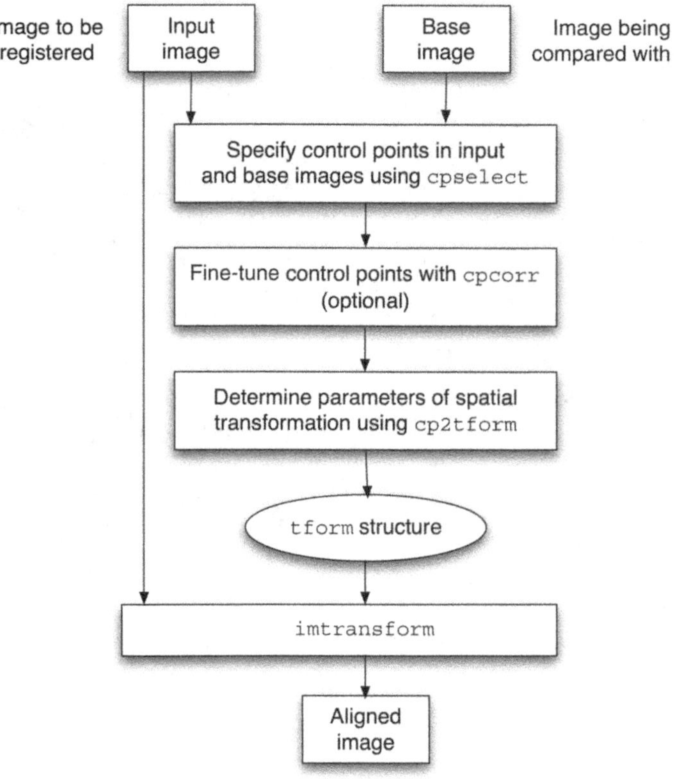

FIGURE 7.8 Image registration using MATLAB and the IPT.

FIGURE 7.9 Interactive image registration: (a) base image; (b) unregistered image.

FIGURE 7.10 The *Control Point Selection* tool.

FIGURE 7.11 Selected points.

16. Inspect the coordinates of the selected control points.

```
base_points
input_points
```

17. Use `cpcorr` to fine-tune the selected control points.

```
input_points_adj = cpcorr(input_points,base_points,...
                          unregistered(:,:,1),base(:,:,1))
```

Question 4 Compare the values for `input_points_adj` with that for `input_points`. Did you notice any changes? Why (not)?

18. This is a critical step. We need to specify the type of transformation we want to apply to the unregistered image based on the type of distortion that it contains. In this case, since the distortion appears to be a combination of translation, rotation, and scaling, we shall use the `'nonreflective similarity'` transformation type. This type requires only two pairs of control points.
19. Once we have selected the type of transformation, we can determine its parameters using `cp2tform`.
20. Use the resulting `tform` structure to align the unregistered image (using `imtransform`).

```
% Select the type of transformation
mytform1 = cp2tform(input_points,base_points,...
           'nonreflective similarity');

% Transform the unregistered image
info = imfinfo('klcc_a.png');
registered = imtransform(unregistered,mytform1,...
             'XData',[1 info.Width], 'YData',[1 info.Height]);
```

21. Display the registered image overlaid on top of the base image.

```
figure, imshow(registered);
hold on
h = imshow(base);
set(h, 'AlphaData', 0.6)
```

Question 5 Are you happy with the results? If you had to do it again, what would you do differently?

WHAT HAVE WE LEARNED?

- Geometric operations modify the geometry of an image by repositioning pixels in a constrained way. They can be used to remove distortions in the image acquisition process or to deliberately introduce a distortion that matches an image with another (e.g., *morphing*).
- Enlarging or reducing a digital image can be done with two different purposes in mind: (1) to actually change the image's dimensions (in pixels), which can be accomplished in MATLAB by function `imresize`; (2) to temporarily change the image size for viewing purposes, through zooming in/out operations, which can be accomplished in MATLAB as part of the functionality of image display primitives such as `imtool` and `imshow`.
- The main interpolation methods used in association with geometric operations are zero-order (or *nearest-neighbor*) interpolation (simple and fast, but leads to low-quality results), first-order (or bilinear) interpolation, and higher-order (e.g., bicubic) interpolation (more sophisticated—and computationally expensive—but leads to best results).
- Affine transformations are a special class of geometric operations, such that once applied to an image, straight lines are preserved and parallel lines remain parallel. Translation, rotation, scaling, and shearing are all special cases of affine transformations. MATLAB's IPT has two functions associated with affine transformations: `maketform` and `imtransform`.
- Image rotation can be performed using the IPT `imrotate` function.
- An image can be flipped horizontally or vertically in MATLAB using simple linear algebra and matrix manipulation instructions.
- The IPT has a function for cropping images, `imcrop`, which crops an image to a specified rectangle and which can be specified either interactively (with the mouse) or via parameter passing.
- Image warping is a technique by which an image's geometry is changed according to a template.
- Image morphing is a geometric transformation technique that converts an image into another in an incremental way. It was popular in TV, movies, and advertisements in the 1980s and 1990s, but has lost impact since then.

LEARN MORE ABOUT IT

- Chapter 5 of [GWE04] contains a MATLAB function for visualizing affine transforms using grids.
- For a deeper coverage of (advanced) interpolation methods, we recommend Chapter 16 of [BB08] and Sections 10.5 and 10.6 of [Jah05].
- Zitová and Flusser [ZF03] have published a survey of image registration methods. For a book-length treatment of the topic, refer to [Gos05].

- The book by Wolberg [Wol90] is a historical reference for image warping.
- For more on warping and morphing, refer to [GDCV99].

ON THE WEB

- MATLAB Image Warping—E. Meyers (MIT)
  http://web.mit.edu/emeyers/www/warping/warp.html
- Image warping using MATLAB GUI (U. of Sussex, England)
  http://www.lifesci.sussex.ac.uk/research/cuttlefish/image_warping_software.htm
- Image morphing with MATLAB
  http://www.stephenmullens.co.uk/image_morphing/

7.8 PROBLEMS

7.1 Use `imrotate` to rotate an image by an arbitrary angle (not multiple of 90°) using the three interpolation methods discussed in Section 7.3. Compare the visual quality of the results obtained with each method and their computational cost (e.g., using MATLAB functions `tic` and `toc`).

7.2 Consider the MATLAB snippet below (assume `X` is a gray-level image) and answer this question: Will `X` and `Z` be identical? Explain.

```
Y = imresize(X,0.5,'nearest');
Z = imresize(Y,2.0,'nearest');
```

7.3 Consider the MATLAB snippet below. It creates an 80×80 black-and-white image (`B`) and uses a simple approach to image interpolation (described in Section 7.3.2) to reduce it to a 40×40 pixel equivalent (`C`). Does the method accomplish its goal? Explain.

```
A = eye(80,80);
A(26:2:54,:)=1;
B = imcomplement(A);
C = B(1:2:end, 1:2:end);
```

CHAPTER 8

GRAY-LEVEL TRANSFORMATIONS

WHAT WILL WE LEARN?

- What does it mean to "enhance" an image?
- How can image enhancement be achieved using gray-level transformations?
- What are the most commonly used gray-level transformations and how can they be implemented using MATLAB?

8.1 INTRODUCTION

This chapter—and also Chapters 9 and 10—will discuss the topic of *image enhancement* in the spatial domain. As discussed in Chapter 1, image enhancement techniques usually have one of these two goals:

1. To improve the subjective quality of an image for human viewing.
2. To modify the image in such a way as to make it more suitable for further analysis and automatic extraction of its contents.

In the first case, the ultimate goal is an improved version of the original image, whose interpretation will be left to a human expert—for example, an enhanced X-ray image that will be used by a medical doctor to evaluate the possibility of a fractured bone. In the second scenario, the goal is to serve as an intermediate step toward an automated solution that will be able to derive the semantic contents of the image—for

Practical Image and Video Processing Using MATLAB®. By Oge Marques.

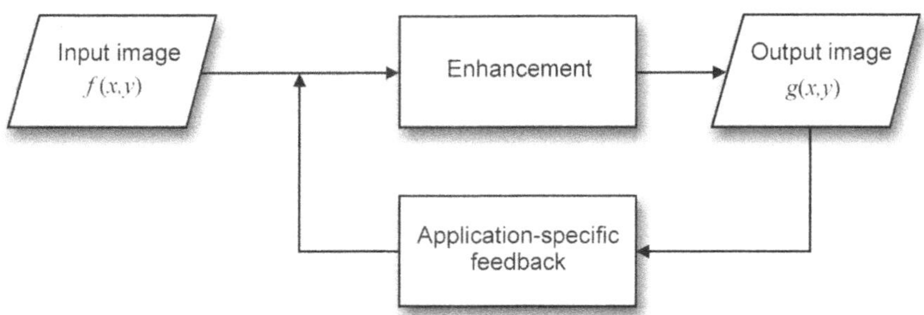

FIGURE 8.1 The image enhancement process. Adapted and redrawn from [Umb05].

example, by improving the contrast between characters and background on a page of text before it is examined by an OCR algorithm. Sometimes these goals can be at odds with each other. For example, sharpening an image to allow inspection of additional fine-grained details is usually desired for human viewing, whereas blurring an image to reduce the amount of irrelevant information is often preferred in the preprocessing steps of a machine vision solution.

Another way to put it is to say that image enhancement techniques are used when either (1) an image needs improvement, or (2) the low-level features must be detected [SS01]. Image enhancement methods "improve the detectability of important image details or objects by man or machine" [SS01]. This is different than attempting to restore a degraded image to its original (or *ideal*) condition, which is the scope of image restoration techniques (Chapter 12).

It is important to mention that image enhancement algorithms are usually *goal specific*, which implies that there is no general theory of image enhancement nor is there a universal enhancement algorithm that always performs satisfactorily. Rather, it is typically an interactive process in which different techniques and algorithms are tried and parameters are fine-tuned, until an acceptable result is obtained (Figure 8.1). Moreover, it is often a subjective process in which the human observer's specialized skills and prior problem-domain knowledge play a significant role (e.g., a radiology expert may find significant differences in the quality of two X-ray images that would go undetected by a lay person).

This chapter focuses on *point operations* whose common goal is to enhance an input image. In some cases, the enhancement results are clearly targeted at a human viewer (e.g., contrast adjustment techniques, Section 8.3.1), while in some other cases the results may be more suitable for subsequent stages of processing in a machine vision system (e.g., image negative, Section 8.3.2).

8.2 OVERVIEW OF GRAY-LEVEL (POINT) TRANSFORMATIONS

Point operations (briefly introduced in Section 2.4.1) are also referred to as *gray-level transformations* or *spatial transformations*. They can be expressed as

$$g(x, y) = T\left[f(x, y)\right] \tag{8.1}$$

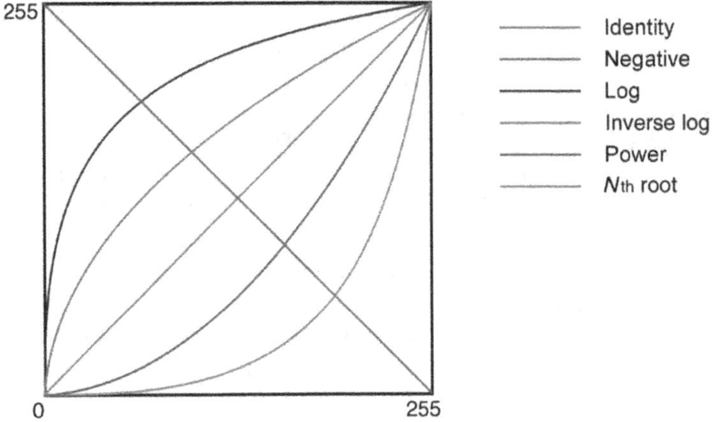

FIGURE 8.2 Basic gray-level transformation functions.

where $g(x, y)$ is the processed image, $f(x, y)$ is the original image, and T is an operator on $f(x, y)$.

Since the actual coordinates do not play any role in the way the transformation function processes the original image,[1] equation (8.1) can be rewritten as

$$s = T[r] \tag{8.2}$$

where r is the original gray level of a pixel and s is the resulting gray level after processing.

Point transformations may be *linear* (e.g., negative), *piecewise linear* (e.g., gray-level slicing), or *nonlinear* (e.g., gamma correction). Figure 8.2 shows examples of basic linear (*identity* and *negative*) and nonlinear (*log*, *inverse log*, *power*—also known as *gamma*—, and nth *root*) transformation functions.

Point operations are usually treated as simple mapping operations whereby the new pixel value at a certain location (x_0, y_0) depends only on the original pixel value at the same location and the mapping function. In other words, the resulting image does not exhibit any change in size, geometry, or local structure if compared with the original image.[2]

In this book, we will call *linear* point transformations those that can be mathematically described by a single linear equation:

$$s = c \cdot r + b \tag{8.3}$$

where r is the original pixel value, s is the resulting pixel value, and c is a constant—responsible for controlling the contrast of the output image—, whereas b is another constant whose value impacts the output image's overall brightness.

[1] If the transformation function $T(.)$ is independent of the image coordinates, the operation is called *homogeneous*.

[2] This is in clear contrast with the operations described in Chapter 7.

From a graphical perspective, a plot of s as a function of r will show a straight line, whose slope (or gradient) is determined by the constant c; the constant term b determines the point at which the line crosses the y-axis. Since the output image should not contain any pixel values outside a certain range (e.g., a range of [0, 255] for monochrome images of class `uint8` in MATLAB), the plot usually also shows a second, horizontal, straight line that is indicative of *clamping* the results to keep them within range (see left and middle columns in Figure 8.4 for examples).

If a point transformation function requires several linear equations, one for each interval of gray-level values, we shall call it a *piecewise linear* function (Section 8.3.5).

Transformation functions that cannot be expressed by one or more linear equations will be called *nonlinear*. The power law (Section 8.3.3) and log (Section 8.3.4) transformations are examples of nonlinear functions.

■ EXAMPLE 8.1

Figure 8.4 shows the results of applying three different linear point transformations to the input image in Figure 8.3. Table 8.1 summarizes the values of c and b for each of the three cases as well as the visual effect on the processed image.

FIGURE 8.3 Linear point transformations example: input image.

TABLE 8.1 Examples of Linear Point Transformations (Images and Curves in Figure 8.4)

Column	c	b	Effect on Image
Left	2	32	Overall brightening, including many saturated pixels
Middle	1	−56	Overall darkening
Right	0.3	0	Significant contrast reduction and overall darkening

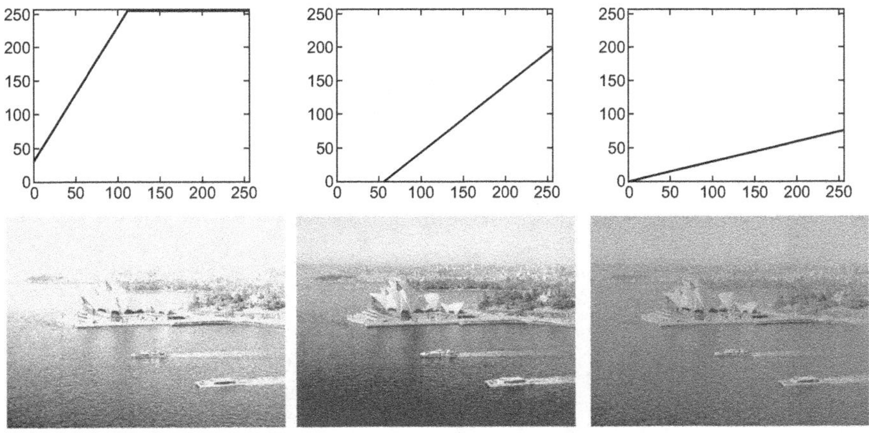

FIGURE 8.4 Linear point transformations and their impact on the overall brightness and contrast of an image: brightening (left), darkening (middle), and contrast reduction (right).

8.3 EXAMPLES OF POINT TRANSFORMATIONS

In this section, we show examples of some of the most widely used point transformations.

8.3.1 Contrast Manipulation

One of the most common applications of point transformation functions is *contrast manipulation* (also known by many other names such as *contrast stretching*, *gray-level stretching*, *contrast adjustment*, and *amplitude scaling*). These functions often exhibit a curve that resembles the curve of a sigmoid function (Figure 8.5a): pixel values of

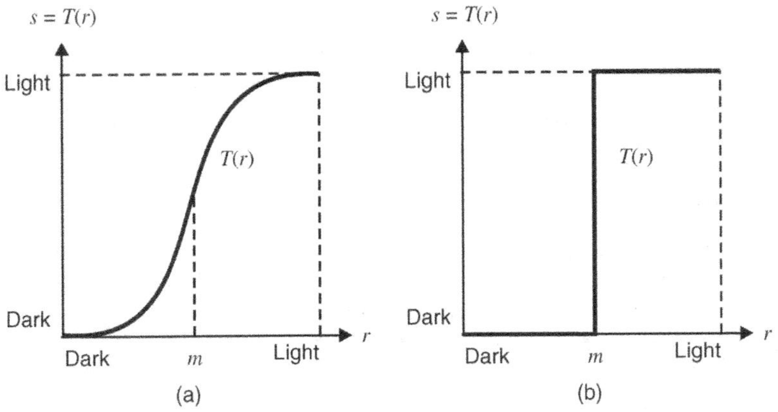

FIGURE 8.5 Examples of gray-level transformations for contrast enhancement. Redrawn from [GW08].

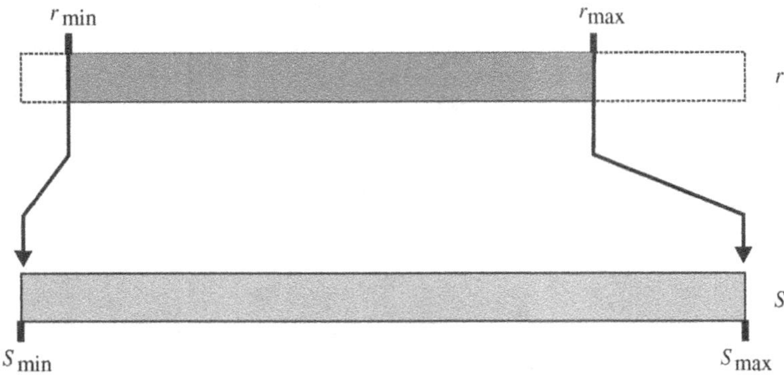

FIGURE 8.6 Autocontrast operation. Redrawn from [BB08].

$f < m$ are compressed toward darker values in the output image, whereas values of $f > m$ are mapped to brighter pixel values in the resulting image. The slope of the curve indicates how dramatic the contrast changes will be; in its most extreme case, a contrast manipulation function degenerates into a binary thresholding[3] function (Figure 8.5b), where pixels in the input image whose value is $f < m$ become black and pixels whose value is $f > m$ are converted to white.

One of the most useful variants of contrast adjustment functions is the *automatic contrast adjustment* (or simply *autocontrast*), a point transformation that—for images of class `uint8` in MATLAB—maps the darkest pixel value in the input image to 0 and the brightest pixel value to 255 and redistributes the intermediate values linearly (Figure 8.6).

The autocontrast function can be described as follows:

$$s = \frac{L-1}{r_{max} - r_{min}} \cdot (r - r_{min}) \tag{8.4}$$

where r is the pixel value in the original image (in the [0, 255] range), r_{max} and r_{min} are the values of its brightest and darkest pixels, respectively, s is the resulting pixel value, and $L - 1$ is the highest gray value in the input image (usually $L = 256$). Figure 8.7 shows an example of an image before and after autocontrast.

In MATLAB

MATLAB's IPT has a built-in function `imadjust` to perform contrast adjustments (including autocontrast). You will learn more about it in Tutorial 9.3 (page 195).

In MATLAB, *interactive* brightness and contrast adjustments can also be performed using `imcontrast` that opens the *Adjust Contrast* tool introduced in Chapter 4.

We shall revisit the topic of contrast adjustments in Chapter 9 when we present techniques such as *histogram equalization* and *histogram stretching*.

[3]We shall study thresholding techniques in Chapter 15.

(a) (b)

FIGURE 8.7 (a) Example of an image whose original gray-level range was [90, 162]; (b) the result of applying the autocontrast transformation (equation (8.4)).

8.3.2 Negative

The negative point transformation function (also known as *contrast reverse* [Pra07]) was described in Section 6.1.2. The negative transformation is used to make the output more suitable for the task at hand (e.g., by making it easier to notice interesting details in the image).

In MATLAB

MATLAB's IPT has a built-in function to compute the negative of an image: `imcomplement`, which was used in Tutorial 6.1 (page 113).

8.3.3 Power Law (Gamma) Transformations

The power law transformation function is described by

$$s = c \cdot r^{\gamma} \tag{8.5}$$

where r is the original pixel value, s is the resulting pixel value, c is a scaling constant, and γ is a positive value. Figure 8.8 shows a plot of equation (8.5) for several values of γ.

■ EXAMPLE 8.2

Figure 8.9 shows the results of applying gamma correction to an input image using two different values of γ. It should be clear from the figure that when $\gamma < 1$, the resulting image is darker than the original one; whereas for $\gamma > 1$, the output image is brighter than the input image.

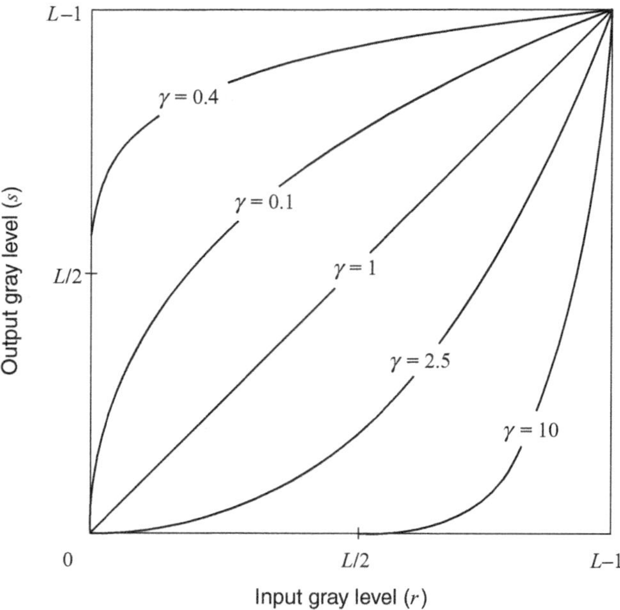

FIGURE 8.8 Examples of power law transformations for different values of γ.

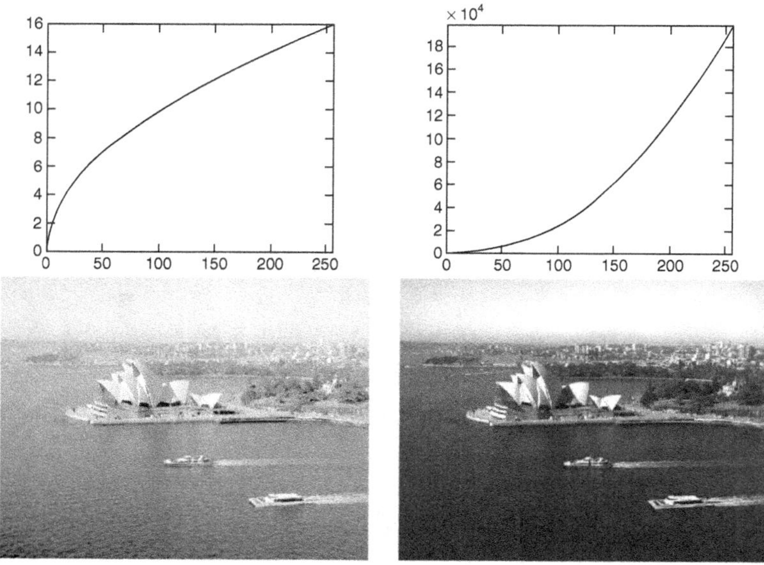

FIGURE 8.9 Examples of gamma correction for two different values of γ: 0.5 (left) and 2.2 (right).

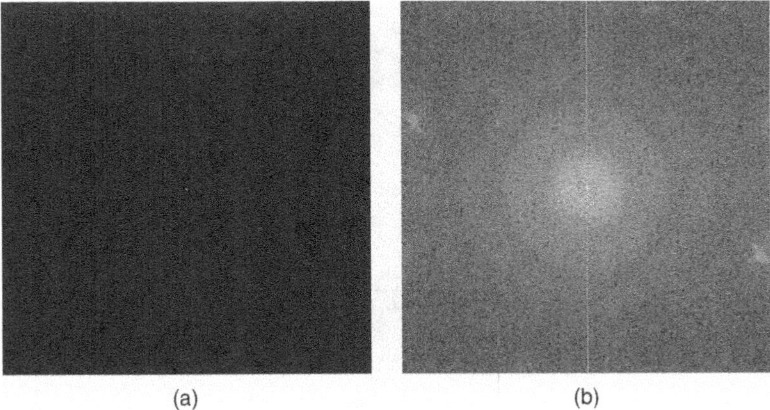

(a) (b)

FIGURE 8.10 Example of using log transformation: (a) Fourier spectrum (amplitude only) of the rice image (available in MATLAB); (b) result of applying equation (8.6) with $c = 1$ followed by autocontrast.

In MATLAB

The `imadjust` function in the IPT can be used to perform gamma correction with the syntax: `g = imadjust(f,[],[],gamma);`

8.3.4 Log Transformations

The log transformation and its inverse are nonlinear transformations used, respectively, when we want to compress or expand the dynamic range of pixel values in an image.

Log transformations can be mathematically described as

$$s = c \cdot \log(1 + r) \tag{8.6}$$

where r is the original pixel value, s is the resulting pixel value, and c is a constant.

Be aware that in many applications of the log transformation, the input "image" is actually a 2D array with values that might lie outside the usual range for gray levels that we usually associate with monochrome images (e.g., [0, 255]).

■ EXAMPLE 8.3

This example uses the log transformation to improve the visualization and display of Fourier transform (FT)[4] results (Figure 8.10). The range of values in the matrix in part (a) is $[0, 2.8591 \times 10^4]$, which—when displayed on a linearly scaled 8-bit system—makes it hard to see anything but the bright spot at the center. Applying a log transform, the dynamic range is compressed to [0, 10.26]. Using the proper

[4]The Fourier transform (FT) and its applications will be presented in Chapter 11.

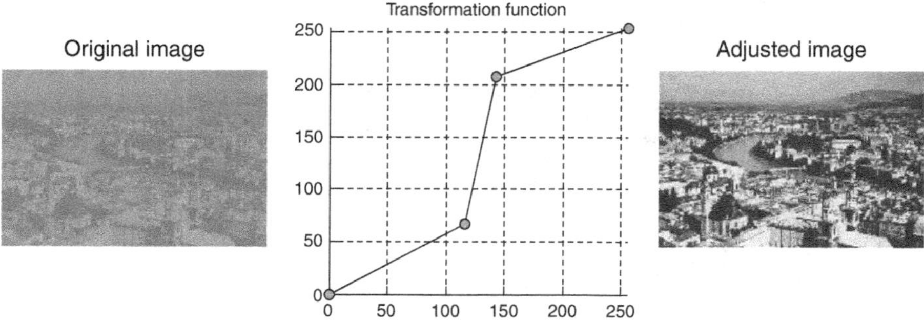

FIGURE 8.11 Piecewise linear transformation using `glsdemo`.

autocontrast transformation to linearly extend the compressed range to [0, 255], we obtain the image in part (b), where significant additional details (e.g., thin vertical line at the center, concentric circles) become noticeable.

8.3.5 Piecewise Linear Transformations

Piecewise linear transformations can be described by several linear equations, one for each interval of gray-level values in the input image. The main advantage of piecewise linear functions is that they can be arbitrarily complex; the main disadvantage is that they require additional user input [GW08], as discussed further in Section 8.4.

■ EXAMPLE 8.4

Figure 8.11 shows an example of an arbitrary piecewise linear transformation function used to improve the contrast of the input image. The function is specified interactively using a GUI-based MATLAB tool `glsdemo` (developed by Jeremy Jacob and available at the book's companion web site).

Figure 8.12 shows an example of *gray-level slicing*, a particular case of piecewise linear transformation in which a specific range of intensity levels (in this case, the

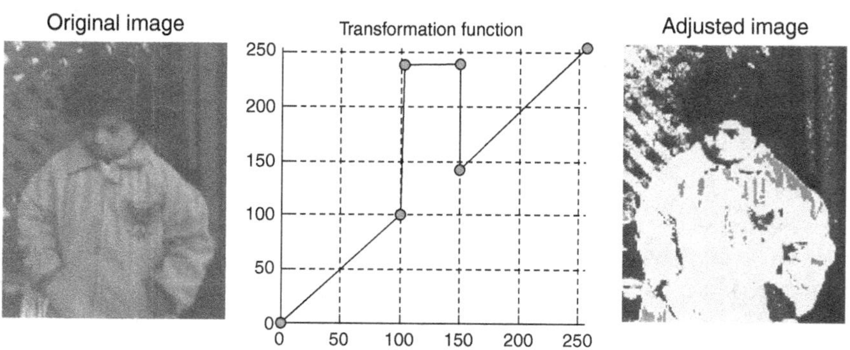

FIGURE 8.12 Gray-level slicing using `glsdemo`. Original image: courtesy of MathWorks.

[100, 150] range) is highlighted in the output image, while all other values remain untouched.[5]

8.4 SPECIFYING THE TRANSFORMATION FUNCTION

All the transformation functions presented in this chapter have been described mathematically in a way that is elegant and appropriate for input variables in the continuous or discrete domain.

In this section, we show that in spite of its elegance, the mathematical formulation is not always useful in practice, for two different—but relevant—reasons:

1. From a user interaction viewpoint, it is often preferable to specify the desired point transformation function interactively using the mouse and a GUI-based application, such as the `glsdemo` in Tutorial 8.1 (page 163).
2. From the perspective of computational efficiency, point operations can be executed at significantly higher speed using lookup tables (LUTs). For images of type `uint8` (i.e., monochrome images with 256 gray levels), the LUT will consist of a 1D array of length 256. LUTs can be easily implemented in MATLAB as demonstrated in the following examples.

■ EXAMPLE 8.5

This example shows how the piecewise linear transformation function specified by equation (8.7) can be implemented using a LUT (and the `intlut` function) in MATLAB. It is important to remember that while arrays (such as the LUT) are 1-based, pixel values vary between 0 and 255. Our MATLAB code must, therefore, include the proper adjustments to avoid off-by-one errors and attempts to access out-of-bounds array elements.

$$s = \begin{cases} 2 \cdot f & \text{for } 0 < r \leq 64 \\ 128 & \text{for } 64 < r \leq 128 \\ f & \text{for } r > 128 \end{cases} \tag{8.7}$$

Using the colon notation (see Chapter 5), we will specify the three linear portions of the transformation function as follows:

```
LUT = uint8(zeros([1 256]));
LUT(1:65) = 2*(0:64);
LUT(66:129) = 128;
LUT(130:256) = (130:256)-1;
```

[5]A variant of the gray-level slicing technique highlights a range of values and maps all other values to a fixed—and usually low—gray level.

Next, we will test the LUT using a 3×3 test image.

```
A = uint8([20 40 0; 178 198 64; 77 128 1])
B = intlut(A, LUT)
```

As expected, if the input array is

$$\begin{bmatrix} 20 & 40 & 0 \\ 178 & 198 & 64 \\ 77 & 128 & 1 \end{bmatrix}$$

the resulting array will be

$$\begin{bmatrix} 40 & 80 & 0 \\ 178 & 198 & 128 \\ 128 & 128 & 2 \end{bmatrix}$$

Finally, we will apply the LUT to a real gray-level image. The result appears in Figure 8.13. Note how many pixels in the original image have been "flattened" to the average gray level (128) in the output image.

```
I = imread('klcc_gray.png');
O = intlut(I,LUT);
figure, subplot(1,2,1), imshow(I), subplot(1,2,2), imshow(O)
```

MATLAB users should be aware that—even though MATLAB makes it extremely simple to apply a transformation function to all pixels using a single line of code—the use of a precomputed LUT enables a much more computationally efficient implementation of point transformations, as demonstrated in the following example.

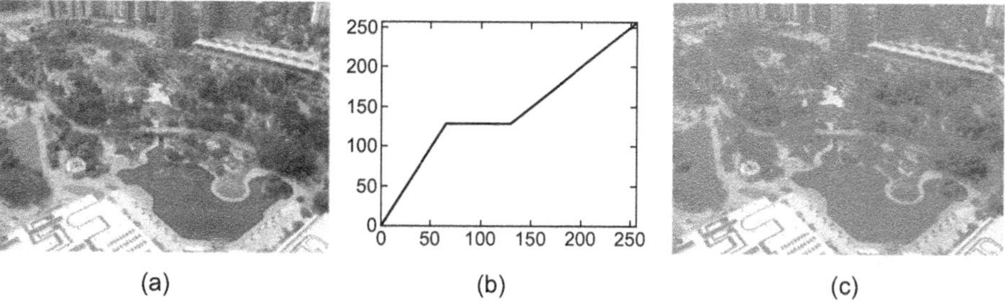

FIGURE 8.13 Example of using a lookup table: (a) input image; (b) transformation function specified by equation (8.7); (c) output image.

■ EXAMPLE 8.6

Let us assume that we want to implement the nonlinear transformation function $s = c\sqrt{r}$, where $c = 5$.

We will first code it using direct implementation and compute the execution times using a 1944 × 2592 pixel monochrome image as input.

```
I = imread('klcc_gray.png');
I2 = double(I);
tic
J = 5*sqrt(I2);
toc
O = uint8(J);
subplot(1,2,1), imshow(I), subplot(1,2,2), imshow(O)
```

Using MATLAB's `tic` and `toc` functions, we can measure the execution time for the transformation step. The value measured using an Apple MacBook with a 2 GHz Intel Core Duo CPU and 1 GB of RAM running MATLAB 7.6.0 (R2008a) was 0.233381 s.

Now we will repeat the process using a precomputed LUT.

```
I = imread('klcc_gray.png');
LUT = double(zeros([1 256]));
LUT(1:256) = 5 * sqrt(0:255);
LUT_int8 = uint8(LUT);
tic
O = intlut(I, LUT_int8);
toc
figure, subplot(1,2,1), imshow(I), subplot(1,2,2), imshow(O)
```

In this case, the measured execution time was 0.030049 s. Even if we move `tic` to the line immediately after the `imread` operation (i.e., if we include the time spent computing the LUT itself), the result would be 0.037198 s, which is more than six times faster than the pixel-by-pixel computation performed earlier.

8.5 TUTORIAL 8.1: GRAY-LEVEL TRANSFORMATIONS

Goal

The goal of this tutorial is to learn how to perform basic point transformations on grayscale images.

Objectives

- Explore linear transformations including the identity function and the negative function.
- Learn how to perform logarithmic grayscale transformations.
- Learn how to perform power law (gamma) grayscale transformations.
- Explore gray (intensity)-level slicing.

What You Will Need

- `radio.tif` image
- `micro.tif` image
- `glsdemo.m` script

Procedure

The most basic transformation function is the identity function, which simply maps each pixel value to the same value.

1. Create an identity transformation function.

```
x = uint8(0:255);
plot(x); xlim([0 255]); ylim([0 255]);
```

2. Use the transformation function on the `moon` image to see how the identity function works.

```
I = imread('moon.tif');
I_adj = x(I + 1);
figure, subplot(1,2,1), imshow(I), title('Original Image');
subplot(1,2,2), imshow(I_adj), title('Adjusted Image');
```

Question 1 Why were we required to use `I+1` when performing the transformation instead of just `I`?

Question 2 How can we show that the adjusted image and the original image are equivalent?

The negative transformation function generates the negative of an image.

3. Create a negative transformation function and show the result after applied to the `moon` image.

```
y = uint8(255:-1:0); I_neg = y(I + 1);
figure, subplot(1,3,1), plot(y), ...
```

```
title('Transformation Function'), xlim([0 255]), ylim([0 255]);
subplot(1,3,2), imshow(I), title('Original Image');
subplot(1,3,3), imshow(I_neg), title('Negative Image');
```

Question 3 How did we create the negative transformation function?

A negative transformation results in the same image as if we were to complement the image logically.

4. Complement the original image and show that it is equivalent to the negative image generated in the previous step.

```
I_cmp = imcomplement(I);
I_dif = imabsdiff(I_cmp,I_neg);
figure, imshow(I_cmp)
figure, imshow(I_dif,[])
```

Logarithmic transformation functions can be used to compress the dynamic range of an image in order to bring out features that were not originally as clear. The log transformation function can be calculated using equation (8.6).

In our case, x represents the value of any particular pixel, and the constant c is used to scale the output within grayscale range [0, 255].

5. Close all open figures and clear all workspace variables.
6. Generate a logarithmic transformation function.

```
x = 0:255; c = 255 / log(256);
y = c * log(x + 1);
figure,  subplot(2,2,1), plot(y), ...
    title('Log Mapping Function'), axis tight, axis square
```

7. Use the transformation function to generate the adjusted image.

```
I = imread('radio.tif');
I_log = uint8(y(I + 1));
subplot(2,2,2), imshow(I), title('Original Image');
subplot(2,2,3), imshow(I_log), title('Adjusted Image');
```

In the second line of code, you will note that we convert the image to uint8. This is necessary because the only way the function imshow will recognize a matrix with a range of [0, 255] as an image is if it is of class uint8. In the next step, we will see that simply brightening the image will *not* show the missing detail in the image.

8. Show a brightened version of the image.

```
I_br = imadd(I,100);
subplot(2,2,4), imshow(I_br), title('Original Image Scaled');
```

Question 4 Why does the log-transformed image display the hidden detail in the radio image, but the brightened image does not?

The inverse of the log function is as follows.

```
y(x) = exp(x/c) - 1;
```

Again, here c is the scaling constant and x is the pixel value. We can demonstrate that applying this equation to the image we previously created (image I_log, which was transformed using the log transformation) will result in the original image.

9. Use inverse log transformation to undo our previous transformation.

```
%
z = exp(x/c) - 1;
I_invlog = uint8(z(I_log + 1));
figure, subplot(2,1,1), plot(z), title('Inverse-log Mapping Function');
subplot(2,1,2), imshow(I_invlog), title('Adjusted Image');
```

Power law transformations include nth root and nth power mapping functions. These functions are more versatile than the log transformation functions because you can specify the value of n, which ultimately changes the shape of the curve to meet your particular needs.

10. Close all open figures and clear all workspace variables.
11. Generate an nth root function where n equals 2.

```
x = 0:255; n = 2; c = 255 / (255 ^ n);
root = nthroot((x/c), n);
figure, subplot(2,2,1), plot(root), ...
    title('2nd-root transformation'), axis tight, axis square
```

Question 5 How does the shape of the curve change if we were to use a different value for n?

12. Use the transformation function to generate the adjusted image.

```
I = imread('drill.tif');
I_root = uint8(root(I + 1));
subplot(2,2,2), imshow(I), title('Original Image');
subplot(2,2,[3 4]), imshow(I_root), title('Nth Root Image');
```

We can see that the adjusted image shows details that were not visible in the original image.

The *n*th power transformation function is the inverse of the *n*th root.

13. Generate an *n*th power transformation function.

```
power = c * (x .^ n);
figure, subplot(1,2,1), plot(power), ...
    title('2nd-power transformation');
axis tight, axis square
```

14. Use the *n*th power transformation to undo our previous transformation.

```
I_power = uint8(power(I_root + 1));
subplot(1,2,2), imshow(I_power), title('Adjusted Image');
```

Question 6 Show that the `I_power` image and the original image `I` are (almost) identical.

The transformation functions we explored thus far have been defined by mathematical equations. During the next steps, we shall explore the creation and application of piecewise linear transformation functions to perform specific tasks. Our first example will be *gray-level slicing*, a process by which we can enhance a particular range of the gray scale for further analysis.

15. Close all open figures and clear all workspace variables.
16. Load the `micro` image and display it.

```
I = imread('micro.tif');
figure, subplot(1,3,1), imshow(I), title('Original Image');
```

17. Create the transformation function.

```
y(1:175) = 0:174;
y(176:200) = 255;
y(201:256) = 200:255;
subplot(1,3,2), plot(y), axis tight, axis square
```

Question 7 Based on the previous step, what do you expect to be the visual effect of applying this transformation function to the original image?

18. Generate the adjusted image.

```
I2 = uint8(y(I + 1));
subplot(1,3,3), imshow(I2), title('Adjusted Image');
```

In the previous steps, we enhanced a particular range of grayscale values while leaving the others unchanged. The adjusted image reflects this change, but it is still difficult to see the enhanced pixels because of the surrounding distractions. We can isolate the enhanced pixels even more by setting all nonenhanced pixels to a constant level.

19. Create a new transformation function and display the adjusted image.

```
z(1:175) = 50;
z(176:200) = 250;
z(201:256) = 50;
I3 = uint8(z(I + 1));
figure, subplot(1,2,1), plot(z), ...
    xlim([0 255]), ylim([0 255]), axis square
subplot(1,2,2), imshow(I3)
```

Although it is possible to create any transformation function in MATLAB by defining a vector of values, as we did above, this can be tedious and time consuming. Through the use of a GUI, we can dynamically generate a function that the user defines visually. The following demo illustrates this.

20. Review the help information for `glsdemo`.
21. Run `glsdemo` with the image `micro.tif` and recreate the transformation functions that we previously used in steps 17 and 19.

WHAT HAVE WE LEARNED?

- Image enhancement is the process of modifying the pixel values within an image in such a way that the resulting image is an improved version of the original image for a particular purpose, whether it is the human perception of subjective quality or further processing by machine vision algorithms.
- Image enhancement can be achieved in many different ways, including the use of certain gray-level transformations, whose chief characteristic is the fact that the resulting gray level of a pixel depends only on the original pixel value and the transformation function. For this reason, gray-level transformations are also referred to as *point transformations*.
- Gray-level transformations can be implemented in MATLAB using the `imadjust` function.
- Gray-level transformations are often used for brightness and contrast adjustments. In MATLAB, interactive brightness and contrast adjustments can also be performed using `imcontrast`.
- Some of the most commonly used point transformations are negative, power law (gamma), logarithmic, and piecewise linear transformations.

- Point transformation functions can be specified interactively using tools such as `glsdemo`.
- The use of a lookup table speeds up the processing of point transformation functions.

LEARN MORE ABOUT IT

- Sections 10.1–10.3 of [Jah05] discuss point transformations and applications such as noise variance equalization, two-point radiometric calibration, and windowing.
- Section 5.4 of [BB08] presents a modified autocontrast operation.
- Section 3.2 of [GWE04] contains MATLAB functions (`intrans` and `gscale`) that extend the functionality of `imadjust`.
- Section 3.8 of [GW08] discusses the use of fuzzy techniques for intensity transformations.

8.6 PROBLEMS

8.1 Write a MATLAB function to perform a piecewise linear brightness and contrast adjustment on monochrome images using the generic method described in equation (8.3). It should take as arguments a monochrome image, the c coefficient (slope), and the b coefficient (offset).

8.2 Write a MATLAB function to perform a simple version of *image solarization* technique (also known as *Sabatier effect*), a point transformation that processes an image by leaving all pixels brighter than a certain value (T) untouched, while extracting the negative of all pixels darker than T.

8.3 Write a MATLAB function to perform a point transformation by which each pixel value in an input image of class `uint8` is replaced by the square of its original value and answer the following questions:

1. Do you have to explicitly make provisions for clamping the results (so that they stay within range)? Why (not)?
2. Is the resulting image brighter or darker than the original image? Explain.

8.4 Repeat Problem 8.3, this time for an input image of class `double`. Does the answer to any of the questions change? Why?

8.5 The sigmoid function used to generate a point transformation similar to Figure 8.5a can be described by the equation

$$s = \frac{1}{1 + (m/r)^S} \tag{8.8}$$

where r is the original pixel value, s is the resulting pixel value, m is a user-specified threshold, and S is a parameter that controls the slope of the curve.

Write a MATLAB script that generates and plots the point transformation function described in equation (8.8).

8.6 Apply the transformation function developed for Problem 8.5 to different images and experiment with different values of m and S. Report a summary of your findings.

CHAPTER 9

HISTOGRAM PROCESSING

WHAT WILL WE LEARN?

- What is the histogram of an image?
- How can the histogram of an image be computed?
- How much information does the histogram provide about the image?
- What is histogram equalization and what happens to an image whose histogram is equalized?
- How can the histogram be modified through direct histogram specification and what happens to an image when we do it?
- What other histogram modification techniques can be applied to digital images and what is the result of applying such techniques?

9.1 IMAGE HISTOGRAM: DEFINITION AND EXAMPLE

The histogram of a monochrome image is a graphical representation of the frequency of occurrence of each gray level in the image. The data structure that stores the frequency values is a 1D array of numerical values, h, whose individual elements store the number (or percentage) of image pixels that correspond to each possible gray level.

Each individual histogram entry can be expressed mathematically as

$$h(k) = n_k = \text{card}\{(x, y) | f(x, y) = k\} \tag{9.1}$$

Practical Image and Video Processing Using MATLAB®. By Oge Marques.

Here, $k = 0, 1, ..., L - 1$, where L is the number of gray levels of the digitized image, and card$\{\cdots\}$ denotes the cardinality of a set, that is, the number of elements in that set (n_k).

A normalized histogram can be mathematically defined as

$$p(r_k) = \frac{n_k}{n} \tag{9.2}$$

where n is the total number of pixels in the image and $p(r_k)$ is the probability (percentage) of the kth gray level (r_k).

Histograms are normally represented using a bar chart, with one bar per gray level, in which the height of the bar is proportional to the number (or percentage) of pixels that correspond to that particular gray level.

In MATLAB

MATLAB's IPT has a built-in function to calculate and display the histogram of a monochrome image: `imhist`. Alternatively, other MATLAB plotting functions such as `bar`, `plot`, and `stem` can also be used to display histograms. In Tutorial 9.1 (page 188), you will have a chance to experiment with all of them.

■ EXAMPLE 9.1

Table 9.1 shows the pixel counts for a hypothetical image containing 128 × 128 pixels, with eight gray levels. The number of pixels that correspond to a given gray level is indicated in the second column and the corresponding percentages (probabilities), $p(r_k)$, are given in the third column. Its bar graph representation is shown in Figure 9.1.

Each value of $p(r_k)$ represents the percentage of pixels in the image whose gray level is r_k. In other words, a histogram can be interpreted as a probability mass function of a random variable (r_k) and as such it follows all the axioms and theorems of elementary probability theory. For instance, it is easy to verify from Table 9.1 that the sum of the values for $p(r_k)$ is 1, as expected.

TABLE 9.1 Example of a Histogram

Gray Level (r_k)	n_k	$p(r_k)$
0	1120	0.068
1	3214	0.196
2	4850	0.296
3	3425	0.209
4	1995	0.122
5	784	0.048
6	541	0.033
7	455	0.028
Total	16,384	1.000

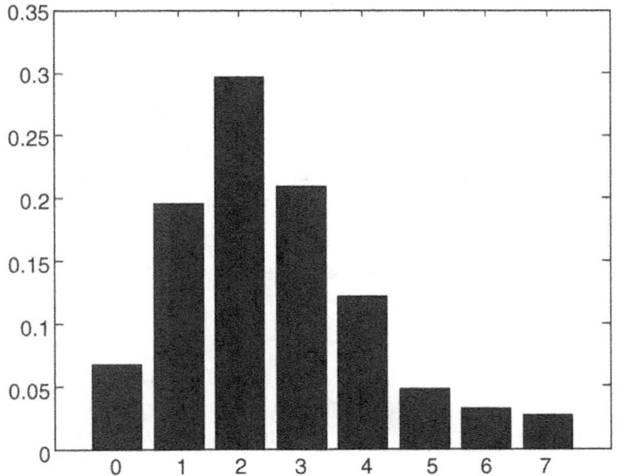

FIGURE 9.1 Example of histogram for an image with eight gray levels.

9.2 COMPUTING IMAGE HISTOGRAMS

To compute the histogram of an 8-bit (256 gray levels) monochrome image, an array of 256 elements (each of which acts as a counter) is created and initialized with zeros. The image is then read, one pixel at a time, and for each pixel the array position corresponding to its gray level is incremented.[1] After the whole image is processed, each array element will contain the number of pixels whose gray level corresponds to the element's index. These values can then be normalized, dividing each of them by the total number of pixels in the image.

For images with more than 8 bits per pixel, it is not practical to map each possible gray level k ($0 \leq k \leq K$) to an array element: the resulting array would be too large and unwieldy. Instead, we use a technique known as *binning*, by which an array of B elements—where B is the number of histogram *bins* (or *buckets*), $B \leq K$—is created and initialized with zeros. In this case, each bin $h(j)$ stores the number of pixels having values within the interval $r_j \leq r < r_{j+1}$ and equation (9.1) can be rewritten as

$$h(j) = \text{card}\{(x, y) | r_j \leq f(x, y) < r_{j+1}\} \quad \text{for } 0 \leq j < B \tag{9.3}$$

The lower limit of each bin can be obtained by the following expression:

$$r_j = j \cdot \frac{K}{B} = j \cdot k_B \tag{9.4}$$

where k_B is the length of each interval.

[1]Keep in mind that arrays in MATLAB are 1 based, and therefore, in MATLAB, the array element with index $k + 1$ will store the number or percentage of pixels whose value is k, where $k \geq 0$.

9.3 INTERPRETING IMAGE HISTOGRAMS

Histograms provide an easy, practical, and straightforward way of evaluating image attributes, such as overall contrast and average brightness. The histogram of a predominantly dark image contains a concentration of bars on the lower end of the gray-level range (Figure 9.2c), whereas the histogram for a bright image is mostly concentrated on the opposite end (Figure 9.2d). For a low contrast image, the histogram is clustered within a narrow range of gray levels (Figure 9.2a), whereas a high contrast image usually exhibits a bimodal histogram with clear separation between the two predominant modes (Figure 9.2b).[2]

■ EXAMPLE 9.2

Figure 9.2 shows four examples of monochrome images and their corresponding histograms (obtained using the `imhist` function in MATLAB).

The histogram in Figure 9.2a shows that the pixels are grouped around intermediate gray-level values (mostly in the [100, 150] range), which corresponds to an image with low contrast. Figure 9.2b shows a typical bimodal histogram, one that has two distinctive hills, a more pronounced one in the dark region (background) and the other smaller one in the light region of the histogram (foreground objects). In these situations, it can be said that the corresponding image has high contrast since the two modes are well spaced from each other.[3] In part (c), the histogram exhibits a large concentration of pixels in the lower gray levels, which corresponds to a mostly dark image. Finally, in Figure 9.2d, the pixel values are grouped close to the higher gray-level values, which corresponds to a bright image.

It is important to note that even though a histogram carries significant qualitative and quantitative information about the corresponding image (e.g., minimum, average, and maximum gray-level values, dominance of bright or dark pixels, etc.), other qualitative conclusions can be reached only upon examining the image itself (e.g., overall quality of the image, presence or absence of noise, etc.). A quick analysis of Figure 9.2 should be enough to prove this statement true. Moreover, although a histogram provides the frequency distribution of gray levels in an image, it tells us nothing about the spatial distribution of the pixels whose gray levels are represented in the histogram (see Problem 9.8).

Histograms have become a popular tool for conveying image statistics and helping determine certain problems in an image. Their usefulness can be demonstrated by the

[2]In many cases, those two modes (hills) correspond to background and one or more objects of interest in the foreground, for example, Figure 9.2b.

[3]It is interesting to note that the concepts of *high* and *low* contrast in this case are only related to the average spacing between groups of histogram bars. The expression "good contrast," on the other hand, should be used with caution, for it often refers to one's subjective opinion about the image quality. Such quality judgments usually cannot be made based upon looking at the histogram alone.

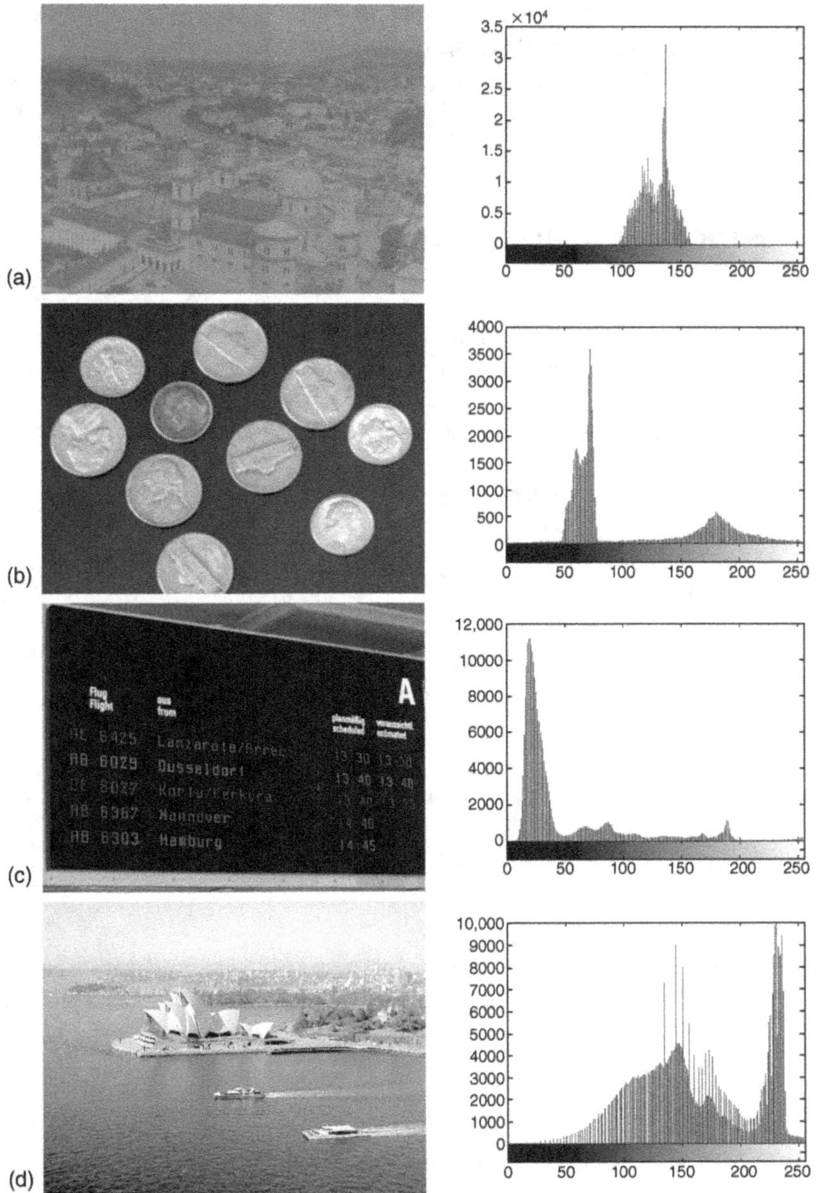

FIGURE 9.2 Examples of images and corresponding histograms. Original image in part (b): courtesy of MathWorks.

fact that many contemporary digital cameras have an optional real-time histogram overlay in the viewfinder to prevent taking underexposed or overexposed pictures.

Histograms can be used whenever a statistical representation of the gray-level distribution in an image or video frame is desired. Histograms can also be used to enhance or modify the characteristics of an image, particularly its contrast. Some of these techniques, generally called *histogram modification (or modeling) techniques*, are histogram equalization, histogram specification (matching), histogram stretching (input cropping), and histogram shrinking (output cropping). They are described in more detail later.

9.4 HISTOGRAM EQUALIZATION

Histogram equalization is a technique by which the gray-level distribution of an image is changed in such a way as to obtain a uniform (flat) resulting histogram, in which the percentage of pixels of every gray level is the same. To perform histogram equalization, it is necessary to use an auxiliary function, called the *transformation function*, $T(r)$. Such transformation function must satisfy two criteria [GW08]:

1. $T(r)$ must be a monotonically increasing function in the interval $0 \leq r \leq L-1$.
2. $0 \leq T(r) \leq L-1$ for $0 \leq r \leq L-1$.

The most usual transformation function is the *cumulative distribution function (cdf)* of the original probability mass function, given by

$$s_k = T(r_k) = \sum_{j=0}^{k} \frac{n_j}{n} = \sum_{j=0}^{k} p(r_j) \tag{9.5}$$

where s_k is the new (mapped) gray level for all pixels whose original gray level used to be r_k.

The inverse of this function is given by

$$r_k = T^{-1}(s_k) \quad \text{for} \quad k = 0, 1, \ldots, L-1 \tag{9.6}$$

Even though the inverse of the cdf is not needed during the equalization process, it will be used during the direct histogram specification method, described later in this chapter.

In MATLAB

MATLAB's IPT has a built-in function to perform histogram equalization of a monochrome image: `histeq`. For the sake of histogram equalization, the syntax for `histeq` is usually `J = histeq(I,n)`, where `n` (whose default value is 64)[4] is the number of desired gray levels in the output image. In addition to histogram equalization, this function can also be used to perform histogram matching (Section 9.5), as it will be seen in Tutorial 9.2.

[4]Caution: This value is *not* the same as the default value for the imhist function that uses $n = 256$ as a default.

■ EXAMPLE 9.3

Assume the histogram data from Table 9.1 and its graphic representation (Figure 9.1). Calculate the equalized histogram using the cdf of the original probability mass function as a transformation function and plot the resulting histogram.

Solution

Using the *cdf* as the transformation function, we can calculate

$$s_0 = T(r_0) = \sum_{j=0}^{0} p(r_j) = p(r_0) = 0.068 \tag{9.7}$$

Similarly,

$$s_1 = T(r_1) = \sum_{j=0}^{1} p(r_j) = p(r_0) + p(r_1) = 0.264 \tag{9.8}$$

and $s_2 = 0.560$, $s_3 = 0.769$, $s_4 = 0.891$, $s_5 = 0.939$, $s_6 = 0.972$, and $s_7 = 1$. The transformation function is plotted in Figure 9.3.

Since the image was quantized with only eight gray levels, each value of s_k must be rounded to the closest valid (multiple of 1/7) value. Thus, $s_0 \simeq 0$, $s_1 \simeq 2$, $s_2 \simeq 4$, $s_3 \simeq 5$, $s_4 \simeq 6$, $s_5 \simeq 7$, $s_6 \simeq 7$, and $s_7 \simeq 7$.

The above values indicate that the original histogram bars must be shifted (and occasionally grouped) according to the following mapping: the original level $r_0 = 0$ should be mapped to the new level $s_0 = 0$, which means the corresponding bar should not change. The 3214 pixels whose original gray level was (1/7) should be mapped to $s_1 = 2(1/7)$. Similarly, those pixels whose gray level was 2 should be mapped to

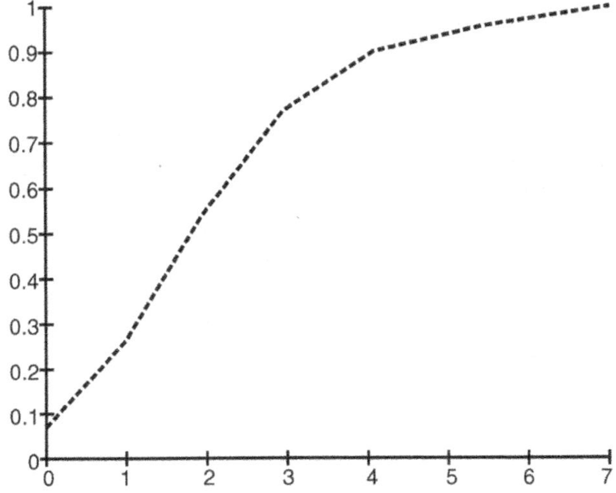

FIGURE 9.3 Transformation function used for histogram equalization.

TABLE 9.2 Equalized Histogram: Values

Gray Level (s_k)	n_k	$p(s_k)$
0	1120	0.068
1	0	0.000
2	3214	0.196
3	0	0.000
4	4850	0.296
5	3425	0.209
6	1995	0.122
7	1780	0.109
Total	16,384	1.000

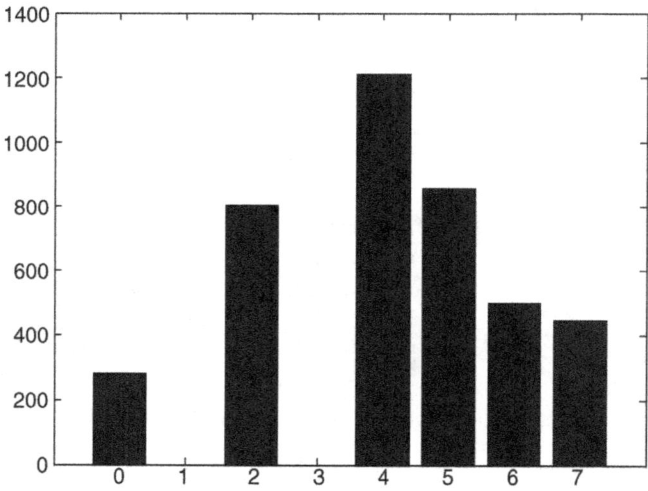

FIGURE 9.4 Equalized histogram—graph.

4, those with $r = 3$ should be mapped to 5, and those with gray level 4 should be mapped to 6. Finally, the three bins that correspond to pixels with gray levels 5, 6, and 7 should be added and mapped to 7.

The resulting (equalized) histogram is shown in Figure 9.4 and the corresponding values are presented in Table 9.2. It should be noted that in the equalized histogram (Figure 9.4), the pixels are more evenly distributed along the gray scale than in the original one (Figure 9.1). Although clearly not a perfectly flat result, it should be interpreted as "the best possible result that can be obtained for this particular image using this transformation function."

■ EXAMPLE 9.4

Figure 9.5 shows an example (obtained using the `histeq` function in MATLAB) of employing histogram equalization to improve the contrast of a 600 × 800

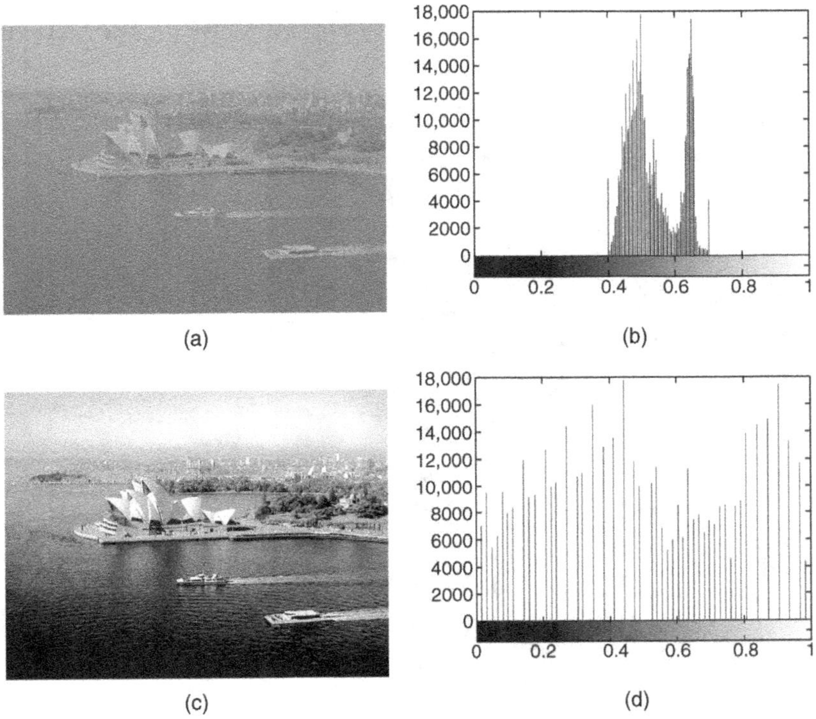

FIGURE 9.5 Use of histogram equalization to improve image contrast.

image with 256 gray levels. Part (a) shows the original image, whose histogram is plotted in Figure 9.5b. Part (d) shows the equalized histogram, corresponding to the image in Figure 9.5c. You may have noticed that the image after equalization exhibits a little bit of *false contouring* (see Chapter 5), particularly in the sky portion of the image. This can be seen as an inevitable "side effect" of the histogram equalization process.

The histogram equalization algorithm described above is *global*; that is, once the mapping (transformation) function is computed, it is applied (as a lookup table) to all pixels of the input image. When the goal is to enhance details in small areas within an image, it is sometimes necessary to apply a *local* version of the algorithm. The local variant of histogram equalization consists of adopting a rectangular (usually square) sliding window (also called a *tile*) and moving it across the entire image. For each image pixel (aligned with the center of the sliding window), the histogram in the neighborhood delimited by the window is computed, the mapping function is calculated, and the reference pixel is remapped to the new value as determined by the transformation function. This process is clearly much more computationally expensive than its global variant and often results in a noisy-looking output image.

■ EXAMPLE 9.5

Figure 9.6 shows a comparison between local and global histogram equalization (obtained using the `adapthisteq` and `histeq` functions in MATLAB, respectively). Parts (a) and (b) show the original image and its histogram, parts (c) and (d) show the results of global histogram equalization, and parts (e) and (f) show the results of local histogram equalization that preserves the bimodal nature of the original histogram while still improving the contrast of the image.

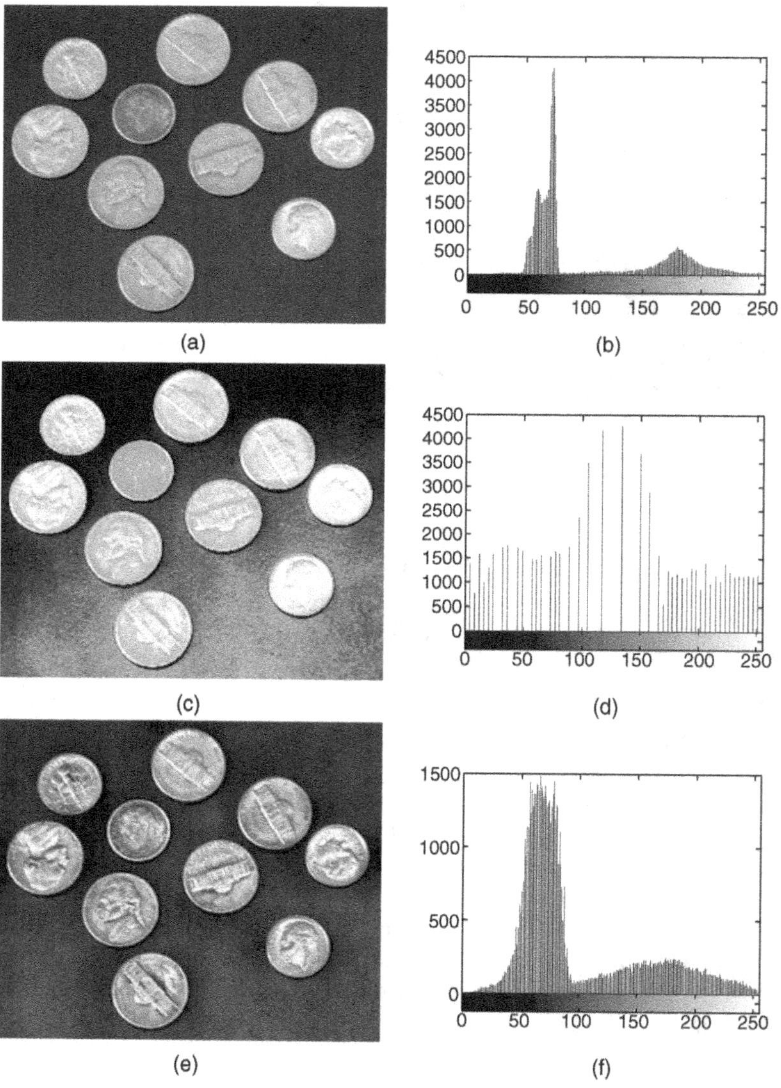

FIGURE 9.6 Global versus local histogram equalization. Original image: courtesy of MathWorks.

In MATLAB

MATLAB's IPT has a built-in function to perform local histogram equalization of a monochrome image, `adapthisteq`, which, unlike `histeq`, operates on small data regions (tiles), rather than the entire image. You will learn how to use this function in Tutorial 9.2.

9.5 DIRECT HISTOGRAM SPECIFICATION

Despite its usefulness in contrast enhancement, histogram equalization is a rather inflexible technique. Its only modifiable parameter is the choice of transformation function, which is normally chosen to be the cdf of the original probability mass function (a convenient choice since the information needed to compute the transformation function can be extracted directly from the pixel values). There might be situations, however, in which one wants to be able to perform specific changes on the original histogram. In these situations, a useful technique is the direct histogram specification, also known as *histogram matching*.

Given an image (and its original histogram) and the desired resulting histogram, the direct histogram specification consists of the following:

1. Equalizing the original image's histogram using the cdf as a transformation function:

$$s_k = T(r_k) = \sum_{j=0}^{k} \frac{n_j}{n} = \sum_{j=0}^{k} p(r_j) \tag{9.9}$$

2. Equalizing the desired probability mass function (in other words, equalizing the desired histogram):

$$v_k = G(z_k) = \sum_{j=0}^{k} p(z_j) \tag{9.10}$$

3. Applying the inverse transformation function

$$z = G^{-1}(s) \tag{9.11}$$

   to the values obtained in step 1.

In MATLAB

The `histeq` function introduced previously can also be used for histogram matching. In this case, the syntax for `histeq` usually changes to `J = histeq(I,h)`, where `h` (a 1D array of integers) represents the specified histogram.

TABLE 9.3 Desired Histogram

Gray Level (s_k)	n_k	$p(s_k)$
0	0	0.0
1	0	0.0
2	0	0.0
3	1638	0.1
4	3277	0.2
5	6554	0.4
6	3277	0.2
7	1638	0.1
Total	16,384	1.0

■ EXAMPLE 9.6

Let us use the histogram from Table 9.1 one more time. Assume that we want to modify this histogram in such a way as to have the resulting pixel distribution as shown in Table 9.3 and plotted in Figure 9.7a. Following the steps outlined above, calculate and plot the resulting histogram that best matches the desired characteristics.

Solution

The equalized histogram has already been calculated before and its results are shown in Table 9.2.

The next step consists in obtaining the cdf of the desired probability mass function. Using equation (9.10), we find the following values:

$v_0 = 0$, $v_1 = 0$, $v_2 = 0$, $v_3 = 0.1$, $v_4 = 0.3$, $v_5 = 0.7$, $v_6 = 0.9$, and $v_7 = 1$.

The last step—and the most difficult to understand when studying this technique for the first time—is obtaining the inverse function. Since we are dealing with discrete values, the inverse function can be obtained simply by searching, for each value of s_k, the closest value of v_k. For instance, for $s_1 = 2/7 \simeq 0.286$, the closest value of v_k is $v_4 = G(z_4) = 0.3$. In inverse function notation, $G^{-1}(0.3) = z_4$. Therefore, pixels

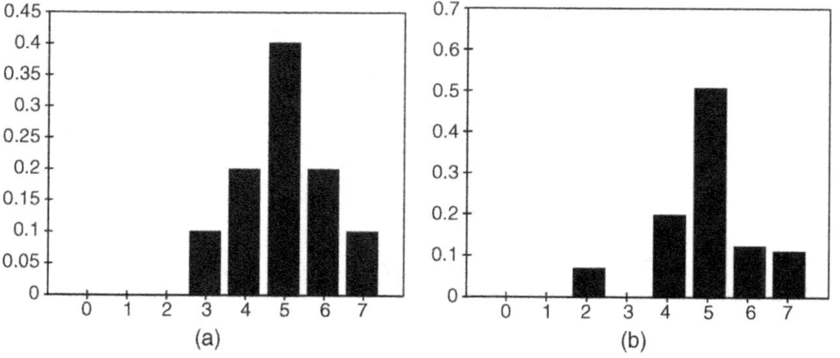

FIGURE 9.7 Histogram matching: (a) desired (specified) histogram; (b) resulting histogram.

TABLE 9.4 Direct Histogram Specification: Summary

k	$p(r_k)$	s_k $(\times 1/7)$	Maps to	v_k	$p(z_k)$
0	0.068	0	z_2	0.00	0.000
1	0.196	2	z_4	0.00	0.000
2	0.296	4	z_5	0.00	0.000
3	0.209	5	z_5	0.10	0.100
4	0.122	6	z_6	0.30	0.200
5	0.048	7	z_7	0.70	0.400
6	0.033	7	z_7	0.90	0.200
7	0.028	7	z_7	1.00	0.100

that were shifted to gray level s_1 after the original histogram equalization should be mapped to gray level z_4. In other words, the 3214 pixels whose original gray level was 1/7 and that were remapped to gray level $s_1 = 2/7$ during the equalization step should now be shifted again to gray level $z_4 = 4/7$. Similarly, for the remaining values of s_k, the following mappings should be performed:

$s_0 = 0 \rightarrow z_2$
$s_1 = 2/7 \simeq 0.286 \rightarrow z_4$
$s_2 = 4/7 \simeq 0.571 \rightarrow z_5$
$s_3 = 5/7 \simeq 0.714 \rightarrow z_5$
$s_4 = 6/7 \simeq 0.857 \rightarrow z_6$
$s_5 = 1 \rightarrow z_7$
$s_6 = 1 \rightarrow z_7$
$s_7 = 1 \rightarrow z_7$

In this case, we have assumed that the algorithm for obtaining the inverse of the transformation function for a given value of s_k would search through the values of v_k and store the index of the most recent (i.e., *last*) value whose absolute difference ($|v_k - s_k|$) is the smallest so far. If the algorithm used another way of breaking ties, s_0 could map to z_0 or z_1 in this example. Table 9.4 summarizes the original and desired histograms, their corresponding cdfs, and the mapping process described above.

Table 9.5 presents the numerical values for the resulting histogram, where $\hat{p}(z_k)$—which was obtained by adding the contents of all rows in column $p(r_k)$ in Table 9.4 that match to each z_k—corresponds to the best possible approximation to the specified $p(z_k)$. For an easy visual comparison between the desired and the resulting histograms, they are plotted side by side in Figure 9.7.

You will probably agree that the resulting histogram approaches, within certain limits, the desired (specified) one.

■ EXAMPLE 9.7

Figure 9.8 shows an example (obtained using the `histeq` function in MATLAB) of direct histogram specification applied to a 600 × 800 image with 256 gray levels.

TABLE 9.5 Resulting Histogram

z_k (×1/7)	$\hat{p}(z_k)$
0	0.000
1	0.000
2	0.068
3	0.000
4	0.196
5	0.505
6	0.122
7	0.109
Total	1

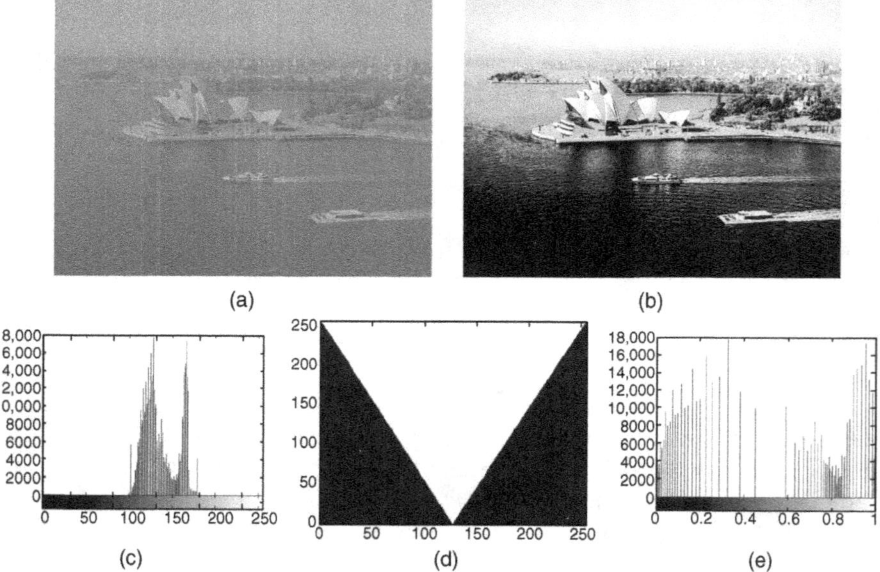

FIGURE 9.8 Histogram matching: (a) original image; (b) resulting image; (c) original histogram; (d) desired histogram; (e) resulting histogram.

9.6 OTHER HISTOGRAM MODIFICATION TECHNIQUES

In this section, we present other techniques by which a histogram can be processed to achieve a specific goal, such as brightness increase (or decrease) or contrast improvement. These techniques parallel some of the point transformation functions presented in Chapter 8, with the main difference being that whereas in Chapter 8 our focus was on the transformation function and its effect on the input image, in this chapter we also inspect the effect on the image's histogram.

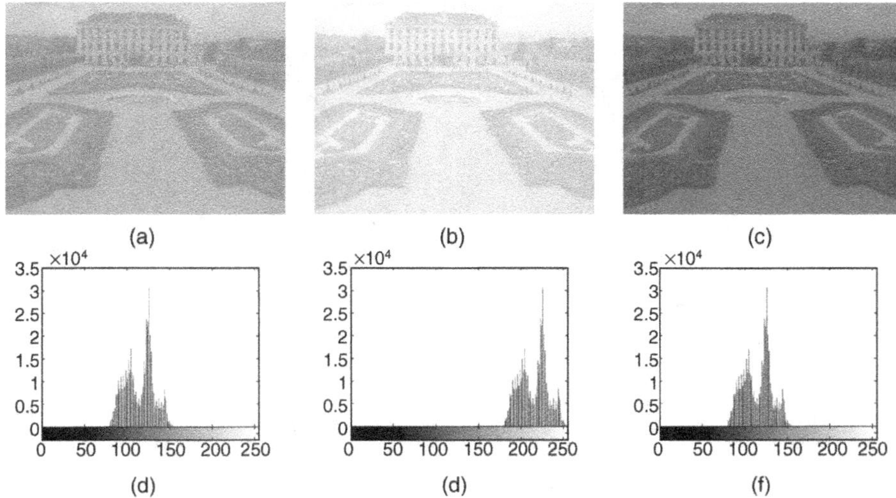

FIGURE 9.9 Histogram sliding: (a) original image; (b) result of sliding to the right by 50; (c) result of sliding to the left by 50; (d–f) histograms corresponding to images in (a)–(c).

9.6.1 Histogram Sliding

This technique consists of simply adding or subtracting a constant brightness value to all pixels in the image. The overall effect will be an image with comparable contrast properties, but higher or lower (respectively) average brightness.

In MATLAB

The `imadd` and `imsubtract` functions introduced earlier in the book can be used for histogram sliding.

■ EXAMPLE 9.8

Figure 9.9 shows an example of histogram sliding using `imadd` and `imsubtract`.

9.6.2 Histogram Stretching

This technique—also known as *input cropping*—consists of a linear transformation that expands (stretches) part of the original histogram so that its nonzero intensity range $[r_{min}, r_{max}]$ occupies the full dynamic gray scale, $[0, L-1]$. Mathematically, each input intensity value, r, is mapped to an output value, s, according to the following linear mapping function:

$$s = \frac{r - r_{min}}{r_{max} - r_{min}} \times (L - 1) \tag{9.12}$$

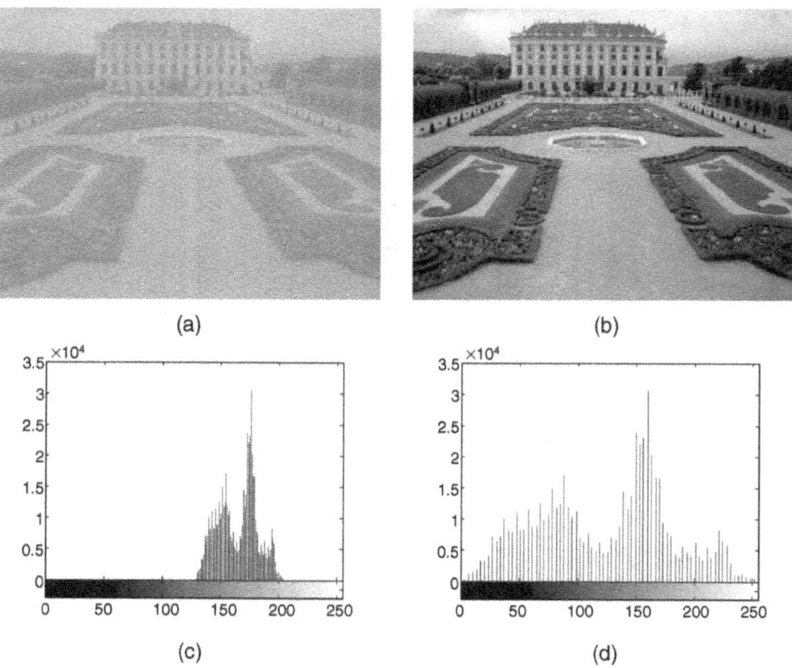

FIGURE 9.10 Example of using histogram stretching to improve contrast: (a) original image ($r_{\text{min}} = 129$, $r_{\text{max}} = 204$); (b) result of stretching using equation (9.12); (c and d) histograms corresponding to images in (a) and (b).

You may have noticed that equation (9.12) is identical to equation (8.4), introduced during our discussion of autocontrast in Chapter 8.

Histogram stretching increases contrast without modifying the shape of the original histogram. It is effective only in cases where the gray-level range of an image is compressed to a narrow range around the middle portion of the dynamic grayscale range. It will not be effective if applied to a poor contrast image whose histogram's lowest and highest occupied bins are close to the minimum and maximum of the full range, such as the one in Figure 9.2c.

■ EXAMPLE 9.9

Figure 9.10 shows an example (using the `imadjust` function in MATLAB) of using the histogram stretching technique to enhance image contrast.

9.6.3 Histogram Shrinking

This technique—also known as *output cropping*—modifies the original histogram in such a way as to compress its dynamic grayscale range, $[r_{\text{min}}, r_{\text{max}}]$, into a narrower

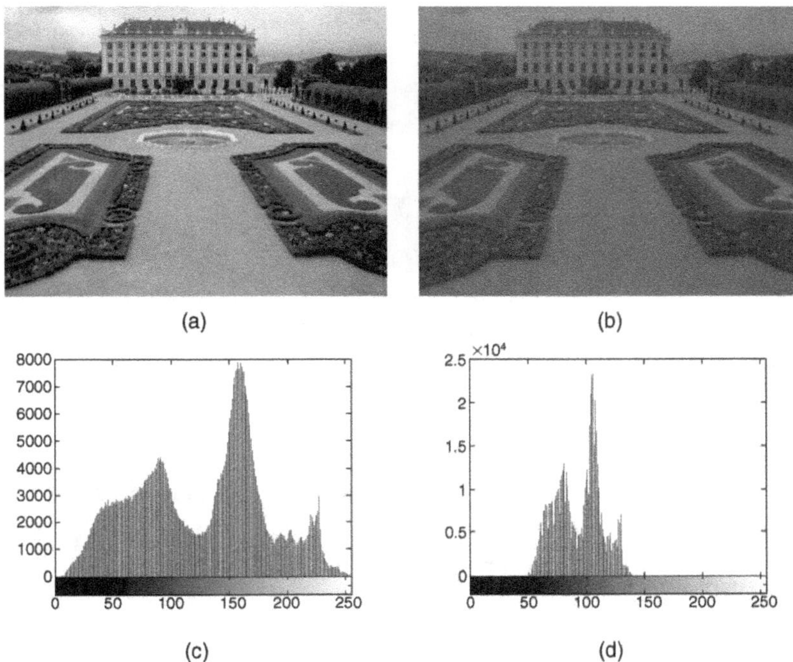

FIGURE 9.11 Example of using histogram shrinking to reduce contrast: (a) original image; (b) result of shrinking using equation (9.13) with $r_{min} = 4$, $r_{max} = 254$, $s_{min} = 49$, and $s_{max} = 140$; (c and d) histograms corresponding to images in (a) and (b).

gray scale, between s_{min} and s_{max}. As a consequence, the resulting image contrast is reduced.

Mathematically,

$$s = \left[\frac{s_{max} - s_{min}}{r_{max} - r_{min}}\right](r - r_{min}) + s_{min} \tag{9.13}$$

■ EXAMPLE 9.10

Figure 9.11 shows an example (using the `imadjust` function in MATLAB) of applying the histogram shrinking technique to a monochrome image.

In MATLAB

MATLAB's IPT has a built-in function to perform histogram stretching and shrinking (among other operations): `imadjust`. You will learn how to use this function in Tutorial 9.3.

9.7 TUTORIAL 9.1: IMAGE HISTOGRAMS

Goal

The goal of this tutorial is to use MATLAB and IPT to calculate and display image histograms.

Objectives

- Learn how to use the IPT function `imhist`.
- Learn how other MATLAB plotting techniques can be used to view and analyze histogram data.

Procedure

Let us begin exploring the `imhist` function that is responsible for computing and displaying the histogram of an image.

1. Display an image and its histogram.

```
I = imread('circuit.tif');
figure, subplot(2,2,1), imshow(I), title('Image')
subplot(2,2,2), imhist(I,256), axis tight, title('Histogram')
```

2. The previous step displayed the default histogram for the image—a histogram with 256 bins. Let us see what happens if we change this value to 64 and 32.

```
subplot(2,2,3), imhist(I,64), axis tight, ...
    title('Histogram with 64 bins')
subplot(2,2,4), imhist(I,32), axis tight, ...
    title('Histogram with 32 bins')
```

You may have noticed that we set the axis to `tight` when displaying histograms. This adjusts the axis limits to the range of the data.

Question 1 Explain the drastic change of the Y-axis values when the histogram is displayed with fewer bins.

There may be a need to postprocess the histogram data or display it using other plotting techniques. To do this, we need the values for each bin of the histogram. The following step illustrates this procedure.

3. Get the values of each bin in the histogram for later use.

```
c = imhist(I,32);
```

We can now use the values in c to display histogram using other plotting techniques. Naturally, the plot of a histogram displays the count of each bin, but it may be more relevant to plot each bin's percentage. This can be done by normalizing the data, as shown in the next step.

4. Normalize the values in c.

```
c_norm = c / numel(I);
```

Question 2 What does the function numel do?

Question 3 Write a one line MATLAB statement that will verify that the sum of the normalized values add to 1.

5. Close any open figures.
6. Display the histogram data using a bar chart.

```
figure, subplot(1,2,1), bar_1 = bar(c);
set(gca, 'XLim', [0 32], 'YLim', [0 max(c)]);
```

In the previous step, we saw how the bar chart can be customized. In MATLAB, almost every object you create can be customized. When we create the bar chart, there is an *axes* object and a bar chart object displayed on the *axes* object. Here, the variable bar_1 is set to the *bar chart* object so that we can reference it later for further customization. The set function allows us to change settings of a particular object. The first parameter of the set function is the object you wish to customize. In this case, the first object we customize is gca, which stands for *get current axes*. Here, we set the limits of the *X* and *Y* axes. Even though the limits have been set, the graph is still ambiguous because the tick marks on the *X* and *Y* axes do not reflect the limits.

7. Set the tick marks to reflect the limits of the graph.

```
set(gca, 'XTick', [0:8:32], 'YTick', ...
    [linspace(0,7000,8) max(c)]);
```

Now the tick marks reflect the limits of the data. We used the set function to change settings of the *current axes*, but we can just as easily use it to customize the bar chart.

8. Use the set function to change the color of the bar chart. Also, give the chart a title.

```
set(bar_1, 'FaceColor', 'r'), title('Bar Chart')
```

Question 4 How would we change the width of the bars in a bar chart?

Notice in the previous step how we used the bar chart object `bar_1` when changing settings. Similarly, we can display the normalized bar chart on the same figure using `subplot`.

9. Display the normalized bar chart and customize its display.

```
subplot(1,2,2), bar_2 = bar(c_norm);
set(gca, 'XTick', [0:8:32], 'YTick', ...
    [linspace(0,0.09,10) max(c_norm)])
xlim([0 32]), ylim([0 max(c_norm)])
title('Normalized Bar Chart')
set(bar_2, 'FaceColor', 'g')
```

Here, we made similar modifications as before. You may have noticed that we used `xlim` and `ylim` functions to set the limits of the axes. Sometimes there is more than one way to accomplish the same task, and this is an example of just that.*Stem charts* are similar to bar charts.

10. Close any open figures.
11. Display stem charts for both standard and normalized histogram data.

```
figure,
subplot(1,2,1), stem(c,'fill','MarkerFaceColor','red'), ...
    axis tight, title('Stem Chart')
subplot(1,2,2), stem(c_norm,'fill','MarkerFaceColor','red'), ...
    axis tight, title('Normalized Stem Chart')
```

In the previous step, we set visual properties of the stem charts by specifying the settings directly in the stem function call—we filled the marker and colored it red. We could have just as easily set a variable equal to the stem plot object and used the `set` function to make the changes.

Question 5 Explore the properties of stem charts. How can we make the lines dotted instead of solid?

Question 6 Alter the axes limits and tick marks to reflect the data being displayed in the stem plot.

The `plot` function will display the data by connecting each point with a straight line.

12. Display a plot graph for both standard and normalized histogram data.

```
figure, subplot(1,2,1), plot(c), axis auto, title('Plot Graph')
subplot(1,2,2), plot(c_norm), axis auto, ...
    title('Normalized Plot Graph')
```

Question 7 Explore the properties of plot graphs. In the above code, the points for each bin are visually lost within the graph line. How can we make the points bolder so that they are more visible?

9.8 TUTORIAL 9.2: HISTOGRAM EQUALIZATION AND SPECIFICATION

Goal

The goal of this tutorial is to learn how to use the IPT for (global and local) histogram equalization and histogram specification (matching).

Objectives

- Explore the process of histogram equalization.
- Learn how to use the `histeq` function.
- Learn how to perform histogram specification (matching).
- Explore the *Interactive Histogram Matching* demo.
- Learn how to perform local histogram equalization with the `adapthisteq` function.

What You Will Need

`ihmdemo.m`—interactive Histogram Matching demo M-file

Procedure

Let us begin by using the function `histeq` to perform histogram equalization on our own images, and by using the `imhist` function, we can view the histogram of the original and the adjusted image.

1. Display the image `pout` and its histogram.

```
I = imread('pout.tif');
figure, subplot(2,2,1), imshow(I), ...
    title('Original Image')
subplot(2,2,2), imhist(I), ...
    title('Original Histogram')
```

2. Use the `histeq` function to perform histogram equalization.

```
I_eq = histeq(I,256);
```

Question 1 Why must we include the second parameter (256) in the `histeq` function call?

3. Display the equalized image and its histogram.

```
subplot(2,2,3), imshow(I_eq), title('Equalized Image')
subplot(2,2,4), imhist(I_eq), title('Equalized Histogram')
```

Question 2 What is the effect of histogram equalization on images with low contrast?

4. Close any open figures and clear all workspace variables.
5. Execute the following code to perform histogram equalization on the `tire` image.

```
I = imread('tire.tif'); I_eq = histeq(I,256);
figure, subplot(2,2,1), imshow(I), title('Original Image')
subplot(2,2,2), imhist(I), title('Original Histogram')
subplot(2,2,3), imshow(I_eq), title('Equalized Image')
subplot(2,2,4), imhist(I_eq), title('Equalized Histogram')
```

Question 3 Based on the tire image's original histogram, what can be said about its overall brightness?

Question 4 How did histogram equalization affect the overall image brightness in this case?

Histogram equalization does not always perform well. As we will see in the next steps, it depends on the original image.

6. Close any open figures and clear all workspace variables.
7. Perform histogram equalization on the `eight` image.

```
I = imread('eight.tif'); I_eq = histeq(I,256);
figure, subplot(2,2,1), imshow(I), title('Original Image')
subplot(2,2,2), imhist(I), title('Original Histogram')
subplot(2,2,3), imshow(I_eq), title('Equalized Image')
subplot(2,2,4), imhist(I_eq), title('Equalized Histogram')
```

Question 5 Why was there such a loss in image quality after histogram equalization?

The transformation function for histogram equalization is simply the cdf of the original image.

8. Display the normalized cdf for the eight.tif image.

```
I_hist = imhist(I); tf = cumsum(I_hist); tf_norm = tf / max(tf);
figure, plot(tf_norm), axis tight
```

Question 6 What does the `cumsum` function do in the previous step?

9. The transformation function can also be obtained without using the `cumsum` function.

```
[newmap, T] = histeq(I);
figure, plot(T)
```

As we have learned, the histogram equalization process attempts to flatten the image histogram. Histogram specification (also known as *histogram matching*) tries to match the image histogram to a specified histogram. The `histeq` function can also be used for this operation.

10. Close any open figures and clear all workspace variables.
11. Prepare a subplot and display original image and its histogram.

```
img1 = imread('pout.tif');
figure, subplot(3,3,1), imshow(img1), title('Original Image')
subplot(3,3,2), imhist(img1), title('Original Histogram')
```

12. Display the image after histogram equalization for comparison.

```
img1_eq = histeq(img1); m1 = ones(1,256)*0.5;
subplot(3,3,4), imshow(img1_eq), title('Equalized Image')
subplot(3,3,5), imhist(img1_eq), title('Equalized Histogram')
subplot(3,3,6), plot(m1), title('Desired Histogram Shape'), ...
    ylim([0 1]), xlim([1 256])
```

13. Display matched image where the desired histogram shape is a straight line from (0, 0) to (1, 1).

```
m2 = linspace(0,1,256); img2 = histeq(img1,m2);
subplot(3,3,7), imshow(img2), title('Matched Image')
subplot(3,3,8), imhist(img2), title('Matched Histogram')
subplot(3,3,9), plot(m2), title('Desired Histogram Shape'), ...
    ylim([0 1]), xlim([1 256])
```

As we can see from the previous steps, performing histogram specification means we must generate a function that represents the shape of the desired histogram. The *Interactive Histogram Matching* demo (developed by Jeremy Jacob and available at the book's companion web site) shows us how creating a desired histogram shape can be an interactive process.

14. Close any open figures and clear all workspace variables.
15. Run the *Interactive Histogram Matching* demo.

```
ihmdemo
```

16. Experiment with creating your own desired histogram shape. To create new points on the function curve, click the curve at the desired location. To move a point, press and drag the point. To delete a point, simply click it.

Question 7 What does the *Continuous Update* checkbox do?

Question 8 How do the different interpolation methods change the shape of the desired histogram curve?

Question 9 How can the demo be loaded with a different image?

Local histogram equalization is performed by the `adapthisteq` function. This function performs contrast limited adaptive histogram equalization (CLAHE) and operates on small data regions (called *tiles*), whose size can be passed as a parameter.

17. Perform local histogram equalization on the `coins` image.

```
I = imread('coins.png');
I_eq = histeq(I,256);
I_leq = adapthisteq(I,'ClipLimit',0.1);
figure, subplot(3,2,1), imshow(I), title('Original Image')
subplot(3,2,2), imhist(I), title('Original Histogram')
subplot(3,2,3), imshow(I_eq), title('Equalized Image')
subplot(3,2,4), imhist(I_eq), title('Equalized Histogram')
subplot(3,2,5), imshow(I_leq), ...
    title('Local Histogram Equalization')
subplot(3,2,6), imhist(I_leq), ...
    title('Local Hist Equalization Histogram')
```

The original image's histogram is clearly bimodal, which separates the pixels of the background from the pixels that make up the coins. We have already seen how images with bimodal distribution of pixel shades do not perform well under (global) histogram equalization.

Question 10 What does the `ClipLimit` setting do in the `adapthisteq` function?

Question 11 What is the default tile size when using `adapthisteq`?

9.9 TUTORIAL 9.3: OTHER HISTOGRAM MODIFICATION TECHNIQUES

Goal

The goal of this tutorial is to learn how to perform other common histogram modification operations.

Objectives

- Learn how to adjust brightness of an image by *histogram sliding.*
- Learn how to use the `imadjust` function.
- Learn how to use the `stretchlim` function.
- Explore adjusting image contrast through *histogram stretching* (also known as *input cropping*).
- Learn how to adjust image contrast with *histogram shrinking* (also known as *output cropping*).

Procedure

Histogram sliding is the process of adding or subtracting a constant brightness value to all pixels in the image. When implementing histogram sliding, we must make sure that pixel values do not go outside the boundaries of the gray scale. Therefore, any pixels that result in values greater than 1 after adjustment will be set to 1. Likewise, any pixels resulting in values less than zero after adjustment will be set to 0.

1. Display original image and prepare subplot.

```
J = imread('pout.tif');
I = im2double(J);
clear J
figure, subplot(3,2,1), imshow(I), title('Original Image')
subplot(3,2,2), imhist(I), axis tight, ...
    title('Original Histogram')
```

2. Obtain a brighter version of the input image by adding 0.1 to each pixel.

```
const = 0.1;
I2 = I + const;
subplot(3,2,3), imshow(I2), title('Original Image + 0.1')
subplot(3,2,4), imhist(I2), axis tight, ...
    title('Original Hist + 0.1')
```

Question 1 How did the histogram change after the adjustment?

3. Produce another brighter image by adding 0.5 to original image.

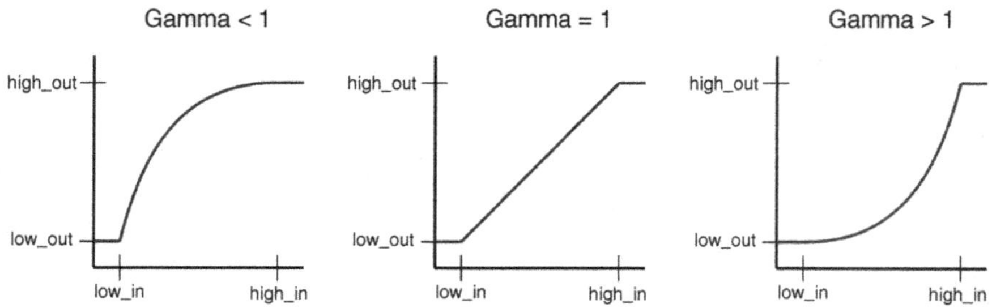

FIGURE 9.12 Gamma transformations for different values of gamma. Redrawn from [GWE04].

```
const = 0.5;
I3 = I + const;
bad_values = find(I3 > 1);
I3(bad_values) = 1;
subplot(3,2,5), imshow(I3), title('Original Image + 0.5')
subplot(3,2,6), imhist(I3), axis tight, ...
    title('Original Hist + 0.5')
```

Question 2 What does the variable `bad_values` contain?

Question 3 Why does the third plot show such an excessive number of pixels with a value of 1?

The brightness of an image can also be modified using the `imadd` function (introduced in Tutorial 6.1), which takes care of truncating and rounding off values outside the desired range in the output image.

Histogram stretching and shrinking can be achieved through use of the `imadjust` function. The syntax for the function is as follows:

```
J = imadjust(I,[low_in; high_in],[low_out; high_out], gamma)
```

Figure 9.12 illustrates what the transformation functions look like when different values of gamma are used. As we already know from Chapter 9, the value of gamma is the exponent in power law transformation.

Any values below `low_in` and above `high_in` are *clipped* or simply mapped to `low_out` and `high_out`, respectively. Only values in between these limits are affected by the curve. Gamma values less than 1 create a weighted curve toward the brighter range, and gamma values greater than 1 weight toward the darker region. The default value of gamma is 1.

Let us explore how to use `imadjust` to perform histogram stretching.

4. Close any open figures.
5. Execute the following code to see histogram stretching on the `pout` image, which is already loaded in variable `I`.

```
img_limits = stretchlim(I);
I_stretch = imadjust(I,img_limits,[]);
figure
subplot(3,2,1), imshow(I), title('Original Image')
subplot(3,2,2), imhist(I), axis tight, ...
    title('Original Histogram')
subplot(3,2,3), imshow(I_stretch), ...
    title('Stretched Image')
subplot(3,2,4), imhist(I_stretch), axis tight, ...
    title('Stretched Histogram')
```

Question 4 How did the histogram change after the adjustment?

Question 5 What is the purpose of using the `stretchlim` function?

In the previous step, we specified the `low_in`, `high_in`, `low_out`, and `high_out` parameters when calling the `imadjust` function when in fact the default operation is histogram stretching—meaning these parameters are not necessary to perform histogram stretching. Notice in the next step how just calling the function and only specifying the image as its parameter will give the same results.

6. Perform histogram stretching with `imadjust` using default parameters and confirm that the results are identical to the ones obtained before.

```
I_stretch2 = imadjust(I);
subplot(3,2,5), imshow(I_stretch2), ...
    title('Stretched Image')
subplot(3,2,6), imhist(I_stretch2), axis tight, ...
    title('Stretched Histogram')
I_stretch_diff = imabsdiff(I_stretch, I_stretch2);
figure, imshow(I_stretch_diff,[])
min(I_stretch_diff(:))
max(I_stretch_diff(:))
```

Question 6 How does the difference image look?

Question 7 What is the purpose of inspecting its maximum and minimum values?

To shrink an image histogram, we must specify the parameters explicitly.

7. Close any open figures and clear all workspace variables.
8. Execute the following code to see the result of histogram shrinking.

```
I = imread('westconcordorthophoto.png');
I_shrink = imadjust(I,stretchlim(I),[0.25 0.75]);
```

```
figure
subplot(2,2,1), imshow(I), title('Original Image')
subplot(2,2,2), imhist(I), axis tight, ...
    title('Original Histogram')
subplot(2,2,3), imshow(I_shrink), ...
    title('Shrunk Image')
subplot(2,2,4), imhist(I_shrink), axis tight, ...
    title('Shrunk Histogram')
```

When we use other techniques to adjust the histogram of an image, we have a means to view the transformation function (i.e., the `histeq` function will return the transformation function as an output parameter if requested). There is no built-in technique for viewing a transformation function when performing histogram sliding, stretching, or shrinking, but we can achieve a visual representation of the transformation function by using the `plot` function. To do so, we specify the original image as the *X* values and the adjusted image as the *Y* values.

9. Display the transformation function for the adjustment performed in the previous step.

```
X = reshape(I,1,prod(size(I)));
Y = reshape(I_shrink,1,prod(size(I_shrink)));
figure, plot(X,Y,'.')
xlim([0 255]); ylim([0 255]);
xlabel('Original Image');
ylabel('Adjusted Image');
```

Question 8 What do the above first two statements in the code do?

Question 9 What does the `xlabel` and `ylabel` functions do?

As noted earlier, gamma values other than 1 will specify the shape of the curve, toward either the bright or the dark region.

10. Close any open figures.
11. Perform histogram shrinking with a gamma value of 2.

```
I_shrink = imadjust(I,stretchlim(I),[0.25 0.75],2);
X = reshape(I,1,prod(size(I)));
Y = reshape(I_shrink,1,prod(size(I_shrink)));
figure
subplot(2,2,1), imshow(I), title('Original Image')
subplot(2,2,2), imhist(I), axis tight, ...
    title('Original Histogram')
subplot(2,2,3), imshow(I_shrink), title('Adjusted Image')
subplot(2,2,4), imhist(I_shrink), axis tight, ...
```

```
    title('Adjusted Histogram')
figure, plot(X,Y,'.'), xlim([0 255]), ylim([0 255])
```

Question 10 The transformation function plot displays a gap from 0 to 12 (on the X axis) where there are no points. Why is this so?

WHAT HAVE WE LEARNED?

- Histograms are a convenient mathematical representation of how many pixels occur at each gray level in a monochrome image.
- In MATLAB, the histogram of an image can be computed and displayed using the `imhist` function.
- Histograms provide valuable quantitative information about an image, such as minimum, maximum and average gray level, global standard deviation, and absolute contrast.
- Histograms also provide valuable qualitative information about an image, such as average brightness level and contrast.
- Histogram equalization is a mathematical technique by which we modify the histogram of the input image in such a way as to approximate a uniform distribution (flat histogram). Histogram equalization is used as a contrast enhancement technique.
- Histograms (and the images to which they correspond) can also be modified through *direct histogram specification*, whose goal is to change the histogram so as to match a desired shape.
- Other histogram modification techniques include histogram sliding (used to modify the brightness properties of an image), histogram stretching (used to increase contrast without modifying the shape of the original histogram), and histogram shrinking (used to reduce the dynamic grayscale range of an image).

LEARN MORE ABOUT IT

- Chapter 4 of [BB08] discusses histograms in detail and provides useful insights into how to use histograms to investigate image acquisition problems and image defects.
- The paper by Hummel [Hum75] provides a review of techniques related to the material in this chapter.
- Many alternative histogram modification techniques have been proposed in the literature. The paper by Stark [Sta00] is a representative recent example.
- Section 5.6 of [BB08] discusses histogram specification in more detail.
- Section 10.2 of [Pra07] provides a deeper discussion on histogram modification techniques.

- Section 2.2 of [SOS00] contains a modified version of the basic histogram expansion algorithm, in which the user can select a cutoff percentage p and a modified version of the general histogram transformation algorithm and the output histogram has the shape of a linear ramp whose slope can be specified as a parameter.
- Many image enhancement techniques based on local and global histogram statistics have been proposed. Section 3.3.4 of [GW08] and Section 8.2.3 of [Umb05] describe representative algorithms under this category.

9.10 PROBLEMS

9.1 The 7×7 image with eight gray levels is given below, where each gray level value is represented in normalized form from 0 (black pixel) to 1 (white pixel).

0	3/7	2/7	2/7	1/7	1/7	4/7
3/7	2/7	1/7	1/7	1/7	1/7	4/7
2/7	0	1	1/7	3/7	0	0
0	5/7	1/7	0	6/7	0	1/7
1/7	1/7	1/7	3/7	6/7	6/7	5/7
1/7	1/7	1/7	1/7	5/7	6/7	4/7
0	1	0	0	0	0	4/7

(a) Calculate the probabilities of each gray level and plot the image's histogram.

(b) Which pixels are predominant in the original image, dark or bright?

(c) Using the cumulative distribution function, equalize the histogram calculated in part (a) and plot the resulting (equalized) histogram.

(d) Show the resulting 7×7 image after histogram equalization.

9.2 Write a MATLAB script to show that a second (or third, fourth, etc.) consecutive application of the histogram equalization algorithm to an image will not produce any significant change in the histogram (and, consequently, in the image).

9.3 Given a 256×256 pixels image with eight gray levels, whose gray-level distribution is given in the following table.

Gray Level (r_k)	n_k	$p(r_k)$
0	2621	0.04
1/7	0	0.00
2/7	0	0.00
3/7	5243	0.08
4/7	7209	0.11
5/7	12,452	0.19
6/7	24,904	0.38
1	13,107	0.20

It is desired that the original histogram is changed to approach the histogram corresponding to the table below.

z_k	$\hat{p}(z_k)$
0	0.27
1/7	0.16
2/7	0.19
3/7	0.16
4/7	0.11
5/7	0.06
6/7	0.03
1	0.02

(a) Which pixels predominate in the original image, dark or bright? Explain.

(b) Assuming the histogram modification will be successful, what will be the probable effect of this modification on the original image?

(c) Equalize the original histogram using the function $s = T(r)$.

(d) Obtain the function $v = G(z)$ and its inverse.

(e) Plot the most relevant histograms: original, desired, equalized, and resulting.

(f) Fill out the table below with the final values for n_k and $\hat{p}(z_k)$ for the eight values of z_k, comparing with the desired values and explaining possible differences.

z_k	n_k	$\hat{p}(z_k)$
0		
1/7		
2/7		
3/7		
4/7		
5/7		
6/7		
1		

9.4 Write a MATLAB script that implements region-based histogram equalization. Your script should allow the user to interactively select (with the mouse) a region of interest (ROI) within an image to which the histogram equalization operation will be applied.

9.5

(a) Write a MATLAB function that creates an 8-bit random image with a uniform distribution of pixel values and takes two parameters: height and width. *Hint*: use MATLAB function `rand`.

(b) Write a MATLAB script that uses the function you have just written to create a 128 × 128 random image.

(c) Inspect the image's histogram. Does it show a uniform (i.e., flat-shaped) distribution as expected? Explain.

9.6

(a) Write a MATLAB function that creates an 8-bit random image with a Gaussian (normal) distribution of pixel values and takes four parameters: height, width, mean value (μ), and standard deviation (σ). *Hint*: use MATLAB function `randn`—which returns a pseudorandom normal distribution with $\mu = 0$ and $\sigma = 1$—and make the necessary adjustments to μ and σ.

(b) Write a MATLAB script that uses the function you have just written to create a 128 × 128 random image with a Gaussian (normal) distribution of pixel values, with mean value $\mu = 128$ and standard deviation $\sigma = 60$.

(c) Inspect the image's histogram. Does it show a Gaussian (i.e., bell-shaped) distribution as expected? Explain.

(d) Repeat the previous two steps for different values of μ and σ.

9.7 An 8-bit image has a minimum gray level of 140 and a maximum gray level of 195. Describe the effect on the histogram of this image after each of these operations is performed (separately):

(a) Subtraction of 130 from all pixel gray levels (histogram sliding).

(b) Histogram stretching.

(c) Histogram equalization.

9.8 Provide empirical evidence of the nonuniqueness of a histogram by writing a MATLAB script that reads a monochrome image, displays its histogram, and generates another gray-level image very different from the original, but whose histogram is identical to the original image's histogram. *Hint*: "Given an image f with a particular histogram H_f, every image that is a spatial shuffling of the gray levels of f has the same histogram H_f" [Bov00a].

CHAPTER 10

NEIGHBORHOOD PROCESSING

WHAT WILL WE LEARN?

- What is neighborhood processing and how does it differ from point processing?
- What is convolution and how is it used to process digital images?
- What is a low-pass linear filter, what is it used for, and how can it be implemented using 2D convolution?
- What is a median filter and what is it used for?
- What is a high-pass linear filter, what is it used for, and how can it be implemented using 2D convolution?

10.1 NEIGHBORHOOD PROCESSING

The underlying theme throughout this chapter is the use of neighborhood-oriented operations for image enhancement. The basics of neighborhood processing were introduced in Section 2.4.2. We call neighborhood-oriented operations those image processing techniques in which the resulting value for a pixel at coordinates (x_0, y_0)—which we shall call the *reference pixel*—is a function of the original pixel value at that point as well as the original pixel value of some of its neighbors. The way by which the neighboring values and the reference pixel value are combined to produce the result can vary significantly among different algorithms. Many algorithms work in a

Practical Image and Video Processing Using MATLAB®. By Oge Marques.

linear way and use 2D convolution (which essentially consists of sums of products, see Section 10.2), while others process the input values in a nonlinear way.

Regardless of the type (linear or nonlinear), neighborhood processing operations follow a sequence of steps [GWE04]:

1. Define a reference point in the input image, $f(x_0, y_0)$.
2. Perform an operation that involves only pixels within a neighborhood around the reference point in the input image.
3. Apply the result of that operation to the pixel of same coordinates in the output image, $g(x_0, y_0)$.
4. Repeat the process for every pixel in the input image.

In this chapter, we provide a representative collection of neighborhood-based image processing techniques for the sake of image enhancement, particularly blurring or sharpening.[1]

The techniques described in this chapter belong to one of these two categories:

- *Linear Filters*: Here the resulting output pixel is computed as a sum of products of the pixel values and mask coefficients in the pixel's neighborhood in the original image. Example: mean filter (Section 10.3.1).
- *Nonlinear Filters*: Here the resulting output pixel is selected from an ordered (ranked) sequence of pixel values in the pixel's neighborhood in the original image. Example: median filter (Section 10.3.4).

10.2 CONVOLUTION AND CORRELATION

Convolution is a widely used mathematical operator that processes an image by computing—for each pixel—a weighted sum of the values of that pixel and its neighbors (Figure 10.1). Depending on the choice of weights, a wide variety of image processing operations can be implemented. Convolution and correlation are the two fundamental mathematical operations involved in linear neighborhood-oriented image processing algorithms. The two operations differ in a very subtle way, which will be explained later in this section.

10.2.1 Convolution in the One-Dimensional Domain

The convolution between two discrete one-dimensional (1D) arrays $A(x)$ and $B(x)$, denoted by $A * B$, is mathematically described by the equation

$$A * B = \sum_{j=-\infty}^{\infty} A(j) \cdot B(x - j) \tag{10.1}$$

[1]We shall see additional examples of neighborhood operations for different purposes—for example, image restoration (Chapter 12) and edge detection (Chapter 14)—later in the book.

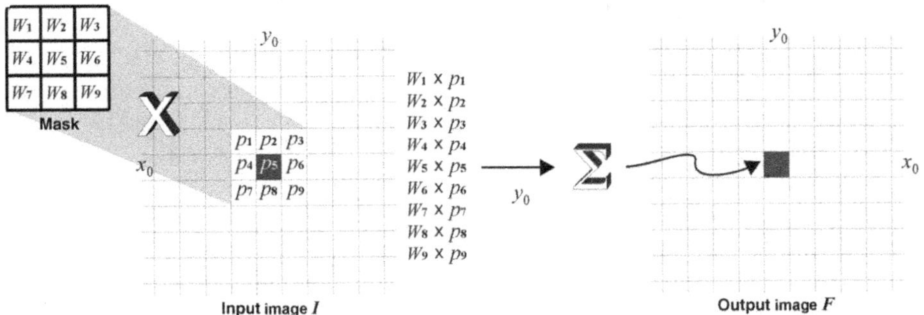

FIGURE 10.1 Neighborhood processing for the case of linear filtering.

■ EXAMPLE 10.1

In this example, we show how the result of a 1D convolution operation can be obtained step-by-step. Let $A = \{0, 1, 2, 3, 2, 1, 0\}$ and $B = \{1, 3, -1\}$. The partial results of multiplying elements in A with corresponding elements in B, as B shifts from $-\infty$ to ∞, are displayed below.

1. Initially, we mirror array B and align its center (reference) value with the first (leftmost) value of array A.[2] The partial result of the convolution calculation $(0 \times (-1)) + (0 \times 3) + (1 \times 1) = 1$ (where empty spots are assumed as zero) is stored in the resulting array $(A * B)$.

A		0	1	2	3	2	1	0
B	−1	3	1					
$A * B$		1						

2. Array B is shifted one position to the right. The partial result of the convolution calculation $(0 \times (-1)) + (1 \times 3) + (2 \times 1) = 5$ is stored in the resulting array $(A * B)$.

A	0	1	2	3	2	1	0
B	−1	3	1				
$A * B$	1	5					

3. Array B is shifted another position to the right. The partial result of the convolution calculation $(1 \times (-1)) + (2 \times 3) + (3 \times 1) = 8$ is stored in the resulting array $(A * B)$.

A	0	1	2	3	2	1	0
B		−1	3	1			
$A * B$	1	5	8				

[2]This is equivalent to saying that all the partial products from $-\infty$ until that point are equal to zero and, therefore, do not contribute to the result of the convolution operation.

4. Array B is shifted another position to the right. The partial result of the convolution calculation $(2 \times (-1)) + (3 \times 3) + (2 \times 1) = 9$ is stored in the resulting array $(A * B)$.

A	0	1	2	3	2	1	0
B			−1	3	1		
$A * B$	1	5	8	8			

5. Array B is shifted another position to the right. The partial result of the convolution calculation $(3 \times (-1)) + (2 \times 3) + (1 \times 1) = 4$ is stored in the resulting array $(A * B)$.

A	0	1	2	3	2	1	0
B				−1	3	1	
$A * B$	1	5	8	8	4		

6. Array B is shifted another position to the right. The partial result of the convolution calculation $(2 \times (-1)) + (1 \times 3) + (0 \times 1) = 1$ is stored in the resulting array $(A * B)$.

A	0	1	2	3	2	1	0
B					−1	3	1
$A * B$	1	5	8	8	4	1	

7. Array B is shifted another position to the right. The partial result of the convolution calculation $(1 \times (-1)) + (0 \times 3) + (0 \times 1) = -1$ is stored in the resulting array $(A * B)$.

A	0	1	2	3	2	1	0	
B						−1	3	1
$A * B$	1	5	8	8	4	1	−1	

The final result of the convolution operation is the array $\{1, 5, 8, 8, 4, 1, -1\}$.

10.2.2 Convolution in the Two-Dimensional Domain

The mathematical definition for 2D convolution is

$$g(x, y) = \sum_{k=-\infty}^{\infty} \sum_{j=-\infty}^{\infty} h(j, k) \cdot f(x - j, y - k) \tag{10.2}$$

In practice, this is rewritten as

$$g(x, y) = \sum_{k=-n_2}^{n_2} \sum_{j=-m_2}^{m_2} h(j, k) \cdot f(x - j, y - k) \tag{10.3}$$

where m_2 is equal to half of the mask's width and n_2 is equal to half of the mask's height, that is,

$$m_2 = \lfloor m/2 \rfloor \tag{10.4}$$

and

$$n_2 = \lfloor n/2 \rfloor \tag{10.5}$$

where $\lfloor x \rfloor$ is the *floor* operator, which rounds a number to the nearest integer less than or equal to x.

The basic mechanism used to understand 1D convolution can be expanded to the 2D domain. In such cases, the 2D array A is usually the input image and B is a small (usually 3×3) mask. The idea of mirroring B and shifting it across A can be adapted to the 2D case as well: mirroring will now take place in both x and y dimensions, and shifting will be done starting from the top left point in the image, moving along each line, until the bottom right pixel in A has been processed.

■ EXAMPLE 10.2

Let

$$A = \begin{bmatrix} 5 & 8 & 3 & 4 & 6 & 2 & 3 & 7 \\ 3 & 2 & 1 & 1 & 9 & 5 & 1 & 0 \\ 0 & 9 & 5 & 3 & 0 & 4 & 8 & 3 \\ 4 & 2 & 7 & 2 & 1 & 9 & 0 & 6 \\ 9 & 7 & 9 & 8 & 0 & 4 & 2 & 4 \\ 5 & 2 & 1 & 8 & 4 & 1 & 0 & 9 \\ 1 & 8 & 5 & 4 & 9 & 2 & 3 & 8 \\ 3 & 7 & 1 & 2 & 3 & 4 & 4 & 6 \end{bmatrix}$$

and

$$B = \begin{bmatrix} 2 & 1 & 0 \\ 1 & 1 & -1 \\ 0 & -1 & -2 \end{bmatrix}$$

The result of the convolution $A * B$ will be

$$A * B = \begin{bmatrix} 20 & 10 & 2 & 26 & 23 & 6 & 9 & 4 \\ 18 & 1 & -8 & 2 & 7 & 3 & 3 & -11 \\ 14 & 22 & 5 & -1 & 9 & -2 & 8 & -1 \\ 29 & 21 & 9 & -9 & 10 & 12 & -9 & -9 \\ 21 & 1 & 16 & -1 & -3 & -4 & 2 & 5 \\ 15 & -9 & -3 & 7 & -6 & 1 & 17 & 9 \\ 21 & 9 & 1 & 6 & -2 & -1 & 23 & 2 \\ 9 & -5 & -25 & -10 & -12 & -15 & -1 & -12 \end{bmatrix}$$

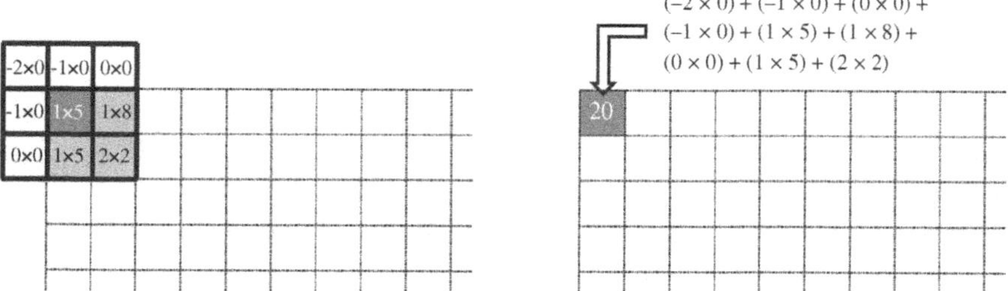

FIGURE 10.2 Two-dimensional convolution example.

Figure 10.2 shows the calculation of the top left pixel in the resulting image in detail. Note how B has been flipped in both dimensions *before* the sum of products was calculated.

Convolution with masks is a very versatile image processing method. Depending on the choice of mask coefficients, entirely different results can be obtained, for example, image blurring, image sharpening, or edge detection.

■ EXAMPLE 10.3

Suppose we apply the three convolution masks from Table 10.1 (one at a time, independently) to the same input image (Figure 10.3a). The resulting images will be a blurred version of the original (part (b)), a sharpened version of the original (part (c)), and an image indicating the presence of horizontal edges in the input image (part (d)).

10.2.3 Correlation

In the context of this chapter, 1D correlation[3] can be mathematically expressed as

$$A \odot B = \sum_{j=-\infty}^{\infty} A(j) \cdot B(x+j) \tag{10.6}$$

whereas the 2D equivalent is given by

$$g(x, y) = \sum_{k=-\infty}^{\infty} \sum_{j=-\infty}^{\infty} h(j, k) \cdot f(x+j, y+k) \tag{10.7}$$

[3]This definition of correlation should not be confused with the most common use of the word, usually to express the degree to which two or more quantities are linearly associated (by means of a *correlation coefficient*).

TABLE 10.1 Examples of Convolution Masks

Low-Pass Filter	High-Pass Filter	Horizontal Edge Detection
$\begin{bmatrix} 1/9 & 1/9 & 1/9 \\ 1/9 & 1/9 & 1/9 \\ 1/9 & 1/9 & 1/9 \end{bmatrix}$	$\begin{bmatrix} 0 & -1 & 0 \\ -1 & 5 & -1 \\ 0 & -1 & 0 \end{bmatrix}$	$\begin{bmatrix} 1 & 1 & 1 \\ 0 & 0 & 0 \\ -1 & -1 & -1 \end{bmatrix}$

In practice, this is rewritten as

$$g(x, y) = \sum_{k=-n_2}^{n_2} \sum_{j=-m_2}^{m_2} h(j, k) \cdot f(x + j, y + k) \tag{10.8}$$

where m_2 and n_2 are as defined earlier.

Simply put, correlation is the same as convolution *without* the mirroring (flipping) of the mask before the sums of products are computed. The difference between using correlation and convolution in 2D neighborhood processing operations is often irrelevant because many popular masks used in image processing are symmetrical around the origin. Consequently, many texts omit this distinction and

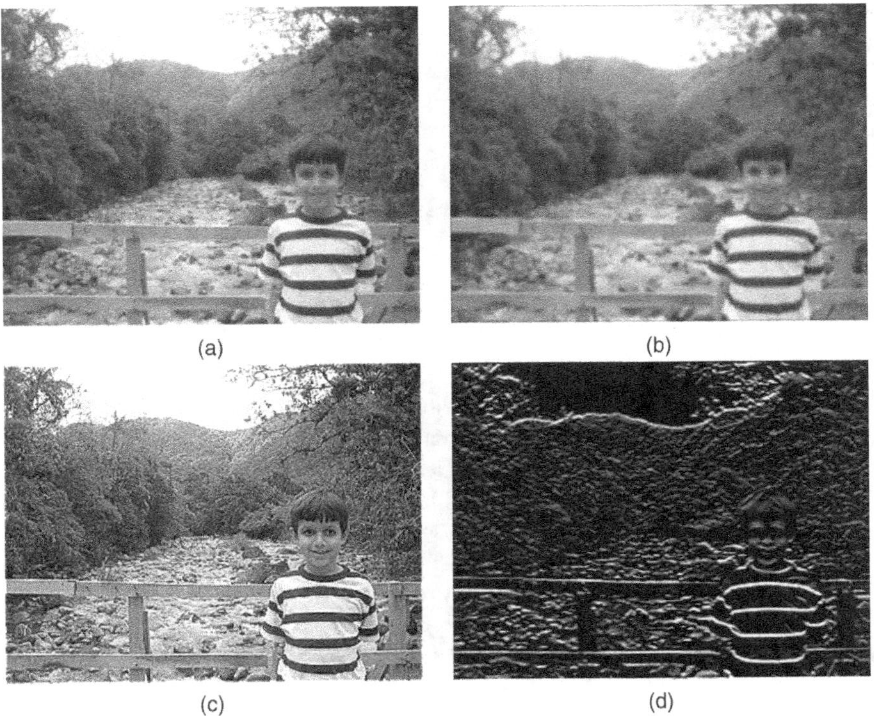

FIGURE 10.3 Applying different convolution masks to the same input image: (a) original image; (b–d) result of 2D convolution using the masks in Table 10.1.

refer to *convolution* or *spatial filtering* when referring to what technically should be called *correlation*.

In MATLAB

MATLAB's Image Processing Toolbox (IPT) has two built-in functions that can be used to implement 2D convolution:

- `conv2`: It computes the 2D convolution between two matrices. In addition to the two matrices, it takes a third parameter that specifies the size of the output:
  - `full`: Returns the full 2D convolution (default).
  - `same`: Returns the central part of the convolution of the same size as A.
  - `valid`: Returns only those parts of the convolution that are computed without the zero-padded edges.
- `filter2`: It rotates the convolution mask (which is treated as a 2D FIR filter) 180° in each direction to create a convolution kernel and then calls `conv2` to perform the convolution operation.

10.2.4 Dealing with Image Borders

Our discussion of convolution and correlation so far has overlooked the need to deal with image borders, that is, those points in the input image for which part of the mask falls outside the image borders (Figure 10.4). There are several ways of handling this:

1. Ignore the borders, that is, apply the mask only to the pixels in the input image for which the mask falls entirely within the image. There are two variants of this approach:
   (a) Keep the pixel values that cannot be reached by the overlapping mask untouched. This will introduce artifacts due to the difference between the processed and unprocessed pixels in the output image.
   (b) Replace the pixel values that cannot be reached by the overlapping mask with a constant fixed value, usually zero (black). If you use this approach, the resulting image will be smaller than the original image. This is unacceptable because the image size will be further decreased by every subsequent filtering operation. Moreover, it will make it harder to combine the input and output images using an arithmetic or logic operation (see Chapter 6) or even to compare them on a pixel-by-pixel basis.
2. Pad the input image with zeros, that is, assume that all values outside the image are equal to zero. If you use this approach, the resulting image will show unwanted artifacts, in this case artificial dark borders, whose width is proportional to the size of the convolution mask. This is implemented in MATLAB by choosing the option `X` (with `X` = 0) for the `boundary_options` parameter for function `imfilter`.

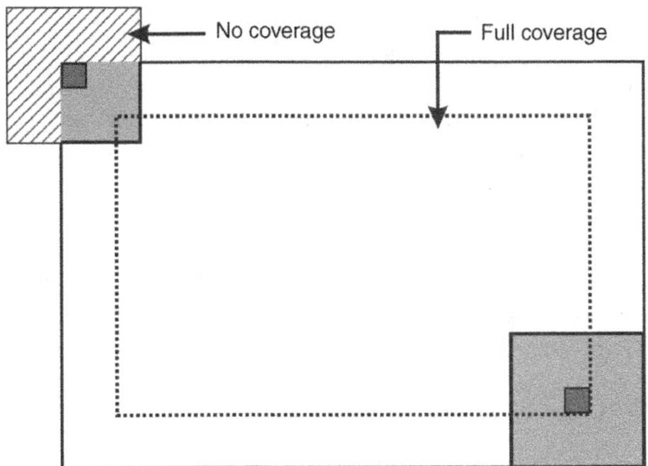

FIGURE 10.4 Border geometry. Redrawn from [BB08].

3. Pad with extended values, that is, assume that the pixel values in the input image extend beyond the image borders. This is equivalent to assuming that the input image has extra rows and columns whose pixel values are identical to the ones of the row/column closest to the border. This is the preferred method because of its simplicity and the relatively low impact of the resulting artifacts on the quality of the output image. This is implemented in MATLAB by choosing the option `'replicate'` for the `boundary_options` parameter for function `imfilter`.
4. Pad with mirrored values, that is, assume that the pixel values in the input image extend beyond the image borders in such a way as to mirror the pixel values in the rows/columns closest to the border. For small mask sizes, the result is comparable to using the *padding with extended values* approach.
5. Treat the input image as a 2D periodic function whose values repeat themselves in both horizontal and vertical directions. This is implemented in MATLAB by choosing the option `'circular'` for the `boundary_options` parameter for function `imfilter`.

10.3 IMAGE SMOOTHING (LOW-PASS FILTERS)

We have seen that different convolution masks can dramatically produce different results when applied to the same input image (Figure 10.3). It is common to refer to these operations as *filtering* operations and to the masks themselves as *spatial filters*. Spatial filters are often named based on their behavior in the *spatial frequency* domain[4]: we call *low-pass* filters (LPFs) those spatial filters whose effect on the output image is equivalent to attenuating high-frequency components (i.e., fine details in the image) and preserving low-frequency components (i.e., coarser details and homogeneous areas in the image). Linear LPFs can be implemented using 2D convolution masks with nonnegative coefficients. Linear LPFs are typically used to either blur an image or reduce the amount of noise present in the image. In this chapter we refer to both uses, but defer a more detailed discussion of noise reduction techniques until Chapter 12.

High-pass filters (HPFs) work in a complementary way to LPFs, that is, they preserve or enhance high-frequency components (with the possible side effect of enhancing noisy pixels as well). HPFs will be discussed in Section 10.4.

In MATLAB

Linear filters are implemented in MATLAB using two functions: `imfilter` and—optionally—`fspecial`.

The syntax for `imfilter` is

```
g = imfilter(f, h, mode, boundary_options, size_options);
```

where

- `f` is the input image.
- `h` is the filter mask.
- `mode` can be either `'conv'` or `'corr'`, indicating, respectively, whether filtering will be done using convolution or correlation (which is the default);
- `boundary_options` refer to how the filtering algorithm should treat border values. There are four possibilities:
    1. `X`: The boundaries of the input array (image) are extended by padding with a value `X`. This is the default option (with `X` = 0).
    2. `'symmetric'`: The boundaries of the input array (image) are extended by mirror-reflecting the image across its border.
    3. `'replicate'`: The boundaries of the input array (image) are extended by replicating the values nearest to the image border.
    4. `'circular'`: The boundaries of the input array (image) are extended by implicitly assuming the input array is periodic, that is, treating the image as one period of a 2D periodic function.
- `size_options`: There are two options for the size of the resulting image: `'full'` (output image is the full filtered result, that is, the size of the extended/padded image) or `'same'` (output image is of the same size as input image), which is the default.
- `g` is the output image.

[4]Frequency-domain image processing techniques will be discussed in Chapter 11.

`fspecial` is an IPT function designed to simplify the creation of common 2D image filters. Its syntax is `h = fspecial(type, parameters)`, where

- `h` is the filter mask.
- `type` is one of the following:
  - `'average'`: Averaging filter
  - `'disk'`: Circular averaging filter
  - `'gaussian'`: Gaussian low-pass filter
  - `'laplacian'`: 2D Laplacian operator
  - `'log'`: Laplacian of Gaussian (LoG) filter
  - `'motion'`: Approximates the linear motion of a camera
  - `'prewitt'` and `'sobel'`: horizontal edge-emphasizing filters
  - `'unsharp'`: unsharp contrast enhancement filter
- `parameters` are optional parameters that vary depending on the `type` of filter, for example, mask size, standard deviation (for `'gaussian'` filter), and so on. See the IPT documentation for full details.

10.3.1 Mean Filter

The *mean* (also known as *neighborhood averaging*) filter is perhaps the simplest and most widely known spatial smoothing filter. It uses convolution with a (usually 3×3) mask whose coefficients have a value of 1 and divides the result by a scaling factor (the total number of elements in the mask). A neighborhood averaging filter in which all coefficients are equal is also referred to as a *box filter*.

The convolution mask for a 3×3 mean filter is given by

$$h(x, y) = \begin{bmatrix} 1/9 & 1/9 & 1/9 \\ 1/9 & 1/9 & 1/9 \\ 1/9 & 1/9 & 1/9 \end{bmatrix} = \frac{1}{9} \begin{bmatrix} 1 & 1 & 1 \\ 1 & 1 & 1 \\ 1 & 1 & 1 \end{bmatrix} \tag{10.9}$$

Figure 10.3b shows the result of applying the mask in equation (10.9) to the image in Figure 10.3a.

The same concept can be applied to larger neighborhoods with a proportional increase in the degree of blurriness of the resulting image. This is illustrated in Figure 10.5 for masks of size 7×7 (part (b)), 15×15 (part (c)), and 31×31 (part (d)).

10.3.2 Variations

Many variations on the basic neighborhood averaging filter have been proposed in the literature. In this section we summarize some of them.

(a) (b) (c) (d)

FIGURE 10.5 Examples of applying the averaging filter with different mask sizes: (a) input image (899 × 675 pixels); (b–d) output images corresponding to averaging masks of size 7 × 7, 15 × 15, and 31 × 31.

Modified Mask Coefficients The mask coefficients from equation (10.9) can be modified, for example, to give more importance to the center pixel and its 4-connected neighbors:

$$h(x, y) = \begin{bmatrix} 0.075 & 0.125 & 0.075 \\ 0.125 & 0.2 & 0.125 \\ 0.075 & 0.125 & 0.075 \end{bmatrix} \tag{10.10}$$

Directional Averaging The square mask can be replaced by a rectangular equivalent to emphasize that the blurring is done in a specific direction.

Selective Application of Averaging Calculation Results Another variation of the basic neighborhood averaging filter consists in applying a decision step between calculating of the neighborhood average for a certain reference pixel and applying the result to the output image. A simple decision step would compare the difference between the original and processed values against a predefined threshold (T): if the difference is less than T, the calculated value is applied; otherwise, the original value is kept. This is usually done to minimize the blurring of important edges in the image.

Removal of Outliers Before Calculating the Average This is the underlying idea of the *average of k nearest-neighbors* technique [DR78], which is a variation of the average filter, whose basic procedure consists of four steps:

1. Sort all pixel values in the neighborhood.
2. Select k values around the median value.
3. Calculate the average gray-level value of the k values selected in step 2.
4. Replace reference (central) pixel in the destination image with the value calculated in step 3.

It was conceived to allow the exclusion of high-contrast or edge pixels from the average calculations and, therefore, reduce the degree of edge blurring in the resulting image. For larger values of k, this filter's performance will approach the conventional average filter.

10.3.3 Gaussian Blur Filter

The Gaussian blur filter is the best-known example of a LPF implemented with a nonuniform kernel. The mask coefficients for the Gaussian blur filter are samples from a 2D Gaussian function (plotted in Figure 10.6):

$$h(x, y) = \exp\left[\frac{-(x^2 + y^2)}{2\sigma^2}\right] \tag{10.11}$$

The parameter σ controls the overall shape of the curve: the larger the value of σ, the flatter the resulting curve.

■ EXAMPLE 10.4

Figure 10.7 shows an example (using the `imfilter` and `fspecial` functions in MATLAB) of applying a Gaussian blur filter to a monochrome image using different kernel sizes and values of σ. You should be able to notice that the Gaussian blur produces a more natural blurring effect than the averaging filter. Moreover, the impact of increasing mask size is less dramatic on the Gaussian blur filter than on the averaging filter.

Some of the most notable properties of the Gaussian blur filter are as follows:

- The kernel is symmetric with respect to rotation; therefore, there is no directional bias in the result.
- The kernel is separable, which can lead to fast computational implementations.
- The kernel's coefficients fall off to (almost) zero at the kernel's edges.

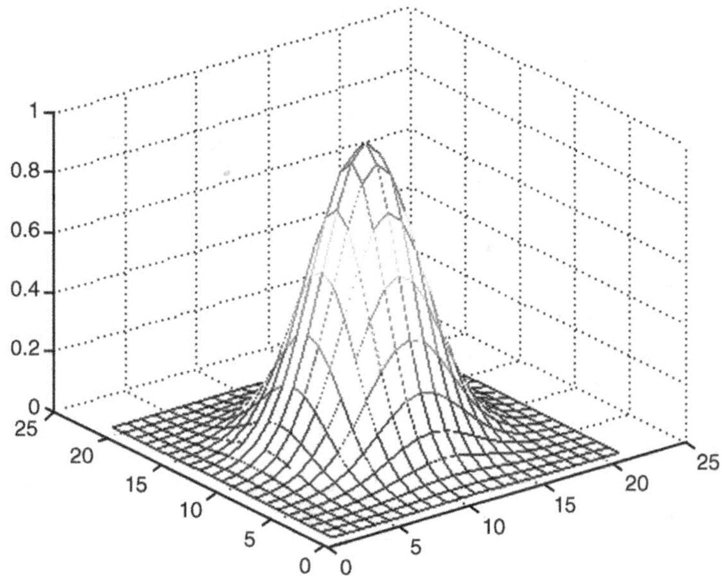

FIGURE 10.6 A 2D Gaussian function (with $\sigma = 3$).

- The Fourier transform (FT) of a Gaussian filter is another Gaussian (this will be explained in Chapter 11).
- The convolution of two Gaussians is another Gaussian.
- The output image obtained after applying the Gaussian blur filter is more pleasing to the eye than the one obtained using other low-pass filters.

10.3.4 Median and Other Nonlinear Filters

As stated earlier in this chapter, nonlinear filters also work at a neighborhood level, but do not process the pixel values using the convolution operator. Instead, they usually apply a ranking (sorting) function to the pixel values within the neighborhood and select a value from the sorted list. For this reason, these are sometimes called *rank filters*. Examples of nonlinear filters include the median filter (described in this section) and the *max* and *min* filters (which will be described in Chapter 12).

The median filter is a popular nonlinear filter used in image processing. It works by sorting the pixel values within a neighborhood, finding the median value, and replacing the original pixel value with the median of that neighborhood (Figure 10.8).

The median filter works very well (and significantly better than an averaging filter with comparable neighborhood size) in reducing "salt and pepper" noise (a type of noise that causes very bright—salt— and very dark—pepper—isolated spots to appear in an image) from images. Figure 10.9 compares the results obtained using median filtering and the averaging filter for the case of an image contaminated with salt and pepper noise.

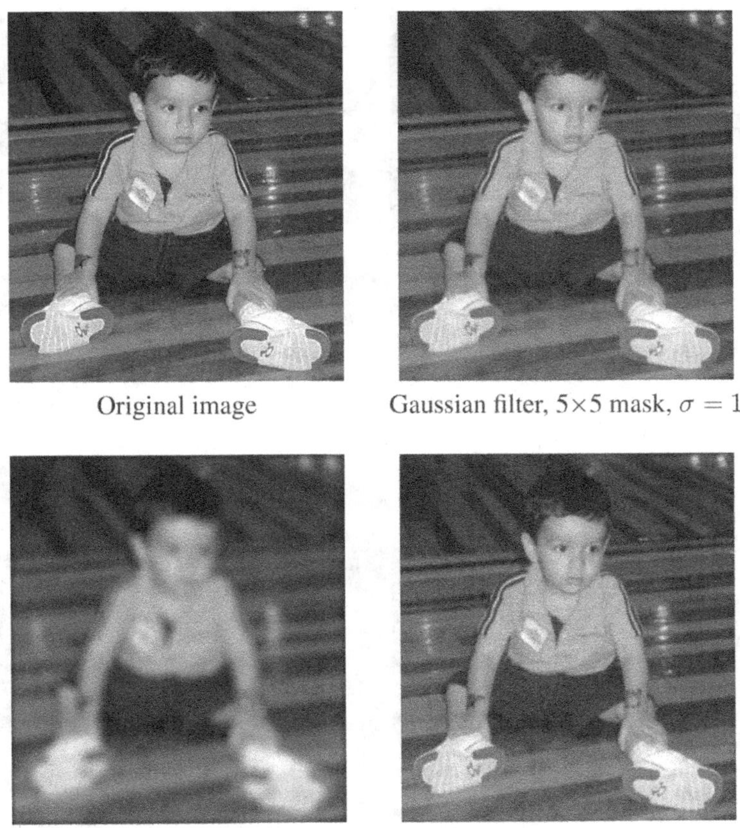

Original image

Gaussian filter, 5×5 mask, $\sigma = 1$

Mean filter, 13×13 mask

Gaussian filter, 13×13 mask, $\sigma = 1$

FIGURE 10.7 Example of using Gaussian blur filters.

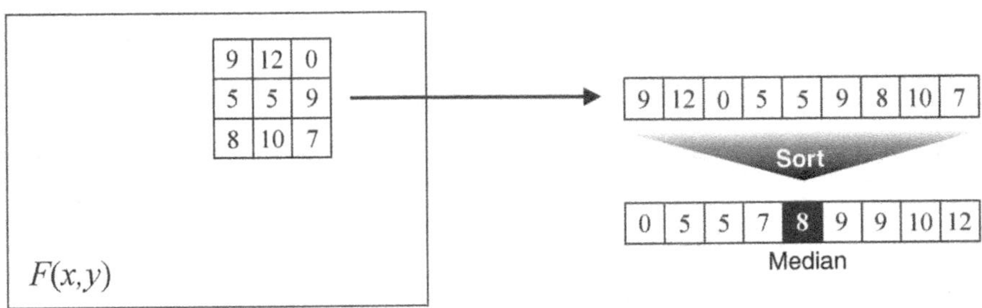

FIGURE 10.8 Median filter. Redrawn from [BB08].

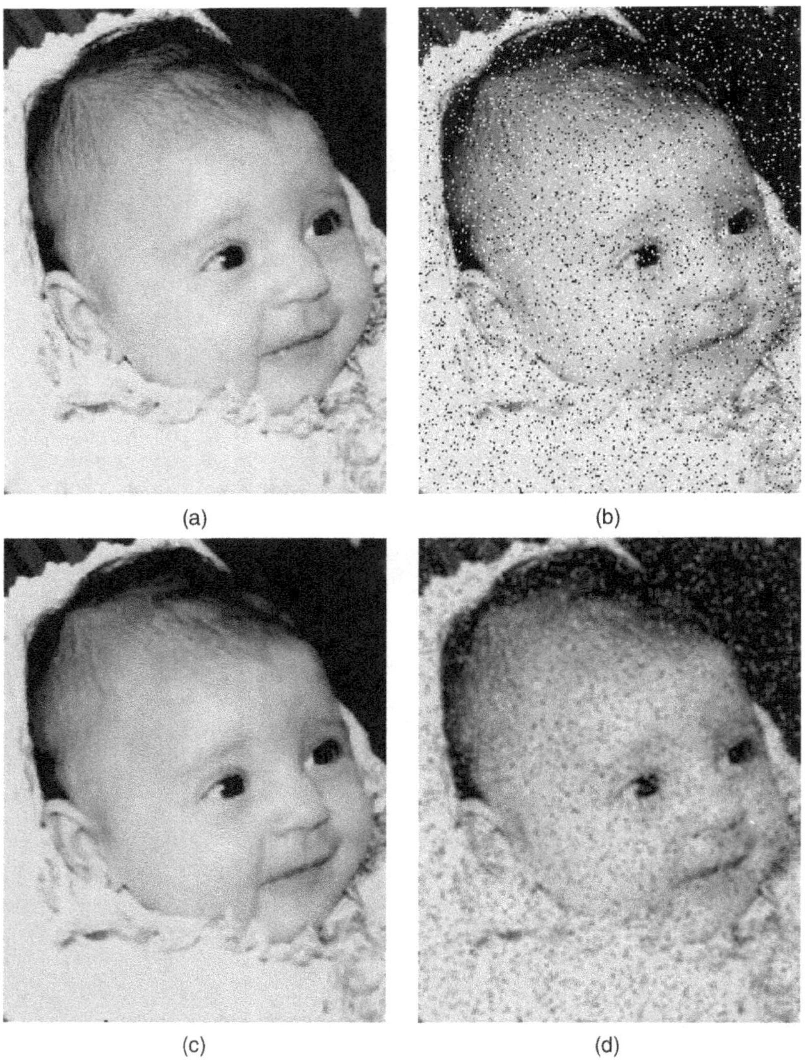

FIGURE 10.9 (a) Original image; (b) image with salt and pepper noise; (c) result of 3 × 3 median filtering; (d) result of 3 × 3 neighborhood averaging.

We shall resume our discussion of median filter, its implementation details, some of its variants, and its use in noise reduction in Chapter 12.

10.4 IMAGE SHARPENING (HIGH-PASS FILTERS)

We call *high-pass* filters those spatial filters whose effect on an image is equivalent to preserving or emphasizing its high-frequency components (i.e., fine details, points, lines, and edges), that is, to highlight transitions in intensity within the image.

Linear HPFs can be implemented using 2D convolution masks with positive and negative coefficients, which correspond to a digital approximation of the *Laplacian*, a simple, *isotropic* (i.e., *rotation invariant*) second-order derivative that is capable of responding to intensity transitions in any direction.

10.4.1 The Laplacian

The Laplacian of an image $f(x, y)$ is defined as

$$\nabla^2(x, y) = \frac{\partial^2(x, y)}{\partial x^2} + \frac{\partial^2(x, y)}{\partial y^2} \tag{10.12}$$

where the second derivatives are usually approximated—for digital signals—as

$$\frac{\partial^2(x, y)}{\partial x^2} = f(x+1, y) + f(x-1, y) - 2f(x, y) \tag{10.13}$$

and

$$\frac{\partial^2(x, y)}{\partial y^2} = f(x, y+1) + f(x, y-1) - 2f(x, y) \tag{10.14}$$

which results in a convenient expression for the Laplacian expressed as a sum of products:

$$\nabla^2(x, y) = f(x+1, y) + f(x-1, y) + f(x, y+1) + f(x, y-1) - 4f(x, y) \tag{10.15}$$

This expression can be implemented by the convolution mask below:

$$\begin{bmatrix} 0 & -1 & 0 \\ -1 & 4 & -1 \\ 0 & -1 & 0 \end{bmatrix}$$

An alternative digital implementation of the Laplacian takes into account all eight neighbors of the reference pixel in the input image and can be implemented by the convolution mask below:

$$\begin{bmatrix} -1 & -1 & -1 \\ -1 & 8 & -1 \\ -1 & -1 & -1 \end{bmatrix}$$

Note that it is common to find implementations in which the signs of each coefficient in the convolution masks above are reversed.

10.4.2 Composite Laplacian Mask

High-pass filters can be implemented using the Laplacian defined in equation (10.15) and combining the result with the original image as follows:

$$g(x, y) = f(x, y) + c[\nabla^2(x, y)] \tag{10.16}$$

where c is a constant used to comply with the sign convention for the particular implementation of the Laplacian mask: $c = 1$ if the center coefficient is positive, while $c = -1$ if the same is negative.

The goal of adding the original image to the results of the Laplacian is to restore the gray-level tonality that was lost in the Laplacian calculations.[5]

It is worth noting that if we want to implement the composite Laplacian mask using the `fspecial` function in MATLAB, we need to figure out the correct value of `alpha`.[6] Choosing `alpha` = 0 will result in the following mask:

$$\begin{bmatrix} 0 & 1 & 0 \\ 1 & -4 & 1 \\ 0 & 1 & 0 \end{bmatrix}$$

It is also common to factor equation (10.16) into the design of the mask, which produces the *composite* Laplacian mask below:

$$\begin{bmatrix} 0 & -1 & 0 \\ -1 & 5 & -1 \\ 0 & -1 & 0 \end{bmatrix}$$

■ **EXAMPLE 10.5**

Figure 10.10 shows an example (using the `imfilter` and `fspecial` functions in MATLAB) of applying a high-pass filter to enhance (sharpen) a monochrome image. Figure 10.10a shows the original image. Figure 10.10b shows the resulting enhanced image obtained by applying equation (10.16) with $c = -1$ and Figure 10.10c shows the result of using the eight-directional Laplacian operator instead. It can be claimed that the results in part (c) are crisper than the ones obtained in part (b).

10.4.3 Directional Difference Filters

Directional difference filters are similar to the Laplacian high-frequency filter discussed earlier. The main difference is that—as their name suggests—directional

[5]The Laplacian mask, as well as any mask whose coefficients add up to zero, tends to produce results centered around zero, which correspond to very dark images, whose only bright spots are indicative of what the mask is designed to detect or emphasize, in this case, omnidirectional edges.

[6]Refer to the IPT documentation for further details.

(a) (b) (c)

FIGURE 10.10 Example of using Laplacian masks to enhance an image.

difference filters emphasize edges in a specific direction. There filters are usually called *emboss filters*. There are four representative masks that can be used to implement the emboss effect:

$$\begin{bmatrix} 0 & 1 & 0 \\ 0 & 0 & 0 \\ 0 & -1 & 0 \end{bmatrix} \quad \begin{bmatrix} 1 & 0 & 0 \\ 0 & 0 & 0 \\ 0 & 0 & -1 \end{bmatrix} \quad \begin{bmatrix} 0 & 0 & 0 \\ 1 & 0 & -1 \\ 0 & 0 & 0 \end{bmatrix} \quad \begin{bmatrix} 0 & 0 & -1 \\ 0 & 0 & 0 \\ 1 & 0 & 0 \end{bmatrix}$$

10.4.4 Unsharp Masking

The unsharp masking technique consists of computing the subtraction between the input image and a blurred (low-pass filtered) version of the input image. The rationale behind this technique is to "increase the amount of high-frequency (fine) detail by reducing the importance of its low-frequency contents." There have been many variants of this basic idea proposed in the literature. In Tutorial 10.3 (page 227), you will use MATLAB to implement unsharp masking in three different ways.

10.4.5 High-Boost Filtering

The high-boost filtering technique—sometimes referred to as *high-frequency emphasis*—emphasizes the fine details of an image by applying a convolution mask:

$$\begin{bmatrix} -1 & -1 & -1 \\ -1 & c & -1 \\ -1 & -1 & -1 \end{bmatrix}$$

where c ($c > 8$) is a coefficient—sometimes called *amplification factor*—that controls how much weight is given to the original image and the high-pass filtered version of that image. For $c = 8$, the results would be equivalent to those seen earlier for the conventional isotropic Laplacian mask (Figure 10.10c). Greater values of c will cause significantly less sharpening.

10.5 REGION OF INTEREST PROCESSING

Filtering operations are sometimes performed only in a small part of an image—known as a *region of interest* (ROI)—which can be specified by defining a (usually binary) *mask* that delimits the portion of the image in which the operation will take place. *Image masking* is the process of extracting such a subimage (or ROI) from a larger image for further processing.

In MATLAB

ROI processing can be implemented in MATLAB using a combination of two functions: `roipoly`—introduced in Tutorial 6.2—for image masking and `roifilt2` for the actual processing of the selected ROI. Selecting a polygonal ROI can be done interactively—clicking on the polygon vertices—or programmatically—specifying the coordinates of the vertices using two separate vectors (one for rows and one for columns).

■ EXAMPLE 10.6

Figure 10.11 shows an example of ROI processing using the `roifilt2` function in MATLAB.

FIGURE 10.11 Example of region of interest processing: (a) original image; (b) result of applying a Gaussian blur to a selected ROI; (c) result of applying a HPF to a selected ROI; (d) result of applying a Laplacian mask to a selected ROI.

10.6 COMBINING SPATIAL ENHANCEMENT METHODS

At the end of this chapter—especially after working on its tutorials—you must have seen a significant number of useful methods and algorithms for processing monochrome images. A legitimate question to ask at this point is as follows "When faced with a practical image processing problem, which techniques should I use and in which sequence?" Naturally, there is no universal answer to this question. Most image processing solutions are problem specific and usually involve the application of several algorithms—in a meaningful sequence—to achieve the desired goal. The choice of algorithms and fine-tuning of their parameters is an almost inevitable trial-and-error process that most image processing solution designers have to go through. Using the knowledge acquired so far and a tool that allows easy experimentation—MATLAB—you should be able to implement, configure, fine-tune, and combine image processing algorithms for a wide variety of real-world problems.

10.7 TUTORIAL 10.1: CONVOLUTION AND CORRELATION

Goal

The goal of this tutorial is to learn how to perform a correlation and convolution calculations in MATLAB.

Objectives

- Learn how to perform a correlation of two (1D and 2D) matrices.
- Learn how to perform a convolution of two (1D and 2D) matrices.
- Explore the `imfilter` function to perform correlation and convolution in MATLAB.

Procedure

We shall start by exploring convolution and correlation in one dimension. This can be achieved by means of the `imfilter` function.

1. Specify the two matrices to be used.

```
a = [0 0 0 1 0 0 0];
f = [1 2 3 4 5];
```

2. Perform convolution, using `a` as the input matrix and `f` as the filter.

```
g = imfilter(a,f,'full','conv')
```

Question 1 What is the relationship between the size of the output matrix, the size of the original matrix, and the length of the filter?

Question 2 How does changing the third parameter from `'full'` to `'same'` affect the output?

3. Perform correlation on the same set of matrices.

```
h = imfilter(a,f,'full','corr')
```

The results from the previous step should confirm that convolution is related to correlation by a reflection of the filter matrix, regardless of the number of dimensions involved.

Let us see how correlation works on a small window of size 3 × 3. Consider the window of values extracted from a larger image in Figure 10.12.

The correlation of two matrices is a sum of products. Numerically, the calculation would be as follows:

$$(140)(-1) + (108)(0) + (94)(1) + (89)(-2) + (99)(0) + (125)(2) \\ + (121)(-1) + (134)(0) + (221)(1) = 126$$

Here, we specify the image (which in our case will be the image region as in Figure 10.12) and the mask from Figure 10.13. We will also explicitly tell the function to use correlation, as it can perform both correlation and convolution.

4. Clear all workspace variables.
5. Use `imfilter` to perform a correlation of the two matrices.

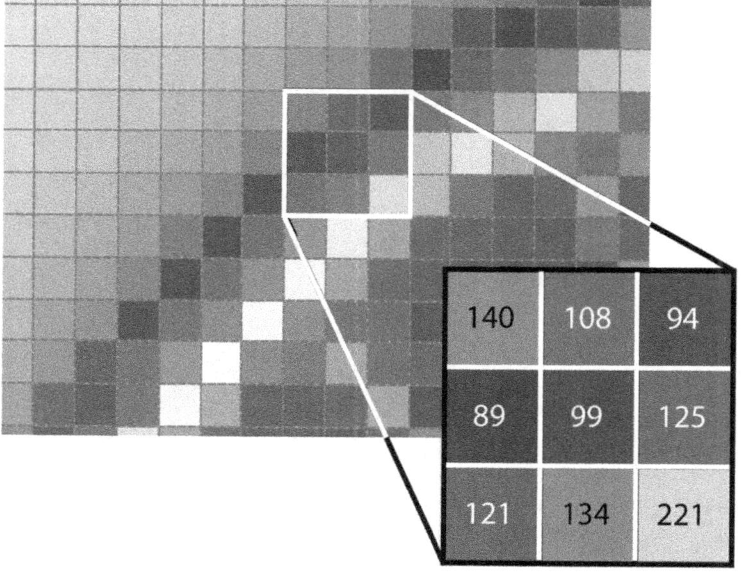

FIGURE 10.12 A 3 × 3 image region.

−1	0	1
−2	0	2
−1	0	1

FIGURE 10.13 A 3 × 3 mask.

```
x = [140 108 94;89 99 125;121 134 221]
y = [-1 0 1;-2 0 2;-1 0 1]
z = imfilter(x,y,'corr')
```

Question 3 In the resulting matrix (`z`), we are interested only in the center value. How does this value compare with our calculation illustrated above?

Question 4 What are the other values in the resulting matrix?

Question 5 Note in the last step we did not specify if the output should be `'full'` or `'same'`. What is the default for this setting if it is not specified?

To perform convolution, we use the same technique as in correlation. The difference here is that the filter matrix is rotated 180° before performing the sum of products. Again, the calculation of the convolution of the given image region and mask is performed as follows:

$$(140)(1) + (108)(0) + (94)(-1) + (89)(2) + (99)(0) + (125)(-2) + (121)(1) \\ +(134)(0) + (221)(-1) = -126$$

6. Use `imfilter` to perform a convolution of the two matrices.

```
z2 = imfilter(x,y,'conv')
```

Question 6 How does the center value of the resulting matrix compare with our calculation above?

10.8 TUTORIAL 10.2: SMOOTHING FILTERS IN THE SPATIAL DOMAIN

Goal

The goal of this tutorial is to learn how to implement smoothing filters in the spatial domain.

Objectives

- Learn how to use the `fspecial` function to generate commonly used kernels.

- Explore applying smoothing filters to images using the `imfilter` function.
- Learn how to implement uniform and nonuniform averaging masks.
- Learn how to implement a Gaussian mask.

Procedure

In the first part of this procedure, we will use the `imfilter` function to implement a 3×3 mean (average) filter. We could easily generate the mask array ourselves (nine values, each equal to 1/9), but the IPT offers a function that will automatically create this and several other commonly used masks.

1. Load the `cameraman` image and prepare a subplot.

```
I = imread('cameraman.tif');
figure, subplot(1,2,1), imshow(I), title('Original Image');
```

2. Create a mean (averaging) filter automatically through the `fspecial` function.

```
fn = fspecial('average')
```

Question 1 Explain what the value of the variable `fn` represents.

Question 2 What other commonly used masks is the `fspecial` function capable of generating?

3. Filter the `cameraman` image with the generated mask.

```
I_new = imfilter(I,fn);
subplot(1,2,2), imshow(I_new), title('Filtered Image');
```

Question 3 What was the effect of the averaging filter?

The mean filter we just implemented was a uniform filter—all coefficients were equivalent. The nonuniform version of the mean filter gives the center of the mask (the pixel in question) a higher weighted value, while all other coefficients are weighted by their distance from the center. This particular mask cannot be generated by the `fspecial` function, so we must create it ourselves.

4. Create a nonuniform version of the mean filter.

```
fn2 = [1 2 1; 2 4 2; 1 2 1]
fn2 = fn2 * (1/16)
```

1/9 ×

1	1	1
1	1	1
1	1	1

1/16 ×

1	2	1
2	4	2
1	2	1

FIGURE 10.14 Uniform and nonuniform averaging masks.

Recall that the uniform mean filter could be created by generating a 3×3 matrix of 1's, and then multiplying each coefficient by a factor of 1/9. In the nonuniform mean filter implantation above, note that the sum of all the original values in the filter equals 16—this is why we divide each coefficient by 16 in the second step. Figure 10.14 illustrates the previous two masks we created.

5. Filter the original image with the new, nonuniform averaging mask.

```
I_new2 = imfilter(I,fn2);
figure, subplot(1,2,1), imshow(I_new), title('Uniform Average');
subplot(1,2,2), imshow(I_new2), title('Non-uniform Average');
```

Question 4 Comment on the subjective differences between using the uniform averaging filter and the nonuniform averaging filter.

The Gaussian filter is similar to the nonuniform averaging filter in that the coefficients are not equivalent. The coefficient values, however, are not a function of their distance from the center pixel, but instead are modeled from the Gaussian curve.

6. Create a Gaussian filter and display the kernel as a 3D plot.

```
fn_gau = fspecial('gaussian',9,1.5);
figure, bar3(fn_gau,'b'), ...
title('Gaussian filter as a 3D graph');
```

7. Filter the `cameraman` image using the Gaussian mask.

```
I_new3 = imfilter(I,fn_gau);
figure
subplot(1,3,1), imshow(I), title('Original Image');
subplot(1,3,2), imshow(I_new), title('Average Filter');
subplot(1,3,3), imshow(I_new3), title('Gaussian Filter');
```

Question 5 Experiment with the size of the Gaussian filter and the value of σ. How can you change the amount of blur that results from the filter?

10.9 TUTORIAL 10.3: SHARPENING FILTERS IN THE SPATIAL DOMAIN

Goal

The goal of this tutorial is to learn how to implement sharpening filters in the spatial domain.

Objectives

- Learn how to implement the several variations of the Laplacian mask.
- Explore different implementations of the unsharp masking technique.
- Learn how to apply a high-boost filtering mask.

Procedure

To implement the Laplacian filter, we can either create our own mask or use the `fspecial` function to generate the mask for us. In the next step, we will use `fspecial`, but keep in mind that you can just as well create the mask on your own.

1. Load the `moon` image and prepare a subplot figure.

```
I = imread('moon.tif');
Id = im2double(I);
figure, subplot(2,2,1), imshow(Id), title('Original Image');
```

We are required to convert the image to doubles because a Laplacian filtered image can result in negative values. If we were to keep the image as class `uint8`, all negative values would be truncated and, therefore, would not accurately reflect the results of having applied a Laplacian mask. By converting the image to `doubles`, all negative values will remain intact.

2. Create a Laplacian kernel and apply it to the image using the `imfilter` function.

```
f = fspecial('laplacian',0);
I_filt = imfilter(Id,f);
subplot(2,2,2), imshow(I_filt), title('Laplacian of Original');
```

Question 1 When specifying the Laplacian filter in the `fspecial` function, what is the second parameter (in the case above, 0) used for?

Question 2 What is the minimum value of the filtered image?

Question 3 Verify that a `uint8` filtered image would not reflect negative numbers. You can use the image `I` that was previously loaded.

You will notice that it is difficult to see details of the Laplacian filtered image. To get a better perspective of the detail the Laplacian mask produced, we can scale the image for display purposes so that its values span the dynamic range of the gray scale.

3. Display a scaled version of the Laplacian image for display purposes.

```
subplot(2,2,3), imshow(I_filt,[]), title('Scaled Laplacian');
```

The center coefficient of the Laplacian mask we created is negative. Recall from the chapter that if the mask center is negative, we subtract the filtered image from the original, and if it is positive, we add. In our case, we will subtract them.

4. Subtract the filtered image from the original image to create the sharpened image.

```
I_sharp = imsubtract(Id,I_filt);
subplot(2,2,4), imshow(I_sharp), title('Sharpened Image');
```

A composite version of the Laplacian mask performs the entire operation all at once. By using this composite mask, we do not need to add or subtract the filtered image—the resulting image *is* the sharpened image.

5. Use the composite Laplacian mask to perform image sharpening in one step.

```
f2 = [0 -1 0; -1 5 -1; 0 -1 0]
I_sharp2 = imfilter(Id,f2);
figure, subplot(1,2,1), imshow(Id), title('Original Image');
subplot(1,2,2), imshow(I_sharp2), title('Composite Laplacian');
```

Question 4 You may have noticed that we created the mask without using the `fspecial` function. Is the `fspecial` function capable of generating the simplified Laplacian mask?

Question 5 Both Laplacian masks used above did not take into account the four corner pixels (their coefficients are 0). Reapply the Laplacian mask, but this time use the version of the mask that accounts for the corner pixels as well. Both the standard and simplified versions of this mask are illustrated in Figure 10.15. How does accounting for corner pixels change the output?

Unsharp Masking Unsharp masking is a simple process of subtracting a blurred image from its original to generate a sharper image. Although the concept is

1	1	1
1	–8	1
1	1	1

–1	–1	–1
–1	9	–1
–1	–1	–1

FIGURE 10.15 Laplacian masks that account for corner pixels (standard and composite).

straightforward, there are three ways it can be implemented. Figures 10.16–10.18 illustrate these processes.

Let us first implement the process described in Figure 10.16.

6. Close all open figures and clear all workspace variables.
7. Load the `moon` image and generate the blurred image.

```
I = imread('moon.tif');
f_blur = fspecial('average',5);
I_blur = imfilter(I,f_blur);
figure, subplot(1,3,1), imshow(I), title('Original Image');
subplot(1,3,2), imshow(I_blur), title('Blurred Image');
```

Question 6 What does the second parameter of the `fspecial` function call mean?

We must now shrink the histogram of the blurred image. The amount by which we shrink the histogram will ultimately determine the level of enhancement in the final result. In our case, we will scale the histogram to range between 0.0 and 0.4, where the full dynamic grayscale range is [0.0 1.0].

8. Shrink the histogram of the blurred image.

```
I_blur_adj = imadjust(I_blur,stretchlim(I_blur),[0 0.4]);
```

9. Now subtract the blurred image from the original image.

```
I_sharp = imsubtract(I,I_blur_adj);
```

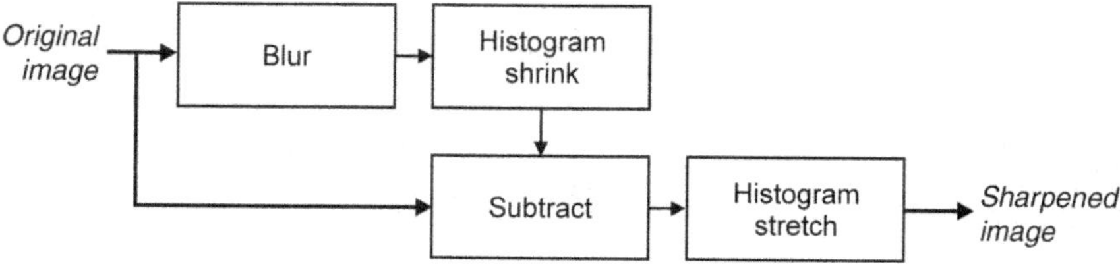

FIGURE 10.16 Unsharp masking process including histogram adjustment.

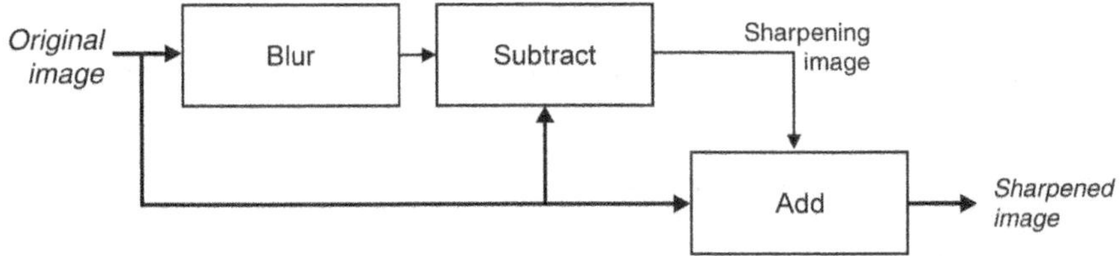

FIGURE 10.17 Unsharp masking process with sharpening image.

We must now perform a histogram stretch on the new image in order to account for previously shrinking the blurred image.

10. Stretch the sharpened image histogram to the full dynamic grayscale range and display the final result.

```
I_sharp_adj = imadjust(I_sharp);
subplot(1,3,3), imshow(I_sharp_adj), title('Sharp Image');
```

Question 7 We learned that by shrinking the blurred image's histogram, we can control the amount of sharpening in the final image by specifying the maximum range value. What other factor can alter the amount of sharpening?

We will now look at the second implementation of the unsharp masking technique, illustrated in Figure 10.17. We have already generated a blurred version of the `moon` image, so we can skip that step.

11. Subtract the blurred image from the original image to generate a sharpening image.

```
I_sharpening = imsubtract(I,I_blur);
```

12. Add sharpening image to original image to produce the final result.

```
I_sharp2 = imadd(I,I_sharpening);
figure, subplot(1,2,1), imshow(I), title('Original Image');
subplot(1,2,2), imshow(I_sharp2), title('Sharp Image');
```

Question 8 How can we adjust the amount of sharpening when using this implementation?

The third implementation uses a convolution mask, which can be generated using the `fspecial` function. This implementation is illustrated in Figure 10.18.

13. Generate unsharp masking kernel using the `fspecial` function.

```
f_unsharp = fspecial('unsharp');
```

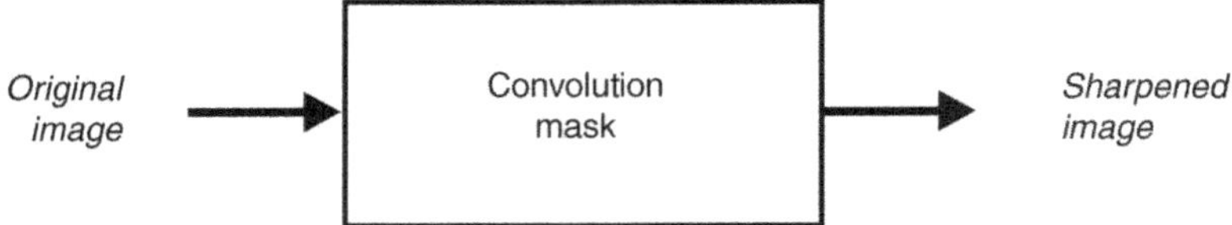

FIGURE 10.18 Unsharp masking process using convolution mask.

14. Apply the mask to the original image to create a sharper image.

```
I_sharp3 = imfilter(I,f_unsharp);
figure, subplot(1,2,1), imshow(I), title('Original Image');
subplot(1,2,2), imshow(I_sharp3), title('Sharp Image');
```

Question 9 How do we control the level of sharpening with this implementation?

High-Boost Filtering High-boost filtering is a sharpening technique that involves creating a sharpening image and adding it to the original image. The mask used to create the sharpening image is illustrated in Figure 10.19. Note that there are two versions of the mask: one that does not include the corner pixels and another that does.

15. Close any open figures.
16. Create a high-boost mask (where `A` = 1) and apply it to the `moon` image.

```
f_hb = [0 -1 0; -1 5 -1; 0 -1 0];
I_sharp4 = imfilter(I,f_hb);
figure, subplot(1,2,1), imshow(I), title('Original Image');
subplot(1,2,2), imshow(I_sharp4), title('Sharp Image');
```

Question 10 What happens to the output image when `A` is less than 1? What about when `A` is greater than 1?

You may have noticed that when `A` = 1, the high-boost filter generalizes to the composite Laplacian mask discussed in step 5. As the value of `A` increases, the output image starts to resemble an image multiplied by a constant.

0	−1	0
−1	A+4	−1
0	−1	0

−1	−1	−1
−1	A+8	−1
−1	−1	−1

FIGURE 10.19 High-boost masks with and without regard to corner pixels.

17. Show that a high-boost mask when A = 3 looks similar to the image simply multiplied by 3.

```
f_hb2 = [0 -1 0; -1 7 -1; 0 -1 0];
I_sharp5 = imfilter(I,f_hb2);
I_mult = immultiply(I,3);
figure, subplot(1,3,1), imshow(I), title('Original Image');
subplot(1,3,2), imshow(I_sharp5), title('High Boost, A = 3');
subplot(1,3,3), imshow(I_mult), title('Multiplied by 3');
```

Question 11 At what value of A does this filter stop being effective (resemble the image multiplied by a constant)?

WHAT HAVE WE LEARNED?

- Neighborhood processing is the name given to image processing techniques in which the new value of a processed image pixel is a function of its original value and the values of (some of) its neighbors. It differs from point processing (Chapters 8 and 9), where the resulting pixel value depends only on the original value and on the transformation function applied to each pixel in the original image.
- Convolution is a widely used mathematical operator that processes an image by computing—for each pixel—a weighted sum of values of that pixel and its neighbors. Depending on the choice of weights, a wide variety of image processing operations can be implemented.
- Low-pass filters are used to smooth an image or reduce the amount of noise in it. Low-pass linear filters can be implemented using 2D convolution masks with nonnegative coefficients. The mean (averaging) filter is the simplest—and most popular—low-pass linear filter. It works by averaging out the pixel values within a neighborhood.
- Low-pass nonlinear filters also work at a neighborhood level, but do not process the pixel values using the convolution operator. The median filter is one of the most popular low-pass linear filters. It works by sorting the pixel values within a neighborhood, finding the median value, and replacing the original pixel value with the median of that neighborhood. The median filter works extremely well in removing salt and pepper noise from images.
- Convolution operators can also be used to detect or emphasize the high-frequency contents of an image, such as fine details, points, lines, and edges. In such cases, the resulting filter is usually called a high-pass filter. High-pass linear filters can be implemented using 2D convolution masks with positive and negative coefficients.
- The spatial-domain image enhancement techniques discussed in Chapters 8–10 can be combined to solve specific image processing problems. The secret of

success consists in deciding which techniques to use, how to configure their parameters (e.g., window size and mask coefficients), and in which sequence they should be applied.

LEARN MORE ABOUT IT

- Section 6.5 of [BB08] and Section 7.2 of [Eff00] discuss implementation aspects associated with spatial filters.
- Chapter 7 of [Pra07] discusses convolution and correlation in greater mathematical depth.
- Chapter 4 of [Jah05] and Section 6.3 of [BB08] describe the formal properties of linear filters.
- Chapter 11 of [Jah05] is entirely devoted to an in-depth discussion of linear filters.
- Section 3.4 of [Dav04] introduces *mode filters* and compares them with mean and median filters in the context of machine vision applications.
- Section 5.3.1 of [SHB08] discusses several *edge-preserving* variants of the basic neighborhood averaging filter. MATLAB implementation for one of these methods—smoothing using a rotating mask—appears in Section 5.4 of [SKH08].
- Section 3.8 of [GW08] discusses the use of fuzzy techniques for spatial filtering.

10.10 PROBLEMS

10.1 Image processing tasks such as blurring or sharpening an image can easily be accomplished using neighborhood-oriented techniques described in this chapter. Can these tasks also be achieved using point operations (such as the ones described in Chapters 8 and 9)? Explain.

10.2 Write MATLAB code to reproduce the results from Example 10.1.

10.3 Write MATLAB code to reproduce the results from Example 10.2.

10.4 Write MATLAB code to implement a linear filter that creates a horizontal blur over a length of 9 pixels, comparable to what would happen to a image if the camera were moved during the exposure interval.

10.5 Write MATLAB code to implement the emboss effect described in Section 10.4.3 and test it with an input image of your choice. Do the results correspond to what you expected?

CHAPTER 11

FREQUENCY-DOMAIN FILTERING

WHAT WILL WE LEARN?

- Which mathematical tools are used to represent an image's contents in the 2D frequency domain?
- What is the Fourier transform, what are its main properties, and how is it used in the context of frequency-domain filtering?
- How are image processing filters designed and implemented in the frequency domain?
- What are the differences between low-pass and high-pass filters (HPFs)?
- What are the differences among ideal, Butterworth, and Gaussian filters?

11.1 INTRODUCTION

This chapter builds upon the ideas introduced in Section 2.4.4, which state that some image processing tasks can be performed by transforming the input images to a different domain, applying selected algorithms in the transform domain, and eventually applying the inverse transformation to the result. In this chapter, we are particularly interested in a special case of operations in the transform domain, which we call *frequency-domain filtering*. Frequency-domain filters work by following a straightforward sequence of steps (Figure 11.1):

Practical Image and Video Processing Using MATLAB®. By Oge Marques.

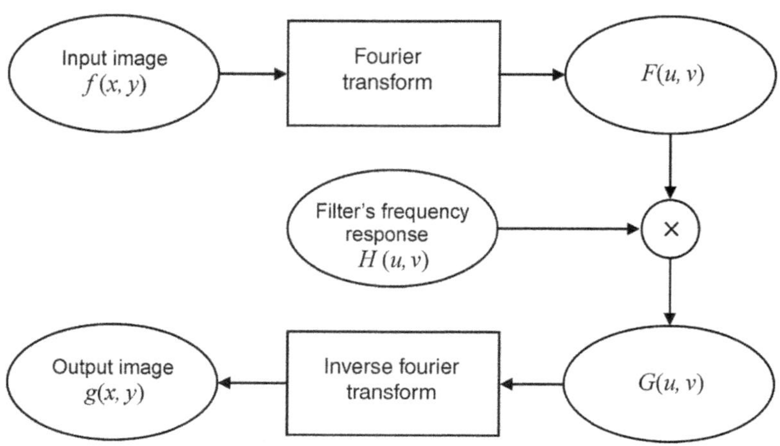

FIGURE 11.1 Frequency-domain operations.

1. The input image is transformed to a 2D frequency-domain representation using the 2D Fourier transform (FT).
2. A filter of specific type (e.g., ideal, Butterworth, Gaussian) and behavior (e.g., low pass, high pass) is specified and applied to the frequency-domain representation of the image.
3. The resulting values are transformed back to the 2D spatial domain by applying the inverse 2D Fourier transform, producing an output (filtered) image.

The mathematical foundation of frequency-domain techniques is the *convolution theorem*. Let $g(x, y)$ be an image obtained by the convolution[1] (denoted by the $*$ symbol) of an image $f(x, y)$ with a linear, position invariant operator $h(x, y)$, that is,

$$g(x, y) = f(x, y) * h(x, y) \tag{11.1}$$

From the convolution theorem, the following frequency-domain relation holds:

$$G(u, v) = F(u, v)H(u, v) \tag{11.2}$$

where G, F, and H are the Fourier transforms of g, f, and h, respectively.

Many image processing problems can be expressed in the form of equation (11.2). In a noise removal application, for instance, given $f(x, y)$, the goal, after computing $F(u, v)$, will be to select $H(u, v)$ such that the desired resulting image,

$$g(x, y) = \mathfrak{F}^{-1}\left[F(u, v)H(u, v)\right] \tag{11.3}$$

where $\mathfrak{F}^{-1}$ is the inverse 2D Fourier transform operation, exhibits a reduction in the noisy contents present in the original image $f(x, y)$. For certain types of noise,

[1]Two-dimensional discrete convolution was introduced in Section 10.2.2.

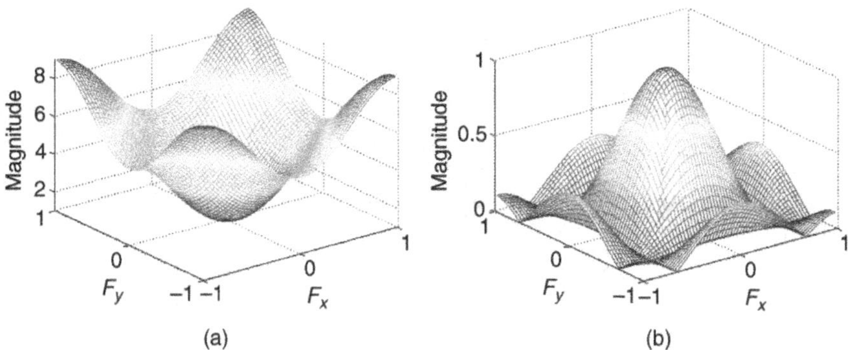

FIGURE 11.2 Two examples of response functions for frequency-domain filters: (a) low-pass filter equivalent to a 3 × 3 average filter in the spatial domain; (b) high-pass filter equivalent to a 3 × 3 composite Laplacian sharpening filter in the spatial domain.

this result could be achieved using a low-pass Butterworth filter (Section 11.3.3), for example.

There are two options for designing and implementing image filters in the frequency domain using MATLAB and IPT:

1. Obtain the frequency-domain filter response function from spatial filter convolution mask. The IPT has a function that does exactly that: `freqz2`. Figure 11.2 shows examples of such response functions for the 3 × 3 average filter described by equation (10.9) and the 3 × 3 composite Laplacian sharpening filter described by equation (10.16).
2. Generate filters directly in the frequency domain. In this case, a meshgrid array (of the same size as the image) is created using the MATLAB function `meshgrid`. This is the method used in the tutorials of this chapter.

11.2 FOURIER TRANSFORM: THE MATHEMATICAL FOUNDATION

The FT is a fundamental tool in signal and image processing.[2] In this section, we discuss the mathematical aspects of 2D transforms in general and then introduce the 2D FT and its main properties.

11.2.1 Basic Concepts

A *transform* is a mathematical tool that allows the conversion of a set of values to another set of values, creating, therefore, a new way of representing the same

[2]A complete, detailed analysis of the FT and associated concepts for 1D signals is beyond the scope of this book. Refer to the section "Learn More About It" for useful pointers.

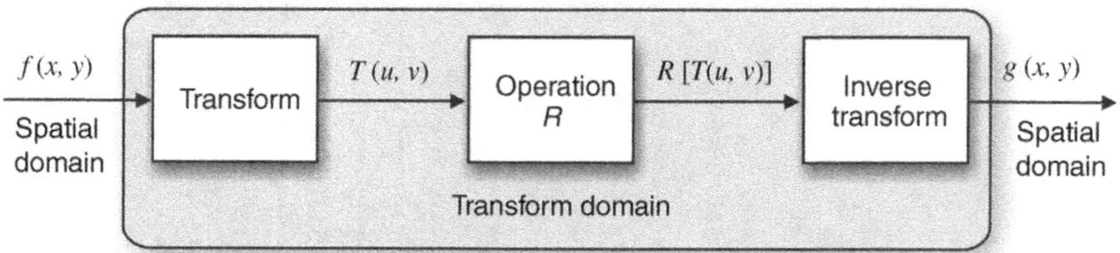

FIGURE 11.3 Operations in a transform domain.

information. In the field of image processing, the original domain is referred to as *spatial domain*, whereas the results are said to lie in the *transform domain*. The motivation for using mathematical transforms in image processing stems from the fact that some tasks are best performed by transforming the input images, applying selected algorithms in the transform domain, and eventually applying the inverse transformation to the result (Figure 11.3).

Linear 2D transforms can be expressed in generic form in terms of a *forward transform* $T(u, v)$ as

$$T(u, v) = \sum_{x=0}^{M-1} \sum_{y=0}^{N-1} f(x, y) \cdot r(x, y, u, v) \tag{11.4}$$

where u ($u = 0, 1, 2, \ldots, M-1$) and v ($v = 0, 1, 2, \ldots, N-1$) are called *transform variables*, $f(x, y)$ is the input image, x and y are the row and column dimensions of $f(x, y)$ (as described in Section 2.1), and $r(x, y, u, v)$ is called the *forward transformation kernel*, sometimes expressed as a collection of *basis images*.

The original image $f(x, y)$ can be recovered by applying the *inverse transform* of $T(u, v)$:

$$f(x, y) = \sum_{u=0}^{M-1} \sum_{v=0}^{N-1} T(u, v) \cdot s(x, y, u, v) \tag{11.5}$$

where $x = 0, 1, 2, \ldots, M-1$, $y = 0, 1, 2, \ldots, N-1$, and $s(x, y, u, v)$ is called the *inverse transformation kernel*.

The combination of equations (11.4) and (11.5) is usually referred to as a *transform pair*. Different mathematical transforms use different transformation kernels. For the sake of this chapter, we shall limit the discussion to the Fourier Transform.[3]

[3]The interested reader should refer to the section "Learn More About It" for useful references on other mathematical transforms.

11.2.2 The 2D Discrete Fourier Transform: Mathematical Formulation

The 2D Fourier transform of an image is a special case of equations (11.4) and (11.5) where

$$r(x, y, u, v) = \exp[-j2\pi(ux/M + vy/N)] \tag{11.6}$$

and

$$s(x, y, u, v) = \frac{1}{MN} \exp[j2\pi(ux/M + vy/N)] \tag{11.7}$$

where $j = \sqrt{-1}$.

Substituting equations (11.6) and (11.7) into equations (11.4) and (11.5), we have

$$F(u, v) = T(u, v) = \sum_{x=0}^{M-1} \sum_{y=0}^{N-1} f(x, y) \cdot \exp[-j2\pi(ux/M + vy/N)] \tag{11.8}$$

and

$$f(x, y) = \mathfrak{F}^{-1}[F(u, v)] = \frac{1}{MN} \sum_{u=0}^{M-1} \sum_{v=0}^{N-1} T(u, v) \cdot \exp[j2\pi(ux/M + vy/N)] \tag{11.9}$$

Remember from our discussion in Chapter 2 that array indices in MATLAB start at 1; therefore, `F(1,1)` and `f(1,1)` in MATLAB correspond to $F(0, 0)$ and $f(0, 0)$ in equations (11.8) and (11.9).

The value of the 2D FT at the origin of the frequency domain (i.e, $F(0, 0)$) is called the *DC component* of the FT, a terminology borrowed from electrical engineering, where *DC* means *direct current*, that is, alternating current of zero frequency. The *DC* component of the FT is the product of the average intensity value of $f(x, y)$ and a factor (MN). The FT is usually implemented using a computationally efficient algorithm known as *fast Fourier transform* (FFT) [Bri74].

The 2D FT of an image is an array of complex numbers, usually expressed in polar coordinates, whose components are *magnitude* (or *amplitude*) ($|F(u, v)|$) and *phase* ($\phi(u, v)$), which can be expressed as

$$|F(u, v)| = \sqrt{[R(u, v)]^2 + [I(u, v)]^2} \tag{11.10}$$

and

$$\phi(u, v) = \arctan\left[\frac{I(u, v)}{R(u, v)}\right] \tag{11.11}$$

where $R(u, v)$ and $I(u, v)$ are the real and imaginary parts of $F(u, v)$, that is, $F(u, v) = R(u, v) + jI(u, v)$.

In MATLAB

The 2D FT and its inverse are implemented in MATLAB by functions `fft2` and `ifft2`, respectively. 2D FT results are usually shifted for visualization purposes in such a way as to position the zero-frequency component at the center of the figure (exploiting the periodicity property of the 2D FT, described in Section 11.2.3). This can be accomplished by function `fftshift`. These functions are extensively used in the tutorials of this chapter. MATLAB also includes function `ifftshift`, whose basic function is to undo the results of `fftshift`.

■ EXAMPLE 11.1

Figure 11.4 shows an image and its frequency spectrum (the log transformation of the amplitude component of its 2D FT coefficients). The transformed version of the image clearly bears no resemblance to the original version. Moreover, it does not provide any obvious visual hints as to what the original image contents were. Do not be discouraged by these facts, though. As you will soon see, these apparent limitations of the transformed version of the image will not prevent us from designing frequency-domain filters to achieve specific goals. The resulting image—obtained after applying the inverse transform—will allow us to judge whether those goals were successfully achieved or not.

MATLAB code:

```
I = imread('Figure11_04_a.png');
Id = im2double(I);
ft = fft2(Id);
ft_shift = fftshift(ft);
imshow(log(1 + abs(ft_shift)), [])
```

(a)

(b)

FIGURE 11.4 (a) Original image (256 × 256 pixels); (b) Fourier spectrum of the image in (a).

11.2.3 Summary of Properties of the Fourier Transform

In this section, we present selected properties of the 2D FT and its inverse that are particularly important in image processing.

Linearity The Fourier transform is a linear operator, that is,

$$\mathfrak{F}[a \cdot f1(x, y) + b \cdot f2(x, y)] = a \cdot F1(u, v) + b \cdot F2(u, v) \tag{11.12}$$

and

$$a \cdot f1(x, y) + b \cdot f2(x, y) = \mathfrak{F}^{-1}[a \cdot F1(u, v) + b \cdot F2(u, v)] \tag{11.13}$$

where a and b are constants.

Translation The translation property of the Fourier transform shows that if an image is moved (translated), the resulting frequency-domain spectrum undergoes a phase shift, but its amplitude remains the same. Mathematically,

$$\mathfrak{F}[f(x - x_0, y - y_0)] = F(u, v) \cdot \exp[-j2\pi(ux_0/M + vy_0/N)] \tag{11.14}$$

and

$$f(x - x_0, y - y_0) = \mathfrak{F}^{-1} F(u, v) \cdot \exp[j2\pi(ux_0/M + vy_0/N)] \tag{11.15}$$

Conjugate Symmetry If $f(x, y)$ is real, its FT is conjugate symmetric about the origin:

$$F(u, v) = F^*(-u, -v) \tag{11.16}$$

where $F^*(u, v)$ is the conjugate of $F(u, v)$; that is, if $F(u, v) = R(u, v) + jI(u, v)$, then $F^*(u, v) = R(u, v) - jI(u, v)$.

Combining equations (11.10) and (11.16), we have

$$|F(u, v)| = |F(-u, -v)| \tag{11.17}$$

Periodicity The FT (and its inverse) are infinitely periodic in both the u and v directions. Mathematically,

$$F(u, v) = F(u + M, v + N) \tag{11.18}$$

and

$$f(x, y) = f(x + M, y + N) \tag{11.19}$$

Separability The Fourier transform is separable; that is, the FT of a 2D image can be computed by two passes of the 1D FT algorithm, one along the rows (columns), and the other along the columns (rows) of the result.

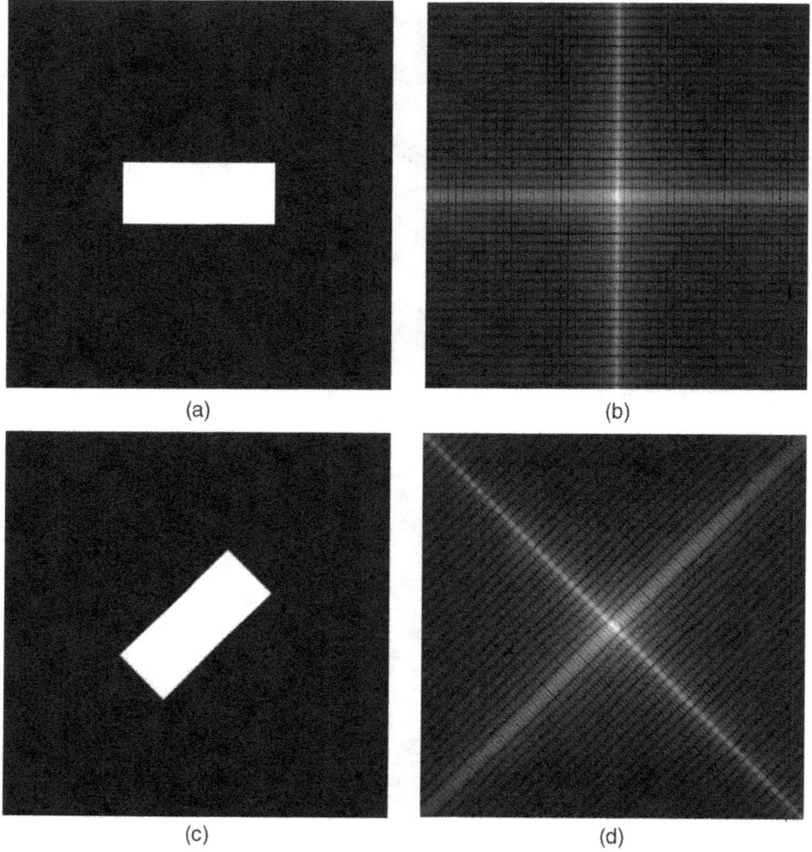

FIGURE 11.5 Original image (a) and its 2D FT spectrum (b); rotated image (c) and its 2D FT spectrum (d).

Rotation If an image is rotated by a certain angle θ, its 2D FT will be rotated by the same angle (Figure 11.5).

11.2.4 Other Mathematical Transforms

In addition to the Fourier transform, there are many other mathematical transforms used in image processing and analysis. Some of those transforms will be described later in the book, whenever needed (e.g., the discrete cosine transform (DCT) in Chapter 17), while many others (e.g., sine, Hartley, Walsh, Hadamard, wavelet, and slant, to mention but a few) will not be discussed in this text. Refer to the Section "Learn More About It" at the end of the chapter for useful pointers and references.

FIGURE 11.6 Example of using LPF to smooth false contours: (a) original image; (b) result of applying a LPF.

FIGURE 11.7 Example of using LPF for noise reduction: (a) original image; (b) result of applying a LPF.

11.3 LOW-PASS FILTERING

Low-pass filters attenuate the high-frequency components of the Fourier transform of an image, while leaving the low-frequency components unchanged. The typical overall effect of applying a low-pass filter (LPF) to an image is a controlled degree of blurring. Figures 11.6 and 11.7 show examples of applications of LPFs for smoothing of false contours (Section 5.4.3) and noise reduction,[4] respectively.

[4]We shall discuss noise reduction in more detail in Chapter 12.

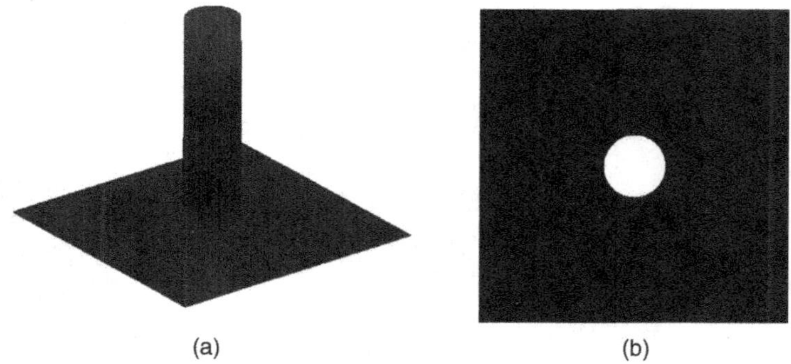

(a) (b)

FIGURE 11.8 Frequency response plot for an ideal LPF: (a) 3D view; (b) 2D view from the top.

11.3.1 Ideal LPF

An ideal low-pass filter enhances all frequency components within a specified radius (from the center of the FT), while attenuating all others. Its mathematical formulation is given as follows:

$$H_I(u,v) = \begin{cases} 1 & \text{if } D(u,v) \leq D_0 \\ 0 & \text{if } D(u,v) > D_0 \end{cases} \tag{11.20}$$

where $D(u,v) = \sqrt{(u^2 + v^2)}$ represents the distance between a point of coordinates (u, v) and the origin of the 2D frequency plan, and D_0 is a nonnegative value, referred to as the *cutoff frequency* (or *cutoff radius*).

Figure 11.8 shows the frequency response plot for an ideal LPF. Figure 11.9 shows an example image and its Fourier spectrum. The rings in Figure 11.9b represent different values for cutoff frequencies (D_0): 8, 16, 32, 64, and 128.

(a) (b)

FIGURE 11.9 (a) Original image (256 × 256 pixels); (b) Fourier spectrum of the image in (a). The rings represent cutoff frequencies for the low-pass filter examples described later.

FIGURE 11.10 (a) Original image (256 × 256 pixels); (b–f) ideal LPF results for filters with cutoff frequency corresponding to the radii in Figure 11.9b, namely, 8, 16, 32, 64, and 128 pixels.

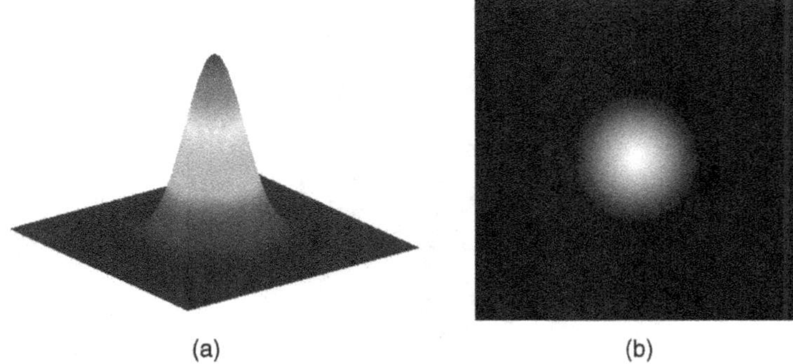

FIGURE 11.11 Frequency response plot for a Gaussian LPF: (a) 3D view; (b) 2D view from the top.

Figure 11.10 shows the results of applying ideal low-pass filters with different cutoff frequencies to the original image: lower values of D_0 correspond to blurrier results. A close inspection of Figure 11.10 shows that the filtered images not only are blurry versions of the input image—an expected outcome, common to all low-pass filters—but also exhibits noticeable *ringing* artifacts that appear because of the sharp transition between passband and stopband in ideal (low-pass) filters.[5]

11.3.2 Gaussian LPF

A Gaussian low-pass filter attenuates high frequencies using a transfer function whose shape is based on a Gaussian curve. The width of the bell-shaped curve can be controlled by specifying the parameter σ, which is functionally equivalent to the cutoff frequency D_0 defined previously: lower values of σ correspond to more strict filtering, resulting in blurrier results. The smooth transition between passband and stopband of the Gaussian LPF guarantees that there will be no noticeable ringing artifacts in the output image. The Gaussian LPF can be mathematically described as

$$H_G(u, v) = e^{-(D(u,v)^2)/2\sigma^2} \tag{11.21}$$

Figure 11.11 shows the frequency response plot for a Gaussian LPF.

Figure 11.12 shows the results of applying Gaussian low-pass filters with different values of σ.

11.3.3 Butterworth LPF

The Butterworth family of filters—widely used in 1D analog signal processing—provides an alternative filtering strategy whose behavior is a function of the cutoff

[5]The fact that the "ideal" LPF produces less than perfect results might seem ironic to the reader; in the context of this discussion, "ideal" refers to abrupt transition between passband and stopband.

FIGURE 11.12 (a) Original image (256×256 pixels); (b–f) Gaussian LPF results for filters with different values for σ: 5, 10, 20, 30, and 75.

frequency, D_0, and the order of the filter, n. The shape of the filter's frequency response, particularly the steepness of the transition between passband and stopband, is also controlled by the value of n: higher values of n correspond to steeper transitions, approaching the ideal filter behavior.

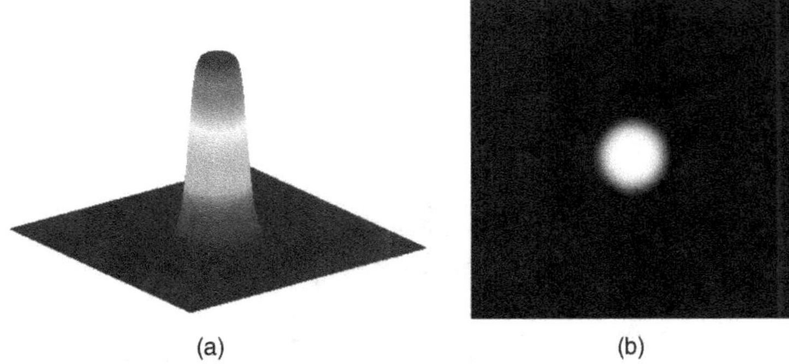

FIGURE 11.13 Frequency response plot for a Butterworth LPF of order $n = 4$: (a) 3D view; (b) 2D view from the top.

The Butterworth filter can be mathematically described as

$$H_B(u, v) = \frac{1}{1 + [D(u, v)/D_0]^{2n}} \tag{11.22}$$

where D_0 is the *cutoff frequency* and n is the *order* of the filter. Figure 11.13 shows the frequency response plot for a Butterworth LPF of order $n = 4$.

Figure 11.14 shows the results of applying Butterworth low-pass filters with $n = 4$ and different cutoff frequencies to the original image. Comparing these results with the ones observed for the ideal LPF (Figure 11.10) for the same cutoff frequency, we can see that they show a comparable amount of blurring, but no ringing effect.

11.4 HIGH-PASS FILTERING

High-pass filters attenuate the low-frequency components of the Fourier transform of an image, while enhancing the high-frequency components (or leaving them unchanged). The typical overall effect of applying a high-pass filter to an image is a controlled degree of sharpening.

11.4.1 Ideal HPF

An ideal high-pass filter attenuates all frequency components within a specified radius (from the center of the FT), while enhancing all others. Its mathematical formulation is given as follows:

$$H_I(u, v) = \begin{cases} 0 & \text{if } D(u, v) \leq D_0 \\ 1 & \text{if } D(u, v) > D_0 \end{cases} \tag{11.23}$$

FIGURE 11.14 (a) Original image (512 × 512 pixels); (b–f) fourth-order Butterworth LPF results for filters with cutoff frequency corresponding to the radii in Figure 11.9b, namely, 8, 16, 32, 64, and 128 pixels.

where $D(u, v) = \sqrt{(u^2 + v^2)}$ represents the distance between a point of coordinates (u, v) and the origin of the 2D frequency plan, and D_0 is a nonnegative value, referred to as the *cutoff frequency* (or *cutoff radius*).

Figure 11.15 shows the frequency response plot for an ideal HPF.

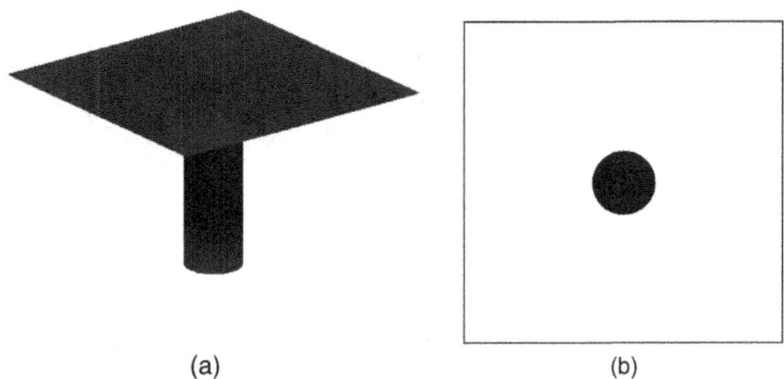

FIGURE 11.15 Frequency response plot for an ideal HPF: (a) 3D view; (b) 2D view from the top.

11.4.2 Gaussian HPF

A Gaussian high-pass filter attenuates low frequencies using a transfer function whose shape is based on a Gaussian curve. The behavior of the filter can be controlled by specifying the parameter σ, which is functionally equivalent to the cutoff frequency D_0. The Gaussian HPF can be mathematically described as

$$H_G(u, v) = 1 - e^{-(D(u,v)^2)/2\sigma^2} \tag{11.24}$$

Figure 11.16 shows the frequency response plot for a Gaussian HPF.

11.4.3 Butterworth HPF

The Butterworth high-pass filter can be mathematically described as

$$H_B(u, v) = \frac{1}{1 + [D_0/D(u, v)]^{2n}} \tag{11.25}$$

where D_0 is the *cutoff frequency* and n is the *order* of the filter.

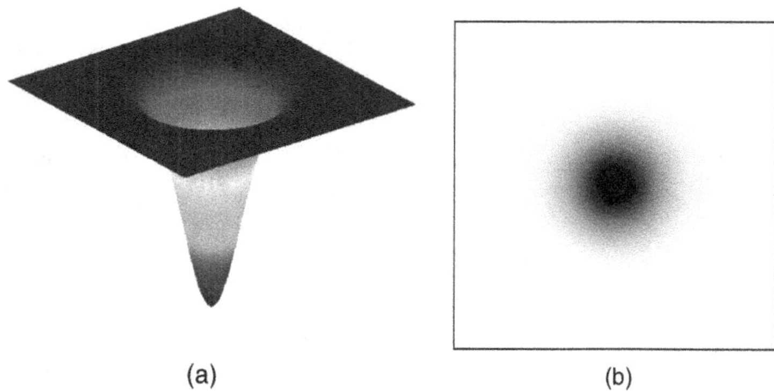

FIGURE 11.16 Frequency response plot for a Gaussian HPF: (a) 3D view; (b) 2D view from the top.

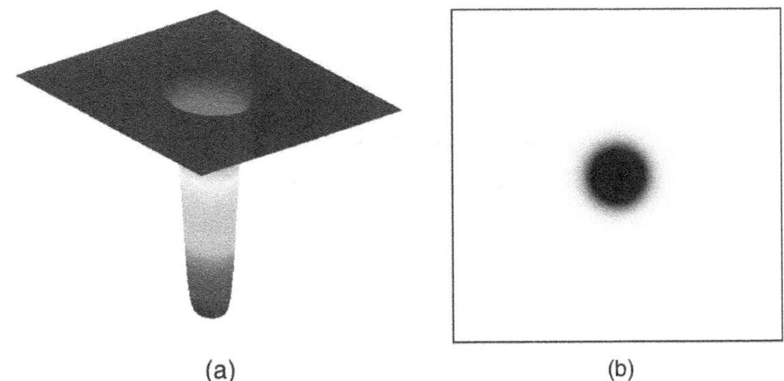

FIGURE 11.17 Frequency response plot for a Butterworth HPF of order $n = 4$: (a) 3D view; (b) 2D view from the top.

Figure 11.17 shows the frequency response plot for a Butterworth HPF of order $n = 4$.

11.4.4 High-Frequency Emphasis

The application of a HPF to an image usually causes a crispening of the high-frequency contents of the image at the expense of the low-frequency contents, which are severely attenuated, leading to a loss of information present in large patches of the original image (its low-frequency components). High-frequency emphasis is a technique that preserves the low-frequency contents of the input image (while enhancing its high-frequency components) by multiplying the high-pass filter function by a constant and adding an offset to the result, that is,

$$H_{\text{hfe}}(u, v) = a + bH(u, v) \tag{11.26}$$

where $H(u, v)$ is the HPF transfer function, $a \geq 0$ and $b > a$. Figure 11.18 shows an example of high-frequency emphasis.

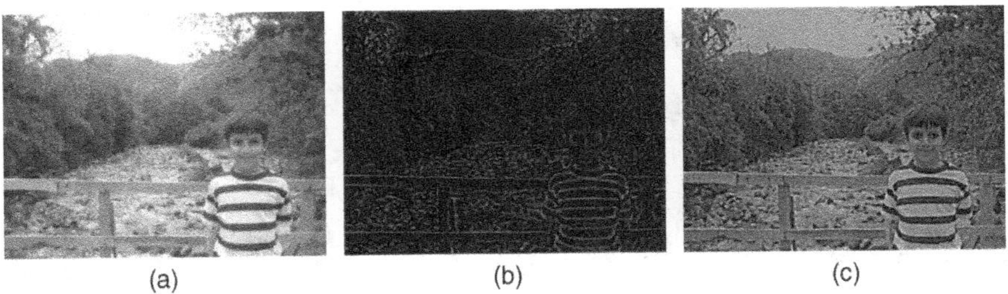

FIGURE 11.18 High-frequency emphasis: (a) input image; (b) result of applying a second-order Butterworth HPF (with $D_0 = 30$) to the input image; (c) result of high-frequency emphasis with $a = 0.5$ and $b = 1$.

11.5 TUTORIAL 11.1: 2D FOURIER TRANSFORM

Goals

The goals of this tutorial are to learn how to compute and display the FT of an image and how to develop filters to be used in the frequency domain.

Objectives

- Learn how to use the `fft2` function to compute the FT of a monochrome image.
- Learn how to visualize the FT results.
- Learn how to generate filters to be used in the frequency domain.

What You Will Need

- `distmatrix.m`

Procedure

To generate the FT of an image (a 2D function), we use the IPT function `fft2`, which implements the FFT algorithm.

1. Load the cameraman image, convert it to `double` (one of the data classes accepted as an input to `fft2`), and generate its FT.

```
I = imread('cameraman.tif');
Id = im2double(I);
ft = fft2(Id);
```

Question 1 What are the minimum and maximum values of the resulting discrete Fourier transform coefficients for the `cameraman` image?

To view the spectrum of the image, that is, the amplitude component of the FT results, we must shift the zero-frequency (DC) component to the center of the image using the `fftshift` function.

2. Shift the FT array of results.

```
ft_shift = fftshift(ft);
```

From your answer to Question 1, you should by now know that the range of values in the FT array of results (`ft`) extends well beyond the typical values of a grayscale image ([0, 255]). Consequently, we will try to display the resulting spectrum as an image using the "scaling for display purposes" option of function `imshow`.

3. Display the FT results, remapped to a grayscale range.

```
figure, subplot(1,2,1), imshow(abs(ft_shift),[]), ...
    title('Direct remap');
```

Question 2 Why are we required to use the `abs` function when displaying the `ft_shift` image?

Question 3 How did we remap the image to a different (narrower) grayscale range?

As you may have noticed, directly remapping the image to the grayscale range does not give us any useful information (only a white pixel at the center of the image—the DC component of the FT results—with all other pixels displayed as black). This suggests that there might be a significant amount of detail in other frequencies that we just cannot see. Recall from Chapter 8 that images with similar attributes (very large dynamic range of gray values) can be brought up by remapping the image with a log transformation.

We can perform the log transformation within the `imshow` function call and then remap the adjusted values to the grayscale range by specifying '[]' in the second parameter.

4. Display the log of the shifted FT image.

```
subplot(1,2,2), imshow(log(1 + abs(ft_shift)), []), ...
    title('Log remap');
```

Question 4 How does the log remap compare to the direct remap?

Distance Matrices

In the second part of this tutorial, we will look at distance matrices. To specify and implement frequency-domain filters, we must first generate a matrix that represents the distance of each pixel from the center of the image. We can create such matrix using the `distmatrix` function. The function takes two parameters, `M` and `N`, and will return the distance matrix of size `M` × `N`.

5. Close any open figures.
6. Generate a distance matrix that is the same size as the image `I`.

```
[M, N] = size(I);
D = distmatrix(M, N);
```

We can visualize the distance matrix in a 3D mesh plot, but we must first shift the image similar to the way we shifted the frequency spectrum earlier.

7. Create a 3D mesh plot of the distance matrix.

```
D_shift = fftshift(D);
figure, mesh(D_shift)
```

Question 5 Explain the shape of the 3D plot.

After obtaining the distance matrix, we can generate the filter of our choice. From that point, filtering in the frequency domain is as simple as multiplying the filter by the FT image and converting back to the spatial domain. These processes will be examined in the remaining tutorials of this chapter.

11.6 TUTORIAL 11.2: LOW-PASS FILTERS IN THE FREQUENCY DOMAIN

Goal

The goal of this tutorial is to demonstrate how to implement low-pass filters in the frequency domain.

Objectives

- Learn how to generate and apply an ideal low-pass filter.
- Learn how to generate and apply a Gaussian low-pass filter.
- Learn how to generate and apply a Butterworth low-pass filter.

What You Will Need

- distmatrix.m
- fddemo.m

Procedure

As we have learned in Tutorial 11.1, the `distmatrix` function returns a two dimensional array, which should be of the same size as the image being processed. The values in the array represent the distance from each pixel to the center of the image. To begin, we will use this matrix to generate an ideal low-pass filter.

1. Load the `eight` image, generate a FT, and display it.

```
I = imread('eight.tif');
Id = im2double(I);
I_dft = fft2(Id);
figure, imshow(Id), title('Original Image');
figure, imshow(log(1 + abs(fftshift(I_dft))),[]), ...
    title('FT of original image');
```

2. Generate a distance matrix with size equal to the input image.

```
[M, N] = size(I);
dist = distmatrix(M, N);
figure, mesh(fftshift(dist)), title('Distance Matrix');
```

Question 1 Verify that the size of the distance matrix is in fact equal to the size of the image.

Question 2 What happens if we display the distance matrix without shifting?

Ideal LPF

To create an ideal low-pass filter, we will start out with a matrix of all zeros and then set specific values to 1 that will represent all frequencies for which we will allow to pass through. Since we are defining an ideal filter, we can simply define the radius of the filter and then set any values within that radius to 1, while all others remain zero.

3. Create initial filter with all values of zero.

```
H = zeros(M, N);
```

4. Create the ideal filter.

```
radius = 35;
ind = dist <= radius;
H(ind) = 1;
Hd = double(H);
```

Question 3 Explain *how* the previous code sets all values within a given radius to 1.

We can visualize the filter's frequency response by displaying it as an image.

5. Display the filter's frequency response.

```
figure, imshow(fftshift(H)), title('Ideal low-pass filter');
```

To apply the filter to the image, we simply multiply each value of the filter by its corresponding frequency value in the FT image.

6. Apply the filter to the FT image.

```
DFT_filt = Hd .* I_dft;
I2 = real(ifft2(DFT_filt));
```

Question 4 Why do we take only the real values when converting the FT of the filtered image back to the spatial domain?

7. Display both the filtered FT image and the final filtered image.

```
figure, imshow(log(1 + abs(fftshift(DFT_filt))),[]), ...
    title('Filtered FT');
figure, imshow(I2), title('Filtered Image');
```

Question 5 How does the filtered image compare to the original image? Can you see any noticeable artifacts?

To see how the choice of radius affects the filtered image, we can use the *frequency-domain demo*, `fddemo` (developed by Jeremy Jacob and available at the book's companion web site).

8. Load the frequency-domain demo.

```
fddemo
```

The default filter is the ideal low pass. To modify the cutoff value, select the magenta circle within the filter profile and drag it to a desired radius. The value of the radius is displayed below the filter profile.

Question 6 Experiment with different values for the radius of the filter. How does your choice of radius affect the amount of ringing in the output image?

9. Close the demo.

Gaussian LPF

The Gaussian low-pass filter is usually specified by providing a value for the standard deviation σ. We can use the distance matrix previously generated to create a Gaussian filter.

10. Create a Gaussian low-pass filter with $\sigma = 30$.

```
sigma = 30;
H_gau = exp(-(dist .^ 2) / (2 * (sigma ^ 2)));
figure, imshow(Id), title('Original Image');
figure, imshow(log(1 + abs(fftshift(I_dft))),[]), ...
    title('DFT of original image');
figure, mesh(fftshift(dist)), title('Distance Matrix');
figure, imshow(fftshift(H_gau)), title('Gaussian low-pass');
```

11. Filter the FT image with the Gaussian low-pass filter and display the filtered image.

```
DFT_filt_gau = H_gau .* I_dft;
I3 = real(ifft2(DFT_filt_gau));
figure, imshow(log(1 + abs(fftshift(DFT_filt_gau))),[]), ...
    title('Filtered FT');
figure, imshow(I3), title('Filtered Image');
```

Question 7 Compare the output images between the ideal filter and the Gaussian filter. What are their similarities? What are their differences?

Question 8 Start the frequency-domain demo (`fddemo`) once again, and this time select the Gaussian low-pass filter. Experiment with different values of sigma (standard deviation). How does this value affect the output image?

Butterworth LPF

The Butterworth low-pass filter is usually specified by providing two parameters: the order of the filter, n, and the cutoff value, D_0. In our implementation of the ideal low-pass filter, earlier in this tutorial, we set the cutoff value to 35. For comparison purposes, we will use the same value for our Butterworth filter.

12. Generate a third-order Butterworth filter, where the cutoff value is 35.

```
D0 = 35; n = 3;
H_but = 1 ./ (1 + (dist ./ D0) .^ (2 * n));
figure, imshow(Id), title('Original Image');
figure, imshow(log(1 + abs(fftshift(I_dft))),[]), ...
    title('FT of original image');
figure, mesh(fftshift(dist)), title('Distance Matrix');
figure, imshow(fftshift(H_but)), title('Butterworth low-pass');
```

13. Filter the image with the Butterworth low-pass filter and display the resulting image.

```
DFT_filt_but = H_but .* I_dft;
I4 = real(ifft2(DFT_filt_but));
figure, imshow(log(1 + abs(fftshift(DFT_filt_but))),[]), ...
    title('Filtered FT');
figure, imshow(I4), title('Filtered Image');
```

Question 9 Compare the ideal-filtered FT image and the Butterworth-filtered FT image.

14. Display all three filters as meshes in 3D and use the *Rotate 3D* option of function `imshow` to explore them in detail.

```
figure, mesh(fftshift(Hd)), title('Ideal low-pass filter');
figure, mesh(fftshift(H_gau)), title('Gaussian low-pass filter');
figure, mesh(fftshift(H_but)), title('Butterworth low-pass filter');
```

Question 10 Implement the Butterworth filter again, but this time using a much higher order, such as 20. How does this output compare to the ideal filter?

Question 11 Experiment with the Butterworth filter by using the frequency-domain demo (`fddemo`). What is the advantage of using this filter over the previous two?

11.7 TUTORIAL 11.3: HIGH-PASS FILTERS IN THE FREQUENCY DOMAIN

Goal

The goal of this tutorial is how to implement high-pass filters in the frequency domain.

Objectives

- Learn how to generate and apply an ideal high-pass filter.
- Learn how to generate and apply a Gaussian high-pass filter.
- Learn how to generate and apply a Butterworth high-pass filter.

What You Will Need

- distmatrix.m
- fddemo.m

Procedure

High-pass filters are conceptually the opposite of low-pass filters and can be implemented in MATLAB using techniques that are very similar to the ones described in Tutorial 11.2. There is, however, a problem when implementing high-pass filters: because a high-pass filter attenuates low frequencies, this means that the zero-frequency term (also known as the DC term) will be set to zero, in turn setting the average value of the image to zero. To compensate for this, we use a technique known as *high-frequency emphasis* filtering, which can be implemented by applying the high-pass filter as we normally would, then multiplying the result (still in the frequency domain) by a constant b, and finally adding an offset constant a.

To begin, let us implement an ideal high-pass filter.

1. Load the `eight` image, generate, and display its FT.

```
I = im2double(imread('eight.tif'));
I_dft = fft2(I);
figure, imshow(I), title('Original Image');
figure, imshow(log(1 + abs(fftshift(I_dft))),[]), ...
    title('FT of original image');
```

Just as in low-pass filtering, we must generate a distance matrix and use the `fftshift` function when displaying it.

2. Generate a distance matrix based on the size of the input image.

```
[M, N] = size(I);
dist = distmatrix(M, N);
```

Ideal HPF

3. Create the ideal high-pass filter.

```
H = ones(M, N);
radius = 30;
ind = dist <= radius;
H(ind) = 0;
```

Question 1 Explain how the previous code generates an ideal high-pass filter.

We will now apply the high-frequency emphasis filtering technique with `a` = `b` = 1.

4. Apply high-frequency emphasis filtering to the high-pass filter.

```
a = 1;   b = 1;
Hd = double(a + (b .* H));
```

5. Apply the filter to the FT image and display the results.

```
DFT_filt = Hd .* I_dft;
I2 = real(ifft2(DFT_filt));
figure, imshow(log(1 + abs(fftshift(DFT_filt))),[]), ...
    title('Filtered FT');
figure, imshow(I2), title('Filtered Image');
```

6. Display the filter as an image and as a 3D mesh in separate figures.

```
figure, imshow(fftshift(Hd),[]), title('Filter as an image');
figure, mesh(fftshift(Hd)), zlim([0 2]), title('Filter as a mesh');
```

Question 2 When displaying the filter as an image, why must we scale the output for display purposes?

Question 3 How does the filtered image compare to the original image?

We can use the frequency-domain demo (`fddemo`) to experiment with different values for the radius of the filter. Note that to use `fddemo`, the M-file `distmatrix.m` must be in the same directory.

7. Start the frequency-domain demo.

```
fddemo
```

8. From the filter pull-down menu, select *Ideal High Pass*.

Notice that the filter profile shows a magenta circle, indicating the current cutoff value for the filter. The cutoff value is displayed below the profile figure. To change the cutoff value, drag the magenta circle in or out. You will notice that as you drag the circle, the displayed cutoff value changes.

9. Drag the cutoff circle so that the cutoff value is 30 to create the same filtered image as we did above.
10. Change the cutoff value to 10.

Question 4 How does the new image compare to the original image?

Question 5 How does the output compare between using a cutoff value of 30 and 10?

Question 6 What happens to the filtered image as we increase the cutoff value (beyond 30) with respect to the original image?

11. Close any open figures or demos.

Gaussian HPF

We can implement the Gaussian high-pass filter using the existing distance matrix (stored in variable `dist`).

12. Generate a Gaussian high-pass filter.

```
sigma = 30;
H_gau = 1 - exp(-(dist .^ 2) / (2 * (sigma ^ 2)));
```

13. Apply high-frequency emphasis filtering to the high-pass filter and display the filter.

```
H_gau_hfe = a + (b .* H_gau);
figure, mesh(fftshift(H_gau_hfe)), zlim([0 2]), ...
    title('Gaussian high-pass filter');
```

We can now apply the filter to the image.

14. Apply the filter and display the results.

```
DFT_filt_gau = H_gau_hfe .* I_dft;
I3 = real(ifft2(DFT_filt_gau));
figure, imshow(I), title('Original Image');
figure, imshow(log(1 + abs(fftshift(I_dft))),[]), ...
    title('FT of original image');
figure, imshow(log(1 + abs(fftshift(DFT_filt_gau))),[]), ...
    title('Filtered FT');
figure, imshow(I3), title('Filtered Image');
```

Question 7 How does the Gaussian-filtered image compare to the ideal-filtered image?

We can see the effects of the value chosen for σ (`sigma`) by using the frequency-domain demo.

15. Start the frequency-domain demo.

```
fddemo
```

16. From the filter pull-down menu, select *Gaussian High Pass*.

Notice that the default value for the standard deviation (σ) is 30, so the filtered image should be equal to what we have implemented above. Let us now change this value to see its effect on both the filter and the resulting image.

17. Change the standard deviation to 10.

Question 8 What happened to the filter?

Question 9 How was the filtered image affected?

Question 10 In general, how does the filter change when the standard deviation of the filter is increased or decreased?

18. Close any open figures or demos.

Butterworth HPF

To implement the Butterworth high-pass filter, we can again use the existing distance matrix (stored in variable `dist`).

19. Generate a Butterworth high-pass filter.

```
cutoff = 30; order = 2;
H_but = 1 ./ (1 + (cutoff ./ dist) .^ (2 * order));
```

20. Apply high-frequency emphasis filtering to the high-pass filter and display the resulting filter.

```
H_but_hfe = a + (b .* H_but);
figure, mesh(fftshift(H_but_hfe)), zlim([0 2]), ...
    title('Butterworth high-pass filter');
```

21. Apply the filter to the input image and display the results.

```
DFT_filt_but = H_but_hfe .* I_dft;
I4 = real(ifft2(DFT_filt_but));
figure, imshow(I), title('Original Image');
figure, imshow(log(1 + abs(fftshift(I_dft))),[]), ...
    title('FT of original image');
figure, imshow(log(1 + abs(fftshift(DFT_filt_but))),[]), ...
    title('Filtered FT');
figure, imshow(I4), title('Filtered Image');
```

In the last portion of this tutorial, we will further explore Butterworth high-pass filters using the `fddemo`.

22. Load `fddemo`.

```
fddemo
```

23. From the filter pull-down menu, select *Butterworth High Pass*.

This filter has two parameters: *cutoff value* and *order*. The cutoff value can be adjusted simply by dragging the magenta circle. To change the order of the filter, type a new value and press *Update*.

Question 11 How does the order parameter change the shape of the filter?

Question 12 For large order values (such as 10), the Butterworth begins to take the shape of what other high-pass filters take?

WHAT HAVE WE LEARNED?

- The *Fourier transform* is the main mathematical tool used to obtain the 2D frequency contents of a digital image. The resulting image is said to be in *Fourier space* or *Fourier domain*. The 2D Fourier transform is implemented in MATLAB by function `fft2` and its results are displayed with function `fftshow`.
- The design of an image processing filter in the frequency domain involves the following steps: (1) determining the desired behavior (low pass, high pass, etc.), (2) choosing the type of filter (ideal, Gaussian, Butterworth, etc.), and (3) specifying the parameters associated with the chosen filter type (e.g, cutoff frequency for ideal filters, filter order for Butterworth filters, etc.).
- Low-pass filters attenuate the high-frequency components of the Fourier transform of an image, while leaving the low-frequency components unchanged. The typical overall effect of applying a low-pass filter to an image is a controlled degree of blurring.
- High-pass filters attenuate the low-frequency components of the Fourier transform of an image, while enhancing the high-frequency components (or leaving them unchanged). The typical overall effect of applying a high-pass filter to an image is a controlled degree of sharpening.
- Ideal (low-pass and high-pass) filters have a sharp transition from passband to stopband, which can lead to undesirable image artifacts, most noticeably ringing. Gaussian filters, on the other hand, show a smooth transition from passband to stopband. Butterworth filters allow the designer to choose their order—which impacts the shape of the transition between passband and stopband—and control how close to an ideal filter they should behave.

LEARN MORE ABOUT IT

- There are entire books devoted to the Fourier transform (e.g., [Pap62]) and its fast implementation (e.g., [Bri74]).
- Recommended references for one- and multidimensional signal analysis and Fourier transform are Chapters 13 and 14 of [BB08], Chapters 2–6 of [vdEV89], Chapters 4 and 5 of [OWY83], and Chapters 1 and 3 of [Lim90], among many others.
- Section 5.1 of [Umb05] provides useful examples of the relationships between mathematical transforms, transform coefficients, and basis images.
- Section 9.3 of [Pra07] discusses computational aspects of 2D FT calculations using the FFT implementation.
- Chapter 2.3 of [Bov00a] provides a friendly and informative explanation of the Fourier transform and image processing in the frequency domain.
- [Pra07] is a good source of additional information for mathematical transforms not covered in this book, for example, the cosine, sine, and Hartley transforms

(Section 8.3), the Hadamard, Haar, and Daubechies (a class of wavelets) transforms (Section 8.4), and the Kahunen–Loeve transform (KLT) (Section 8.5).

- Chapter 5 of [Umb05] also covers mathematical transforms not included in this chapter, such as cosine, Walsh–Hadamard, wavelet, Haar, and the principal components transform (PCT).
- A homomorphic filter is a special type of frequency-domain filter that can be particularly useful in situations where the input image shows significant variations in illumination. Homomorphic filtering is discussed in Section 4.9.6 of [GW08] and Section 7.9 of [McA04], among many other references.
- Other types of filters with image processing applications include bandpass, bandreject, and notch filters. These selective filters will be discussed in Section 12.4 in the context of noise reduction.
- The discussion on the computational cost of spatial-domain filtering versus frequency-domain filtering (and other related implementation implications) is beyond the scope of this book. The interested reader may consult Chapter 7 of [SOS00].

11.8 PROBLEMS

11.1 Write a MATLAB script to generate a 256 $\times$ 256 test image consisting of a white circle against a black background, use this image as an input to the `fft2` function, and display the resulting spectrum. You should be able to notice the presence of "ringing" artifacts in the resulting spectrum, owing to the sharp transitions between the circle and the background.

11.2 Implement the solution to the ringing artifact in Problem 11.1, while keeping the input image similar to the original (white circle against a black background) as much as possible.

11.3 In this problem,

(a) Design a spatial-domain averaging filter whose output is the average of the four neighbors of the center pixel in a 3 $\times$ 3 neighborhood.

(b) Use the MATLAB function `freqz2` to obtain its frequency-domain equivalent and plot the resulting filter function.

(c) Apply the two filters, one at a time, to an input image of your choice and observe the results. Are there any noticeable differences? Explain.

CHAPTER 12

IMAGE RESTORATION

WHAT WILL WE LEARN?

- What is noise (in the context of image processing) and how can it be modeled?
- What are the main types of noise that may affect an image?
- What is blurring (in the context of image processing) and how can it be modeled?
- Which noise removal techniques are typically used in image processing?
- Which deblurring techniques are typically used in image processing?

12.1 MODELING OF THE IMAGE DEGRADATION AND RESTORATION PROBLEM

This chapter presents techniques used to improve the appearance of an image that has been subject to degradation and noise. Figure 12.1 shows a diagram of the degradation and restoration processes. In this diagram, it is assumed that an original image $f(x, y)$ has been subject to some sort of quality degradation (e.g., blurring caused by lack of focus or camera motion, atmospheric disturbances, or geometric distortions caused by imperfect lenses) that can be modeled by a function $h(x, y)$. The image may (also) have been contaminated by additive noise ($n(x, y)$). The resulting degraded image, $g(x, y)$, is the input for the image restoration algorithms described later. These algorithms are usually implemented as restoration filters that are able to undo—to some extent—the

Practical Image and Video Processing Using MATLAB®. By Oge Marques.

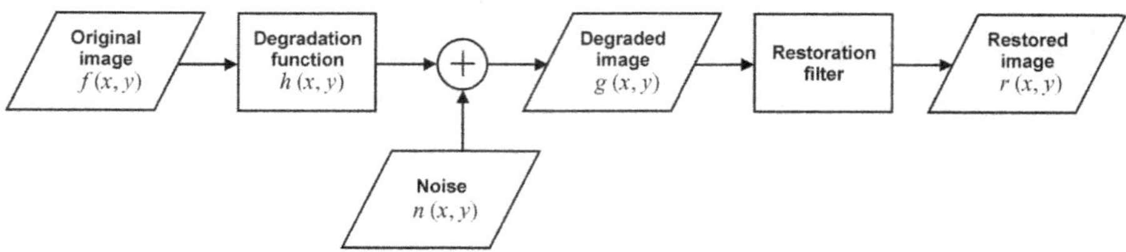

FIGURE 12.1 Image degradation and restoration.

degradation process, resulting in a restored image $r(x, y)$ (which should be interpreted as an estimate of the original image $f(x, y)$ and is often referred to as $\hat{f}(x, y)$). In other words, the goal of restoration techniques is to obtain an image that is as close to the original image as possible.

Mathematically, the degradation and restoration problem can be described as

$$g(x, y) = f(x, y) * h(x, y) + n(x, y) \tag{12.1}$$

where $*$ denotes convolution.

From the convolution theorem, the following frequency-domain relation among the Fourier transform of $f(x, y)$, $g(x, y)$, $h(x, y)$, and $n(x, y)$ holds:

$$G(u, v) = F(u, v)H(u, v) + N(u, v) \tag{12.2}$$

The *restoration filter* block in Figure 12.1 is typically designed following these steps:

1. Collect knowledge about the degradation process (usually through examples of degraded images and knowledge of the image acquisition process).
2. Use that knowledge to develop a degradation model.
3. Develop the *inverse* degradation process and model it as a filter.

Restoration filters are, therefore, specifically designed to solve one particular type of degradation. They can work in the spatial domain or in the frequency domain.

Note that the goal of image *restoration* techniques is clearly distinct from the goal of image *enhancement* techniques—such as the ones described in Chapters 8–11—where no mathematical modeling of (the inverse of) a degradation process is needed. This chapter presents representative examples of restoration filters, particularly for noise reduction and deblurring, and their application to digital images.

12.2 NOISE AND NOISE MODELS

Noise can be defined as any undesired artifact that contaminates an image. The presence of noise in an image can be due to several sources, resulting in different types of noise, from thermal noise in acquisition devices to periodic noise in the communication channel used to transmit an image from a remote sensing location to a base station, among many others.

In this section, we present an overview of the main types of noise that may be found in degraded digital images. Our main goals are threefold: (1) to describe the main noise models in a mathematical and graphical way, (2) to remind the reader that different types of noise will require different noise reduction techniques, and (3) to introduce the problem of noise estimation.

We treat noise as a random variable whose probability density function (PDF), or histogram, describes its shape and distribution across the range of gray levels. In addition to that spatial-domain representation, noise patterns can sometimes also be represented in the frequency domain (through their frequency spectrum).

12.2.1 Selected Noise Probability Density Functions

Gaussian Noise The PDF of a Gaussian random variable z is given by

$$p_{\mathrm{g}}(z) = \frac{1}{\sqrt{2\pi}\sigma} e^{-(z-\bar{z})^2/2\sigma^2} \tag{12.3}$$

where z represents the gray level, $\bar{z}$ is its mean, and σ is its standard deviation (σ^2 is called the *variance* of z). A plot of this function is shown in Figure 12.2a.

Impulse (Salt and Pepper) Noise The PDF of (bipolar) impulse noise is given by

$$p_{\mathrm{sp}}(z) = \begin{cases} P_p & \text{for } z = p \\ P_s & \text{for } z = s \\ 0 & \text{otherwise} \end{cases} \tag{12.4}$$

where P_p and P_s are the probability of occurrence of pixel whose values are equal to p (pepper) or s (salt), respectively. A plot of this function is shown in Figure 12.2b.

Uniform Noise The histogram of uniform noise is given by

$$p_{\mathrm{u}}(z) = \begin{cases} \frac{1}{b-a} & \text{if } a \leq z \leq b \\ 0 & \text{otherwise} \end{cases} \tag{12.5}$$

where $a \geq 0$ and $0 < a < b$.

The mean of the uniform noise is given by

$$\bar{z} = \frac{a+b}{2} \tag{12.6}$$

whereas the variance is given by

$$\sigma^2 = \frac{(b-a)^2}{12} \tag{12.7}$$

A plot of this function is shown in Figure 12.2c.

Rayleigh Noise The PDF of Rayleigh noise is given by

$$p_{\mathrm{r}}(z) = \begin{cases} \frac{2}{b}(z-a)e^{-(z-a)^2/b} & \text{for } z \geq a \\ 0 & \text{for } z < a \end{cases} \tag{12.8}$$

where $a \geq 0$ and $0 < a < b$.

The mean and the variance of the Rayleigh PDF are given by

$$\bar{z} = a + \sqrt{\pi b/4} \tag{12.9}$$

and

$$\sigma^2 = \frac{b(4-\pi)}{4} \tag{12.10}$$

A plot of this function is shown in Figure 12.2d.

Gamma (Erlang) Noise The histogram of gamma (Erlang) noise is given by

$$p_{\mathrm{E}}(z) = \begin{cases} \frac{a^b z^{b-1}}{(b-1)!}e^{-az} & \text{for } z \geq 0 \\ 0 & \text{for } z < 0 \end{cases} \tag{12.11}$$

where $a > 0$, b is a positive integer, and "!" indicates factorial.

The mean and the variance of the gamma PDF are given by

$$\bar{z} = \frac{b}{a} \tag{12.12}$$

and

$$\sigma^2 = \frac{b}{a^2} \tag{12.13}$$

A plot of this function is shown in Figure 12.2e.

Exponential Noise The PDF of exponential noise (a special case of the Erlang PDF, with $b = 1$) is given by

$$p_{\exp}(z) = \begin{cases} ae^{-az} & \text{for } z \geq 0 \\ 0 & \text{for } z < 0 \end{cases} \tag{12.14}$$

The mean and the variance of the exponential PDF are given by

$$\bar{z} = \frac{1}{a} \tag{12.15}$$

and

$$\sigma^2 = \frac{1}{a^2} \tag{12.16}$$

A plot of this function is shown in Figure 12.2f.

■ **EXAMPLE 12.1**

Figures 12.3 and 12.4 show a test image contaminated by different types of noise and the resulting histograms. Since the test image contains only two gray levels, it is possible to map the histograms to the plots in Figure 12.2.[1]

12.2.2 Noise Estimation

It is often necessary to estimate the type of noise (and the main parameters of its PDF) before designing a solution for noise reduction. In cases where the acquisition device (sensor) is the primary source of noise, it is common to generate test images consisting of large patches of homogeneous gray-level values and observe the resulting histograms. If creating such test images is not possible, a widely used alternative is to crop a relatively large homogeneous region from an image and inspect its histogram: even though the histogram of the entire image may not provide an adequate hint as to the type of noise present in the image, the histogram of the cropped portion will.

■ **EXAMPLE 12.2**

Figure 12.5 shows the process of noise estimation. The addition of noise to the original image (a) causes dramatic changes to its histogram (from (c) to (d)), without, however, providing a reliable hint as to the type of noise present in the image (b). Inspecting the histogram (e) of the cropped portion of the image (indicated by a rectangle in (b)) allows us to conclude that the noise is of Gaussian type.

In MATLAB

MATLAB's IPT has a built-in function to add noise to an image: `imnoise`. It allows the generation of noisy versions of input images. It supports several types of noise, including Gaussian and salt and pepper additive noise and speckle multiplicative noise. It does not, however, support other types of noise described in this section (e.g., Rayleigh, uniform, Erlang, and exponential) (see Problem 12.1). You will learn how to use this function in Tutorial 12.1.

12.3 NOISE REDUCTION USING SPATIAL-DOMAIN TECHNIQUES

Noise reduction filters work under the assumption that the only degradation present in an image is additive noise.[2] Mathematically, equations (12.1) and (12.2) become

$$g(x, y) = f(x, y) + n(x, y) \tag{12.17}$$

and

$$G(u, v) = F(u, v) + N(u, v) \tag{12.18}$$

[1]Some of the histograms show a peak at the rightmost value because the corresponding noise PDF has a "long tail" and, consequently, many values get truncated at 1.

[2]From the perspective of Figure 12.1, you can think of the degradation function as an *all-pass* filter.

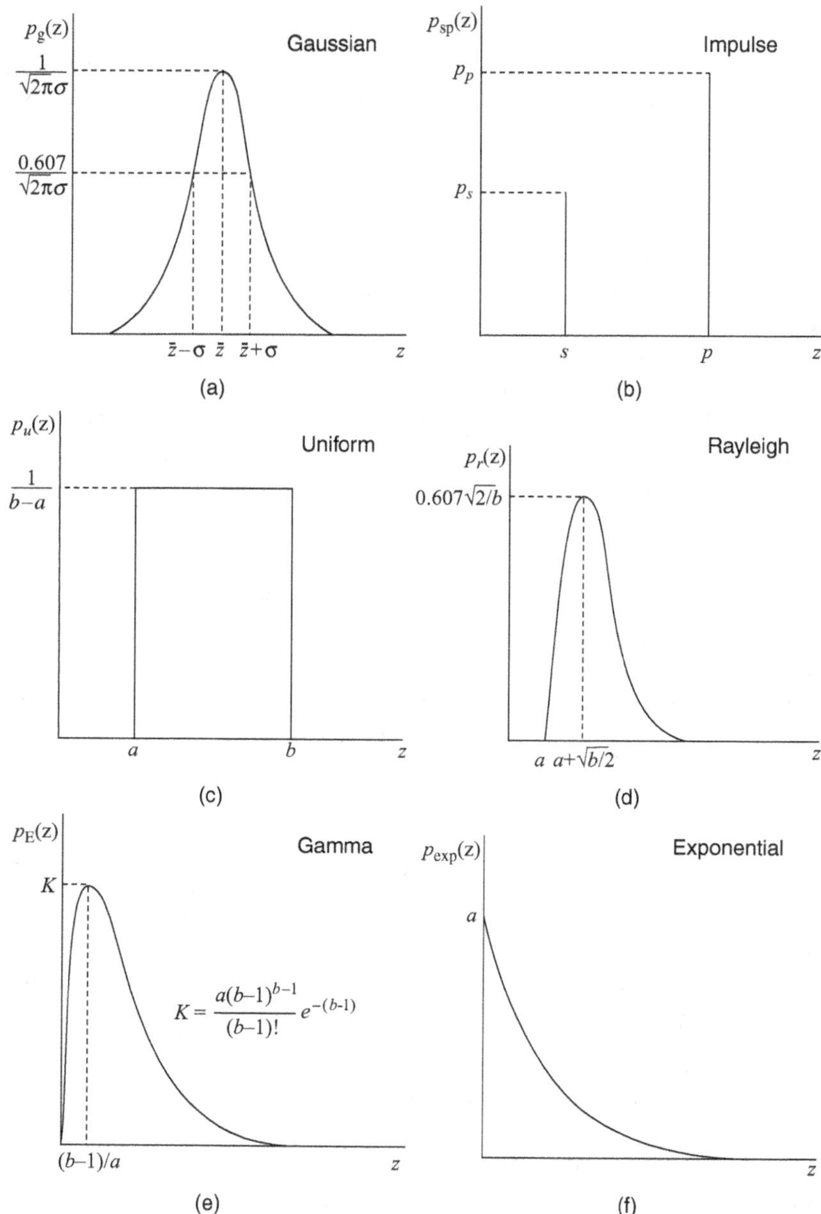

FIGURE 12.2 Histograms of representative noise types: (a) Gaussian, (b) impulse (salt and pepper), (c) uniform, (d) Rayleigh, (e) gamma (Erlang), and (e) exponential. Redrawn from [Pra07].

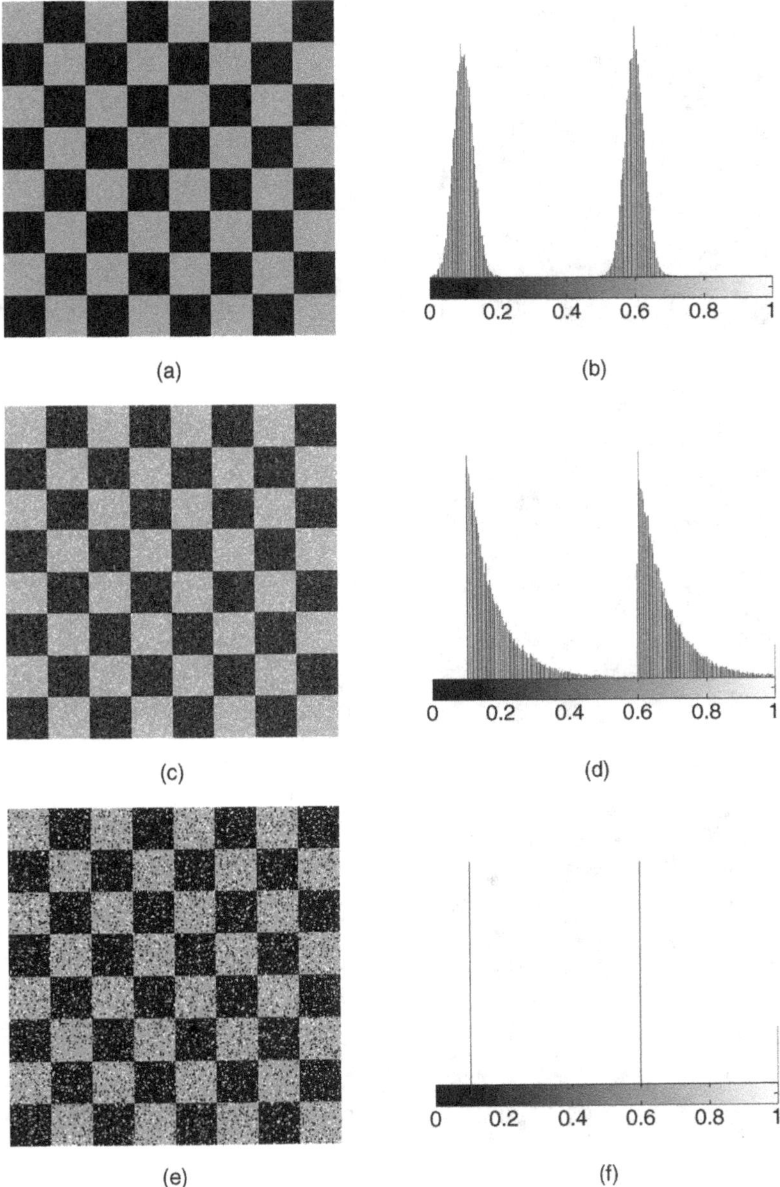

FIGURE 12.3 Test images and corresponding histograms for different types of noise: (a and b) Gaussian; (c and d) exponential; (e and f) salt and pepper.

This section presents the most popular spatial-domain techniques for noise reduction. Although some of them have been introduced previously (Chapter 10), they are being discussed again in this section. There are two main groups of spatial-domain noise reduction techniques: mean filters (Section 12.3.1) and order statistic (or *rank*)

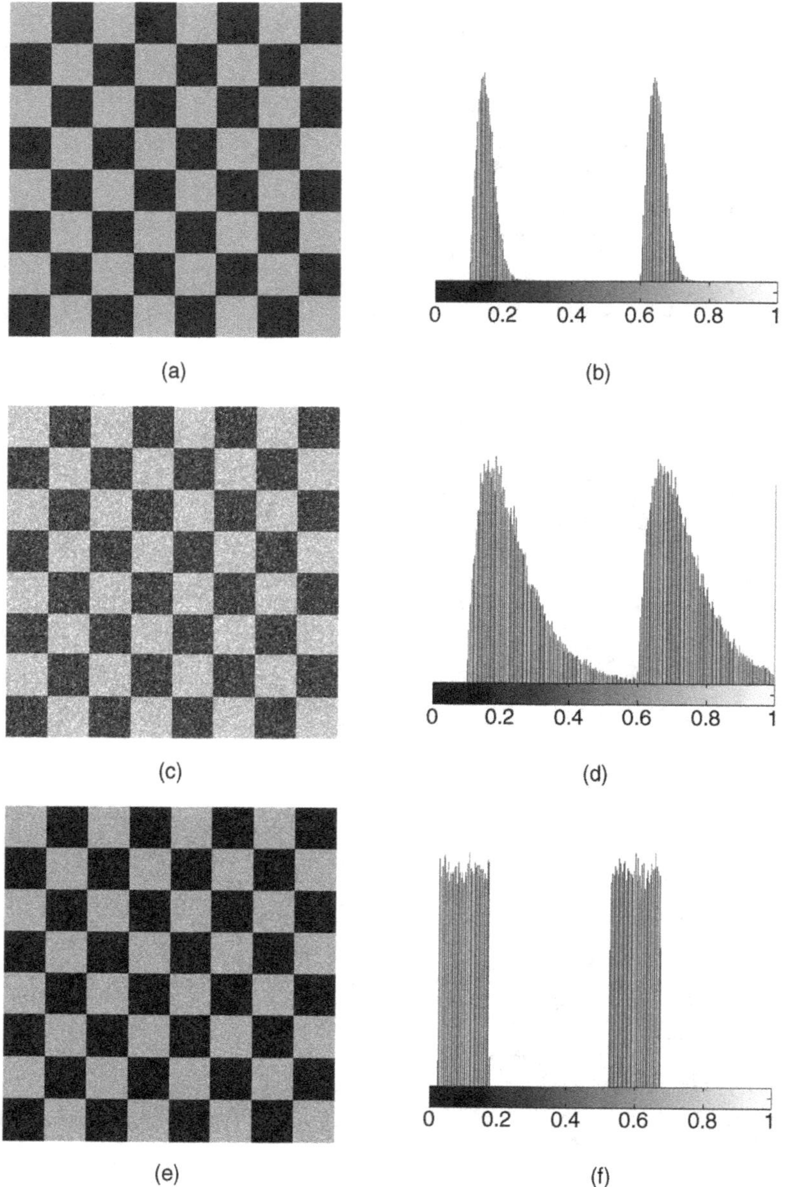

FIGURE 12.4 Test images and corresponding histograms for different types of noise: (a and b) Rayleigh; (c and d) Gamma; (e and f) uniform.

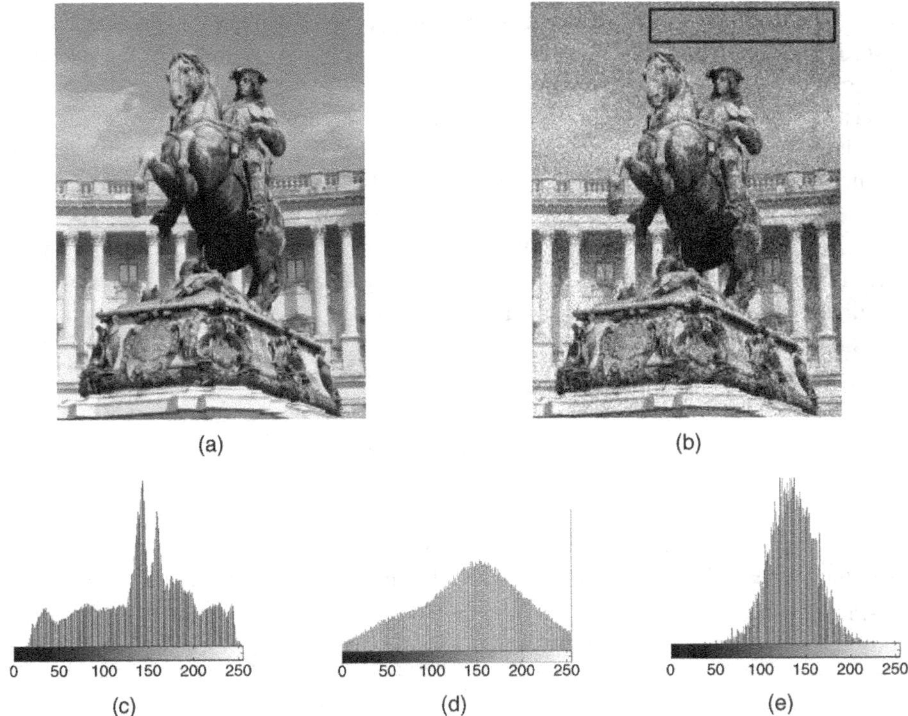

FIGURE 12.5 Estimating noise type from a homogeneous patch within an image: (a) original image; (b) noisy image (where the rectangle indicates a manually selected patch); (c) histogram of the original image; (d) histogram of the noisy image; (e) histogram of selected patch showing clearly that the noise is of Gaussian type in this case.

filters (Section 12.3.2). At the end of this section, we will briefly introduce the third category of spatial domain noise reduction techniques: adaptive filters.

12.3.1 Mean Filters

In this section, we revisit the arithmetic mean filter introduced in Section 10.3.1—from the perspective of its noise reduction capabilities—and expand the discussion to include other related filters and discuss their performance.

Arithmetic Mean Filter The arithmetic mean filter, also known as averaging filter, operates on an $m \times n$ sliding window by calculating the average of all pixel values within the window and replacing the center pixel value in the destination image with the result. Its mathematical formulation is given as follows:

$$\hat{f}(x, y) = \frac{1}{mn} \sum_{(r,c)\in W} g(r, c) \tag{12.19}$$

where g is the noisy image, $\hat{f}$ is the restored image, and r and c are the row and column coordinates, respectively, within a window W of size $m \times n$ where the operation takes place.

The arithmetic mean filter causes a certain amount of blurring (proportional to the window size) to the image, thereby reducing the effects of noise. It can be used to reduce noise of different types, but works best for Gaussian, uniform, or Erlang noise.

Geometric Mean Filter The geometric mean filter is a variation of the arithmetic mean filter and is primarily used on images with Gaussian noise. This filter is known to retain image detail better than the arithmetic mean filter. Its mathematical formulation is as follows:

$$\hat{f}(x, y) = \left[\prod_{(r,c) \in W} g(r, c) \right]^{1/mn} \tag{12.20}$$

Harmonic Mean Filter The harmonic mean filter is yet another variation of the arithmetic mean filter and is useful for images with Gaussian or salt noise. Black pixels (pepper noise) are not filtered. The filter's mathematical formulation is as follows:

$$\hat{f}(x, y) = \frac{mn}{\sum_{(r,c) \in W} (1/g(r, c))} \tag{12.21}$$

Contraharmonic Mean Filter The contra-harmonic mean filter is another variation of the arithmetic mean filter and is primarily used for filtering salt *or* pepper noise (but not both). Images with salt noise can be filtered using negative values of R, whereas those with pepper noise can be filtered using positive values of R. The filter's mathematical formulation is

$$\hat{f}(x, y) = \frac{\sum_{(r,c) \in W} g(r, c)^{R+1}}{\sum_{(r,c) \in W} g(r, c)^{R}} \tag{12.22}$$

where R is called the *order* of the filter.

■ EXAMPLE 12.3

Figure 12.6 shows the effects of applying different filters to an image corrupted by Gaussian noise of zero mean and variance 0.01. The result obtained with any of the four filters contains less noise than that with the image in Figure 12.6b. Moreover, you may also have noticed that the geometric and harmonic mean filters generate more dark pixels in the resulting image. Finally, the figure also helps to confirm the trade-off involved when choosing the mask size for a mean averaging filter: the result for Figure 12.6d (5×5 window) contains less noise—but is significantly more blurry—than the one for Figure 12.6c (3×3 window).

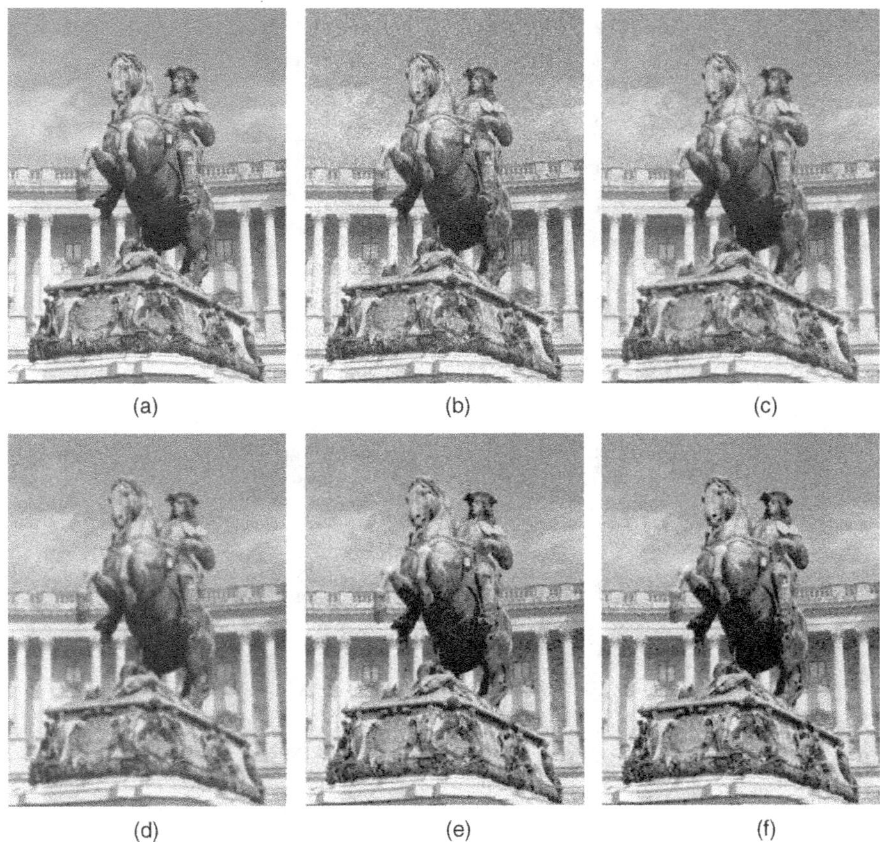

FIGURE 12.6 (a) Original image; (b) image with Gaussian noise; (c) result of 3 × 3 arithmetic mean filtering; (d) result of 5 × 5 arithmetic mean filtering; (e) result of 3 × 3 geometric mean filtering; (f) result of 3 × 3 harmonic mean filtering.

■ EXAMPLE 12.4

Figure 12.7 shows the effects of applying different filters to an image corrupted by salt and pepper noise. The result obtained with the mean averaging filter is the only acceptable one. The geometric and the harmonic mean filters perform very poorly for the pepper portion of the noise, whereas the performance of the contraharmonic filter confirms what we had stated earlier: depending on the choice of R, it will reduce the amount of salt *or* pepper noise, but not both.

12.3.2 Order Statistic Filters

Order statistic filters (also known as *rank filters*, or simply *order filters*) operate on a neighborhood around a reference pixel by ordering (i.e., ranking) the pixel values and then performing an operation on those ordered values to obtain the new value for

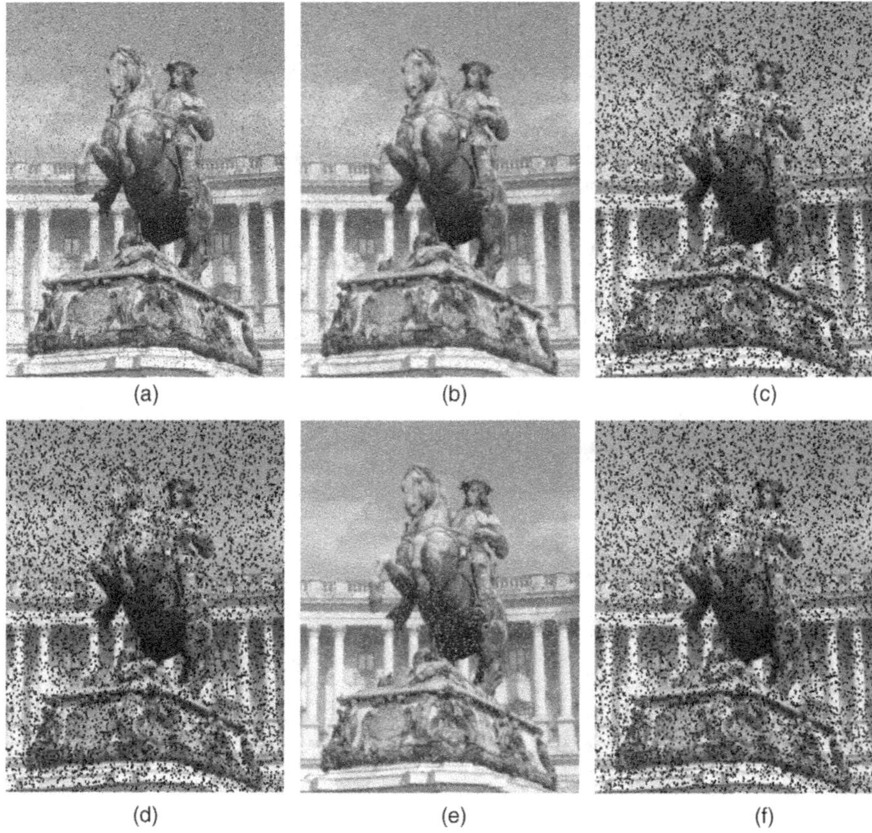

FIGURE 12.7 (a) Image with salt and pepper noise; (b) result of 3×3 arithmetic mean filtering; (c) result of 3×3 geometric mean filtering; (d) result of 3×3 harmonic mean filtering; (e) result of 3×3 contraharmonic mean filtering with $R = 0.5$; (f) result of 3×3 contraharmonic mean filtering with $R = -0.5$.

the reference pixel. Order statistic filters perform very well in the presence of salt and pepper noise but are computationally more expensive than mean filters.

In this section, we revisit the median filter introduced in Section 10.3.4—from the perspective of its noise reduction capabilities—and expand the discussion to include other related filters (the min filter, the max filter, the midpoint filter, and the alpha-trimmed mean filter) and discuss their performance.

The Median Filter The most popular and useful of the rank filters is the median filter. It works by selecting the middle pixel value from the ordered set of values within the $m \times n$ neighborhood (W) around the reference pixel. If mn is an even number (which is not common), the arithmetic average of the two values closest to the middle of the ordered set is used instead.

Mathematically,

$$\hat{f}(x, y) = \text{median}\,\{g(r, c)|(r, c) \in W\} \tag{12.23}$$

There have been many variants, extensions, and optimized implementations of the median filter proposed in the literature. Please refer to the "Learn More About It" section at the end of the chapter for useful pointers.

The Min and Max Filters The min and max filters also work on a ranked set of pixel values. Contrary to the median filter—which replaces the reference pixel with the median of the ordered set—the *min* filter, also known as the zeroth percentile filter, replaces it with the lowest value instead.

Mathematically,

$$\hat{f}(x, y) = \min\,\{g(r, c)|(r, c) \in W\} \tag{12.24}$$

Similarly, the *max* filter, also known as the 100th percentile filter, replaces the reference pixel within the window with the highest value, that is,

$$\hat{f}(x, y) = \max\,\{g(r, c)|(r, c) \in W\} \tag{12.25}$$

The min filter is useful for reduction of salt noise, whereas the max filter can help remove pepper noise.

The Midpoint Filter The midpoint filter calculates the average of the highest and lowest pixel values within a window, thereby combining order statistics and averaging into one filter. It is used to reduce Gaussian and uniform noise in images.

Mathematically,

$$\hat{f}(x, y) = \frac{1}{2}\left[\max\,\{g(r, c)|(r, c) \in W\} + \min\,\{g(r, c)|(r, c) \in W\}\right] \tag{12.26}$$

The Alpha-Trimmed Mean Filter The alpha-trimmed filter uses another combination of order statistics and averaging, in this case an average of the pixel values closest to the median, after the D lowest and D highest values in an ordered set have been excluded. The rationale behind this filter is to allow its user to control its behavior by specifying the parameter D: for $D = 0$, the filter behaves as a regular arithmetic mean filter; for $D = (mn - 1)/2$, it is equivalent to the median filter. It is used in cases where the image is corrupted by more than one type of noise, for example, salt and pepper (where the median filter performs well) and Gaussian (where the arithmetic mean filter shows satisfactory performance).

The mathematical description of the alpha-trimmed filter is as follows:

$$\hat{f}(x, y) = \frac{1}{mn - 2D} \sum_{(r,c)\in W} g(r, c) \tag{12.27}$$

where D is the number of pixel values excluded at each end of the ordered set, which can range from 0 to $(mn - 1)/2$.

■ **EXAMPLE 12.5**

Figure 12.8 shows the effects of applying different rank filters to an image corrupted by salt and pepper noise. The results obtained with the median filter (part (c)) are clearly better than the ones obtained with the average filter of same window size (part (b)) as expected. The midpoint filter not only is ineffective in removing this type of noise, but also makes it worse by replacing the noisy pixels (and their immediate neighbors) with the average between the minimum—pepper—and maximum—salt—values.

In MATLAB

Sliding window neighborhood operations are implemented in the IPT using one of these two functions: `nlfilter` or `colfilt`. Both functions accept a (user-defined) function as a parameter. Such function can perform linear (e.g., averaging) or nonlinear (e.g., median) operations on the pixels within a window. Both functions might take long to process results on an image with hundreds (or thousands) of pixels in each dimension; they both provide a progress bar indicator to inform to the user that the processing is taking place, but `colfilt` is considerably faster than `nlfilter`. You will have a chance to use `nlfilter` in Tutorial 12.1 (page 289).

For rank filters, the IPT function `ordfilt2` makes it very easy to create the min, max, and median filters, as demonstrated in Tutorial 12.1. Since the median filter is by far the most popular rank filter, the IPT has another function with a more familiar name, `medfilt2`, that implements it.

12.3.3 Adaptive Filters

The basic idea behind adaptive filters is to design the filter in such a way that its behavior changes depending on the pixel values of the neighborhood currently being processed. A classical use of adaptive filters is found under the category of *edge-preserving smoothing filters*: in this case, the goal is to apply a low-pass filter to an image in a selective way, minimizing the edge blurring effect that would be present if a standard LPF had been applied to the image. There are many variants of adaptive filters for noise reduction and image restoration in the literature. Refer to the "Learn More About It" section at the end of the chapter for useful pointers.

12.4 NOISE REDUCTION USING FREQUENCY-DOMAIN TECHNIQUES

In Chapter 11, we studied frequency-domain filters commonly used in image enhancement tasks, notably high-pass and low-pass filters. In this section, we introduce three additional types of frequency-domain filters—bandpass, bandreject,

FIGURE 12.8 (a) Image with salt and pepper noise; (b) result of 3 × 3 arithmetic mean filtering (for comparison); (c) result of 3 × 3 median filtering; (d) result of 3 × 3 midpoint filtering.

and notch—whose main application is in the reduction or removal of periodic noise.

12.4.1 Periodic Noise

This is a type of noise that usually arises as a result of electrical or electromechanical interference during the image acquisition process.

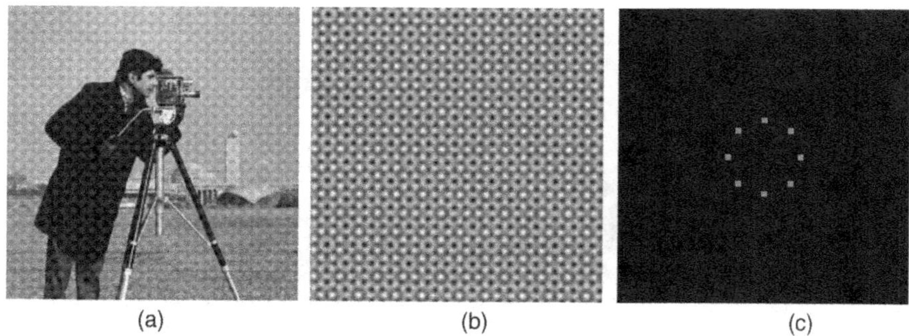

(a) (b) (c)

FIGURE 12.9 Example of an image corrupted by periodic noise: (a) noisy image; (b) periodic noise component; (c) the Fourier spectrum of the noise component (bright dots were enlarged for viewing purposes).

■ EXAMPLE 12.6

Figure 12.9 shows an example of an image corrupted by periodic noise, the noise component, and its Fourier spectrum. The bright dots in the spectrum indicate four pairs of impulse functions, each pair corresponding to a sinusoidal noise source.

12.4.2 Bandreject Filter

A bandreject filter, as its name suggests, attenuates frequency components within a certain range (the *stopband* of the filter), while leaving all other frequency components untouched (or amplifying them by a certain gain). The mathematical formulation for an *ideal* bandreject filter is as follows:

$$H_{\mathrm{br}}^{\mathrm{i}}(u,v) = \begin{cases} 1 & \text{if } D(u,v) < D_0 - \frac{W}{2} \\ 0 & \text{if } D_0 - \frac{W}{2} \leq D(u,v) \leq D_0 + \frac{W}{2} \\ 1 & \text{if } D(u,v) > D_0 + \frac{W}{2} \end{cases} \tag{12.28}$$

where $D(u, v)$ is the distance from the origin of the frequency spectrum, W is the width of the band, and D_0 is the radius of the circle-shaped band.

A *Butterworth* bandreject filter of order n is described mathematically as

$$H_{\mathrm{br}}^{\mathrm{b}}(u,v) = \frac{1}{1 + \left[\frac{D(u,v)W}{D^2(u,v) - D_0^2}\right]^{2n}} \tag{12.29}$$

whereas the transfer function for a *Gaussian* bandreject filter is given by the following equation:

$$H_{\mathrm{br}}^{\mathrm{g}}(u,v) = 1 - e^{-\frac{1}{2}\left[\frac{D^2(u,v) - D_0^2}{D(u,v)W}\right]^2} \tag{12.30}$$

FIGURE 12.10 Example of using a bandreject filter to reduce periodic noise: (a) noisy image; (b) noisy image spectrum (the eight spots corresponding to the noise have been made brighter and bigger for visualization purposes); (c) the Fourier spectrum of the image after applying the bandreject filter; (d) resulting image.

■ EXAMPLE 12.7

Figure 12.10 shows an example of an ideal bandreject filter (of radius $D_0 = 32$ and width $W = 6$) used to reduce periodic noise. Although the periodic noise has been successfully removed, the resulting image shows an undesirable ringing effect as a result of the sharp transition between passband and stopband in the bandreject filter.

12.4.3 Bandpass Filter

A bandpass filter allows certain frequencies (within its passband) to be preserved while attenuating all others. It is, in effect, the opposite of a bandreject filter (Section 12.4.2), and its mathematical formulation can be simply stated as follows:

$$H_{\mathrm{bp}}(u, v) = 1 - H_{\mathrm{br}}(u, v) \tag{12.31}$$

where $H_{\mathrm{br}}(u, v)$ is the transfer function of a bandreject filter.

Applying equation (12.31) to equations (12.28)–(12.30) leads to the mathematical formulations for the different types of bandpass filters:

Ideal BPF

$$H_{\text{bp}}^{i}(u, v) = \begin{cases} 0 & \text{if } D(u, v) < D_0 - \frac{W}{2} \\ 1 & \text{if } D_0 - \frac{W}{2} \leq D(u, v) \leq D_0 + \frac{W}{2} \\ 0 & \text{if } D(u, v) > D_0 + \frac{W}{2} \end{cases} \tag{12.32}$$

Butterworth BPF

$$H_{\text{bp}}^{\text{b}}(u, v) = \frac{\left[\frac{D(u,v)W}{D^2(u,v)-D_0^2}\right]^{2n}}{1 + \left[\frac{D(u,v)W}{D^2(u,v)-D_0^2}\right]^{2n}} \tag{12.33}$$

Gaussian BPF

$$H_{\text{bp}}^{\text{g}}(u, v) = e^{-\frac{1}{2}\left[\frac{D^2(u,v)-D_0^2}{D(u,v)W}\right]^2} \tag{12.34}$$

12.4.4 Notch Filter

The *notch* filter is a special kind of frequency-domain filter that attenuates (or allows) frequencies within a neighborhood around a center frequency. Owing to the symmetry property of the Fourier transform (Section 11.2.3), the spectrum of notch filters shows symmetric pairs around the origin (except for a notch filter located at the origin, of course).

The mathematical formulation for an ideal notch filter of radius D_0 with center at (u_0, v_0) (and by symmetry at $(-u_0, -v_0)$) that rejects frequencies within a predefined neighborhood is

$$H_{\text{nr}}^{\text{i}}(u, v) = \begin{cases} 0 & \text{if } D_1(u, v) < D_0 \text{ or } D_2(u, v) < D_0 \\ 1 & \text{otherwise} \end{cases} \tag{12.35}$$

where

$$D_1(u, v) = \left[(u - M/2 - u_0)^2 + (v - N/2 - v_0)^2\right]^{1/2} \tag{12.36}$$

and

$$D_2(u, v) = \left[(u - M/2 + u_0)^2 + (v - N/2 + v_0)^2\right]^{1/2} \tag{12.37}$$

It is assumed that the spectrum has been shifted by $(M/2, N/2)$[3] and the values of (u_0, v_0) are relative to the shifted center.

A Butterworth notch filter of order n can be mathematically described as

$$H_{\mathrm{nr}}^{\mathrm{b}}(u, v) = \frac{1}{1 + \left[\frac{D_0^2}{D_1(u,v)D_2(u,v)}\right]^n} \tag{12.38}$$

where $D_1(u, v)$ and $D_2(u, v)$ are given by equations (12.36) and (12.37), respectively.

A Gaussian notch filter is mathematically described by the following equation:

$$H_{\mathrm{nr}}^{\mathrm{g}}(u, v) = 1 - e^{-\frac{1}{2}\left[\frac{D_1(u,v)D_2(u,v)}{D_0^2}\right]} \tag{12.39}$$

To convert a notch filter that rejects certain frequencies into one that allows those same frequencies to pass, one has to simply compute

$$H_{\mathrm{np}}(u, v) = 1 - H_{\mathrm{nr}}(u, v) \tag{12.40}$$

where $H_{\mathrm{np}}(u, v)$ is the transfer function of the notch pass filter that corresponds to the notch reject filter whose transfer function is $H_{\mathrm{nr}}(u, v)$.

12.5 IMAGE DEBLURRING TECHNIQUES

The goal of image deblurring techniques is to process an image that has been subject to blurring caused, for example, by camera motion during image capture or poor focusing of the lenses. The simplest image deblurring filtering technique, *inverse filtering*, operates in the frequency domain, according to the model in Figure 12.1, and assuming that there is no significant noise in the degraded image,

$$G(u, v) = F(u, v)H(u, v) + 0 \tag{12.41}$$

which leads to

$$F(u, v) = \frac{G(u, v)}{H(u, v)} = G(u, v)\frac{1}{H(u, v)} \tag{12.42}$$

where the term $1/H(u, v)$ is the FT of the restoration filter, which will be denoted by $R_{\mathrm{inv}}(u, v)$.

The simplicity of this formulation hides some of its pitfalls. If there are any points in $H(u, v)$ that are zero, a divide by zero exception will be generated.[4] Even worse, if the assumption of no additive noise is correct, the degraded image $H(u, v)$ will

[3]The shift can be performed using MATLAB function *fftshift*, similarly to what we did in the tutorials for Chapter 11.

[4]Even a milder version of this problem—where the values of $H(u, v)$ are close to zero—will lead to unacceptable results, as the following example demonstrates.

exhibit zeros at the same points, which will lead to a 0/0 indeterminate form. On the other hand, if the image is contaminated by noise—although we assumed it had not—the zeros will not coincide but the result of the inverse filter calculations will be heavily biased by the noise term. The latter problem will be handled by Wiener filters (Section 12.5.1).

There are two common solutions to the former problem:

1. Apply a low-pass filter with transfer function $L(u, v)$ to the division, thus limiting the restoration to a range of frequencies below the *restoration cutoff frequency*, that is,

$$F(u, v) = \frac{G(u, v)}{H(u, v)} L(u, v) \tag{12.43}$$

   For frequencies within the filter's passband, the filter's gain is set to any desired positive value (usually the gain is set to 1); for frequencies within the filter's stopband, the gain should be zero. If the transition between the passband and stopband is too sharp (as in the case of the ideal LPF), however, ringing artifacts may appear in the restored image; other filter types (e.g., Butterworth with high order) may be preferred in this case.

2. Use *constrained division* where a threshold value T is chosen such that if $|H(u, v)| < T$, the division does not take place and the original value is kept untouched:

$$F(u, v) = \begin{cases} \frac{G(u,v)}{H(u,v)} & \text{if } |H(u, v)| \geq T \\ G(u, v) & \text{otherwise} \end{cases} \tag{12.44}$$

■ EXAMPLE 12.8

Figure 12.11 shows an example of image restoration using inverse filtering. Part (a) shows the input (blurry) image. Part (b) shows the result of naively applying inverse filtering (equation (12.42)), which is completely unacceptable due to the division by very small values of $H(u, v)$. Parts (c) and (d) show the results of applying a 10th-order Butterworth low-pass filter to the division, with different cutoff frequencies. Parts (e) and (f) show the results of using constrained division, with different values for the threshold T.

Motion deblurring can be considered a special case of inverse filtering. Figure 12.12 shows the results of applying inverse filtering with constrained division to a blurry image (generated using `fspecial('motion',10,0)` to simulate a horizontal displacement equivalent to 10 pixels).

FIGURE 12.11 Example of image restoration using inverse filtering: (a) input (blurry) image; (b) result of naive inverse filtering; (c) applying a 10th-order Butterworth low-pass filter with cutoff frequency of 20 to the division; (d) same as (c), but with cutoff frequency of 50; (e) results of using constrained division, with threshold $T = 0.01$; (f) same as (e), but with threshold $T = 0.001$.

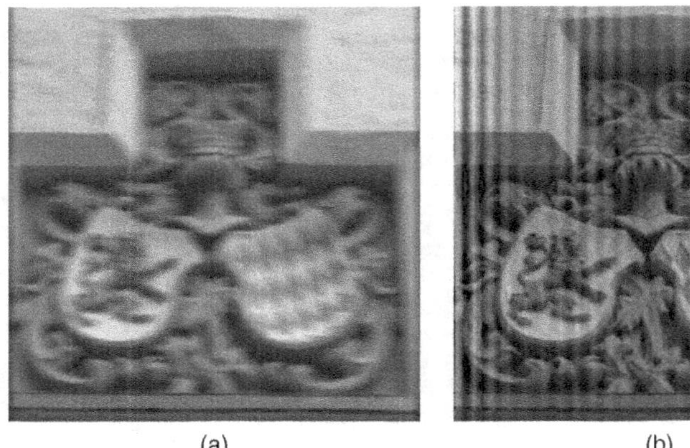

(a) (b)

FIGURE 12.12 Example of motion deblurring using inverse filtering: (a) input image; (b) result of applying inverse filtering with constrained division and threshold $T = 0.05$: the motion blurred has been removed at the expense of the appearance of vertical artifacts.

12.5.1 Wiener Filtering

The Wiener filter (developed by Norbert Wiener in 1942) is an image restoration solution that can be applied to images that have been subject to a degradation function *and* also contain noise, which corresponds to the worst-case scenario for the degraded image $g(x, y)$ in Figure 12.1.

The design of the Wiener filter is guided by an attempt to model the error in the restored image through statistical methods, particularly the *minimum mean square estimator*: once the error is modeled, the average error is mathematically minimized. Assuming a degraded version $g(x, y)$ of some original image $f(x, y)$, and a restored version $r(x, y)$, if we compute the sum of the squared differences between each and every pixel in $f(x, y)$ and the corresponding pixel in $r(x, y)$, we have a figure of merit that captures how well the restoration algorithm worked: smaller values mean better results.

The transfer function of a Wiener filter is given by

$$R(u, v) = \left[\frac{1}{H(u, v)} \frac{|H(u, v)|^2}{|H(u, v)|^2 + K}\right] G(u, v) \tag{12.45}$$

where $H(u, v)$ is the degradation function and K is a constant used to approximate the amount of noise. When $K = 0$, equation (12.45) reduces to equation (12.42).

In MATLAB

The IPT has a function that implements image deblurring using Wiener filter: `deconvwnr`.

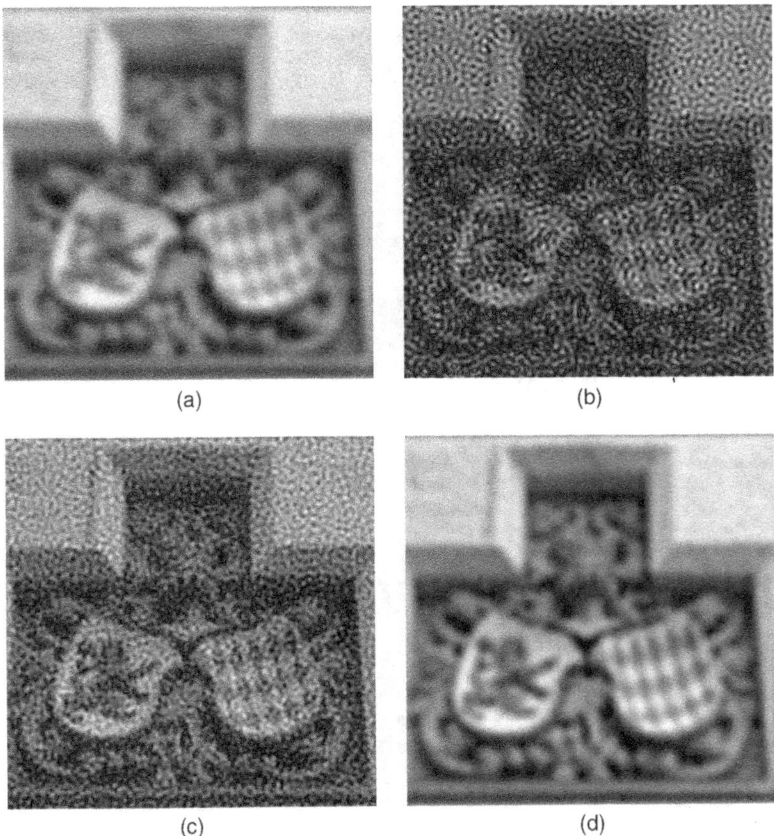

FIGURE 12.13 Example of image restoration using Wiener filtering: (a) input image (blurry and noisy); (b) result of inverse filtering, applying a 10th-order Butterworth low-pass filter with cutoff frequency of 50 to the division; (c) results of Wiener filter, with $K = 10^{-3}$; (d) same as (c), but with $K = 0.1$.

■ EXAMPLE 12.9

Figure 12.13 shows an example of image restoration using Wiener filtering. Part (a) shows the input image, which has been degraded by blur and noise. Part (b) shows the results of applying inverse filtering, using a 10th-order Butterworth low-pass filter and limiting the restoration cutoff frequency to 50. Part (c) shows the results of using Wiener filter with different values of K, to illustrate the trade-off between noise reduction (which improves for higher values of K) and deblurring (which is best for lower values of K).

The Wiener filter can also be used to restore images in the presence of blurring only (i.e., without noise), as shown in Figure 12.14. In such cases, the best results are obtained for lower values of K (Figure 12.14c), but the resulting image is still not as crisp as the one obtained with the inverse filter (Figure 12.14b).

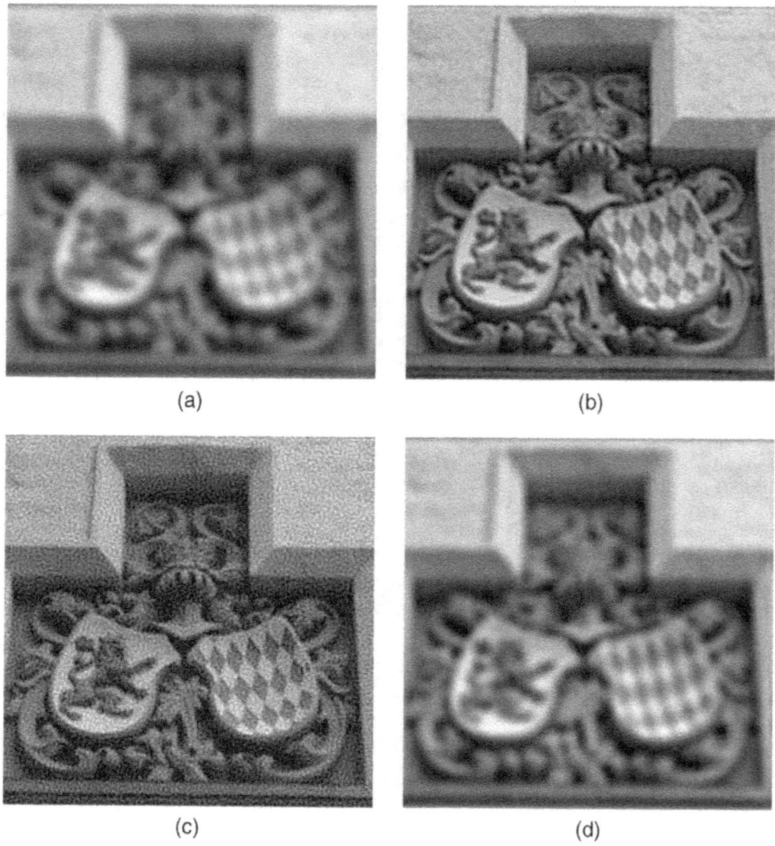

FIGURE 12.14 Example of image restoration using Wiener filtering: (a) input (blurry) image; (b) result of inverse filtering, applying a 10th-order Butterworth low-pass filter with cutoff frequency of 50 to the division; (c) results of Wiener filter, with $K = 10^{-5}$; (d) same as (c), but with $K = 0.1$.

In MATLAB

The IPT has several other built-in functions for image deblurring, namely,

- `deconvreg`: deblur image using regularized filter
- `deconvlucy`: deblur image using Lucy–Richardson method
- `deconvblind`: deblur image using blind deconvolution

The complexity of these functions (in terms of both their mathematical formulation and the number of parameters that need to be adjusted for better performance) is beyond the scope of this book. The reader should refer to the corresponding MATLAB documentation and demos for additional information.

12.6 TUTORIAL 12.1: NOISE REDUCTION USING SPATIAL-DOMAIN TECHNIQUES

Goal

The goal of this tutorial is to learn how to perform noise reduction using spatial-domain techniques.

Objectives

- Learn how to implement the arithmetic mean filter, as well as some of its variations, such as the contraharmonic mean, the harmonic mean, and the geometric mean filters.
- Learn how to perform order statistic filtering, including median, min, max, midpoint, and alpha-trimmed mean filters.

What You Will Need

- `atmean.m`
- `geometric.m`
- `harmonic.m`
- `c_harmonic.m`

Procedure

Arithmetic Mean Filter

The *arithmetic mean filter*, also known as an *averaging* or *low-pass* (from its frequency-domain equivalent) filter, is a simple process of replacing each pixel value with the average of an $N \times N$ window surrounding the pixel. The averaging filter can be implemented as a convolution mask. As in previous tutorials, we will use the function `fspecial` to generate the averaging convolution mask.

1. Load the `eight` image, add (Gaussian) noise to it, and display the image before and after adding noise.

```
I = imread('eight.tif');
In = imnoise(I,'gaussian',0,0.001);
figure, subplot(2,2,1), imshow(I), title('Original Image');
subplot(2,2,2), imshow(In), title('Noisy Image');
```

2. Apply an averaging filter to the image using the default kernel size (3×3).

```
f1 = fspecial('average');
I_blur1 = imfilter(In,f1);
```

```
subplot(2,2,3), imshow(I_blur1), ...
    title('Averaging with default kernel size');
```

Question 1 What is the general effect of the arithmetic mean filter?

3. Implement an averaging kernel with a 5 × 5 mask.

```
f2 = fspecial('average',[5 5]);
I_blur2 = imfilter(In,f2);
subplot(2,2,4), imshow(I_blur2), ...
    title('Averaging with 5x5 kernel');
```

Question 2 How does the size of the kernel affect the resulting image?

Contraharmonic Mean Filter

The *contraharmonic mean filter* is used for filtering an image with either salt or pepper noise (but not both). When choosing a value for `r`, it is important to remember that negative values are used for salt noise and positive values are used for pepper noise. As we will see, using the wrong sign will give undesired results. This filter does not have a convolution mask equivalent, so we must implement it as a sliding neighborhood operation using the `nlfilter` function. This function allows us to define how we want to operate on the window, which can be specified within a function of our own.

4. Close any open figures.
5. Load two noisy versions of the eight image: one with salt noise and the other with pepper. Also, display the original image along with the two affected images.

```
I_salt = im2double(imread('eight_salt.tif'));
I_pepper = im2double(imread('eight_pepper.tif'));
figure
subplot(2,3,1), imshow(I), title('Original Image');
subplot(2,3,2), imshow(I_salt), title('Salt Noise');
subplot(2,3,3), imshow(I_pepper), title('Pepper Noise');
```

The contraharmonic function requires that images be of class `double`. This is why we convert the image when loading it.

6. Filter the salt noise affected image using −1 for the value of `r`.

```
I_fix1 = nlfilter(I_salt,[3 3],@c_harmonic,-1);
subplot(2,3,5), imshow(I_fix1), title('Salt Removed, r = -1');
```

Our function `c_harmonic` takes two parameters: the current window matrix and a value for `r`. The window matrix, which gets stored into variable `x`, is passed implicitly by the `nlfilter` function. Note how we specified the `c_harmonic` function as the third parameter of the `nlfilter` function call. When using the `nlfilter` function, if you want to pass any additional parameters to your function, you can specify those parameters after the function handle (labeled with a '@' in front of it). Notice above how we specified the value of `r` directly after the function handle.

7. Filter the pepper noise affected image using 1 for the value of `r`.

```
I_fix2 = nlfilter(I_pepper,[3 3],@c_harmonic,1);
subplot(2,3,6), imshow(I_fix2), title('Pepper Removed, r = 1');
```

As mentioned previously, using the wrong sign for the value of `r` can lead to unwanted results.

8. Filter the pepper noise image using the wrong sign for `r`.

```
I_bad = nlfilter(I_pepper,[3 3],@c_harmonic,-1);
subplot(2,3,4), imshow(I_bad), title('Using wrong sign for r');
```

Question 3 What is the effect of using the wrong sign when filtering with the contraharmonic mean filter?

Harmonic Mean Filter

The *harmonic mean filter* is another variation of the mean filter and is good for salt and Gaussian noise. It fails, however, when used on pepper noise.

9. Close any open figures.
10. Filter the salt noise affected image with the harmonic filter.

```
I_fix4 = nlfilter(I_salt,[3 3],@harmonic);
figure
subplot(2,3,1), imshow(I), title('Original Image');
subplot(2,3,2), imshow(I_salt), title('Salt Noise');
subplot(2,3,3), imshow(I_pepper), title('Pepper Noise');
subplot(2,3,5), imshow(I_fix4), title('Harmonic Filtered (salt)');
```

11. Filter the pepper noise image and display the result.

```
I_bad2 = nlfilter(I_pepper,[3 3],@harmonic);
subplot(2,3,6), imshow(I_bad2), title('Harmonic Filtered (pepper)');
```

Question 4 Why does the harmonic mean filter fail for images with pepper noise?

12. Try to filter the `In` image (`I` with additive Gaussian noise) with the harmonic mean filter. The image must be converted to `double` first.

```
In_d = im2double(In);
I_fix5 = nlfilter(In_d,[3 3],@harmonic);
figure
subplot(1,3,1), imshow(I), title('Original Image');
subplot(1,3,2), imshow(In_d), title('Image w/ Gaussian Noise');
subplot(1,3,3), imshow(I_fix5), title('Filtered w/ Harmonic Mean');
```

Question 5 How does the size of the window affect the output image?

Geometric Mean Filter

The last variation of the mean filters we will look at is the *geometric mean filter*. This filter is known to preserve image detail better than the arithmetic mean filter and works best on Gaussian noise.

13. Close any open figures.
14. Perform a geometric mean filter on the `eight` image with Gaussian noise (currently loaded in the variable `In_d`).

```
I_fix6 = nlfilter(In_d,[3 3],@geometric);
figure
subplot(1,3,1), imshow(I), title('Original Image');
subplot(1,3,2), imshow(In_d), title('Gaussian Noise');
subplot(1,3,3), imshow(I_fix6), title('Geometric Mean Filtered');
```

Question 6 Filter the salt and pepper noise images with the geometric mean filter. How does the filter perform?

Order Statistic Filters

The *median filter* is the most popular example of an *order statistic filter*. This filter simply sorts all values within a window, finds the median value, and replaces the original pixel value with the median value. It is commonly used for salt and pepper noise. Because of its popularity, the median filter has its own function (`medfilt2`) provided by the IPT.

15. Close any open figures and clear all workspace variables.
16. Load the `coins` image and apply salt and pepper noise.

```
I = imread('coins.png');
I_snp = imnoise(I,'salt & pepper');
figure
subplot(1,3,1), imshow(I), title('Original Image');
subplot(1,3,2), imshow(I_snp), title('Salt & Pepper Noise');
```

17. Filter the image using the `medfilt2` function.

```
I_filt = medfilt2(I_snp,[3 3]);
subplot(1,3,3), imshow(I_filt), title('Filtered Image');
```

Question 7 How does the size of the window affect the output image?

18. Apply the filter to an image with Gaussian noise.

```
I_g = imnoise(I,'gaussian');
I_filt2 = medfilt2(I_g,[3 3]);
figure
subplot(1,3,1), imshow(I), title('Original Image');
subplot(1,3,2), imshow(I_g), title('Gaussian Noise');
subplot(1,3,3), imshow(I_filt2), title('Filtered');
```

Question 8 Why do you think the median filter works on salt and pepper noise but not Gaussian noise?

A quick way to get rid of salt noise in an image is to use the *min filter*, which simply takes the minimum value of a window when the values are ordered. Recall that we previously used the `imfilter` function when dealing with convolution masks and `nlfilter` for sliding neighborhood operations that could not be implemented as a convolution mask. Similarly, the `ordfilt2` function is used for order statistic operations.

19. Close any open figures and clear all workspace variables.
20. Use the `ordfilt2` function to implement a min filter on an image with salt noise.

```
I_s = imread('eight_salt.tif');
I2 = ordfilt2(I_s, 1, ones(3,3));
figure
subplot(1,2,1), imshow(I_s), title('Salt Noise');
subplot(1,2,2), imshow(I2), title('Min Filter');
```

Question 9 Why would this filter not work on pepper noise?

The first parameter specified in the `ordfilt2` function is the image we wish to filter. The second parameter specifies the index of the value to be useed after all values in the window have been ordered. Here, we specified this parameter as 1, which means we want the first value after reordering, that is, the minimum value. The last parameter defines the size of the window as well as which values in that window will be used in the ordering. A 3×3 matrix of 1's would indicate a 3×3 window and to use all values when ordering. If we instead specified a 3×3 matrix where only the first row was 1's and the last two rows were zeros, then the sliding window would consist of a 3×3 matrix, but only the top three values would be considered when ordering. In addition, keep in mind that even though we used a special function to implement the median filter, it is still an order statistic filter, which means we could have implemented it using the `ordfilt2` function.

Question 10 Implement the median filter using the `ordfilt2` function.

The *max filter* is used for filtering pepper noise, similar to the technique of the min filter.

21. Filter a pepper noise affected image with the max filter.

```
I_p = imread('eight_pepper.tif');
I3 = ordfilt2(I_p, 9, ones(3,3));
figure
subplot(1,2,1), imshow(I_p), title('Pepper Noise');
subplot(1,2,2), imshow(I3), title('Max Filter');
```

Although the *midpoint filter* is considered an order statistic filter, it cannot be directly implemented using the `ordfilt2` function because we are not selecting a particular element from the window, but instead performing a calculation on its values—namely, the minimum and maximum values. Rather, we will implement it using the familiar `nlfilter` function. This noise removal technique is best used on Gaussian or uniform noise.

22. Filter an image contaminated with Gaussian noise using the midpoint filter.

```
I = imread('coins.png');
I_g = imnoise(I,'gaussian',0,0.001);
midpoint = inline('0.5 * (max(x(:)) + min(x(:)))');
I_filt = nlfilter(I_g,[3 3],midpoint);
figure
subplot(1,2,1), imshow(I_g), title('Gaussian Noise');
subplot(1,2,2), imshow(I_filt), title('Midpoint Filter');
```

You may have noticed that we have used an inline function instead of creating a separate function, as we did in previous steps. Inline functions are good for quick tests, but—as you may have realized—they are much slower than regular functions.

Alpha-Trimmed Mean Filters

The *alpha-trimmed mean filter* is basically an averaging filter whose outlying values are removed before averaging. To do this, we sort the values in the window, discard elements on both ends, and then take the average of the remaining values. This has been defined in the function `atmean`.

23. Close any open figures and clear all workspace variables.
24. Generate a noisy image with Gaussian noise and salt and pepper noise.

```
I = imread('cameraman.tif');
Id = im2double(I);
In = imnoise(Id,'salt & pepper');
In2 = imnoise(In,'gaussian');
```

25. Filter the image using the alpha-trimmed mean filter.

```
I_filt = nlfilter(In2,[5 5],@atmean,6);
figure
subplot(1,3,1), imshow(I), title('Original Image');
subplot(1,3,2), imshow(In2), title('S&P and Gaussian Noise');
subplot(1,3,3), imshow(I_filt), title('Alpha Trimmed Mean');
```

Question 11 When filtering an image with both types of noise, how does the alpha-trimmed mean filter compare to the arithmetic mean filter?

WHAT HAVE WE LEARNED?

- In the context of image processing, *noise* is a general term used to express deviations from a pixel's expected (or *true*) value. When these deviations are offsets from the true value, the noise is said to be of *additive* type. When the true value is rescaled as a result of noise, such noise is said to be of *multiplicative* type.
- The statistical properties of noise are usually modeled in a way that is independent of the actual causes of the noise. Common probability distribution functions associated with noise are Gaussian, exponential, uniform, and gamma (Erlang).
- Two of the most common types of noise in image processing are the Gaussian noise and the *salt and pepper* noise. Both are forms of additive noise. The Gaussian noise follows a zero-mean normal distribution. The salt and pepper noise is a type of impulsive noise that appears as black and white specks on the image.

- The most common noise removal techniques in the spatial domain are the mean filter, the median filter, and variants and combinations of them.
- The most common noise removal techniques in the frequency domain are low-pass, bandpass, bandreject, and notch filters.
- *Blurring* is the loss of sharpness in an image. It can be caused by poor focusing, relative motion between sensor and scene, and noise, among other factors.
- Deblurring techniques typically consist of applying inverse filtering techniques with the goal of "undoing" the degradation. The best-known approach to image deblurring (even in the presence of noise) is the Wiener filter, which is implemented in MATLAB by the `deconvwnr` function.

LEARN MORE ABOUT IT

- Chapter 4.5 of [Bov00a] provides additional information on noise sources and noise models.
- Chapters 11 and 12 of [Pra07] discuss image restoration models and techniques in more detail.
- Section 5.2 of [GWE04] discusses the generation of spatial random noise with specified distributions and extends the functionality of the IPT function `imnoise`.
- There have been many variants, extensions, and optimized implementations of the median filter proposed in the literature, such as
  - the pseudomedian [PCK85], also described in Section 10.3 of [Pra07];
  - weighted median filters, described in Chapter 3.2 of [Bov00a];
  - faster implementations, such as the one proposed in [HYT79].
- Nonlinear filters have been the subject of entire chapters, for example, [Dou94], and book-length treatment, for example, [PV90].
- To learn more about adaptive filters, we recommend Section 5.3.3 of [GW08] and Chapter 11 of [MW93].
- Sections 5.7–5.10 of [GWE04] discuss the main IPT functions for image deblurring.
- Chapter 4.2 of [SOS00] presents a technique for noise removal in binary images, the *k*Fill filter.
- Chapter 3.5 of [Bov00a] provides additional information on image restoration, image deblurring, and blur identification techniques.

12.7 PROBLEMS

12.1 Write a modified and expanded version of the IPT function `imnoise`. Your function should allow the specification of other types of noise, currently not supported by `imnoise`, for example, Rayleigh, Erlang, uniform, and exponential.

12.2 Write a MATLAB function to generate periodic noise (and its spectrum), given a set of coordinate pairs corresponding to different values of frequencies (u, v) in the 2D spatial frequency domain. Use your function to generate the same periodic noise as displayed in Figure 12.9 (where the value of D_0 is 32).

12.3 What is the effect on an image if we apply a contraharmonic mean filter with the following values of R:

(a) $R = 0$

(b) $R = -1$

12.4 Write a MATLAB function to implement an ideal bandreject filter of radius D_0 and width W and use it to reduce periodic noise on an input noisy image.

12.5 Write a MATLAB function to implement a Butterworth bandreject filter of order n, radius D_0, and width W and use it to reduce periodic noise on an input noisy image.

12.6 Write a MATLAB function to implement a Gaussian bandreject filter of radius D_0 and width W and use it to reduce periodic noise on an input noisy image.

12.7 Test your solutions to Problems 12.4–12.6 using the image from Figure 12.9a (available at the book web site) as an input.

12.8 Write a MATLAB function to implement an ideal bandpass filter of radius D_0 and width W.

12.9 Write a MATLAB function to implement a Butterworth bandpass filter of order n, radius D_0, and width W.

12.10 Write a MATLAB function to implement a Gaussian bandpass filter of radius D_0 and width W.

12.11 Test your solutions to Problems 12.8–12.10 using the image from Figure 12.9a (available at the book web site) as an input and comparing the result produced by your function with the image in Figure 12.9b.

12.12 Write a MATLAB function to implement an ideal notch reject filter of radius D_0, centered at (u_0, v_0), and use it to reduce periodic noise on an input noisy image.

12.13 Write a MATLAB function to implement a Butterworth notch reject filter of order n and radius D_0, centered at (u_0, v_0), and use it to reduce periodic noise on an input noisy image.

12.14 Write a MATLAB function to implement a Gaussian notch reject filter of radius D_0, centered at (u_0, v_0), and use it to reduce periodic noise on an input noisy image.

12.15 Test your solutions to Problems 12.12–12.14 using a test image with one sinusoidal noise component created with your solution to Problem 12.2.

12.16 Write a MATLAB function to implement a *range filter* whose output is the difference between the maximum and the minimum gray levels in a neighborhood centered on a pixel [Eff00], test it, and answer the following questions:

(**a**) Is the range filter a linear or nonlinear one? Explain.

(**b**) What can it be used for?

CHAPTER 13

MORPHOLOGICAL IMAGE PROCESSING

WHAT WILL WE LEARN?

- What is mathematical morphology and how is it used in image processing?
- What are the main morphological operations and what is the effect of applying them to binary and grayscale images?
- What is a structuring element (SE) and how does it impact the result of a morphological operation?
- What are some of the most useful morphological image processing algorithms?

13.1 INTRODUCTION

Mathematical morphology is a branch of image processing that has been successfully used to provide tools for representing, describing, and analyzing shapes in images. It was initially developed by Jean Serra in the early 1980s [SC82] and—because of its emphasis on studying the geometrical structure of the components of an image—named after the branch of biology that deals with the form and structure of animals and plants. In addition to providing useful tools for extracting image components, morphological algorithms have been used for pre- or postprocessing the images containing shapes of interest.

Practical Image and Video Processing Using MATLAB®. By Oge Marques.

The basic principle of mathematical morphology is the extraction of geometrical and topological information from an unknown set (an image) through transformations using another, well-defined, set known as *structuring element*. In morphological image processing, the design of SEs, their shape and size, is crucial to the success of the morphological operations that use them.

The IPT in MATLAB has an extensive set of built-in morphological functions, which will be introduced throughout the chapter. You will have a chance to work with many of them in the tutorials at the end of the chapter.

13.2 FUNDAMENTAL CONCEPTS AND OPERATIONS

The basic concepts of mathematical morphology can be introduced with the help of set theory and its standard operations: *union* (∪), *intersection* (∩), and *complement*, defined as

$$A^c = \{z | z \notin A\} \tag{13.1}$$

and the *difference* of two sets A and B:

$$A - B = \{z | z \in A, z \notin B\} = A \cap B^c \tag{13.2}$$

Let A be a set (of pixels in a binary image) and $w = (x, y)$ be a particular coordinate point. The *translation* of set A by point w is denoted by A_w and defined as

$$A_w = \{c | c = a + w, \quad \text{for} \quad a \in A\} \tag{13.3}$$

The *reflection* of set A relative to the origin of a coordinate system, denoted $\hat{A}$, is defined as

$$\hat{A} = \{z | z = -a, \quad \text{for} \quad a \in A\} \tag{13.4}$$

Figure 13.1 shows a graphical representation of the basic set operations defined above. The black dot represents the origin of the coordinate system.

Binary mathematical morphology theory views binary images as a set of its foreground pixels (whose values are assumed to be 1), the elements of which are in $\mathcal{Z}^2$. Classical image processing refers to a binary image as a function of x and y, whose only possible values are 0 and 1. To avoid any potential confusion that this dual view may cause, here is an example of how a statement expressed in set theory notation can be translated into a set of logical operations applied to binary images:

The statement $C = A \cap B$, from a set theory perspective, means

$$C = \{(x, y) | (x, y) \in A \text{ and } (x, y) \in B\} \tag{13.5}$$

The equivalent expression using conventional image processing notation would be

$$C(x, y) = \begin{cases} 1 & \text{if } A(x, y) \text{ and } B(x, y) \text{ are both 1} \\ 0 & \text{otherwise} \end{cases} \tag{13.6}$$

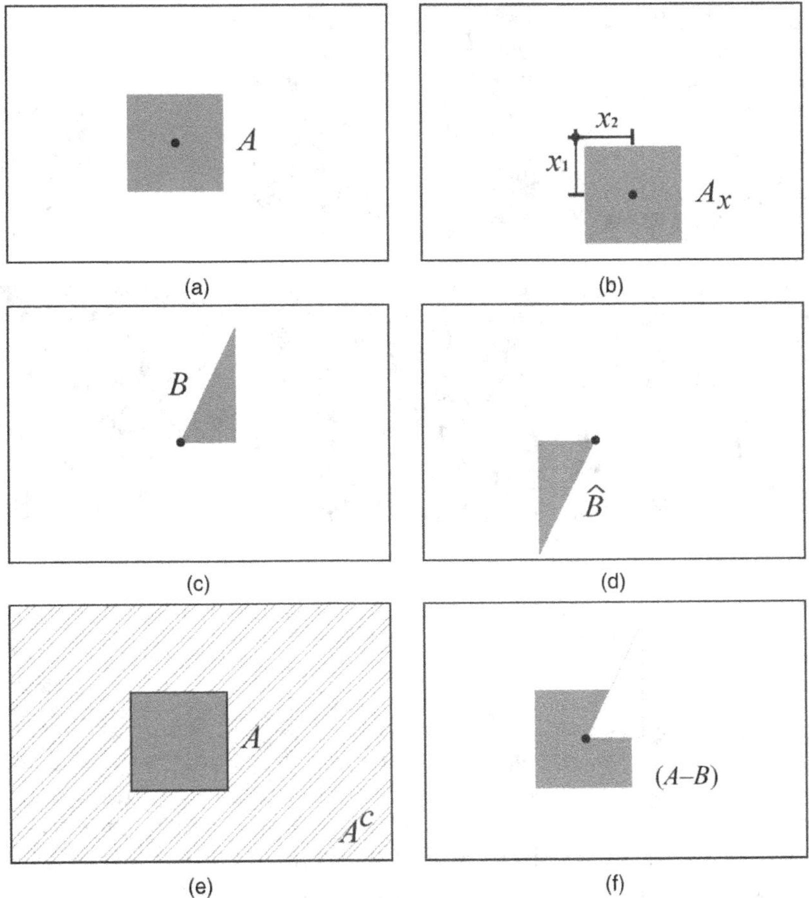

FIGURE 13.1 Basic set operations: (a) set A; (b) translation of A by $x = (x_1, x_2)$; (c) set B; (d) reflection of B; (e) set A and its complement A^c; (f) set difference $(A-B)$.

This expression leads quite easily to a single MATLAB statement that performs the intersection operation using the logical operator AND (&). Similarly, complement can be obtained using the unary NOT (~) operator, set union can be implemented using the logical operator OR (|), and set difference $(A - B)$ can be expressed as `(A & ~B)`. Figure 13.2 shows representative results for two binary input images. Note that we have followed the IPT convention, representing foreground (1-valued) pixels as white pixels against a black background.

13.2.1 The Structuring Element

The structuring element is the basic neighborhood structure associated with morphological image operations. It is usually represented as a small matrix, whose shape

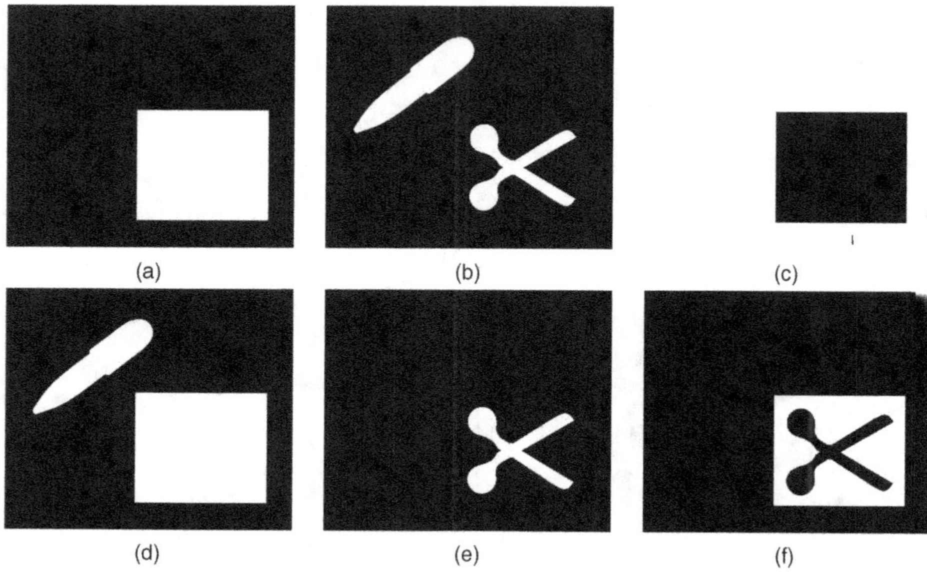

FIGURE 13.2 Logical equivalents of set theory operations: (a) Binary image (A); (b) Binary image (B); (c) Complement (A^c); (d) Union ($A \cup B$); (e) Intersection ($A \cap B$); (f) Set difference ($A-B$).

and size impact the results of applying a certain morphological operator to an image. Figure 13.3 shows two examples of SEs and how they will be represented in this chapter: the black dot corresponds to their origin (reference point), the gray squares represent 1 (true), and the white squares represent 0 (false). Although a structuring element can have any shape, its implementation requires that it should be converted to a rectangular array. For each array, the shaded squares correspond to the members of the SE, whereas the empty squares are used for padding, only.

In MATLAB

MATLAB's IPT provides a function for creating structuring elements, `strel`, which supports arbitrary shapes, as well as commonly used ones, such as square, diamond, line, and disk. The result is stored as a variable of class `strel`.

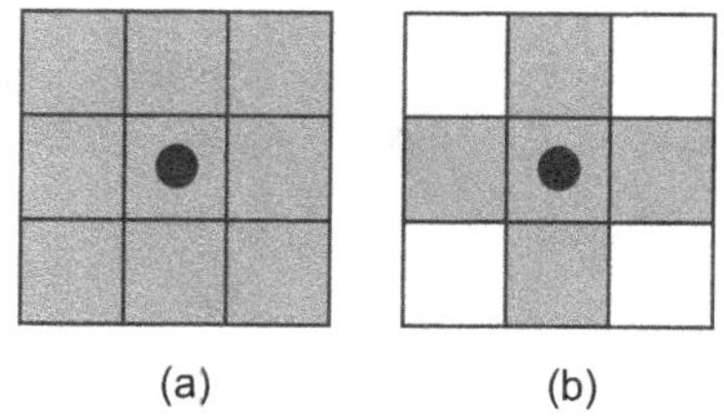

FIGURE 13.3 Examples of structuring elements: (a) square; (b) cross.

■ EXAMPLE 13.1

This example shows the creation of a square SE using `strel` and interpretation of its results.

```
>> se1 = strel('square',4)

se1 =

Flat STREL object containing 16 neighbors.
Decomposition: 2 STREL objects containing a total of 8 neighbors

Neighborhood:
     1     1     1     1
     1     1     1     1
     1     1     1     1
     1     1     1     1
```

The 4×4 square SE has been stored in a variable `se1`. The results displayed on the command window also indicate that the STREL object contains 16 neighbors, which can be decomposed (for faster execution) into 2 STREL objects of 8 elements each.

We can then use the `getsequence` function to inspect the decomposed structuring elements.

```
>> decomp = getsequence(se1)

decomp =

2x1 array of STREL objects

>> decomp(1)

ans =

Flat STREL object containing 4 neighbors.

Neighborhood:
     1
     1
     1
     1

>> decomp(2)
```

```
ans =

Flat STREL object containing 4 neighbors.

Neighborhood:
     1     1     1     1
```

■ **EXAMPLE 13.2**

This example shows the creation of two additional SEs of different shape and size using `strel`. Note that the rectangular SE does not require decomposition.

```
>> se2 = strel('diamond',4)

se2 =

Flat STREL object containing 41 neighbors.
Decomposition: 3 STREL objects containing a total of 13 neighbors

Neighborhood:
     0     0     0     0     1     0     0     0     0
     0     0     0     1     1     1     0     0     0
     0     0     1     1     1     1     1     0     0
     0     1     1     1     1     1     1     1     0
     1     1     1     1     1     1     1     1     1
     0     1     1     1     1     1     1     1     0
     0     0     1     1     1     1     1     0     0
     0     0     0     1     1     1     0     0     0
     0     0     0     0     1     0     0     0     0

>> se3 = strel('rectangle',[1 3])

se3 =

Flat STREL object containing 3 neighbors.

Neighborhood:
     1     1     1
```

13.3 DILATION AND EROSION

In this section, we discuss the two fundamental morphological operations upon which all other operations and algorithms are built: dilation and erosion.

13.3.1 Dilation

Dilation is a morphological operation whose effect is to "grow" or "thicken" objects in a binary image. The extent and direction of this thickening are controlled by the size and shape of the structuring element.

Mathematically, the dilation of a set A by B, denoted $A \oplus B$, is defined as

$$A \oplus B = \left\{ z | (\hat{B})_z \cap A \neq \emptyset \right\} \tag{13.7}$$

Figure 13.4 illustrates how dilation works using the same input object and three different structuring elements: rectangular SEs cause greater dilation along their longer dimension, as expected.

In MATLAB

Morphological dilation is implemented by function `imdilate`, which takes two parameters: an image and a structuring element. In Tutorial 13.1, you will have an opportunity to learn more about this function and apply it to binary images.

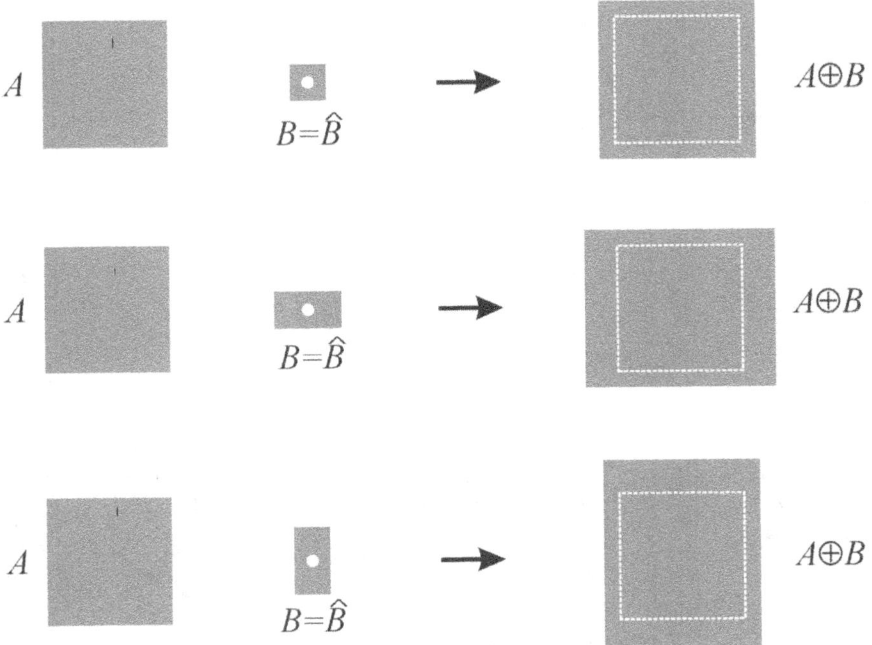

FIGURE 13.4 Example of dilation using three different rectangular structuring elements.

■ EXAMPLE 13.3

This example shows the application of `imdilate` to a small binary test image with two different SEs.

```
>> a = [ 0 0 0 0 0 ; 0 1 1 0 0; 0 1 1 0 0; 0 0 1 0 0; 0 0 0 0 0]

a =

     0     0     0     0     0
     0     1     1     0     0
     0     1     1     0     0
     0     0     1     0     0
     0     0     0     0     0

>> se1 = strel('square',2)

se1 =

Flat STREL object containing 4 neighbors.

Neighborhood:
     1     1
     1     1


>> b = imdilate(a,se1)

b =

     0     0     0     0     0
     0     1     1     1     0
     0     1     1     1     0
     0     1     1     1     0
     0     0     1     1     0
```

The dilation of a with SE `se1` (whose reference pixel is the top left corner) caused every pixel in a whose value was 1 to be replaced with a 2×2 square of pixels equal to 1.

```
>> se2 = strel('rectangle', [1 2])

se2 =
```

```
Flat STREL object containing 2 neighbors.

Neighborhood:
     1     1


>> c = imdilate(a,se2)

c =

     0     0     0     0     0
     0     1     1     1     0
     0     1     1     1     0
     0     0     1     1     0
     0     0     0     0     0
```

The dilation of a with SE se2 (whose reference pixel is the leftmost one) caused every pixel in a whose value was 1 to be replaced with a 1×2 rectangle of pixels equal to 1, that is, a predominantly horizontal dilation.

13.3.2 Erosion

Erosion is a morphological operation whose effect is to "shrink" or "thin" objects in a binary image. The direction and extent of this thinning is controlled by the shape and size of the structuring element.

Mathematically, the erosion of a set A by B, denoted $A \ominus B$, is defined as

$$A \ominus B = \left\{ z | (\hat{B})_z \cup A^c \neq \emptyset \right\} \tag{13.8}$$

Figure 13.5 illustrates how erosion works using the same input object and three different structuring elements. Erosion is more severe in the direction of the longer dimension of the rectangular SE.

In MATLAB

Morphological erosion is implemented by function imerode, which takes two parameters: an image and a structuring element. In Tutorial 13.1, you will have an opportunity to learn more about this function.

■ EXAMPLE 13.4

This example shows the application of imerode to a small binary test image with two different SEs.

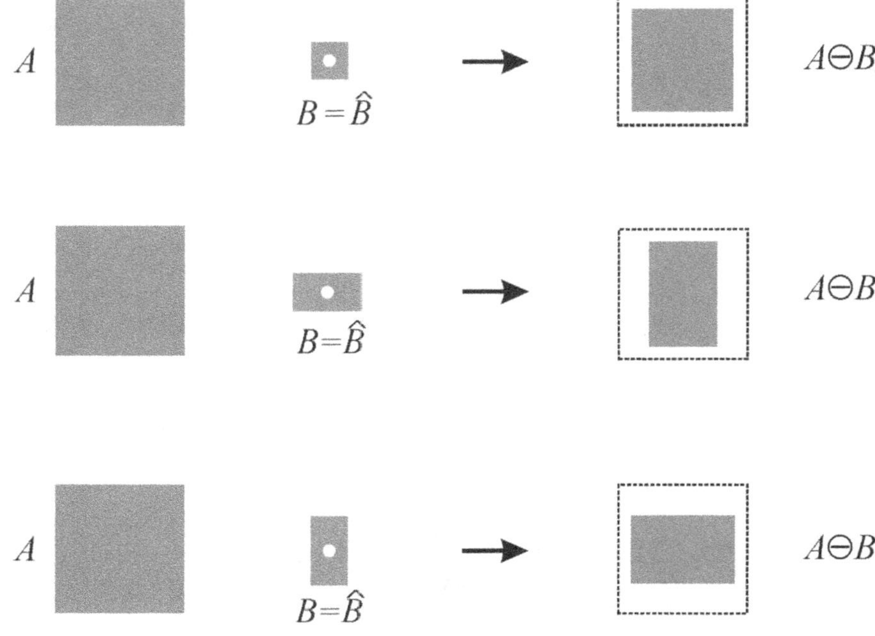

FIGURE 13.5 Example of erosion using three different rectangular structuring elements.

```
>> a = [ 0 0 0 0 0 ; 0 1 1 1 0; 1 1 1 0 0; 0 1 1 1 1; 0 0 0 0 0]

a =

     0     0     0     0     0
     0     1     1     1     0
     1     1     1     0     0
     0     1     1     1     1
     0     0     0     0     0

>> se1 = strel('square',2)

se1 =

Flat STREL object containing 4 neighbors.

Neighborhood:
     1     1
     1     1


>> b = imerode(a,se1)
```

```
b =

     0     0     0     0     0
     0     1     0     0     0
     0     1     0     0     0
     0     0     0     0     0
     0     0     0     0     0
```

The erosion of a with SE `se1` (whose reference pixel is the top left corner) caused the disappearance of many pixels in a; the ones that remained were the ones corresponding to the top left corner of a 2 × 2 square of pixels equal to 1.

```
>> se2 = strel('rectangle', [1 2])

se2 =

Flat STREL object containing 2 neighbors.

Neighborhood:
     1     1


>> c = imerode(a,se2)

c =

     0     0     0     0     0
     0     1     1     0     0
     1     1     0     0     0
     0     1     1     1     1
     0     0     0     0     0
```

The erosion of a with SE `se2` (whose reference pixel is the leftmost one) caused the disappearance of many pixels in a; the ones that remained were the ones corresponding to the leftmost pixel of a 1×2 rectangle of pixels equal to 1.

Erosion is the dual operation of dilation and vice versa:

$$(A \ominus B)^c = A^c \oplus \hat{B} \tag{13.9}$$

$$A \oplus B = (A^c \ominus \hat{B})^c \tag{13.10}$$

Erosion and dilation can also be interpreted in terms of whether a SE *hits* or *fits* an image (region), as follows [Eff00].

For dilation, the resulting image $g(x, y)$, given an input image $f(x, y)$ and a SE *se*, will be

$$g(x, y) = \begin{cases} 1 & \text{if } se \text{ hits } f \\ 0 & \text{otherwise} \end{cases} \tag{13.11}$$

for all x and y.

For erosion, the resulting image $g(x, y)$, given an input image $f(x, y)$ and a SE *se*, will be

$$g(x, y) = \begin{cases} 1 & \text{if } se \text{ fits } f \\ 0 & \text{otherwise} \end{cases} \tag{13.12}$$

for all x and y.

13.4 COMPOUND OPERATIONS

In this section, we present morphological operations that combine the two fundamental operations (erosion and dilation) in different ways.

13.4.1 Opening

The morphological opening of set A by B, represented as $A \circ B$, is the erosion of A by B followed by the dilation of the result by B. Mathematically,

$$A \circ B = (A \ominus B) \oplus B \tag{13.13}$$

Alternatively, the opening operation can be expressed using set notation as

$$A \circ B = \bigcup \{(B)_z | (B)_z \subseteq A\} \tag{13.14}$$

where $\bigcup \{\bullet\}$ represents the union of all sets within the curly braces, and the symbol $\subseteq$ means "is a subset of."

The opening operation is *idempotent*, that is, once an image has been opened with a certain SE, subsequent applications of the opening algorithm with the same SE will not cause any effect on the image. Mathematically,

$$(A \circ B) \circ B = A \circ B \tag{13.15}$$

Morphological opening is typically used to remove thin protrusions from objects and to open up a gap between objects connected by a thin bridge without shrinking the objects (as erosion would have done). It also causes a smoothening of the object's contour (Figure 13.6).

The geometric interpretation of the opening operation is straightforward: $A \circ B$ is the union of all translations of B that fit entirely within A (Figure 13.7).

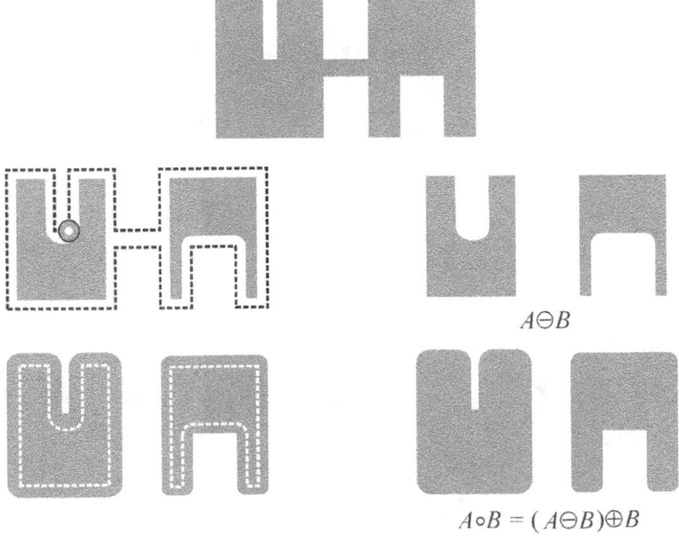

FIGURE 13.6 Example of morphological opening.

In MATLAB

Morphological opening is implemented by function `imopen`, which takes two parameters: input image and structuring element. In Tutorial 13.1, you will have an opportunity to learn more about this function.

13.4.2 Closing

The morphological closing of set A by B, represented as $A \bullet B$, is the dilation of A by B followed by the erosion of the result by B. Mathematically,

$$A \bullet B = (A \oplus B) \ominus B \tag{13.16}$$

Just as with opening, the closing operation is *idempotent*, that is, once an image has been closed with a certain SE, subsequent applications of the opening algorithm

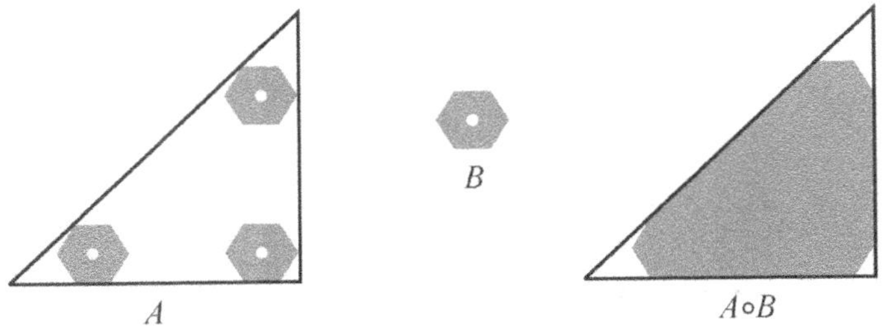

FIGURE 13.7 Geometric interpretation of the morphological opening operation.

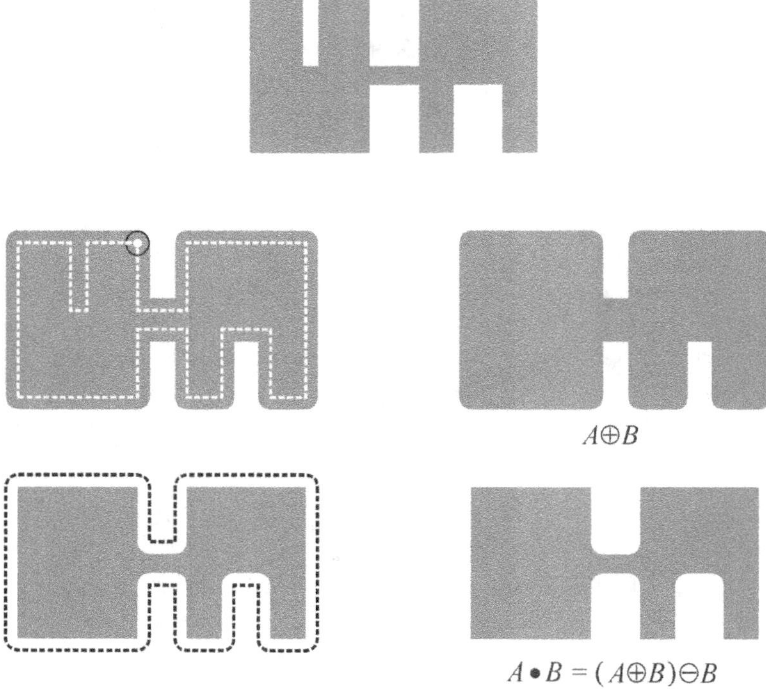

FIGURE 13.8 Example of morphological closing.

with the same SE will not cause any effect on the image. Mathematically,

$$(A \bullet B) \bullet B = A \bullet B \tag{13.17}$$

Morphological closing is typically used to fill small holes, fuse narrow breaks, and close thin gaps in the objects within an image, without changing the objects' size (as dilation would have done). It also causes a smoothening of the object's contour (Figure 13.8).

The geometric interpretation of the closing operation is as follows: $A \bullet B$ is the complement of the union of all translations of B that do not overlap A (Figure 13.9).

Closing is the dual operation of opening and vice versa:

$$A \bullet B = (A^c \circ B)^c \tag{13.18}$$

$$A \circ B = (A^c \bullet B)^c \tag{13.19}$$

In MATLAB

Morphological closing is implemented by function `imclose`, which takes two parameters: input image and structuring element. In Tutorial 13.1, you will have an opportunity to learn more about this function.

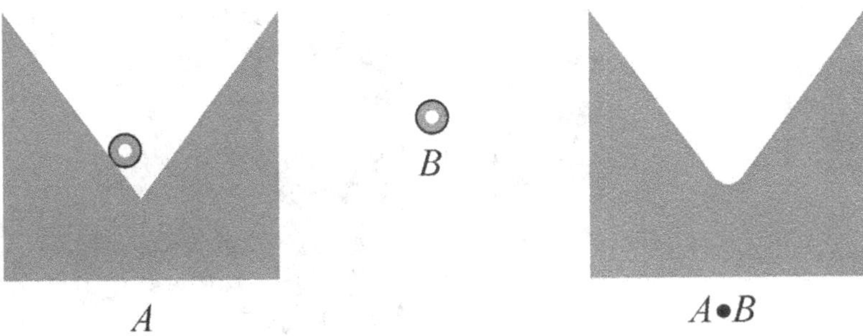

FIGURE 13.9 Geometric interpretation of the morphological closing operation. Adapted and redrawn from [GW08].

13.4.3 Hit-or-Miss Transform

The hit-or-miss (HoM) transform is a combination of morphological operations that uses two structuring elements (B_1 and B_2) designed in such a way that the output image will consist of all locations that match the pixels in B_1 (a *hit*) and that have none of the pixels in B_2 (a *miss*). Mathematically, the HoM transform of image A by the structuring element set B ($B = (B_1, B_2)$), denoted $A \otimes B$, is defined as

$$A \otimes B = (A \ominus B_1) \cap (A^c \ominus B_2) \tag{13.20}$$

Alternatively, the HoM transform can be expressed as

$$A \otimes B = (A \ominus B_1) - (A \oplus \hat{B}_2) \tag{13.21}$$

Figure 13.10[1] shows how the HoM transform can be used to locate squares of a certain size in a binary image. Figure 13.10a shows the input image (A), consisting of two squares against a background, and Figure 13.10b shows its complement (A^c). The two structuring elements, in this case, have been chosen to be a square of the same size as the smallest square in the input image, B_1 (Figure 13.10c), and a square of the larger size, indicating that the smallest square should be surrounded by pixels of opposite color, B_2 (Figure 13.10d). Figure 13.10e shows the partial result after the first erosion: the smallest square is hit at a single spot, magnified, and painted red for viewing purposes, whereas the largest square is hit at many points. Figure 13.10f shows the final result, containing the set of all points[2] for which the HoM transform found a hit for B_1 in A *and* a hit for B_2 in A^c (which is equivalent to a miss for B_2 in A).

In MATLAB

The binary hit-or-miss transformation is implemented by function `bwhitmiss`, whose basic syntax is `J = bwhitmiss(I, B1, B2)`, where `I` is the input

[1]The thin black borders in parts (a), (d), (e), and (f) were added for viewing purposes only.

[2]In this case, this set contains only one point, which was enlarged and painted red for viewing purposes.

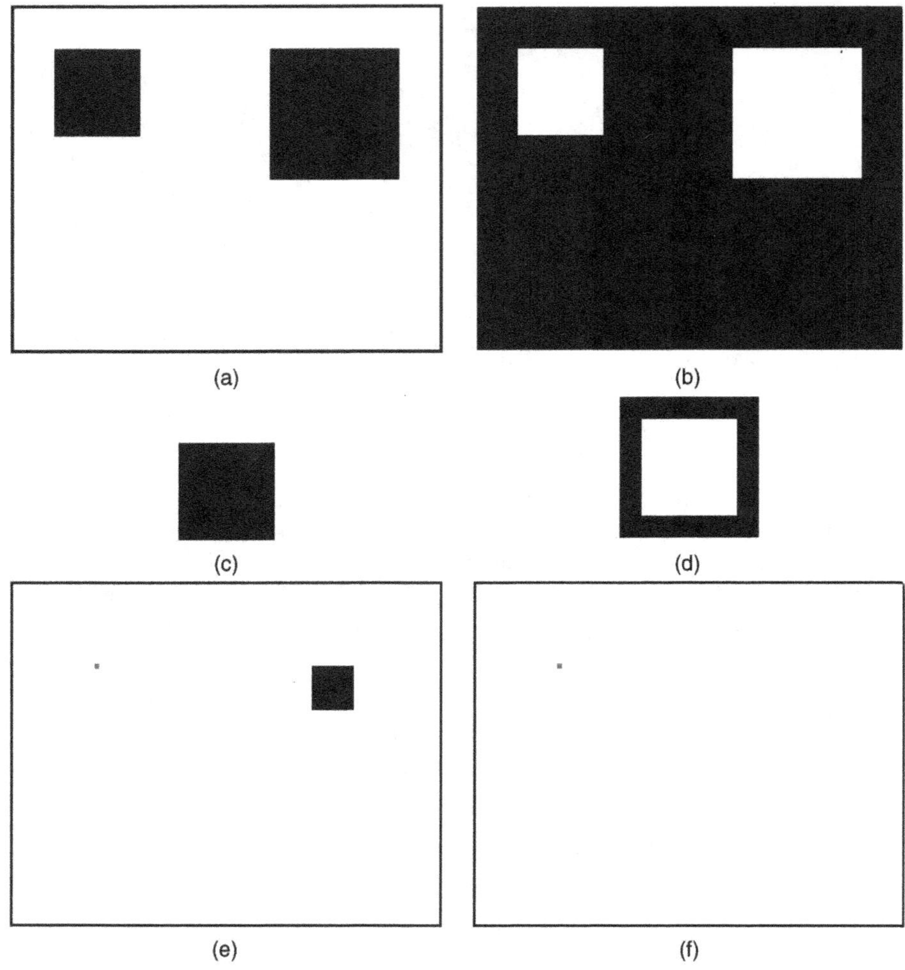

FIGURE 13.10 Example of HoM transform.

image, `B1` and `B2` are the structuring elements, and `J` is the resulting image. You will learn how to use this function in Tutorial 13.1.

13.5 MORPHOLOGICAL FILTERING

Morphological filters are Boolean filters that apply a many-to-one binary (or Boolean) function h within a window W in the binary input image $f(x, y)$, producing at the output an image $g(x, y)$ given by

$$g(x, y) = h\left[Wf(x, y)\right] \tag{13.22}$$

Examples of Boolean operations (denoted as h in equation (13.22)) are as follows:

- OR: This (as we have seen in Section 13.2) is equivalent to a morphological dilation with a square SE of the same size as W.
- AND: This (as we have seen in Section 13.2) is equivalent to a morphological erosion with a square SE of the same size as W.
- MAJ (Majority): This is the morphological equivalent to a median filter (introduced in Section 12.3.2) applicable to binary images.

Morphological filters can also be used in noise reduction. Let A be a binary image corrupted by impulse (salt and pepper) noise. The application of a morphological opening operation followed by a morphological closing will remove a significant amount of noise present in the input image, resulting in an image C given by

$$C = ((A \circ B) \bullet B) \tag{13.23}$$

where B is the chosen SE.

■ EXAMPLE 13.5

Figure 13.11 shows an example of morphological filtering for noise removal on a binary image using a circular SE of radius equal to 2 pixels. Part (a) shows the (641 × 535) input image, contaminated by salt and pepper noise. Part (b) shows the partial result (after the opening operation): at this stage, all the "salt" portion of the noise has been removed, but the "pepper" noise remains. Finally, part (c) shows the result of applying closing to the result of the opening operation. The noise has been completely removed, at the expense of imperfections at the edges of the objects. Such imperfections would be even more severe for larger SEs, as demonstrated in part (d), where the radius of the circular SE is equal to 4 pixels.

13.6 BASIC MORPHOLOGICAL ALGORITHMS

In this section, we present a collection of simple and useful morphological algorithms and show how they can be implemented using MATLAB and the IPT. Several of these algorithms will also be covered in Tutorial 13.2.

In MATLAB

The IPT function `bwmorph` implements a number of useful morphological operations and algorithms, listed in Table 13.1. This function takes three arguments: input image, desired operation, and the number of times the operation is to be repeated.

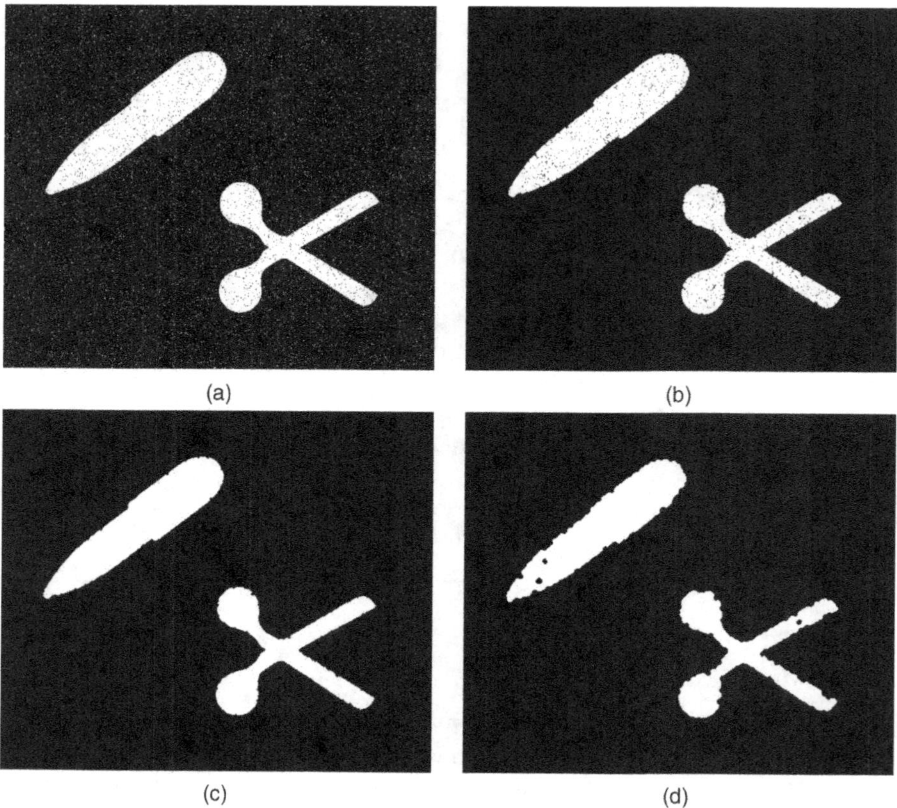

FIGURE 13.11 Morphological filtering. (a) input (noisy) image; (b) partial result (after opening) with SE of radius = 2 pixels; (c) final result with SE of radius = 2 pixels; (d) final result with SE of radius = 4 pixels.

■ EXAMPLE 13.6

In this example, we use `bwmorph` to apply different operations to the same small test image (Figure 13.12a).[3]

Here is the sequence of steps used to obtain the results in Figure 13.12(b–f):

```
B = bwmorph(A,'skel', Inf);
C = bwmorph(B,'spur',Inf);
D = bwmorph(A,'remove');
E = bwmorph(D,'thicken',3);
F = bwmorph(E,'thin',3);
```

[3]The goal is to give you a quick glance at some of the most useful morphological algorithms and their effect on an input test image. A detailed formulation of each operation is beyond the scope of this book. Refer to "Learn More About It" section at the end of the chapter.

TABLE 13.1 Operations Supported by bwmorph

Operation	Description
`bothat`	Subtract the input image from its closing
`bridge`	Bridge previously unconnected pixels
`clean`	Remove isolated pixels (1s surrounded by 0s)
`close`	Perform binary closure (dilation followed by erosion)
`diag`	Diagonal fill to eliminate 8-connectivity of background
`dilate`	Perform dilation using the structuring element 1s(3)
`erode`	Perform erosion using the structuring element 1s(3)
`fill`	Fill isolated interior pixels (0s surrounded by 1s)
`hbreak`	Remove H-connected pixels
`majority`	Set a pixel to 1 if five or more pixels in its 3×3 neighborhood are 1s
`open`	Perform binary opening (erosion followed by dilation)
`remove`	Set a pixel to 0 if its 4-connected neighbors are all 1s, thus leaving only boundary pixels
`shrink`	With N = Inf, shrink objects to points; shrink objects with holes to connected rings
`skel`	With N = Inf, remove pixels on the boundaries of objects without allowing objects to break apart
`spur`	Remove endpoints of lines without removing small objects completely
`thicken`	With N = Inf, thicken objects by adding pixels to the exterior of objects without connecting previously unconnected objects
`thin`	With N = Inf, remove pixels so that an object without holes shrinks to a minimally connected stroke, and an object with holes shrinks to a ring halfway between the hold and outer boundary
`tophat`	Subtract the opening from the input image

13.6.1 Boundary Extraction

Dilation, erosion, and set difference operations can be combined to perform boundary extraction of a set A, denoted by $\mathcal{BE}(A)$, as follows:

- *Internal Boundary*: Consists of the pixels in A that sit at the edge of A.

$$\mathcal{BE}(A) = A - (A \ominus B) \tag{13.24}$$

- *External Boundary*: Consists of the pixels outside A that sit immediately next to A.

$$\mathcal{BE}(A) = (A \oplus B) - A \tag{13.25}$$

- *Morphological Gradient*: Consists of the combination of internal and external boundaries.

$$\mathcal{BE}(A) = (A \oplus B) - (A \ominus B) \tag{13.26}$$

where B is a suitable structuring element.

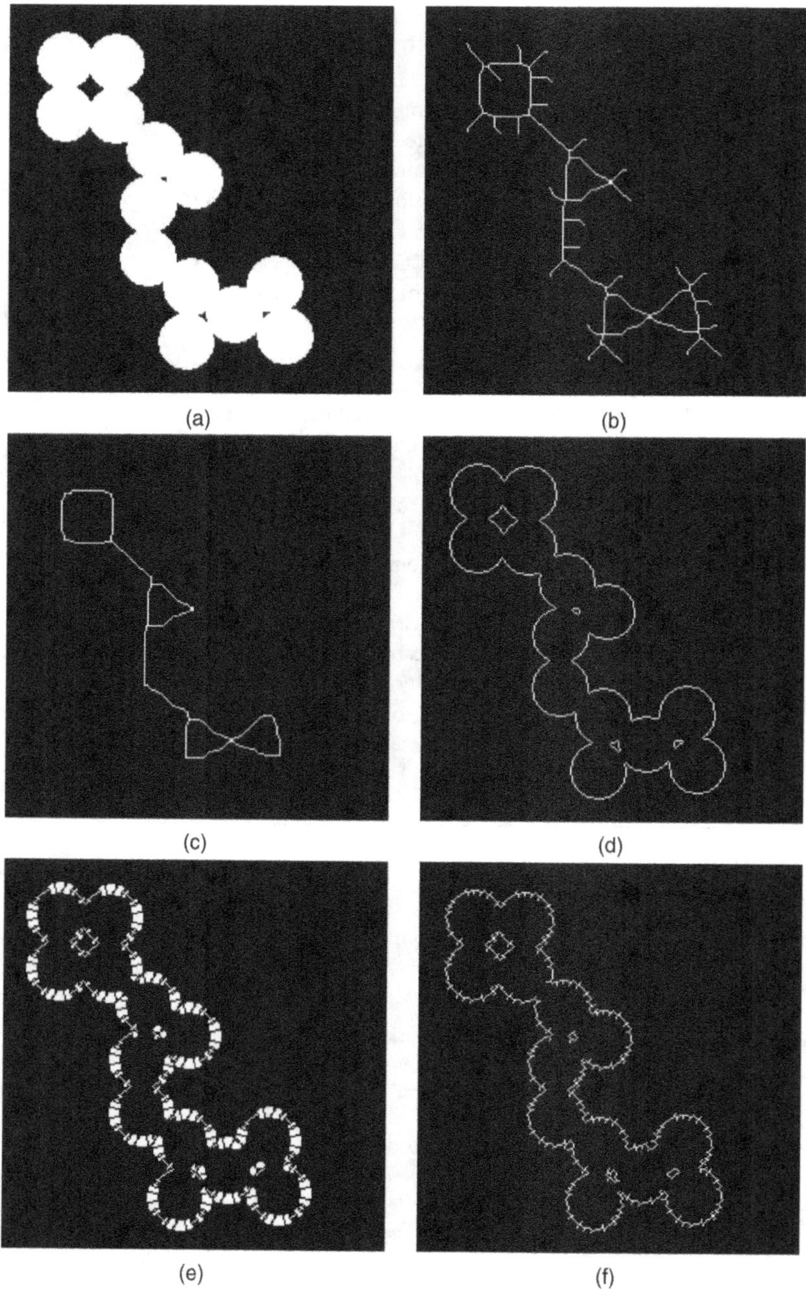

FIGURE 13.12 Morphological algorithms. (a) input image; (b) skeleton of (a); (c) pruning spurious pixels from (b); (d) removing interior pixels from (a); (e) thickening the image in (d); (f) thinning the image in (e). Original image: courtesy of MathWorks.

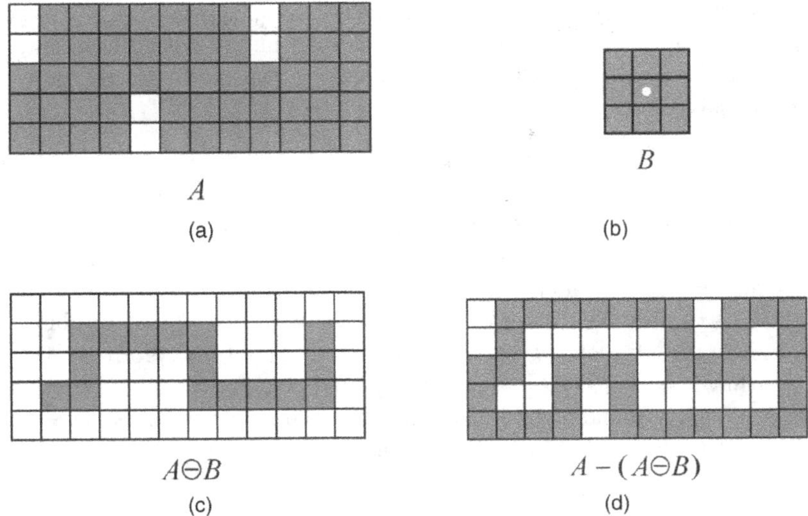

FIGURE 13.13 Boundary extraction.

Figure 13.13 shows an example of internal boundary extraction of a small (5×12) object using a 3×3 square as SE.

In MATLAB

The IPT function `bwperim` returns a binary image containing only the perimeter pixels of objects in the input image.

■ EXAMPLE 13.7

This example shows the application of `bwperim` (with 8-connectivity, to be consistent with the SE in Figure 13.13b) to extract the internal boundary of a small binary test image identical to the image in Figure 13.13a.

```
a = ones(5,12)
a (1:2,1)=0
a (1:2,9)=0
a (4:5,5)=0
b = bwperim(a,8)
```

When you execute the steps above, you will confirm that the result (stored in variable `b`) is identical to the image in Figure 13.13d.

13.6.2 Region Filling

In this section, we present an algorithm that uses morphological and set operations to fill regions (which can be thought of as *holes*) in a binary image.

Let p be a pixel in a region surrounded by an 8-connected boundary, A. The goal of a region filling algorithm is to fill up the entire region with 1s using p as a starting point (i.e., setting it as 1). Region filling can be accomplished using an iterative procedure, mathematically expressed as follows:

$$X_k = (X_{k-1} \oplus B) \cap A^c \quad k = 1, 2, 3, \ldots \tag{13.27}$$

where $X_0 = p$ and B is the cross-shaped structuring element. The algorithm stops at the kth iteration if $X_k = X_{k-1}$. The union of X_k and A contains the original boundary (A) and all the pixels within it labeled as 1.

Figure 13.14 illustrates the process. Part (a) shows the input image, whose complement is in part (b). Part (c) shows the partial results of the algorithm after each iteration (the number inside the square): the initial pixel (p, top left) corresponds to iteration 0. This example requires six iterations to complete and uses the SE shown in part (e). Part (d) shows the union of X_6 and A.

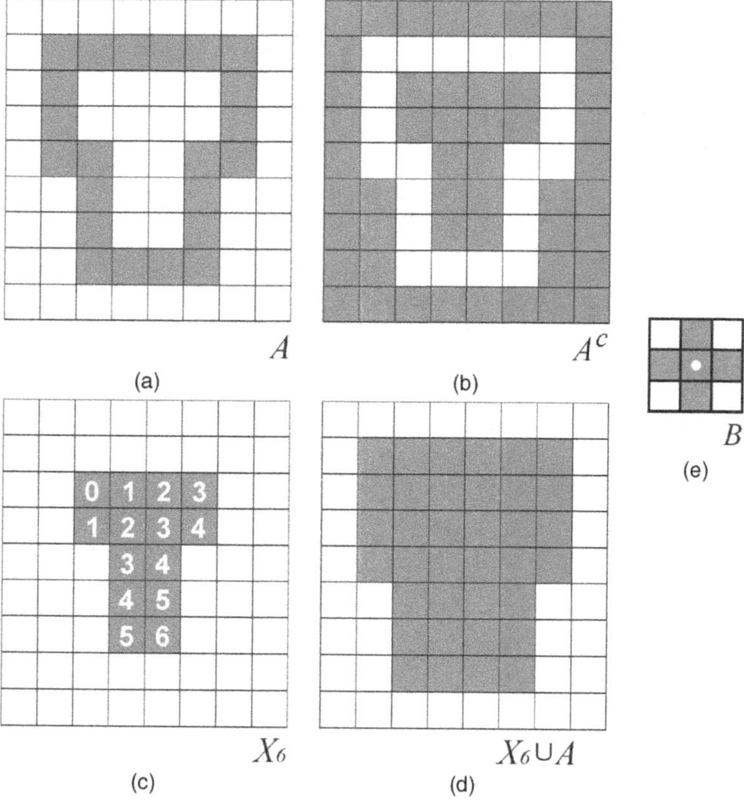

FIGURE 13.14 Region filling: (a) input image; (b) complement of (a); (c) partial results (numbered according to the iteration in the algorithm described by equation (13.27); (d) final result; (e) structuring element.

In MATLAB

The IPT function `imfill` implements region filling. It can be used in an interactive mode (where the user clicks on the image pixels that should be used as starting points) or by passing the coordinates of the starting points. You will use `imfill` in Tutorial 13.2.

13.6.3 Extraction and Labeling of Connected Components

Morphological concepts and operations can be used to extract and label connected components[4] in a binary image.

The morphological algorithm for extraction of connected components is very similar to the region filling algorithm described in Section 13.6.2. Let A be a set containing one or more connected components and p ($p \in A$) be a starting pixel. The process of finding all other pixels in a component can be accomplished using an iterative procedure, mathematically expressed as follows:

$$X_k = (X_{k-1} \oplus B) \cap A, \quad k = 1, 2, 3, \ldots \tag{13.28}$$

where $X_0 = p$ and B is a suitable structuring element: cross-shaped for 4-connectivity, 3×3 square for 8-connectivity.

The algorithm stops at the kth iteration if $X_k = X_{k-1}$.

Figure 13.15 shows an example of extraction of connected components. Part (a) shows the initial set A and the starting pixel p, represented by the number 0. Part (e) shows the SE used in this case. Parts (b) and (c) show the result of the first and second iterations, respectively. The final result (after six iterations) is shown in Figure 13.15d.

In MATLAB

The IPT has a very useful function for computing connected components in a binary image: `bwlabel`. The function takes two parameters (input image and connectivity criterion: 4 or 8 (default)) and returns a matrix of the same size as the input image, containing labels for the connected objects in the image. Pixels labeled 0 correspond to the background; pixels labeled 1 and higher correspond to the connected components in the image. You will learn how to use this function in Tutorial 13.2.

Visualization of the connected components extracted by `bwlabel` is often achieved by pseudocoloring the label matrix (assigning a different color to each component) using the `label2rgb` function.

The IPT also has a function for selecting objects, that is, connected components, in a binary image: `bwselect`. It can be used in an interactive mode (where the user clicks on the image pixels that should be used as starting points) or by passing the coordinates of the starting points. You will use `bwselect` in Tutorial 13.2.

[4]The concept of *connected component* (a set of pixels that are 4- or 8- connected to each other) was introduced in Section 2.3.

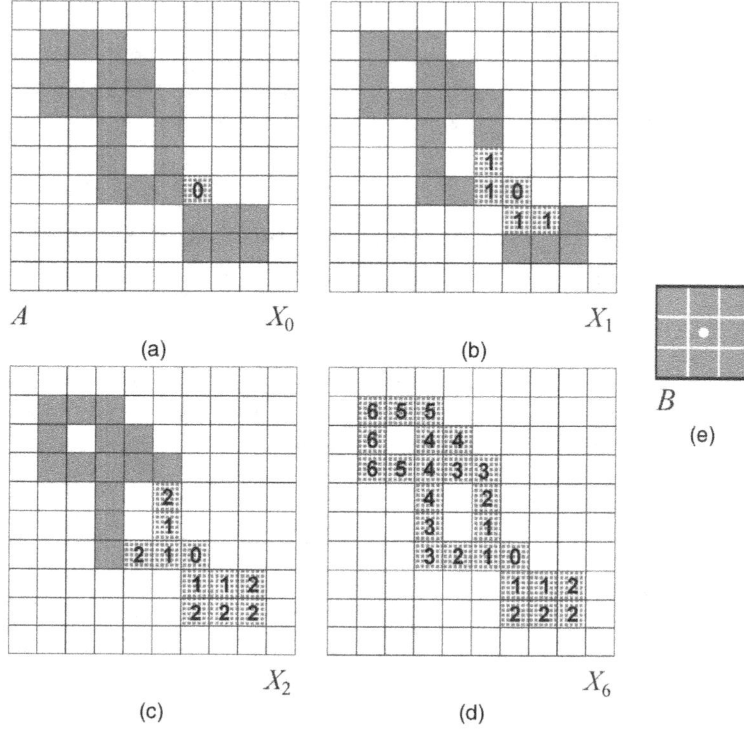

FIGURE 13.15 Extraction of connected components: (a) input image; (b) first iteration; (c) second iteration; (d) final result, showing the contribution of each iteration (indicated by the numbers inside the squares); (e) structuring element.

13.7 GRAYSCALE MORPHOLOGY

Many morphological operations originally developed for binary images can be extended to grayscale images. This section presents the grayscale version of the basic morphological operations (erosion, dilation, opening, and closing) and introduces the top-hat and bottom-hat transformations.

The mathematical formulation of grayscale morphology uses an input image $f(x, y)$ and a structuring element (SE) $b(x, y)$. Structuring elements in grayscale morphology come in two categories: nonflat and flat.

In MATLAB

Nonflat SEs can be created with the same function used to create flat SEs: `strel`. In the case of nonflat SEs, we must also pass a second matrix (containing the height values) as a parameter.

13.7.1 Dilation and Erosion

The dilation of an image $f(x, y)$ by a flat SE $b(x, y)$ is defined as

$$[A \oplus B](x, y) = \max\{f(x+s, y+t) | (s, t) \in D_{\mathrm{b}}\} \tag{13.29}$$

where D_{b} is called the *domain* of b.

For a nonflat SE, $b_N(x, y)$, equation (13.29) becomes

$$[A \oplus B](x, y) = \max\{f(x+s, y+t) + b_N(s, t) | (s, t) \in D_{\mathrm{b}}\} \tag{13.30}$$

Some references use $f(x-s, y-t)$ in equations (13.29) and (13.30), which just requires the SE to be rotated by 180° or mirrored around its origin, that is, $\hat{B}(x, y) = b(-x, -y)$.

The erosion of an image $f(x, y)$ by a flat SE $b(x, y)$ is defined as

$$[A \ominus B](x, y) = \min\{f(x+s, y+t) | (s, t) \in D_{\mathrm{b}}\} \tag{13.31}$$

where D_{b} is called the *domain* of b.

For a nonflat SE, $b_N(x, y)$, equation (13.31) becomes

$$[A \ominus B](x, y) = \min\{f(x+s, y+t) - b_N(s, t) | (s, t) \in D_{\mathrm{b}}\} \tag{13.32}$$

■ **EXAMPLE 13.8**

Figure 13.16 shows examples of grayscale erosion and dilation with a nonflat ball-shaped structuring element with radius 5 (`se = strel('ball',5,5);`).

13.7.2 Opening and Closing

Grayscale opening and closing are as defined exactly as they were for binary morphology.

The opening of grayscale image $f(x, y)$ by SE $b(x, y)$ is given by

$$f \circ b = (f \ominus b) \oplus b \tag{13.33}$$

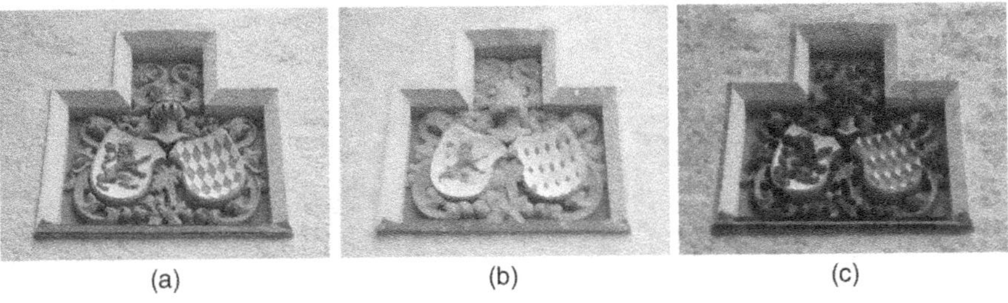

FIGURE 13.16 Grayscale erosion and dilation with a nonflat ball-shaped structuring element with radius 5: (a) input image; (b) result of dilation; (c) result of erosion.

The closing of grayscale image $f(x, y)$ by SE $b(x, y)$ is given by

$$f \bullet b = (f \oplus b) \ominus b \tag{13.34}$$

Grayscale opening and closing are used in combination for the purpose of smoothing and noise reduction, similar to what we saw in Section 13.5 for binary images. This technique, known as *morphological smoothing*, is discussed in the following example.

■ EXAMPLE 13.9

Figure 13.17 shows examples of applying grayscale opening and closing with a flat disk-shaped structuring element with radius 3 (`se = strel('disk',3);`) for noise reduction. The upper part of the figure refers to an input image corrupted by

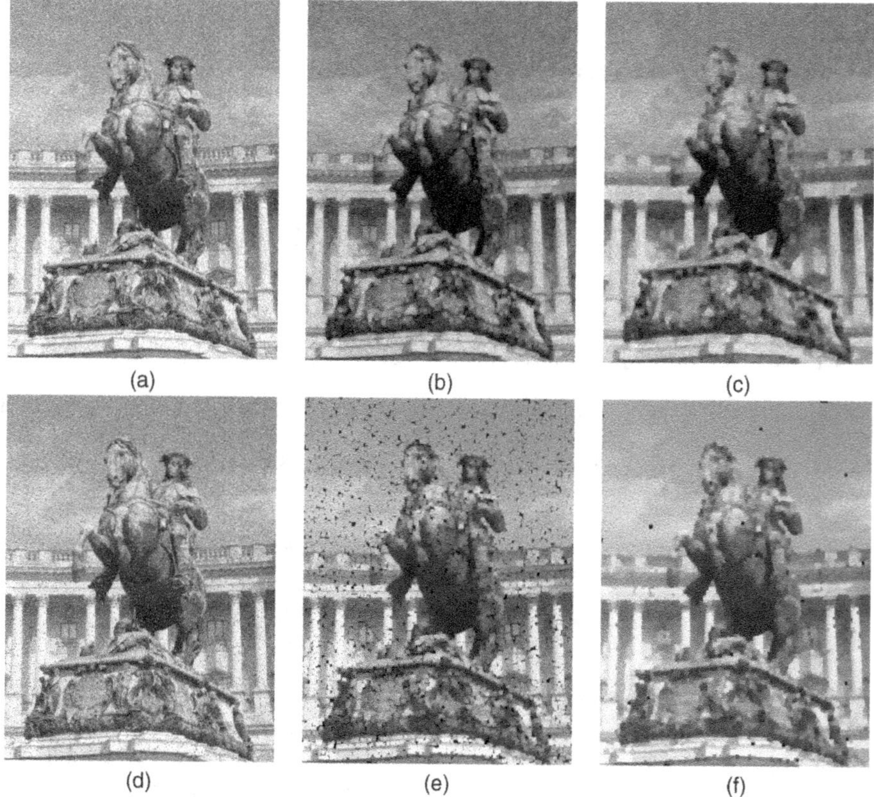

FIGURE 13.17 Grayscale opening and closing with a flat disk-shaped structuring element with radius 3: (a) input image (Gaussian noise); (b) result of opening image (a); (c) result of closing image (b); (d) input image (salt and pepper noise); (e) result of opening image (d); (f) result of closing image (e).

Gaussian noise of mean 0 and variance 0.01, whereas the bottom part refers to an input image corrupted by salt and pepper noise.

13.7.3 Top-Hat and Bottom-Hat Transformations

The top-hat and bottom-hat transformations are operations that combine morphological openings and closings with image subtraction. The top-hat transformation is often used in the process of *shading correction*, which consists in compensating for nonuniform illumination of the scene. Top-hat and bottom-hat operations can be combined for contrast enhancement purposes.

Mathematically, the top-hat transformation of a grayscale image f is defined as the subtraction of f from its opening with SE b:

$$\text{Top-hat}\,(f) = f - (f \circ b) \tag{13.35}$$

and the bottom-hat transformation of f is given by the subtraction of the closing of f with SE b from f:

$$\text{Bottom-hat}\,(f) = (f \bullet b) - f \tag{13.36}$$

In MATLAB

IPT functions `imtophat` and `imbothat` implement top-hat and bottom-hat filtering, respectively.

■ EXAMPLE 13.10

Figure 13.18 shows an example of applying `imtophat` and `imbothat` to improve the contrast of an input image `I` using a flat disk-shaped SE of radius 3. The output image was obtained as follows:

```
J = imsubtract(imadd(I,imtophat(I,se)), imbothat(I,se));
```

13.8 TUTORIAL 13.1: BINARY MORPHOLOGICAL IMAGE PROCESSING

Goal

The goal of this tutorial is to learn how to implement the basic binary morphological operations in MATLAB.

Objectives

- Learn how to dilate an image using the `imdilate` function.
- Learn how to erode an image using the `imerode` function.

(a)

(b)

FIGURE 13.18 Example of using top-hat and bottom-hat filtering for contrast improvement: (a) input image; (b) output image.

- Learn how to open an image using the `imopen` function.
- Learn how to close an image using the `imclose` function.
- Explore the hit-or-miss transformation using the `bwhitmiss` function.

Procedure

1. Load and display the `blobs` test image.

```
I = imread('blobs.png');
figure, imshow(I), title('Original Image');
```

Dilation

2. Create a 3×3 structuring element with all coefficients equal to 1.

```
SE_1 = strel('square',3);
```

3. Perform dilation using the generated SE and display the results.

```
I_dil_1 = imdilate(I,SE_1);
figure,  imshow(I_dil_1), title('Dilated with 3x3');
```

Question 1 What happened when we dilated the image with a square 3×3 SE?

Let us now see what happens when using a SE shape other than a square.

4. Create a 1×7 SE with all elements equal to 1 and dilate the image.

```
SE_2 = strel('rectangle', [1 7]);
I_dil_2 = imdilate(I, SE_2);
figure, imshow(I_dil_2), title('Dilated with 1x7');
```

Question 2 What is the difference in results between the dilation using the 3×3 SE and the 1×7 SE?

Question 3 How would the results change if we use a 7×1 SE? Verify your prediction.

Question 4 What other SE shapes does the `strel` function support?

Erosion

The procedure for erosion is similar to that for dilation. First, we create a structuring element with the `strel` function, followed by eroding the image using the `imerode` function. We will use the same two SEs already created in the previous steps.

5. Erode the original image with a 3×3 structuring element and display the results.

```
I_ero_1 = imerode(I, SE_1);
figure, imshow(I), title('Original Image');
figure, imshow(I_ero_1), title('Eroded with 3x3');
```

6. Erode the original image with a 1×7 structuring element.

```
I_ero_2 = imerode(I, SE_2);
figure, imshow(I_ero_2), title('Eroded with 1x7');
```

Question 5 What is the effect of eroding the image?

Question 6 How does the size and shape of the SE affect the results?

Opening

7. Perform morphological opening on the original image using the square 3×3 SE created previously.

```
I_open_1 = imopen(I, SE_1);
figure, imshow(I), title('Original Image');
figure, imshow(I_open_1), title('Opening the image');
```

Question 7 What is the overall effect of opening a binary image?

8. Compare opening with eroding.

```
figure, subplot(2,2,1), imshow(I), title('Original Image');
subplot(2,2,2), imshow(I_ero_1), title('Result of Erosion');
subplot(2,2,3), imshow(I_open_1), title('Result of Opening (3x3)');
```

Question 8 How do the results of opening and erosion compare?

9. Open the image with a 1×7 SE.

```
I_open_2 = imopen(I, SE_2);
subplot(2,2,4), imshow(I_open_2), title('Result of Opening (1x7)');
```

Question 9 How does the shape of the SE affect the result of opening?

Closing

10. Create a square 5×5 SE and perform morphological closing on the image.

```
SE_3 = strel('square',5);
I_clo_1 = imclose(I, SE_3);
figure,  imshow(I), title('Original Image');
figure,  imshow(I_clo_1), title('Closing the image');
```

Question 10 What was the overall effect of closing this image?

11. Compare closing with dilation.

```
figure, imshow(I), title('Original Image');
figure, imshow(I_dil_1), title('Dilating the image');
figure, imshow(I_clo_1), title('Closing the image');
```

Question 11 How does closing differ from dilation?

Hit-or-Miss Transformation

The HoM transformation is implemented through the `bwhitmiss` function, which takes three variables: an image and two structuring elements. Technically, the operation will keep pixels whose neighbors match the first structuring element, and will also keep any pixels whose neighbors do not match the second structuring element. This justifies the name of the operations: a hit for the first structuring element or a miss for the second structuring element.

Normally we would use `strel` to generate the structuring element; but in this case, we are not simply generating a matrix of 1s. Therefore, we will define the structuring elements manually.

12. Close any open windows.
13. Define the two structuring elements.

```
SE1 = [     0 0 0 0 0
         0 0 0 0 0
         0 1 1 0 0
         0 0 1 0 0
         0 0 0 0 0]

SE2 = [ 0 0 0 0 0
         1 1 1 1 0
         0 0 0 1 0
         0 0 0 1 0
         0 0 0 1 0]
```

14. Apply the HoM operation on the original image.

```
I_hm = bwhitmiss(I,SE1,SE2);
figure,  imshow(I), title('Original Image');
figure,  imshow(I_hm), title('Hit-or-miss operation');
```

Question 12 What was the result of applying the HoM operation to the image with the given structuring elements?

Question 13 How could we define the structuring elements so that the bottom left corner pixels of objects are shown in the result?

The `bwhitmiss` function offers a shortcut for creating structuring elements like the one above. Because the two structuring elements above do not have any values in common, we can actually use one array to represent both structuring elements.

In the array in Figure 13.19, 1 refers to ones in the first structuring element, -1 refers to ones in the second structuring element, and all 0s are ignored. This array is called an *interval*.

15. Define an equivalent interval to that of the two structuring elements previously defined.

```
interval = [ 0 0 0 0 0
             -1 -1 -1 -1 0
             0 1 1 -1 0
             0 0 1 -1 0
             0 0 0 -1 0]
```

Interval =

0	0	0	0	0
−1	−1	−1	−1	0
0	1	1	−1	0
0	0	1	−1	0
0	0	0	−1	0

FIGURE 13.19 Combining two structuring elements into one for the HoM transformation.

```
I_hm2 = bwhitmiss(I,interval);
figure, imshow(I_hm), title('Using two SEs');
figure, imshow(I_hm2), title('Using interval');
```

Question 14 Will this technique work if the two structuring elements have elements in common? Explain.

13.9 TUTORIAL 13.2: BASIC MORPHOLOGICAL ALGORITHMS

Goal

The goal of this tutorial is to learn how to implement basic morphological algorithms in MATLAB.

Objectives

- Learn how to perform boundary extraction.
- Explore the `bwperim` function.
- Learn how to fill object holes using the `imfill` function.
- Explore object selection using the `bwselect` function.
- Learn how to label objects in a binary image using the `bwlabel` function.
- Explore the `bwmorph` function to perform thinning, thickening, and skeletonization.

What You Will Need

- `morph.bmp`

Procedure

Boundary Extraction

1. Load and display the test image.

```
I = imread('morph.bmp');
figure, imshow(I), title('Original image');
```

2. Subtract the original image from its eroded version to get the boundary image.

```
se = strel('square',3);
I_ero = imerode(I,se);
I_bou = imsubtract(I,I_ero);
figure, imshow(I_bou), title('Boundary Extraction');
```

3. Perform boundary extraction using the `bwperim` function.

```
I_perim = bwperim(I,8);
figure, imshow(I_perim), title('Boundary using bwperim');
```

Question 1 Show that the `I_perim` image is exactly the same as `I_bou`.

Question 2 If we specify 4-connectivity in the `bwperim` function call, how will this affect the output image?

Region Filling

We can use the `imfill` function to fill holes within objects (among other operations).

4. Close any open figures.
5. Fill holes in the image using the `imfill` function.

```
I_fill1 = imfill(I,'holes');
figure, imshow(I_fill1), title('Holes filled');
```

Function `imfill` can also be used in an interactive mode.

6. Pick two of the three holes interactively by executing this statement. After selecting the points, press Enter.

```
I_fill2 = imfill(I);
imshow(I_fill2), title('Interactive fill');
```

Question 3 What other output parameters can be specified when using `imfill`?

Selecting and Labeling Objects

The `bwselect` function allows the user to interactively select connected components—which often correspond to objects of interest—in a binary image.

7. Close any open figures.
8. Select any of the white objects and press Enter.

```
bwselect(I);
```

Question 4 In the last step, we did not save the output image into a workspace variable. Does this function allow us to do this? If so, what is the syntax?

In many cases, we need to label the connected components so that we can reference them individually. We can use the function `bwlabel` for this purpose.

9. Label the objects in an image using `bwlabel`.

```
I_label = bwlabel(I);
figure, imshow(I_label,[]), title('Labeled image');
```

Question 5 What do the different shades of gray represent when the image is displayed?

Question 6 What other display options can you choose to make it easier to tell the different labeled regions apart?

Thinning

Thinning is one of the many operations that can be achieved through the use of the `bwmorph` function (see Table 13.1 for a complete list).

10. Use `bwmorph` to thin the original image with five iterations.

```
I_thin = bwmorph(I,'thin',5);
figure, imshow(I_thin), title('Thinning, 5 iterations');
```

Question 7 What happens if we specify 10, 15, or `Inf` (infinitely many) iterations instead?

Thickening

11. Thicken the original image with five iterations.

```
I_thick = bwmorph(I,'thicken',5);
figure, imshow(I_thick), title('Thicken, 5 iterations');
```

Question 8 What happens when we specify a higher number of iterations?

Question 9 How does MATLAB know when to stop thickening an object? (*Hint*: Check to see what happens if we use `Inf` (infinitely many) iterations?

Skeletonization

12. Close any open figures.
13. Generate the skeleton using the `bwmorph` function.

```
I_skel = bwmorph(I,'skel',Inf);
figure, imshow(I_skel), title('Skeleton of image');
```

Question 10 How does skeletonization compare with thinning? Explain.

WHAT HAVE WE LEARNED?

- Mathematical morphology is a branch of image processing that has been successfully used to represent, describe, and analyze shapes in images. It offers useful tools for extracting image components, such as boundaries and skeletons, as well as pre- or postprocessing images containing shapes of interest.
- The main morphological operations are erosion (implemented by the `imerode` function in MATLAB), dilation (`imdilate`), opening (`imopen`), closing (`imclose`), and the hit-or-miss transform (`bwhitmiss`).
- The structuring element is the basic neighborhood structure associated with morphological image operations. It is usually represented as a small matrix, whose shape and size impact the results of applying a certain morphological operator to an image.
- Morphological operators can be combined into useful algorithms for commonly used image processing operations, for example, boundary extraction, region filling, and extraction (and labeling) of connected components within an image.

LEARN MORE ABOUT IT

- The books by Serra ([SC82], [Ser88], and [SS96]) are considered the primary references in the field of mathematical morphology.
- Book-length treatment of the topics discussed in this chapter can also be found in [GD88] and [Dou92], among others.
- Section 14.2 of [Pra07] and Section 9.3.3 of [GWE04] discuss the implementation of the HoM transform using lookup tables (LUTs).
- Chapter 2.2 of [Bov00a] provides examples to illustrate the operation of basic morphological filters.

- Maragos and Pessoa (Chapter 3.3 of [Bov00a]) provide an extensive coverage of morphological filtering and applications.
- For additional discussions and examples of morphological algorithms, refer to Sections 9.5 and 9.6.3 of [GW08] and Section 10.6 of [McA04].
- The topic of morphological reconstruction [Vin93] is discussed in Sections 9.5.9 and 9.6.4 of [GW08] and Sections 9.5 and 9.6.3 of [GWE04].
- The paper by Jang and Chin [JC90] discusses morphological thinning algorithms.
- Section 3.4 of [SS01] and Section 11.1 of [BB08] contain additional algorithms for extraction and labeling of connected components.
- Chapter 13 of [SKH08] has MATLAB implementations of morphological algorithms for object detection, thinning, and granulometry applications, among others.

13.10 PROBLEMS

13.1 Use `strel` and `getsequence` to create SEs of different shape and size and inspect their decomposed structuring elements.

13.2 Use the equations and definitions from Section 13.2 to prove that equation (13.21) is, indeed, equivalent to equation (13.20).

13.3 Write a MATLAB script to demonstrate that erosion and dilation are dual operations.

13.4 Repeat Example 13.7 using MATLAB commands that correspond to equation (13.26).

13.5 Write a MATLAB script (or function) that extracts the connected components of a binary image and displays the results using different colors for each component and overlays a cross-shaped symbol on top of each component's center of gravity.

13.6 How many connected components would the `bwlabel` function find, if presented with the negative (i.e., the image obtained by swapping white and black pixels) of the image in Figure 2.8a as an input image? Explain.

13.7 Explore the MATLAB IPT and try to find at least two other functions—besides `bwlabel`—whose results depend on whether 4- or 8-connectivity has been selected.

CHAPTER 14

EDGE DETECTION

WHAT WILL WE LEARN?

- What is edge detection and why is it so important to computer vision?
- What are the main edge detection techniques and how well do they work?
- How can edge detection be performed in MATLAB?
- What is the Hough transform and how can it be used to postprocess the results of an edge detection algorithm?

14.1 FORMULATION OF THE PROBLEM

Edge detection is a fundamental image processing operation used in many computer vision solutions. The goal of edge detection algorithms is to find the most relevant edges in an image or scene. These edges should then be connected into meaningful lines and boundaries, resulting in a segmented image[1] containing two or more regions. Subsequent stages in a machine vision system will use the segmented results for tasks such as object counting, measuring, feature extraction, and classification.

The need for edge detection algorithms as part of a vision system also has its roots in biological vision: there is compelling evidence that the very early stages of the human visual system (HVS) contain edge-sensitive cells that respond strongly

[1] Image segmentation will be discussed in Chapter 15.

Practical Image and Video Processing Using MATLAB®. By Oge Marques.

(i.e., exhibit a higher firing rate) when presented with edges of certain intensity and orientation. Edge detection algorithms, therefore, attempt to emulate an ability present in the human visual system.

Such an ability is, according to many vision theorists, essential to all processing steps that take place afterward in the HVS. David Marr, in his very influential theory of vision [Mar82], speaks of *primal sketches*, which can be understood as the outcome of the very early steps in the transition from an image to a symbolic representation of its features, such as edges, lines, blobs, and terminations. In Marr's terminology, the result of edge detection algorithms compose the *raw* primal sketch; the outcome of additional processing, linking the resulting edges together, constitute the *full* primal sketch, which will be used by subsequent stages in the human visual processing system.

Edge detection is a hard image processing problem. Most edge detection solutions exhibit limited performance in the presence of images containing real-world scenes, that is, images that have not been carefully controlled in their illumination, size and position of objects, and contrast between objects and background. The impacts of shadows, occlusion among objects and parts of the scene, and noise—to mention just a few—on the results produced by an edge detection solution are often significant. Consequently, it is common to precede the edge detection stage with preprocessing operations such as noise reduction and illumination correction.

14.2 BASIC CONCEPTS

An *edge* can be defined as a boundary between two image regions having distinct characteristics according to some feature (e.g., gray level, color, or texture). In this chapter, we focus primarily on edges in grayscale 2D images, which are usually associated with a sharp variation of the intensity function across a portion of the image. Figure 14.1 illustrates this concept and shows the difference between an ideal edge (sharp and abrupt transition) and a ramp edge (gradual transition between dark and bright areas in the image).

Edge detection methods usually rely on calculations of the first or second derivative along the intensity profile. The first derivative has the desirable property of being directly proportional to the difference in intensity across the edge; consequently, the magnitude of the first derivative can be used to detect the presence of an edge at a certain point in the image. The sign of second derivative can be used to determine whether a pixel lies on the dark or on the bright side of an edge. Moreover, the zero crossing between its positive and negative peaks can be used to locate the center of thick edges.

Figure 14.2 illustrates these concepts. It shows an image with a ramp edge and the corresponding intensity profile, first and second derivatives, and zero crossing for any horizontal line in the image. It shows that the first derivative of the intensity function has a peak at the center of the luminance edge, whereas the second derivative—which is the slope of the first derivative function—has a zero crossing at the center

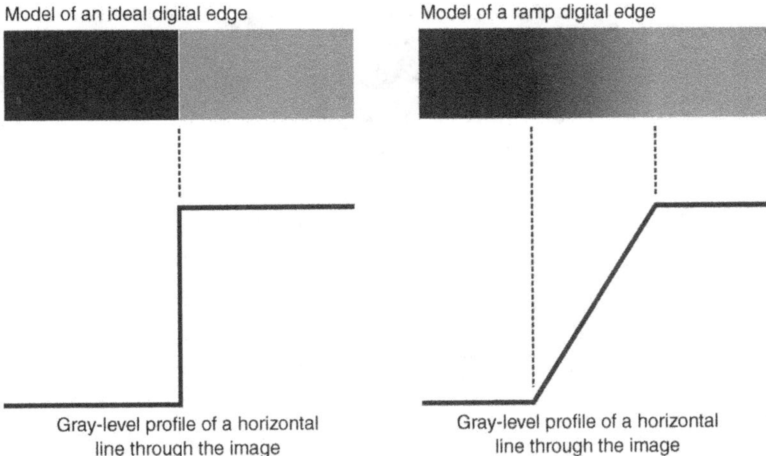

FIGURE 14.1 Ideal and ramp edges: (a) ideal edge on a digital image and corresponding profile along a horizontal line; (b) ramp edge and corresponding profile.

of the luminance edge, with a positive value on one side and a negative value on the other.

The edges in Figures 14.1 and 14.2 were noise-free. When the input image is corrupted by noise, the first and second derivatives respond quite differently. Even modest noise levels—barely noticeable when you look at the original image or its profile—can render the second derivative results useless, whereas more pronounced noise levels will also impact the first derivative results to a point that they cannot be used for edge detection. Figure 14.3 illustrates this problem.

In summary, the process of edge detection consists of three main steps:

1. *Noise Reduction*: Due to the first and second derivative's great sensitivity to noise, the use of image smoothing techniques (see Chapters 10–12) before applying the edge detection operator is strongly recommended.
2. *Detection of Edge Points*: Here local operators that respond strongly to edges and weakly elsewhere (described later in this chapter) are applied to the image, resulting in an output image whose bright pixels are candidates to become edge points.
3. *Edge Localization*: Here the edge detection results are postprocessed, spurious pixels are removed, and broken edges are turned into meaningful lines and boundaries, using techniques such as the Hough transform (Section 14.6.1).

In MATLAB

The IPT has a function for edge detection (`edge`), whose variants and options will be explored throughout this chapter and its tutorial.

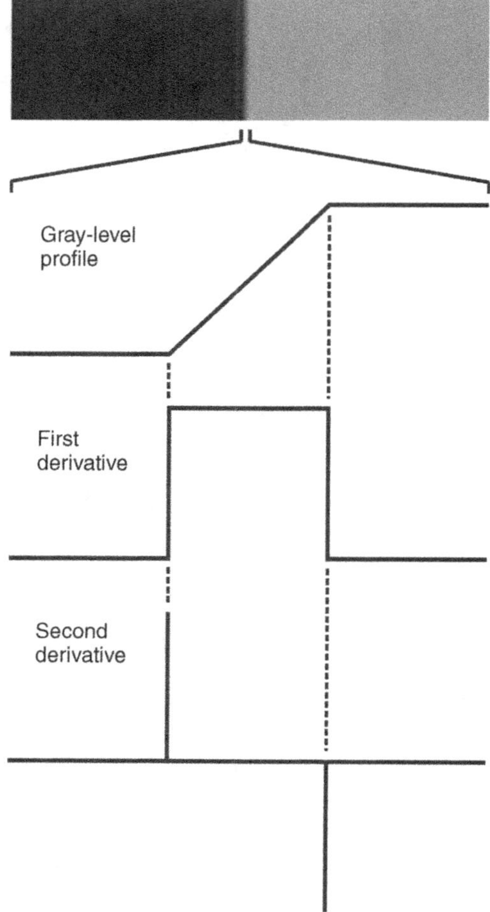

FIGURE 14.2 Grayscale image containing two regions separated by a ramp edge: intensity profile and, first and second derivative results.

14.3 FIRST-ORDER DERIVATIVE EDGE DETECTION

The simplest edge detection methods work by estimating the gray-level gradient at a pixel, which can be approximated by the digital equivalent of the first-order derivative as follows:

$$g_x(x, y) \approx f(x+1, y) - f(x-1, y) \tag{14.1}$$

$$g_y(x, y) \approx f(x, y+1) - f(x, y-1) \tag{14.2}$$

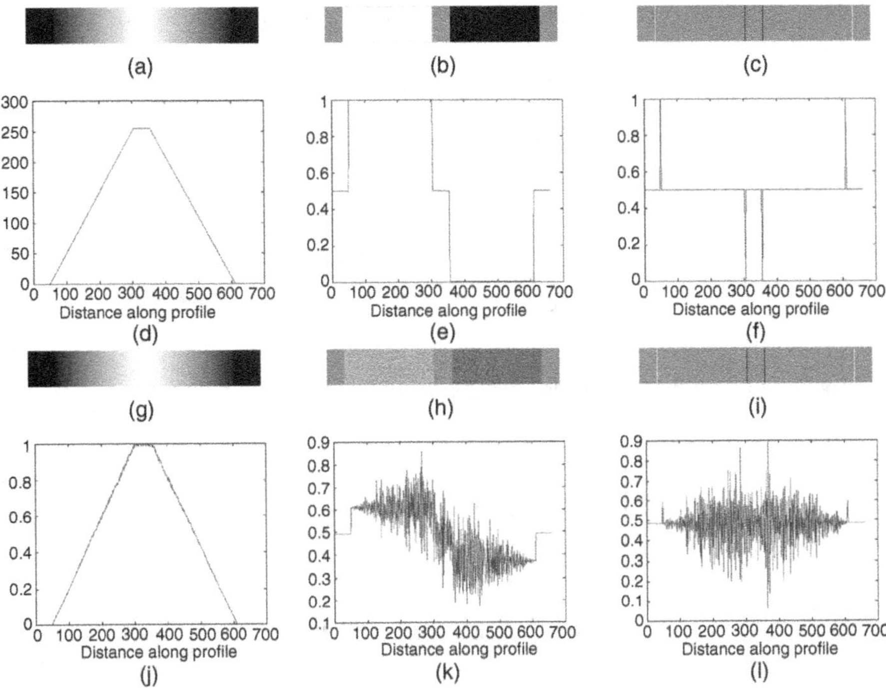

FIGURE 14.3 First- and second-order edge detectors with and without noise: (a) original image; (b) first derivative; (c) second derivative; (d–f) horizontal profiles for images (a)–(c); (g–i) noisy versions of images (a)–(c); (j–l) horizontal profiles for images (g)–(i).

The 2×2 approximations of the first-order derivative above are also known as Roberts operators and can be represented using a 2×2 matrix notation as

$$g_x = \begin{bmatrix} 0 & -1 \\ 1 & 0 \end{bmatrix} \tag{14.3}$$

$$g_y = \begin{bmatrix} -1 & 0 \\ 0 & 1 \end{bmatrix} \tag{14.4}$$

These gradients are often computed within a 3×3 neighborhood using convolution:

$$g_x(x, y) = h_x * f(x, y) \tag{14.5}$$

$$g_y(x, y) = h_y * f(x, y) \tag{14.6}$$

where h_x and h_y are appropriate convolution masks (kernels).

The simplest pair of kernels, known as the Prewitt [Pre70] edge detector (operator), are as follows:

$$h_x = \begin{bmatrix} -1 & 0 & 1 \\ -1 & 0 & 1 \\ -1 & 0 & 1 \end{bmatrix} \tag{14.7}$$

$$h_y = \begin{bmatrix} -1 & -1 & -1 \\ 0 & 0 & 0 \\ 1 & 1 & 1 \end{bmatrix} \tag{14.8}$$

A similar pair of kernels, which gives more emphasis to on-axis pixels, is the Sobel edge detector, given by the following:

$$h_x = \begin{bmatrix} -1 & 0 & 1 \\ -2 & 0 & 2 \\ -1 & 0 & 1 \end{bmatrix} \tag{14.9}$$

$$h_y = \begin{bmatrix} -1 & -2 & -1 \\ 0 & 0 & 0 \\ 1 & 2 & 1 \end{bmatrix} \tag{14.10}$$

As you may have noticed, despite their differences, the 3×3 masks presented so far share two properties:

- They have coefficients of opposite signs (across a row or column of coefficients equal to zero) in order to obtain a high response in image regions with variations in intensity (possibly due to the presence of an edge).
- The sum of the coefficients is equal to zero, which means that when applied to perfectly homogeneous regions in the image (i.e., a patch of the image with constant gray level), the result will be 0 (black pixel).

In MATLAB

The IPT function `edge` has options for both Prewitt and Sobel operators. Edge detection using Prewitt and Sobel operators can also be achieved by using `imfilter` with the corresponding 3×3 masks (which can be created using `fspecial`).

■ EXAMPLE 14.1

Figure 14.4 shows an example of using `imfilter` to apply the Prewitt edge detector to a test image. Due to the fact that the Prewitt kernels have both positive and negative coefficients, the resulting array contains negative and positive values. Since negative

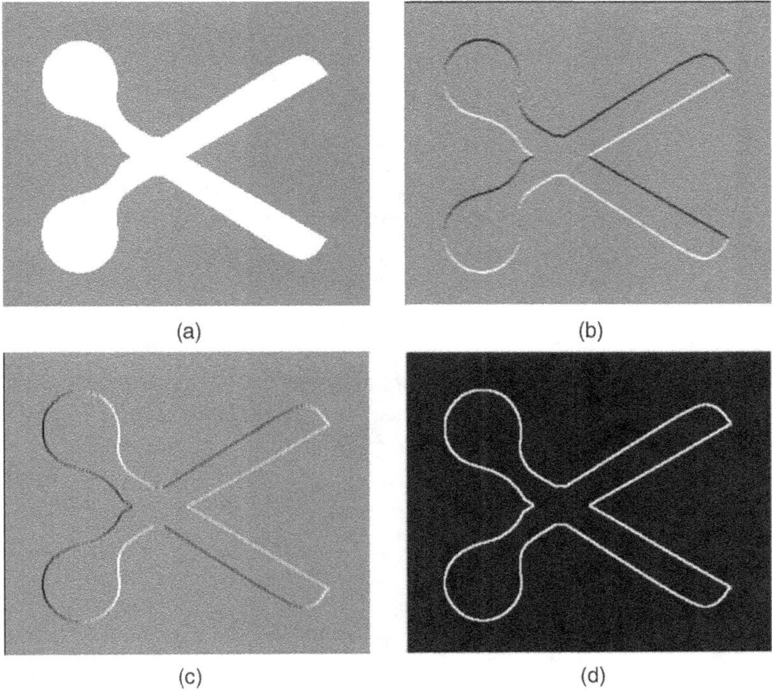

FIGURE 14.4 Edge detection example: (a) original image; (b) result of Prewitt horizontal kernel; (c) result of Prewitt vertical kernel; (d) combination of (b) and (c).

values are usually truncated when displaying an image in MATLAB, the partial results (Figure 14.4b and c) have been mapped to a modified gray-level range (where the highest negative value becomes black, the highest positive value is displayed as white, and all zero values are shown with a midlevel gray). This display strategy also gives us insight into which side of the edge corresponds to a dark pixel and which side corresponds to a bright one. Moreover, it is worth observing that both detectors can find edges in this test image, despite the fact that it does not contain any purely vertical or horizontal edges of significant length.

The combined final result (Figure 14.4d) was obtained by computing the magnitude of the gradient, originally defined as

$$g = \sqrt{g_x^2 + g_y^2} \tag{14.11}$$

which can be approximated by

$$g = |g_x| + |g_y| \tag{14.12}$$

■ EXAMPLE 14.2

Figure 14.5 shows examples of edge detection using `imfilter` to apply the Sobel operator on a grayscale image. All results are displayed in their negative (using `imcomplement`) for better viewing on paper.

The idea of using horizontal and vertical masks used by the Prewitt and Sobel operators can be extended to include all eight compass directions: north, northeast, east, southeast, south, southwest, west, and northwest. The Kirsch [Kir71] (Figure 14.6) and the Robinson [Rob77] (Figure 14.7) kernels are two examples of *compass masks*. You will have a chance to design and apply the Kirsch and the Robinson masks to grayscale images in Tutorial 14.1.

Edge detection results can be thresholded to reduce the number of false positives (i.e., pixels that appear in the output image although they do not correspond to actual edges). Figure 14.8 shows an example of using `edge` to implement the Sobel operator with different threshold levels. The results range from unacceptable because of too many spurious pixels (part (a)) to unacceptable because of too few edge pixels (part (d)). Part (c) shows the result using the best threshold value, as determined by the

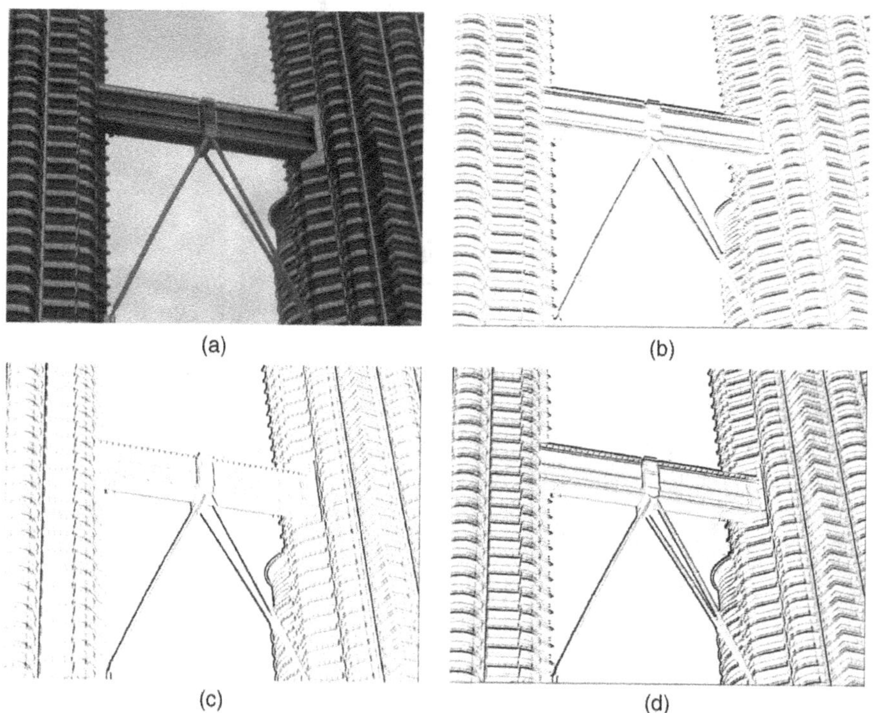

FIGURE 14.5 Edge detection using Sobel operator: (a) original image; (b) result of Sobel horizontal kernel; (c) result of Sobel vertical kernel; (d) combination of (b) and (c).

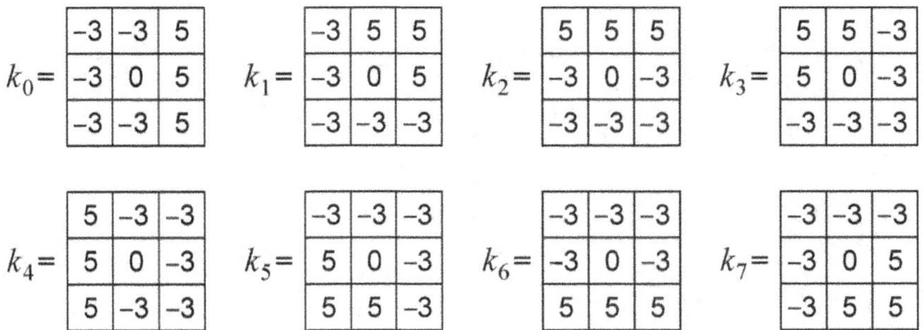

FIGURE 14.6 Kirsch compass masks.

`edge` function using the syntax: `[BW, thresh] = edge(I,'sobel');`, where `I` is the input image.

14.4 SECOND-ORDER DERIVATIVE EDGE DETECTION

The Laplacian operator (originally introduced in Section 10.4.1) is a straightforward digital approximation of the second-order derivative of the intensity. Although it has the potential for being employed as an isotropic (i.e., omnidirectional) edge detector, it is rarely used in isolation because of two limitations (commented earlier in this chapter):

- It generates "double edges," that is, positive and negative values for each edge.
- It is extremely sensitive to noise.

In MATLAB

Edge detection using the Laplacian operator can be implemented using the `fspecial` function (to generate the Laplacian 3×3 convolution mask) and the

$$r_0 = \begin{bmatrix} -1 & 0 & 1 \\ -2 & 0 & 2 \\ -1 & 0 & 1 \end{bmatrix} \quad r_1 = \begin{bmatrix} 0 & 1 & 2 \\ -1 & 0 & 1 \\ -2 & 1 & 0 \end{bmatrix} \quad r_2 = \begin{bmatrix} 1 & 2 & 1 \\ 0 & 0 & 0 \\ -1 & -2 & -1 \end{bmatrix} \quad r_3 = \begin{bmatrix} 2 & 1 & 0 \\ 1 & 0 & -1 \\ 0 & -1 & -2 \end{bmatrix}$$

$$r_4 = \begin{bmatrix} 1 & 0 & -1 \\ 2 & 0 & -2 \\ 1 & 0 & -1 \end{bmatrix} \quad r_5 = \begin{bmatrix} 0 & -1 & -2 \\ 1 & 0 & -1 \\ 2 & 1 & 0 \end{bmatrix} \quad r_6 = \begin{bmatrix} -1 & -2 & -1 \\ 0 & 0 & 0 \\ 1 & 2 & 1 \end{bmatrix} \quad r_7 = \begin{bmatrix} -2 & -1 & 0 \\ -1 & 0 & 1 \\ 0 & 1 & 2 \end{bmatrix}$$

FIGURE 14.7 Robinson compass masks.

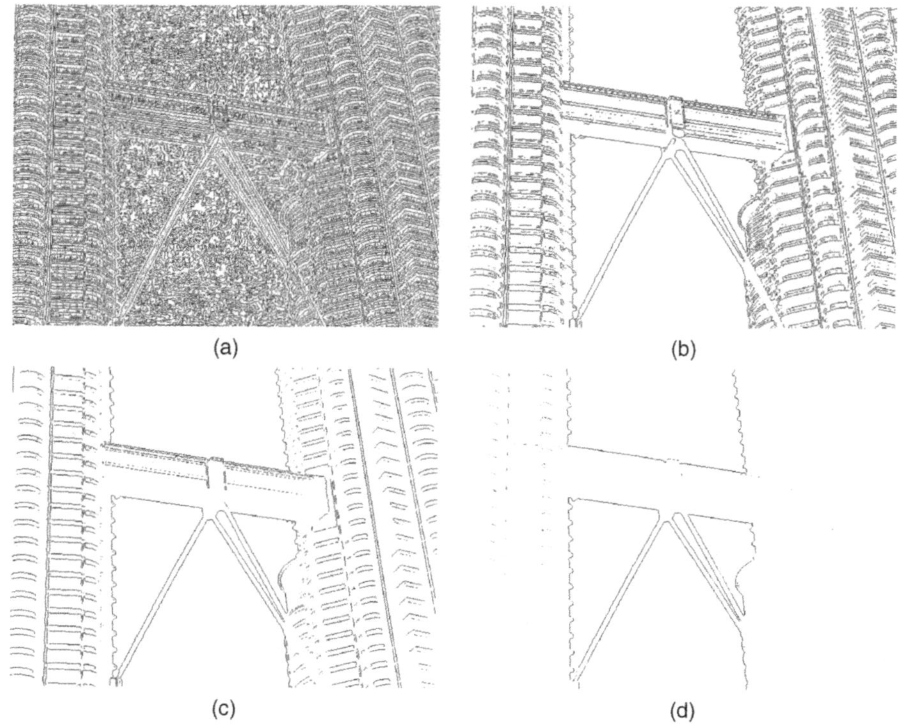

FIGURE 14.8 Edge detection using Sobel operator and thresholding (the original image is the same as Figure 14.5a): (a) threshold of 0; (b) threshold of 0.05; (c) threshold of 0.1138 (the best value); (d) threshold of 0.2.

`zerocross` option in function `edge` as follows:

```
h = fspecial('laplacian',0);
J = edge(I,'zerocross',t,h);
```

where `t` is a user-provided sensitivity threshold.

■ EXAMPLE 14.3

Figure 14.9 shows the results of applying the zero-cross edge detector to an image and the impact of varying the thresholds. Part (a) shows a clean input image; part (b) shows the results of edge detection in (a) using default parameters, whereas part (c) shows the effects of reducing the threshold to 0. Clearly the result in (b) is much better than (c).

Part (d) is a noisy version of (a) (with zero-mean Gaussian noise with $\sigma = 0.0001$). Parts (e) and (f) are the edge detection results using the noisy image in part (d) as an input and the same options as in (b) and (c), respectively. In this case, although the amount of noise is hardly noticeable in the original image (d), both edge detection results are unacceptable.

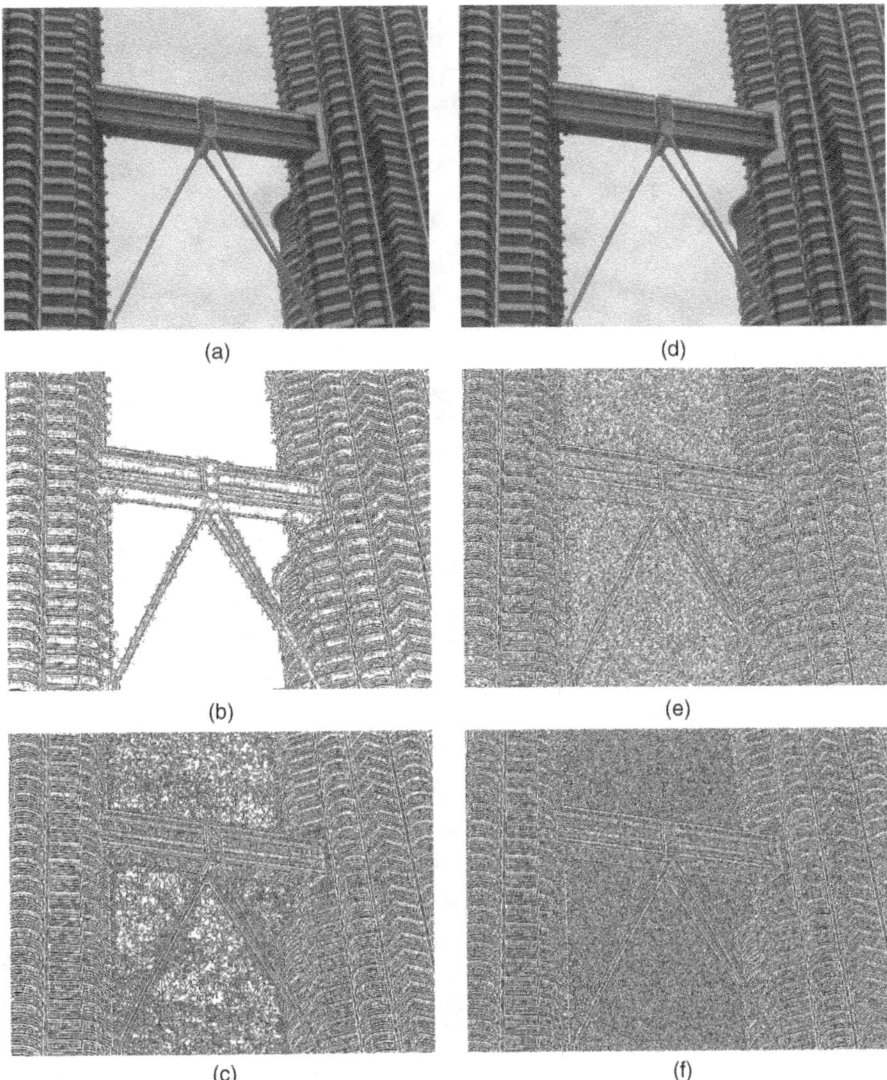

FIGURE 14.9 Edge detection using the zero-cross edge detector: (a) input image (without noise); (b) results using default values; (c) results using threshold zero; (d) noisy input image; (e) results using default values; (f) results using threshold zero. Edge results have been inverted for clarity.

14.4.1 Laplacian of Gaussian

The Laplacian of Gaussian (LoG) edge detector works by smoothing the image with a Gaussian low-pass filter (LPF), and then applying a Laplacian edge detector to the result. The resulting transfer function (which resembles a Mexican hat in its 3D view) is represented in Figure 14.10. The LoG filter can sometimes be approximated by

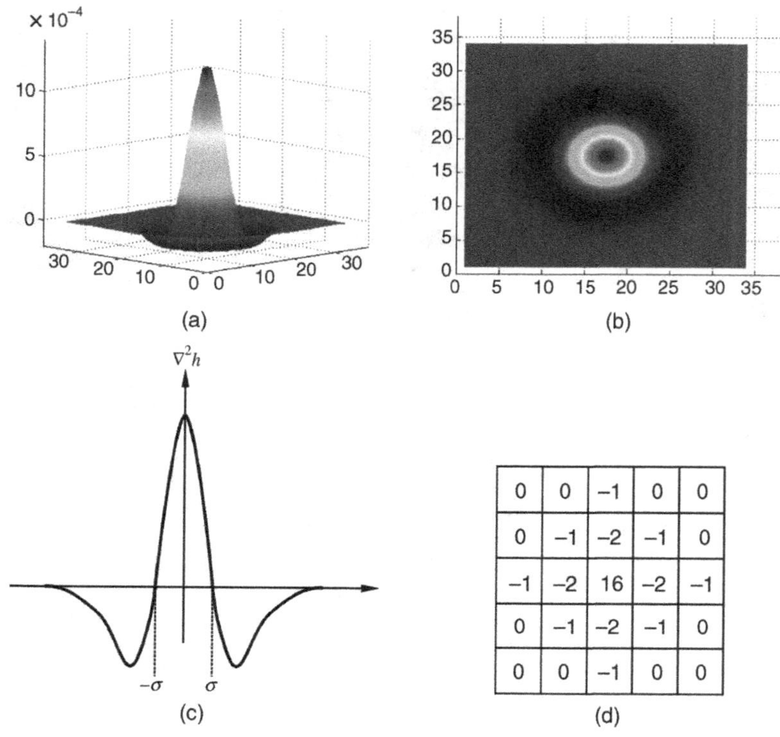

0	0	−1	0	0
0	−1	−2	−1	0
−1	−2	16	−2	−1
0	−1	−2	−1	0
0	0	−1	0	0

FIGURE 14.10 Laplacian of Gaussian: (a) 3D plot; (b) 2D intensity plot; (c) cross section of (a).

taking the differences of two Gaussians of different widths in a method known as *difference of Gaussians* (DoG).

In a landmark paper [MH80], David Marr and Ellen Hildreth proposed that LoG filtering explains much of the low-level behavior of the human vision system, since the response profile of an LoG filter approximates the receptive field of retinal cells tuned to respond to edges. They proposed an architecture based on LoG filters with four or five different spreads of σ to derive a primal sketch from a scene. The Marr–Hildreth zero-crossing algorithm was eventually supplanted by the Canny edge detector as the favorite edge detection solution among image processing and computer vision practitioners, but Marr's ideas continue to influence researchers in both human as well as computer vision.

In MATLAB

Edge detection using the LoG filter can be implemented using the `log` option in function `edge`.

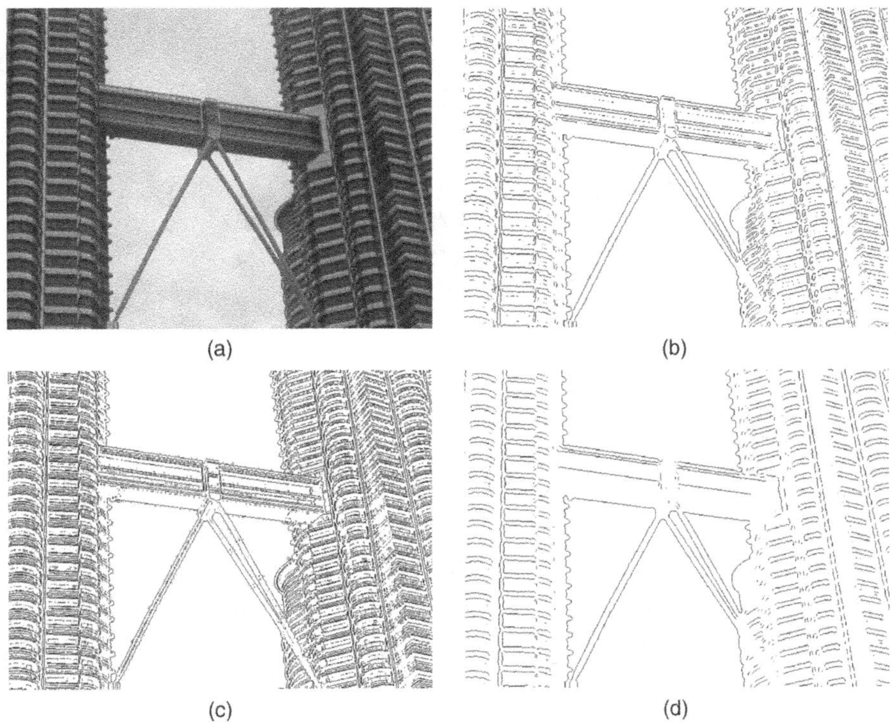

FIGURE 14.11 Edge detection using the LoG edge detector: (a) input image; (b) results using default values; (c) results using $\sigma = 1$; (d) results using $\sigma = 3$. Edge results have been inverted for clarity.

■ EXAMPLE 14.4

Figure 14.11 shows the results of applying the LoG edge detector to an image, and the impact of varying σ. Part (a) shows the input image; part (b) shows the results of edge detection in (a) using default parameters (i.e., $\sigma = 2$), whereas parts (c) and (d) show the effects of reducing or increasing sigma (to 1 and 3, respectively). Reducing σ causes the resulting image to contain more fine details, whereas an increase in σ leads to a coarser edge representation, as expected.

14.5 THE CANNY EDGE DETECTOR

The Canny edge detector [Can86] is one of the most popular, powerful, and effective edge detection operators available today. Its algorithm can be described as follows:

1. The input image is smoothed using a Gaussian low-pass filter (Section 10.3.3), with a specified value of σ: large values of σ will suppress much of the noise at the expense of weakening potentially relevant edges.

2. The local gradient (intensity and direction) is computed for each point in the smoothed image.
3. The edge points at the output of step 2 result in wide ridges. The algorithm thins those ridges, leaving only the pixels at the top of each ridge, in a process known as *nonmaximal suppression*.
4. The ridge pixels are then thresholded using two thresholds T_{low} and T_{high}: ridge pixels with values greater than T_{high} are considered *strong* edge pixels; ridge pixels with values between T_{low} and T_{high} are said to be *weak* pixels. This process is known as *hysteresis thresholding*.
5. The algorithm performs edge linking, aggregating weak pixels that are 8-connected[2] to the strong pixels.

In MATLAB

The `edge` function includes the Canny edge detector, which can be invoked using the following syntax:

```
J = edge(I, 'canny', T, sigma);
```

where `I` is the input image, `T = [T_low T_high]` is a 1×2 vector containing the two thresholds explained in step 4 of the algorithm, `sigma` is the standard deviation of the Gaussian smoothing filter, and `J` is the output image.

■ EXAMPLE 14.5

Figure 14.12 shows the results of applying the Canny detector to an image (Figure 14.5a), and the impact of varying σ and the thresholds. Part (a) uses the syntax `BW = edge(J,'canny');`, which results in `t = [0.0625 0.1563]` and $\sigma = 1$. In part (b), we change the value of σ (to 0.5) leaving everything else unchanged. In part (c), we change the value of σ (to 2) leaving everything else unchanged. Changing σ causes the resulting image to contain more (part (b)) or fewer (part (c)) edge points (compared to part (a)), as expected. Finally, in part (d), we keep σ in its default value and change the thresholds to `t = [0.01 0.1]`. Since both T_{low} and T_{high} were lowered, the resulting image contains more strong and weak pixels, resulting in a larger number of edge pixels (compared to part (a)), as expected.

14.6 EDGE LINKING AND BOUNDARY DETECTION

The goal of edge detection algorithms should be to produce an image containing only the edges of the original image. However, due to the many technical challenges discussed earlier (noise, shadows, and occlusion, among others), most edge detection algorithms will output an image containing fragmented edges. In order to turn

[2]In some implementations, only the neighbors along a line normal to the gradient orientation at the edge pixel are considered, not the entire 8-neighborhood.

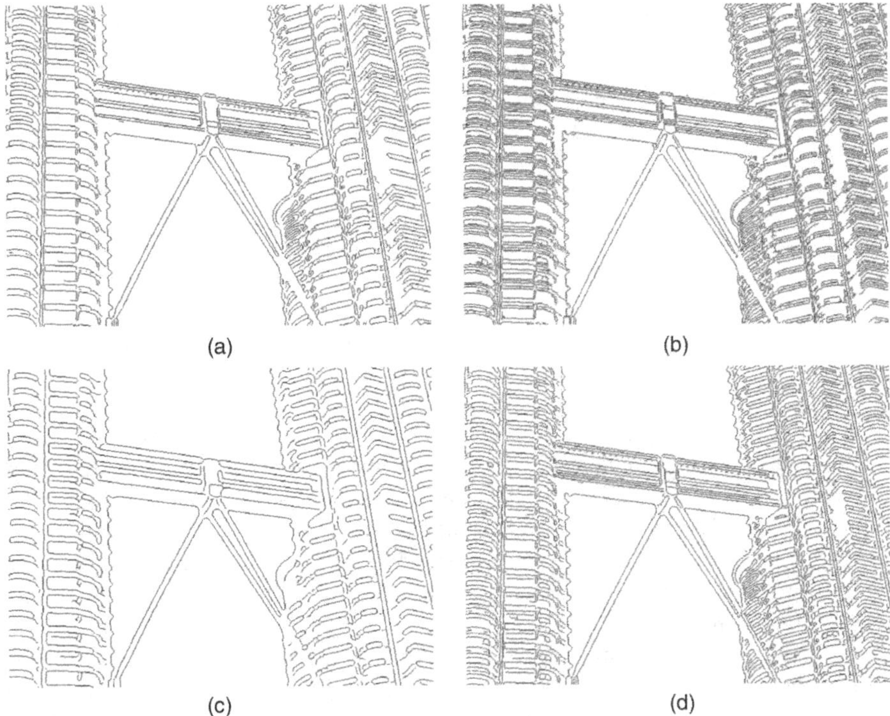

FIGURE 14.12 Edge detection using the Canny edge detector: (a) default values ($\sigma = 1$, $T_{\text{low}} = 0.0625$, $T_{\text{high}} = 0.1563$); (b) $\sigma = 0.5$; (c) $\sigma = 2$; (d) $\sigma = 1$, $T_{\text{low}} = 0.01$, $T_{\text{high}} = 0.1$.

these fragmented edge segments into useful lines and object boundaries, additional processing is needed. In this section, we discuss a global method for edge linking and boundary detection: the Hough transform.[3]

14.6.1 The Hough Transform

The Hough transform [Hou] is a mathematical method designed to find lines in images. It can be used for linking the results of edge detection, turning potentially sparse, broken, or isolated edges into useful lines that correspond to the actual edges in the image.

Let (x, y) be the coordinates of a point in a binary image.[4] The Hough transform stores in an *accumulator array* all pairs (a, b) that satisfy the equation $y = ax + b$. The (a, b) array is called the *transform array*. For example, the point $(x, y) = (1, 3)$ in the input image will result in the equation $b = -a + 3$, which can be plotted as a line that represents all pairs (a, b) that satisfy this equation (Figure 14.13).

[3]Pointers to other edge linking and boundary detection techniques can be found in "Learn More About It" section at the end of the chapter.

[4]This binary image could, of course, be an image containing thresholded edge detection results.

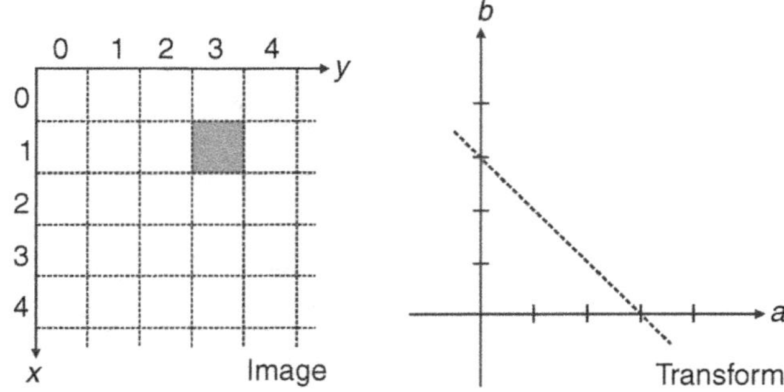

FIGURE 14.13 The Hough transform maps a point into a line.

Since each point in the image will map to a line in the transform domain, repeating the process for other points will result in many intersecting lines, one per point (Figure 14.14). The meaning of two or more lines intersecting in the transform domain is that the points to which they correspond are aligned in the image. The points with the greatest number of intersections in the transform domain correspond to the longest lines in the image.

Describing lines using the equation $y = ax + b$ (where a represents the gradient) poses a problem, since vertical lines have infinite gradient. This limitation can be circumvented by using the *normal representation* of a line, which consists of two parameters: ρ (the perpendicular distance from the line to the origin) and θ (the angle between the line's perpendicular and the horizontal axis). In this new representation (Figure 14.15), vertical lines will have $\theta = 0$. It is common to allow ρ to have negative values, therefore restricting θ to the range $-90° < \theta \leq 90°$.

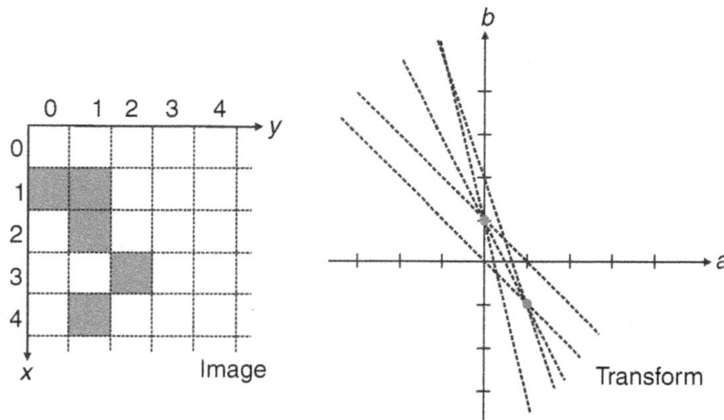

FIGURE 14.14 The Hough transform: intersections in the transform domain correspond to aligned points in the image.

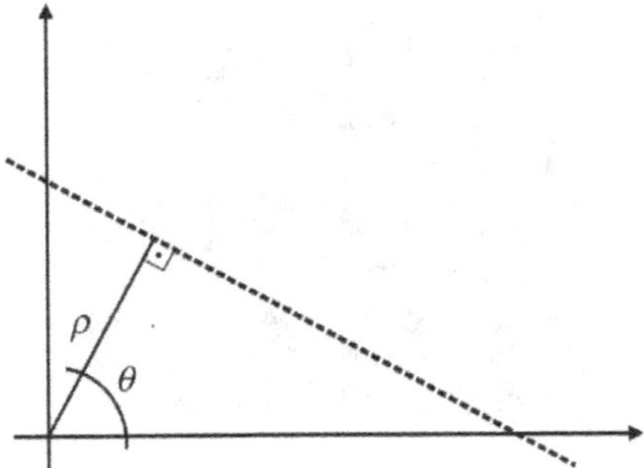

FIGURE 14.15 The Hough transform: a line and its parameters in the polar coordinate system.

The relationship between ρ, θ, and the original coordinates (x, y) is

$$\rho = x \cos\theta + y \sin\theta \tag{14.13}$$

Under the new set of coordinates, the Hough transform can be implemented as follows:

1. Create a 2D array corresponding to a discrete set of values for ρ and θ. Each element in this array is often referred to as an *accumulator cell.*
2. For each pixel (x, y) in the image and for each chosen value of θ, compute $x \cos\theta + y \sin\theta$ and write the result in the corresponding position—(ρ, θ)—in the accumulator array.
3. The highest values in the (ρ, θ) array will correspond to the most relevant lines in the image.

IN MATLAB

The IPT contains a function for Hough transform calculations, `hough`, which takes a binary image as an input parameter, and returns the corresponding Hough transform matrix and the arrays of ρ and θ values over which the Hough transform was calculated. Optionally, the resolution of the discretized 2D array for both ρ and θ can be specified as additional parameters.

■ EXAMPLE 14.6

In this example, we use the `hough` function to find the strongest lines in a binary image obtained as a result of an edge detection operator (`BW`), using the

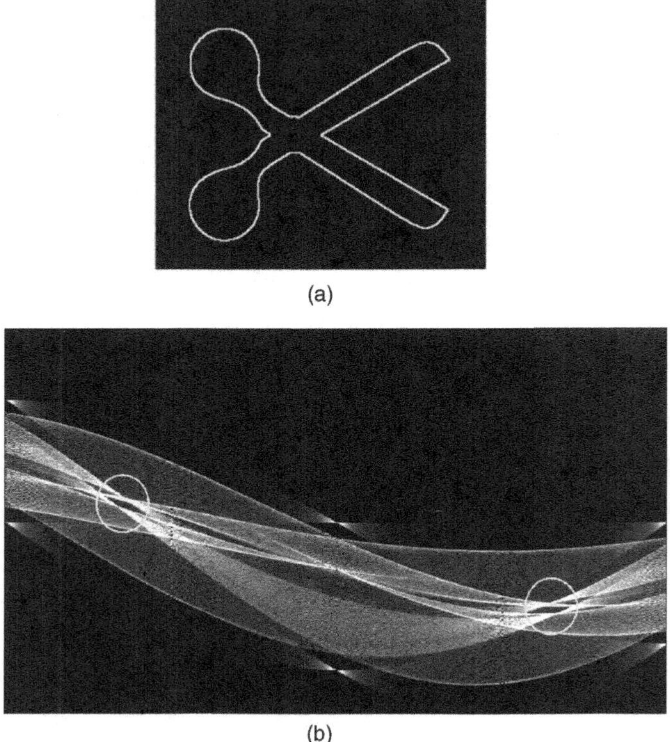

(a)

(b)

FIGURE 14.16 Hough transform example: (a) input image; (b) results of Hough transform, highlighting the intersections corresponding to the predominant lines in the input image.

following steps:

```
[H,T,R] = hough(BW,'RhoResolution',0.5,'ThetaResolution',0.5);
```

Figure 14.16 shows the original image and the results of the Hough transform calculations. You will notice that some of the highest peaks in the transform image (approximately at $\theta = -60^\circ$ and $\theta = 60^\circ$) correspond to the main diagonal lines in the scissors shape.

IN MATLAB

The IPT also includes two useful companion functions for exploring and plotting the results of Hough transform calculations: `houghpeaks` (which identifies the k most salient peaks in the Hough transform results, where k is passed as a parameter) and `houghlines`, which draws the lines associated with the highest peaks on top of the original image.

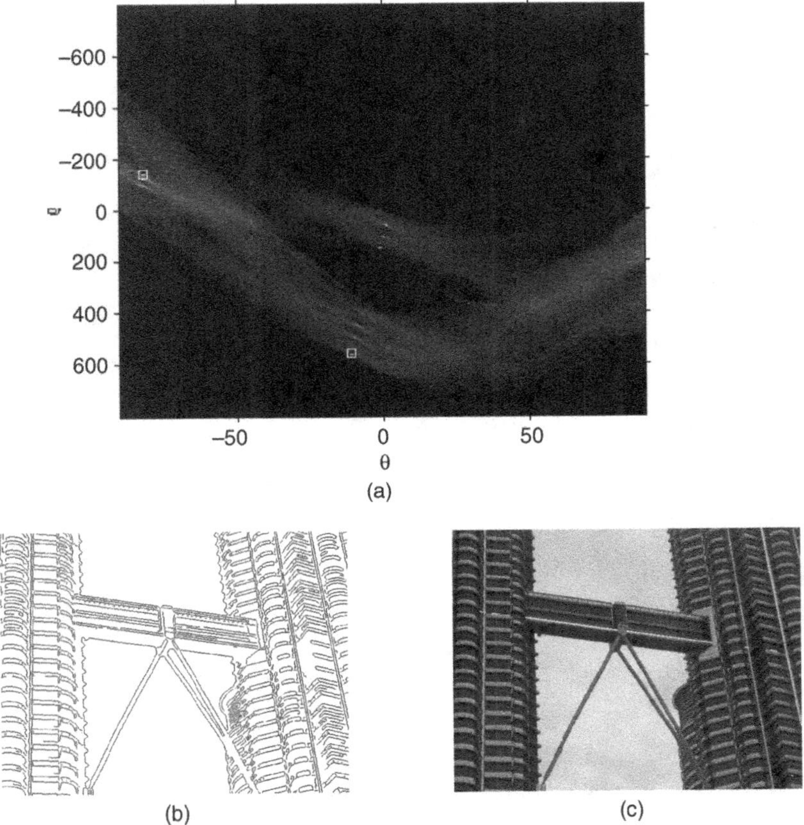

FIGURE 14.17 Hough transform example: (a) results of Hough transform highlighting the two highest peaks; (b) (negative of) edge detection results; (c) lines corresponding to the longest peaks overlaid on top of original image.

■ EXAMPLE 14.7

In this example, we use `hough`, `houghpeaks`, and `houghlines` on a grayscale test image whose edges have been extracted using the Canny edge detector.[5]

Figure 14.17a shows the results of the Hough transform calculations with two small squares indicating the two highest peaks. Figure 14.17b shows the result of the Canny edge detector (displayed with black pixels against a white background for better visualization). Figure 14.17c displays the original image with the highest (cyan) and second highest (yellow) lines overlaid.

The Hough transform can be extended and generalized to find other shapes in images. Refer to "Learn More About It" section at the end of the chapter for useful references.

[5]The complete sequence of MATLAB commands is available at the book web site.

14.7 TUTORIAL 14.1: EDGE DETECTION

Goal

The goal of this tutorial is to learn how to implement edge detection and associated techniques in MATLAB.

Objectives

- Learn how to use the IPT `edge` function.
- Explore the most popular first-derivative edge detectors: Roberts, Sobel, and Prewitt.
- Explore the Marr–Hildreth Laplacian of Gaussian edge detector.
- Explore the Canny edge detector.
- Learn how to implement edge detection with compass masks (Kirsch and Robinson).

What You Will Need

- `lenna.tif`
- `mandrill.tif`

Procedure

First-order edge detection methods, such as Prewitt and Sobel, are defined as convolution kernels. As we have seen in previous tutorials, convolution can be executed in MATLAB using the `imfilter` function. Although we could use this function to implement edge detection, we would be required to perform additional tasks, such as convolving the image twice (once for horizontal edges, and another for vertical ones) and adding the absolute values of these results to yield the final image. In practice, this image might also then be thresholded to produce a binary image where white pixels would represent edges. The function `edge` will do all this for us and will even determine a threshold value if we choose not to specify one.

Edge Detection Using the Prewitt Operator

1. Load and display the test image.

```
I = imread('lenna.tif');
figure, subplot(2,2,1), imshow(I), title('Original Image');
```

2. Extract the edges in the image using the Prewitt operator.

```
[I_prw1,t1] = edge(I,'prewitt');
subplot(2,2,2), imshow(I_prw1), title('Prewitt, default thresh');
```

Question 1 What does the `t1` variable represent?

Edge detection methods are often compared by their ability to detect edges in noisy images. Let us perform the Prewitt operator on the Lenna image with additive Gaussian noise.

3. Add noise to the test image and extract its edges.

```
I_noise = imnoise(I,'gaussian');
[I_prw2,t2] = edge(I_noise,'prewitt');
subplot(2,2,3), imshow(I_noise), title('Image w/ noise');
subplot(2,2,4), imshow(I_prw2), title('Prewitt on noise');
```

Question 2 How did the Prewitt edge detector perform in the presence of noise (compared to no noise)?

Question 3 Did MATLAB use a different threshold value for the noisy image?

Question 4 Try using different threshold values. Do these different values affect the operator's response to noise? How does the threshold value affect the edges of the object?

Edge Detection Using the Sobel Operator

4. Extract the edges from the test image using the Sobel edge detector.

```
[I_sob1,t1] = edge(I,'sobel');
figure, subplot(2,2,1), imshow(I), title('Original Image');
subplot(2,2,2), imshow(I_sob1), title('Sobel, default thresh');
```

5. Extract the edges from the test image with Gaussian noise using the Sobel edge detector.

```
[I_sob2,t2] = edge(I_noise,'sobel');
subplot(2,2,3), imshow(I_noise), title('Image w/ noise');
subplot(2,2,4), imshow(I_sob2), title('Sobel on noise');
```

Question 5 How does the Sobel operator compare with the Prewitt operator with and without noise?

Another feature of the edge function is `'thinning'`, which reduces the thickness of the detected edges. Although this feature is turned on by default, it can be turned off, which results in faster edge detection.

6. Extract the edges from the test image with the Sobel operator with no thinning.

```
I_sob3 = edge(I,'sobel','nothinning');
figure, subplot(1,2,1), imshow(I_sob1), title('Thinning');
subplot(1,2,2), imshow(I_sob3), title('No Thinning');
```

As you already know, the Sobel operator actually performs two convolutions (horizontal and vertical). These individual images can be obtained by using additional output parameters.

7. Display the horizontal and vertical convolution results from the Sobel operator.

```
[I_sob4,t,I_sobv,I_sobh] = edge(I,'sobel');
figure
subplot(2,2,1), imshow(I), title('Original Image');
subplot(2,2,2), imshow(I_sob4), title('Complete Sobel');
subplot(2,2,3), imshow(abs(I_sobv),[]), title('Sobel Vertical');
subplot(2,2,4), imshow(abs(I_sobh),[]), title('Sobel Horizontal');
```

Question 6 Why do we display the absolute value of the vertical and horizontal images? *Hint*: Inspect the minimum and maximum values of these images.

Question 7 Change the code in step 7 to display thresholded (binarized), not thinned, versions of all images.

As you may have noticed, the `edge` function returns the vertical and horizontal images before any thresholding takes place.

Edge Detection with the Roberts Operator

Similar options are available with the `edge` function when the Roberts operator is used.

8. Extract the edges from the original image using the Roberts operator.

```
I_rob1 = edge(I,'roberts');
figure
subplot(2,2,1), imshow(I), title('Original Image');
subplot(2,2,2), imshow(I_rob1), title('Roberts, default thresh');
```

9. Apply the Roberts operator to a noisy image.

```
[I_rob2,t] = edge(I_noise,'roberts');
subplot(2,2,3), imshow(I_noise), title('Image w/ noise');
subplot(2,2,4), imshow(I_rob2), title('Roberts on noise');
```

Question 8 Compare the Roberts operator with the Sobel and Prewitt operators. How does it hold up to noise?

Question 9 If we were to adjust the threshold, would we get better results when filtering the noisy image?

Question 10 Suggest a method to reduce the noise in the image before performing edge detection.

Edge Detection with the Laplacian of a Gaussian Operator

The LoG edge detector can be implemented with the `edge` function as well. Let us see its results.

10. Extract the edges from the original image using the LoG edge detector.

```
I_log1 = edge(I,'log');
figure
subplot(2,2,1), imshow(I), title('Original Image');
subplot(2,2,2), imshow(I_log1), title('LoG, default parameters');
```

11. Apply the LoG edge detector to the noisy image.

```
[I_log2,t] = edge(I_noise,'log');
subplot(2,2,3), imshow(I_noise), title('Image w/ noise');
subplot(2,2,4), imshow(I_log2), title('LoG on noise');
```

Question 11 By default, the LoG edge detector uses a value of 2 for σ (the standard deviation of the filter). What happens when we increase this value?

Edge Detection with the Canny Operator

12. Extract the edges from the original image using the Canny edge detector.

```
I_can1 = edge(I,'canny');
figure
subplot(2,2,1), imshow(I), title('Original Image');
subplot(2,2,2), imshow(I_log1), title('Canny, default parameters');
```

13. Apply the filter to the noisy image.

```
[I_can2,t] = edge(I_noise,'canny', [], 2.5);
subplot(2,2,3), imshow(I_noise), title('Image w/ noise');
subplot(2,2,4), imshow(I_can2), title('Canny on noise');
```

As you know, the Canny detector first applies a Gaussian smoothing function to the image, followed by edge enhancement. To achieve better results on the noisy image, we can increase the size of the Gaussian smoothing filter through the `sigma` parameter.

14. Apply the Canny detector on the noisy image where `sigma` = 2.

```
[I_can3,t] = edge(I_noise,'canny', [], 2);
figure
subplot(1,2,1), imshow(I_can2), title('Canny, default parameters');
subplot(1,2,2), imshow(I_can3), title('Canny, sigma = 2');
```

Question 12 Does increasing the value of `sigma` give us better results when using the Canny detector on a noisy image?

Another parameter of the Canny detector is the threshold value, which affects the sensitivity of the detector.

15. Close any open figures and clear all workspace variables.
16. Load the `mandrill` image and perform the Canny edge detector with default parameters.

```
I = imread('mandrill.tif');
[I_can1,thresh] = edge(I,'canny');
figure
subplot(2,2,1), imshow(I), title('Original Image');
subplot(2,2,2), imshow(I_can1), title('Canny, default parameters');
```

17. Inspect the contents of variable `thresh`.
18. Use a threshold value higher than the one in variable `thresh`.

```
[I_can2,thresh] = edge(I, 'canny', 0.4);
subplot(2,2,3), imshow(I_can2), title('Canny, thresh = 0.4');
```

19. Use a threshold value lower than the one in variable `thresh`.

```
[I_can2,thresh] = edge(I, 'canny', 0.08);
subplot(2,2,4), imshow(I_can2), title('Canny, thresh = 0.08');
```

Question 13 How does the sensitivity of the Canny edge detector change when the threshold value is increased?

Edge Detection with the Kirsch Operator

The remaining edge detection techniques discussed in this tutorial are not included in the current implementation of the `edge` function, so we must implement them as they are defined. We will begin with the Kirsch operator.

20. Close any open figures and clear all workspace variables.
21. Load the `mandrill` image and convert it to `double` format.

```
I = imread('mandrill.tif');
I = im2double(I);
```

Previously, when we were using the `edge` function, we did not need to convert the image to class `double` because the function took care of this for us automatically. Since now we are implementing the remaining edge detectors on our own, we must perform the class conversion to properly handle negative values (preventing unwanted truncation).

Next we will define the eight Kirsch masks. For ease of implementation, we will store all eight masks in a $3 \times 3 \times 8$ matrix. Figure 14.18 illustrates this storage format.

22. Create the Kirsch masks and store them in a preallocated matrix.

```
k = zeros(3,3,8);
k(:,:,1) = [-3 -3 5; -3 0 5; -3 -3 5];
k(:,:,2) = [-3 5 5; -3 0 5; -3 -3 -3];
k(:,:,3) = [5 5 5; -3 0 -3; -3 -3 -3];
k(:,:,4) = [5 5 -3; 5 0 -3; -3 -3 -3];
k(:,:,5) = [5 -3 -3; 5 0 -3; 5 -3 -3];
k(:,:,6) = [-3 -3 -3; 5 0 -3; 5 5 -3];
k(:,:,7) = [-3 -3 -3; -3 0 -3; 5 5 5];
k(:,:,8) = [-3 -3 -3; -3 0 5; -3 5 5];
```

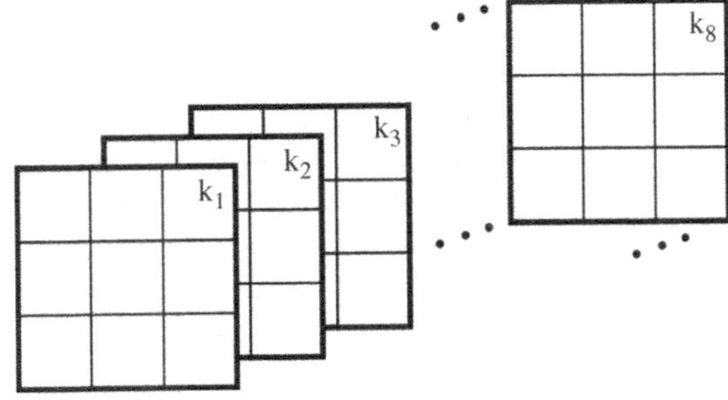

FIGURE 14.18 Kirsch masks stored in a $3 \times 3 \times 8$ matrix.

Next we must convolve each mask on the image, generating eight images. We will store these images in a three-dimensional matrix just as we did for the masks. Because all the masks are stored in one matrix, we can use a `for` loop to perform all eight convolutions with less lines of code.

23. Convolve each mask with the image using a `for` loop.

```
I_k = zeros(size(I,1), size(I,2), 8);
for i = 1:8
    I_k(:,:,i) = imfilter(I,k(:,:,i));
end
```

24. Display the resulting images.

```
figure
for j = 1:8
    subplot(2,4,j), imshow(abs(I_k(:,:,j)),[]), ...
        title(['Kirsch mask ', num2str(j)]);
end
```

Question 14 Why are we required to display the absolute value of each mask? *Hint*: Inspect the minimum and maximum values.

Question 15 How did we dynamically display the mask number when displaying all eight images?

Next we must find the maximum value of all the images for each pixel. Again, because they are all stored in one matrix, we can do this in one line of code.

25. Find the maximum values.

```
I_kir = max(I_k,[],3);
figure, imshow(I_kir,[]);
```

Question 16 When calculating the maximum values, what does the last parameter in the `max` function call mean?

Question 17 Why are we required to scale the image when displaying it?

In the previous step we scaled the result for display purposes. If we wish to threshold the image (as we did with all previous edge detectors), we must first scale the image so that its values are within the range [0, 255] as well as convert to class `uint8`. To do so, we can create a linear transformation function that maps all current values to values within the range we want.

26. Create a transformation function to map the image to the grayscale range and perform the transformation.

```
m = 255 / (max(I_kir(:)) - min(I_kir(:)));
I_kir_adj = uint8(m * I_kir);
figure, imshow(I_kir_adj);
```

Question 18 Why is it not necessary to scale this image (`I_kir_adj`) when displaying it?

Question 19 Make a copy of the `mandrill` image and add Gaussian noise to it. Then perform the Kirsch edge detector on it. Comment on its performance when noise is present.

Edge Detection with the Robinson Operator

The Robinson edge detector can be implemented in the same manner as the Kirsch detector. The only difference is the masks.

27. Generate the Robinson masks.

```
r = zeros(3,3,8);
r(:,:,1) = [-1 0 1; -2 0 2; -1 0 1];
r(:,:,2) = [0 1 2; -1 0 1; -2 -1 0];
r(:,:,3) = [1 2 1; 0 0 0; -1 -2 -1];
r(:,:,4) = [2 1 0; 1 0 -1; 0 -1 -2];
r(:,:,5) = [1 0 -1; 2 0 -2; 1 0 -1];
r(:,:,6) = [0 -1 -2; 1 0 -1; 2 1 0];
r(:,:,7) = [-1 -2 -1; 0 0 0; 1 2 1];
r(:,:,8) = [-2 -1 0; -1 0 1; 0 1 2];
```

28. Filter the image with the eight Robinson masks and display the output.

```
I_r = zeros(size(I,1), size(I,2), 8);
for i = 1:8
    I_r(:,:,i) = imfilter(I,r(:,:,i));
end
figure
for j = 1:8
    subplot(2,4,j), imshow(abs(I_r(:,:,j)),[]), ...
        title(['Robinson mask ', num2str(j)]);
end
```

29. Calculate the `max` of all eight images and display the result.

```
I_rob = max(I_r,[],3);
figure, imshow(I_kir,[]);
```

Question 20 How does the Robinson edge detector compare with the Kirsch detector?

WHAT HAVE WE LEARNED?

- Edge detection is a fundamental image processing operation that attempts to emulate an ability present in the human visual system. *Edges* in grayscale 2D images are usually defined as a sharp variation of the intensity function. In a more general sense, an *edge* can be defined as a boundary between two image regions having distinct characteristics according to some feature (e.g., gray level, color, or texture). Edge detection is a fundamental step in many image processing techniques: after edges have been detected, the regions enclosed by these edges are segmented and processed accordingly.
- There are numerous edge detection techniques in the image processing literature. They range from simple convolution masks (e.g., Sobel and Prewitt) to biologically inspired techniques (e.g., the Marr–Hildreth method) and the quality of the results they provide vary widely. The Canny edge detector is allegedly the most popular contemporary edge detection method.
- MATLAB has a function `edge` that implements several edge detection methods such as Prewitt, Sobel, Laplacian of Gaussian, and Canny.
- The results of the edge detection algorithm are typically postprocessed by an edge linking algorithm that typically eliminates undesired points, bridges gaps, and results in cleaner edges that are then used in subsequent stages of an edge-based image segmentation solution. The Hough transform is a commonly used technique to find long straight edges (i.e., line segments) within the edge detection results.

LEARN MORE ABOUT IT

- David Marr's book [Mar82] is one of the most influential books ever written in the field of vision science. Marr's theories remain a source of inspiration to computer vision scientists more than 25 years after they were published.
- Mlsna and Rodriguez discuss first- and second-order derivative edge methods, as well as the Canny edge detector in Chapter 4.11 of [Bov00a].
- Section 15.2 of [Pra07] presents additional first-order derivative edge operators, for example, Frei–Chen, boxcar, truncated pyramid, Argyle, Macleod, and first derivative of Gaussian (FDOG) operators.
- Section 15.5 of [Pra07] discusses several performance criteria that might be used for a comparative analysis of edge detection techniques.

- Other methods for edge linking and boundary detection are discussed in Section 10.2.7 of [GW08], Section 17.4 of [Pra07], and Section 10.3 of [SS01], among other references.
- Chapter 9 of [Dav04] describes in detail the standard Hough transform and its application to line detection.
- Chapters 11, 12, and 14 of [Dav04] discuss the generalized Hough transform (GHT) and its applications to line, ellipse, polygon, and corner detection.
- Computer vision algorithms usually require that other types of relevant primitive properties of the images, such as lines, corners, and points, be detected. For line detection, refer to Section 5.3.9 of [SHB08]. Section 5.4 of [SOS00] covers critical point detection. For corner detection, we recommend Chapter 8 of [BB08] and Section 5.3.10 of [SHB08].

14.8 PROBLEMS

14.1 Write a MATLAB script to generate a test image containing an ideal edge and plot the intensity profile and the first and second derivatives along a horizontal line of the image.

14.2 Repeat Problem 14.1 for a ramp edge.

14.3 Show that the LoG edge detector can be implemented using `fspecial` and `imfilter` (instead of `edge`) and provide a reason why this implementation may be preferred.

CHAPTER 15

IMAGE SEGMENTATION

WHAT WILL WE LEARN?

- What is image segmentation and why is it relevant?
- What is image thresholding and how is it implemented in MATLAB?
- What are the most commonly used image segmentation techniques and how do they work?

15.1 INTRODUCTION

Segmentation is one of the most crucial tasks in image processing and computer vision. As you may recall from our discussion in Chapter 1 (Section 1.5), image segmentation is the operation that marks the transition between *low-level image processing* and *image analysis*: the input of a segmentation block in a machine vision system is a preprocessed image, whereas the output is a representation of the regions within that image. This representation can take the form of the boundaries among those regions (e.g., when edge-based segmentation techniques are used) or information about which pixel belongs to which region (e.g., in clustering-based segmentation). Once an image has been segmented, the resulting individual regions (or objects) can be described, represented, analyzed, and classified with techniques such as the ones presented in Chapters 18 and 19.

Practical Image and Video Processing Using MATLAB®. By Oge Marques.

Segmentation is defined as the process of partitioning an image into a set of nonoverlapping regions whose union is the entire image. These regions should ideally correspond to objects and their meaningful parts, and background. Most image segmentation algorithms are based on one of two basic properties that can be extracted from pixel values—discontinuity and similarity—or a combination of them.

Segmentation of nontrivial images is a very hard problem—made even harder by nonuniform lighting, shadows, overlapping among objects, poor contrast between objects and background, and so on—that has been approached from many different angles, with limited success to this date. Many image segmentation techniques and algorithms have been proposed and implemented during the past 40 years and yet, except for relatively "easy" scenes, the problem of segmentation remains unsolved.

Figure 15.1 illustrates the problem. At the top, it shows the color and grayscale versions of a *hard* test image that will be used later in this chapter. Segmenting this image into its four main objects (Lego bricks) and the background is not a simple task for contemporary image segmentation algorithms, due to uneven lighting, projected shadows, and occlusion among objects. Attempting to do so without resorting to color information makes the problem virtually impossible to solve for the techniques described in this chapter.

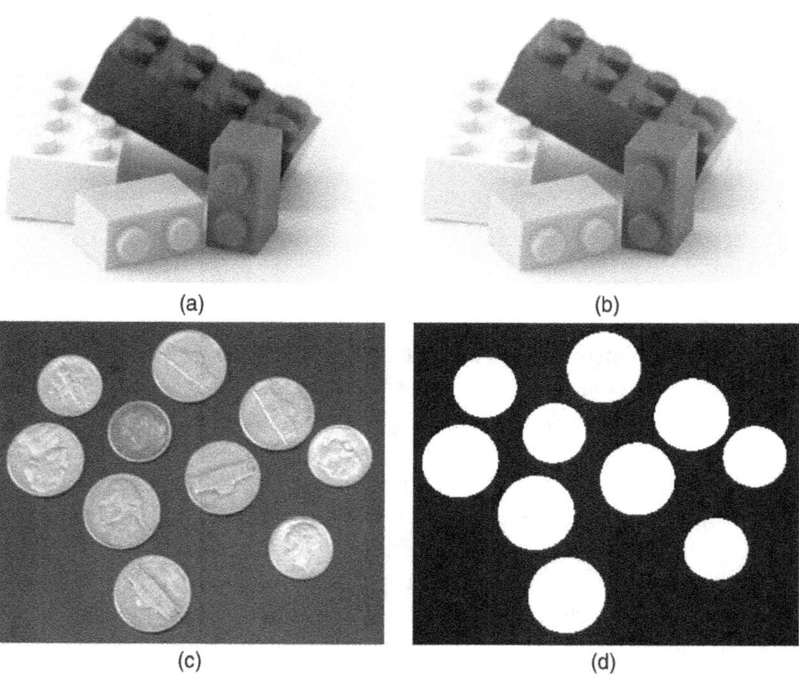

FIGURE 15.1 Test images for segmentation algorithms: (a) a *hard* test image and (b) its grayscale equivalent; (c) an easier test image (courtesy of MathWorks) and (d) the result of morphological preprocessing and thresholding.

The bottom part of Figure 15.1 shows another test image, which is considerably simpler and will probably lead to perfect segmentation with even the simplest techniques. Although the original image has a few imperfections (particularly on one coin that is significantly darker than the others), simple preprocessing operations such as region filling (Section 13.6.2) using `imfill` will turn it into an image suitable for global thresholding (Section 15.2.1) and subsequent labeling of the individual regions.

There is no underlying theory of image segmentation, only *ad hoc* methods, whose performance is often evaluated indirectly, based on the performance of the larger system to which they belong. Even though they share the same goal, image segmentation techniques can vary widely according to the type of image (e.g., binary, gray, color), choice of mathematical framework (e.g., morphology, image statistics, graph theory), type of features (e.g., intensity, color, texture, motion), and approach (e.g., top-down, bottom-up, graph-based).[1]

There is no universally accepted taxonomy for classification of image segmentation algorithms either. In this chapter, we have organized the different segmentation methods into the following categories:

- Intensity-based methods (Section 15.2), also known as *noncontextual* methods, work based on pixel distributions (i.e., histograms). The best-known example of intensity-based segmentation technique is *thresholding*.
- Region-based methods (Section 15.3), also known as *contextual* methods, rely on adjacency and connectivity criteria between a pixel and its neighbors. The best-known examples of region-based segmentation techniques are *region growing* and *split and merge*.
- Other methods, where we have grouped relevant segmentation techniques that do not belong to any of the two categories above. These include segmentation based on texture, edges, and motion, among others.[2]

15.2 INTENSITY-BASED SEGMENTATION

Intensity-based methods are conceptually the simplest approach to segmentation. They rely on pixel statistics—usually expressed in the form of a histogram (Chapter 9)—to determine which pixels belong to foreground objects and which pixels should be labeled as background. The simplest method within this category is *image thresholding*, which will be described in detail in the remaining part of this section.

[1]The field of image segmentation research is still very active. Most recently published algorithms are far too complex to be included in this text, and their computational requirements often push MATLAB to its limits. Refer to "Learn More About It" section at the end of the chapter for useful pointers.

[2]Segmentation of color images will be discussed in Chapter 16.

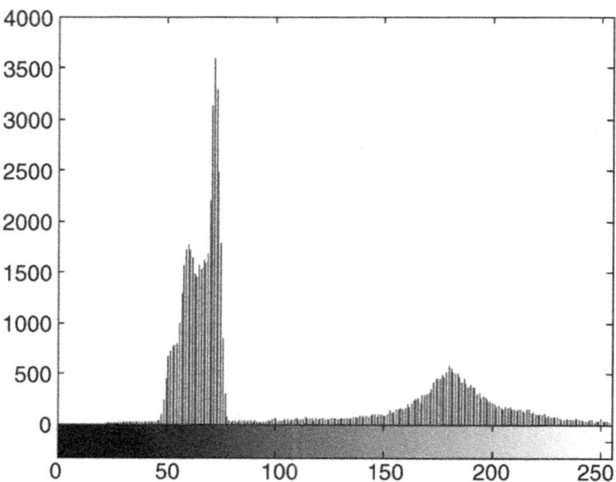

FIGURE 15.2 The histogram for the image in Figure 15.1c: an example of histogram suitable for partitioning using a single threshold.

15.2.1 Image Thresholding

The basic problem of thresholding is the conversion of an image with many gray levels into another image with fewer gray levels, usually only two. This conversion is usually performed by comparing each pixel intensity with a reference value (*threshold*, hence the name) and replacing the pixel with a value that means "white" or "black" depending on the outcome of the comparison. Thresholding is a very popular image processing technique, due to its simplicity, intuitive properties, and ease of implementation.

Thresholding an image is a common preprocessing step in machine visual systems in which there are relatively few objects of interest whose shape (silhouette) is more important than surface properties (such as texture) and whose average brightness is relatively higher or lower than the other elements in the image. The test image in Figure 15.1c is an example of image suitable for image thresholding. Its histogram (Figure 15.2) has two distinct modes, the narrowest and most prominent one (on the left) corresponding to background pixels, the broadest one (on the right) reflecting the intensity distribution of pixels corresponding to the coins.

Mathematically, the process of thresholding an input image $f(x, y)$ and producing a binarized version of it, $g(x, y)$, can be described as

$$g(x, y) = \begin{cases} 1 & \text{if } f(x, y) > T \\ 0 & \text{otherwise} \end{cases} \tag{15.1}$$

where T is the threshold. If the same value of T is adopted for the entire image, the process is called *global thresholding*. When the choice of value for T at a point of coordinates (x, y) depends on statistical properties of pixel values in a neighborhood around (x, y), it will be referred to as *local* or *regional thresholding*.[3]

In MATLAB

The IPT has a function to convert a grayscale image into a binary (black-and-white) image, `im2bw`, that takes an image and a threshold value as input parameters. You will learn how to use it in Tutorial 15.1.

15.2.2 Global Thresholding

When the intensity distribution of an image allows a clear differentiation between pixels of two distinct predominant average gray levels, the resulting histogram has a bimodal shape (such as the one in Figure 15.2), which suggests that there may be a single value of T that can be used as a threshold for the entire image. For the case of a single image, the choice of the actual value of T can be done manually, in typical trial and error fashion, as follows:

1. Inspecting the image's histogram (using `imhist`).
2. Select an appropriate value for T.
3. Apply the selected value (using `im2bw`) to the image.
4. Inspect the results: if they are acceptable, save resulting image. Otherwise, make adjustments and repeat steps 2–4.

For the cases where many images need to be segmented using global thresholding, a manual, labor-intensive approach such as described above is not appropriate. An automated procedure for selecting T has to be employed. Gonzalez and Woods [GW08] proposed an iterative algorithm for this purpose, whose MATLAB implementation (based on [GWE04]) follows:

```
Id = im2double(I); % I is a uint8 grayscale image
T = 0.5*(min(Id(:)) + max(Id(:)));
deltaT = 0.01; % convergence criterion
done = false;
while ~done
     g = Id >= T;
     Tnext = 0.5*(mean(Id(g)) + mean(Id(~g)));
     done = abs(T - Tnext) < deltaT;
     T = Tnext;
end
```

[3]Techniques that rely on the spatial coordinates (x, y), often called *dynamic* or *adaptive thresholding*, have also been proposed in the literature. Since they cannot be considered purely "intensity-based," they have been left out of this discussion.

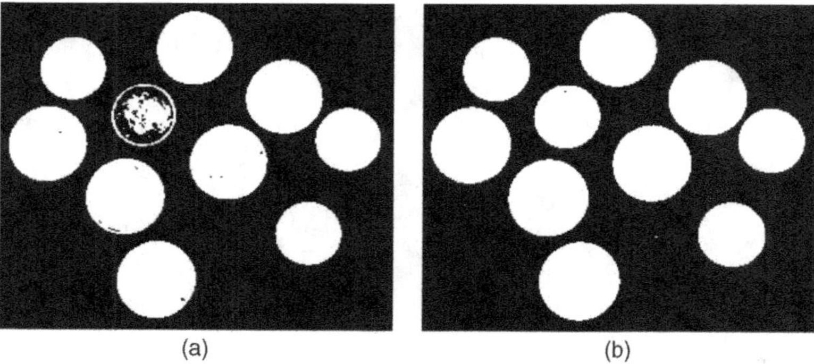

FIGURE 15.3 Image thresholding results for the image in Figure 15.1c using iterative threshold selection algorithm (a) and manually selected threshold (b).

Figure 15.3a shows the result of applying the threshold value to the image in Figure 15.1c. In this case, the maximum and minimum gray values in `Id` are 1 and 0.0902, respectively, `deltaT` was chosen to be 0.01, and it took the algorithm three iterations to arrive at the final result: $T = 0.4947$. For the sake of comparison, the results obtained with a manually selected $T = 0.25$ are shown in Figure 15.3b.

Optimal Thresholding Many *optimal* strategies for selecting threshold values have been suggested in the literature. These strategies usually rely on assumed statistical models and consist of modeling the thresholding problem as a statistical inference problem. Unfortunately, such statistical models usually cannot take into account important factors such as borders and continuity, shadows, nonuniform reflectance, and other perceptual aspects that would impact a human user making the same decision. Consequently, for most of the cases, manual threshold selection by humans will produce better results than statistical approaches would [BD00].

The most popular approach under this category was proposed by Otsu[4] in 1979 [Ots79] and implemented as an IPT function: `graythresh`. Applying that function to the image in Figure 15.1c results in an optimal value for $T = 0.4941$, which—in this particular case—is remarkably close to the one obtained with the (much simpler) iterative method described earlier ($T = 0.4947$). However, as shown in Figure 15.3, neither of these methods produce a better (from a visual interpretation standpoint) result than the manually chosen threshold.

15.2.3 The Impact of Illumination and Noise on Thresholding

Illumination and reflectance patterns play a critical role in thresholding. Even an easy input image (such as the `coins` image), which could be successfully segmented using global thresholding, poses a much harder challenge if the illumination pattern changes

[4] A detailed description of the approach is beyond the scope of this text. See Section 10.3.3 of [GW08] or Section 3.8.2 of [SS01] for additional information.

FIGURE 15.4 An example of uneven illumination pattern used to generate the image in Figure 15.5a.

from constant (uniform) to gradual (Figure 15.4). The resulting image (Figure 15.5a) is significantly darker overall and the corresponding histogram (Figure 15.5b) shows an expected shift to the left. Consequently, using the same value of threshold ($T = 25$) that produced very good results before (Figure 15.3b) will lead to an unacceptable binarized image (Figure 15.5c).

Noise can also have a significant impact on thresholding, as illustrated in Figure 15.5(d–f). In this case a Gaussian noise of mean zero and variance 0.03 has been applied to the image, resulting in the image in Figure 15.5d, whose histogram, shown in Figure 15.5e, has lost its original bimodal shape. The result of segmenting the image using $T = 0.25$ is shown in Figure 15.5f. Although not as bad as one could expect, it would need postprocessing (noise reduction) to be truly useful.

In summary, in both cases, the images changed significantly, their histograms lost their bimodal shape, and the originally chosen value for global threshold ($T = 0.25$) was no longer adequate. In addition, no other value could be easily chosen just by inspecting the histogram and following the trial and error procedure suggested earlier in this chapter.

15.2.4 Local Thresholding

Local (also called *adaptive*) thresholding uses block processing to threshold blocks of pixels, one at a time. The size of the block is usually specified by the user, with the two extreme conditions being avoided: blocks that are too small may require an enormous amount of processing time to compute, whereas large blocks may produce results that are not substantially better than the ones obtained with global thresholding.

In MATLAB

The block processing technique is implemented using the `blkproc` function. You will learn how to use this function in the context of local thresholding in Tutorial 15.1.

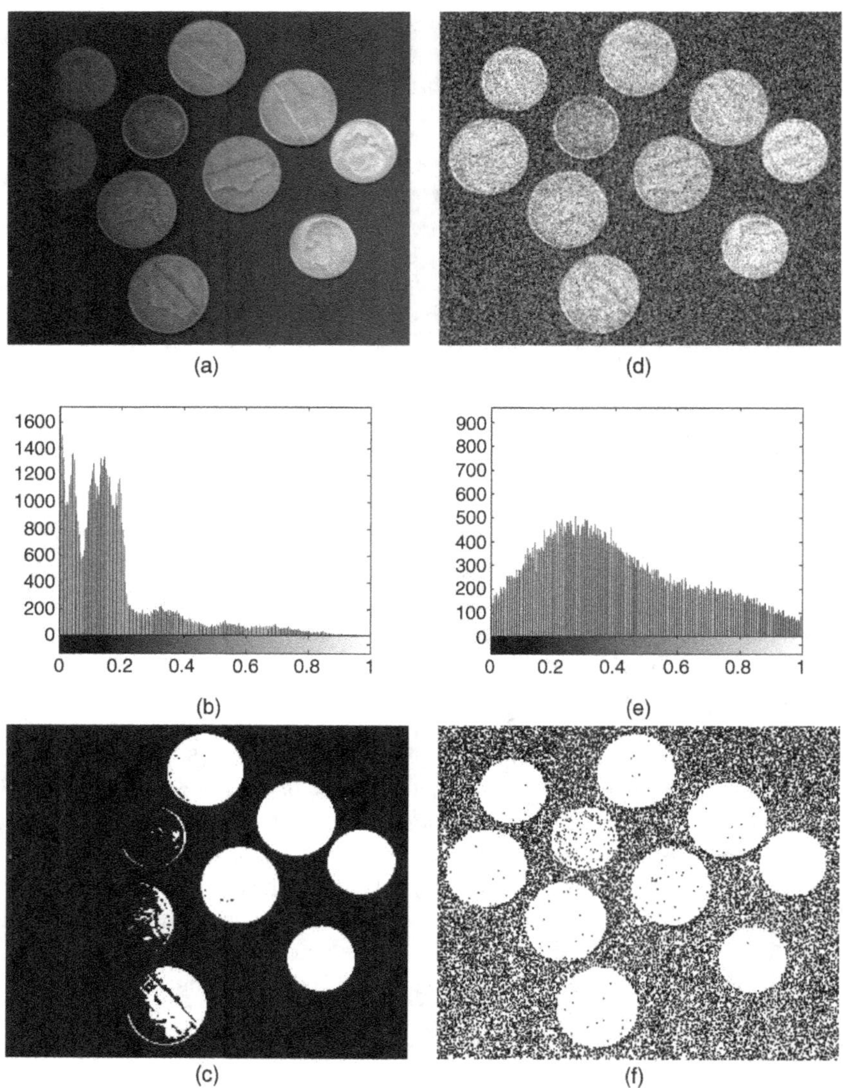

FIGURE 15.5 Effect of illumination (left) and noise (right) on thresholding. See text for details.

■ EXAMPLE 15.1

In this example, we divide the image in Figure 15.5a into six vertical slices, treating each of them separately and using `graythresh` to choose a different value of threshold for each slice. The resulting image (Figure 15.6b) is significantly better than the one we would have obtained using a single value ($T = 0.3020$, also calculated using

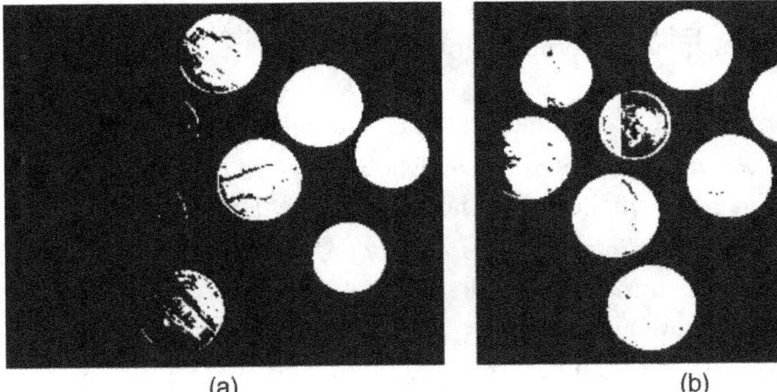

FIGURE 15.6 Local thresholding. Using a single threshold for the entire image (a) and dividing it up into six slices and choosing a different threshold for each vertical slice (b).

`graythresh`) for the entire image (Figure 15.6a). The left portion of Figure 15.6b gives away some hints at where the slices were placed.

15.3 REGION-BASED SEGMENTATION

Region-based segmentation methods are based on the fact that a pixel cannot be considered a part of an object or not based solely on its gray value (as intensity-based methods do). They incorporate measures of connectivity among pixels in order to decide whether these pixels belong to the same region (or object) or not.

Mathematically, region-based segmentation methods can be described as a systematic way to partition an image I into n regions, $R_1, R_2, \cdots, R_n$, such that the following properties hold [GW08]:

1. $\bigcup_{i=1}^{n} R_i = I$.
2. R_i is a connected region, $i = 1, 2, \cdots, n$.
3. $R_i \cap R_j = \emptyset$ for all i and j, $i \neq j$.
4. $P(R_i) = \text{TRUE}$ for $i = 1, 2, \cdots, n$.
5. $P(R_i \cup R_j) = \text{FALSE}$ for any adjacent regions R_i and R_j.

Here $P(R_i)$ is a logical predicate defined over the points in set R_i and $\emptyset$ is the empty set.

The first property states that the segmentation will be complete, that is, each pixel in the image will be labeled as belonging to one of the n regions. Property 2 requires that all points within a region be 4- or 8- connected. Property 3 states that the regions cannot overlap. Property 4 states which criterion must be satisfied so that a pixel is granted membership in a certain region, for example, all pixel values must be within a certain range of intensities. Finally, property 5 ensures that two adjacent regions are different in the sense of predicate P.

These logical predicates are also called *homogeneity criteria*, $H(R_i)$. Some of the most common homogeneity criteria for grayscale images are as follows [Umb05]:

- *Pure Uniformity*: All pixel values in a region are the same.
- *Local Mean Relative to Global Mean*: The average intensity in a region is significantly greater (or smaller) than the average gray level in the whole image.
- *Local Standard Deviation Relative to Global Mean*: The standard deviation of the pixel intensities in a region is less than a small percentage of the average gray level in the whole image.
- *Variance*: At least a certain percentage of the pixels in a region are within two standard deviations of the local mean.
- *Texture*: All four quadrants within a region have comparable texture.

15.3.1 Region Growing

The basic idea of region growing methods is to start from a pixel and grow a region around it, as long as the resulting region continues to satisfy a homogeneity criterion. It is, in that sense, a bottom-up approach to segmentation, which starts with individual pixels (also called *seeds*) and produces segmented regions at the end of the process.

The key factors in region growing are as follows:

- *The Choice of Similarity Criteria*: For monochrome images, regions are analyzed based on intensity levels (either the gray levels themselves or the measures that can easily be calculated from them, for example, moments and texture descriptors[5]) and connectivity properties.
- *The Selection of Seed Points*: These can be determined interactively (if the application allows) or based on a preliminary cluster analysis of the image, used to determine groups of pixels that share similar properties, from which a seed (e.g., corresponding to the centroid of each cluster) can be chosen.
- *The Definition of a Stopping Rule*: A region should stop growing when there are no further pixels that satisfy the homogeneity and connectivity criteria to be included in that region.

It can be described in algorithmic form as follows [Eff00]:

```
Let f(x,y) be the input image
Define a set of regions R1, R2, ..., Rn, each consisting of a
          single seed pixel
repeat
   for i = 1 to n do
      for each pixel p at the border of Ri do
         for all neighbors of p do
```

[5]These will be discussed in Chapter 18.

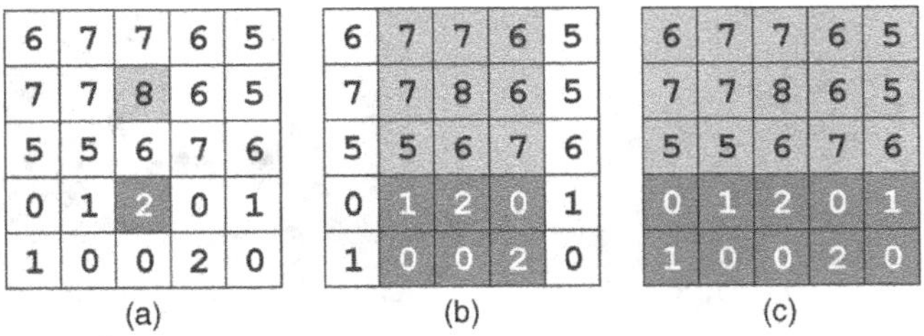

6	7	7	6	5
7	7	8	6	5
5	5	6	7	6
0	1	2	0	1
1	0	0	2	0

(a)

6	7	7	6	5
7	7	8	6	5
5	5	6	7	6
0	1	2	0	1
1	0	0	2	0

(b)

6	7	7	6	5
7	7	8	6	5
5	5	6	7	6
0	1	2	0	1
1	0	0	2	0

(c)

FIGURE 15.7 Region growing: (a) seed pixels; (b) first iteration; (c) final iteration.

```
          Let (x,y) be the neighbor's coordinates
          Let Mi be the mean gray level of pixels in Ri
          if the neighbor is unassigned and
                    |f(x,y) - Mi| <= Delta then
             Add neighbor to Ri
             Update Mi
          end if
       end for
    end for
 end for
until no more pixels can be assigned to regions
```

■ EXAMPLE 15.2

Figure 15.7 shows an example of region growing on small test image, where the logical predicate for uniformity is given by

$$P(R_i) = \begin{cases} \text{TRUE} & \text{if } |f(x, y) - \mu_i| \leq \Delta \\ \text{FALSE} & \text{otherwise} \end{cases} \tag{15.2}$$

where μ_i is the average intensity of all pixels in R_i except the reference pixel at (x, y) and Δ is a user-selected threshold. In this example, $\Delta = 3$. Figure 15.7 shows the seed pixels (a), and the results of the first (b) and last (c) iterations.

■ EXAMPLE 15.3

Figure 15.8 shows the results of applying a region growing algorithm[6] to the two test images originally introduced in Figure 15.1. Part (a) shows the *hard* input image with the seed points (specified interactively by the user) overlaid. Part (b) shows the results of using region growing (pseudocolored for easier visualization): four regions plus background. A quick inspection shows that the results are virtually useless, primarily

[6]The region growing algorithm used in this example appears in [GWE04].

FIGURE 15.8 Region growing results for two test images. See text for details.

due to poor contrast between the two darker and two brighter bricks and the influence of projected shadows.

For comparison purposes, we ran the same algorithm with an *easy* input image (Figure 15.8c) in *unsupervised mode*, that is, without specifying any points to be used as seeds (or even the total number of regions that the algorithm should return). The results (Figure 15.8d) are good, comparable to the ones obtained using global thresholding earlier in this chapter.

Limitations of Region Growing The basic region growing algorithm described in this section has several limitations, listed below [Eff00]:

- It is not very stable: significantly different results are obtained when switching between 4-connectivity and 8-connectivity criteria.
- Segmentation results are very sensitive to choice of logical uniformity predicate.
- The number of seeds provided by the user may not be sufficient to assign every pixel to a region.
- If two or more seeds that should belong to the same region are incorrectly provided to the algorithm, it will be forced to create distinct regions around them although only one region should exist.

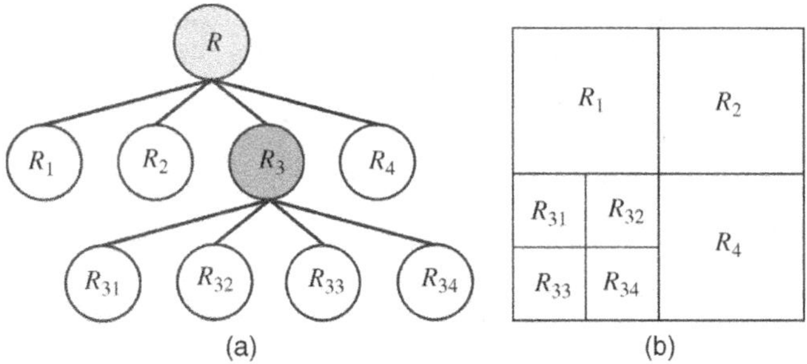

FIGURE 15.9 The quadtree data structure used in the split and merge segmentation algorithm (a) and the corresponding regions in the image (b).

15.3.2 Region Splitting and Merging

Region splitting is a top-down approach to image segmentation. It starts from the entire image and partitions it into smaller subimages until each resulting region is considered homogeneous by some criteria. At the end of the process, it is guaranteed that the resulting regions satisfy the homogeneity criterion. It is possible, however, that two or more adjacent regions are similar enough that should be combined into one. This is the goal of the *merging* step: to merge two or more adjacent regions into one if they satisfy a homogeneity criterion.

The data structure most commonly used for this algorithm is the *quadtree*, a special type of tree in which each node (except for the leaves) has four children. Each leaf node in the quadtree corresponds to a region in the segmented image (Figure 15.9).

The split and merge segmentation algorithm can be described in algorithmic form as follows:

1. Define a logical uniformity predicate $P(R_i)$.
2. Compute $P(R_i)$ for each region.
3. Split into four disjoint quadrants any region R_i for which $P(R_i) = \text{FALSE}$.
4. Repeat steps 2 and 3 until all resulting regions satisfy the uniformity criterion, that is, $P(R_i) = \text{TRUE}$.
5. Merge any adjacent regions R_j and R_k for which $P(R_j \cup R_k) = \text{TRUE}$.
6. Repeat step 5 until no further merging is possible.

15.4 WATERSHED SEGMENTATION

In this section, we describe a popular application of the morphological watershed transform in image segmentation. The watershed transform is a morphological technique that derives its name from an expression in geography, where *watershed* is defined as the ridge that divides areas drained by different river systems. A related term, *catchment* (or *drainage*) *basin*, is used to represent the geographical area that drains into a river or reservoir.

In morphological image processing, the watershed transform is used to represent regions in a segmented image (equivalent to catchment basins) and the boundaries among them (analogous to the ridge lines).

In MATLAB

The IPT function `watershed` implements the watershed transform. It takes an input image and (optionally) a connectivity criterion (4- or 8-connectivity) as input parameters and produces a labeled matrix (of the same size as the input image) as a result. Elements labeled 1 and higher belong to a unique watershed region, identified by their number, whereas elements labeled 0 do not belong to any watershed region.

15.4.1 The Distance Transform

The distance transform is a useful tool employed in conjunction with the watershed transform. It computes the distance from every pixel to the nearest nonzero-valued pixel. It is implemented in MATLAB by function `bwdist`, which allows specification of the distance method (Euclidean distance being the default) to be used.

■ EXAMPLE 15.4

This example shows the creation of a test matrix of size 5×5 and the results of computing the distance transform using `bwdist` and two different distance calculations: Euclidean and city block.

```
>> a = [0 1 1 0 1; 1 1 1 0 0; 0 0 0 1 0; 0 0 0 0 0; 0 1 0 0 0]

a =

     0     1     1     0     1
     1     1     1     0     0
     0     0     0     1     0
     0     0     0     0     0
     0     1     0     0     0

>> b = bwdist(a)
```

```
b =

    1.0000         0         0    1.0000         0
         0         0         0    1.0000    1.0000
    1.0000    1.0000    1.0000         0    1.0000
    1.4142    1.0000    1.4142    1.0000    1.4142
    1.0000         0    1.0000    2.0000    2.2361

>> b = bwdist(a,'cityblock')

b =

     1     0     0     1     0
     0     0     0     1     1
     1     1     1     0     1
     2     1     2     1     2
     1     0     1     2     3
```

■ **EXAMPLE 15.5**

Figure 15.10 shows an example of segmentation using watershed. It uses a binarized and postprocessed version of the coins test image as input (a). Part (b) shows the results of the distance transform calculations. Part (c) shows the ridge lines obtained as a result of applying the watershed transform. Finally, part (d) shows the overlap between parts (a) and (c), indicating how the watershed transform results lead to an excellent segmentation result in this particular case.

15.5 TUTORIAL 15.1: IMAGE THRESHOLDING

Goal

The goal of this tutorial is to learn to perform image thresholding using MATLAB and the IPT.

Objectives

- Learn how to visually select a threshold value using a heuristic approach.
- Explore the `graythresh` function for automatic threshold value selection.
- Learn how to implement adaptive thresholding.

What You Will Need

- `gradient_with_text.tif`

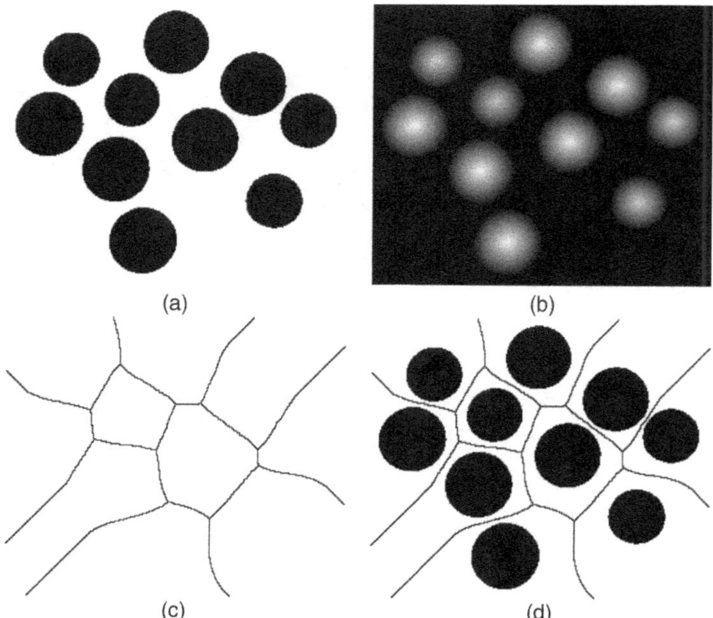

FIGURE 15.10 Segmentation using the morphological watershed transform: (a) complement of the image shown in Figure 15.3; (b) distance transform; (c) watershed ridge lines; (d) result of segmentation.

Procedure

Global Thresholding

The first method of thresholding that we will explore involves visually analyzing the histogram of an image to determine the appropriate value of `T` (the threshold value).

1. Load and display the test image.

```
I = imread('coins.png');
figure, imshow(I), title('Original Image');
```

2. Display a histogram plot of the `coins` image to determine what threshold level to use.

```
figure, imhist(I), title('Histogram of Image');
```

Question 1 Which peak of the histogram represents the background pixels and which peak represents the pixels associated with the coins?

The histogram of the image suggests a bimodal distribution of grayscale values. This means that the objects in the image are clearly separated from the background.

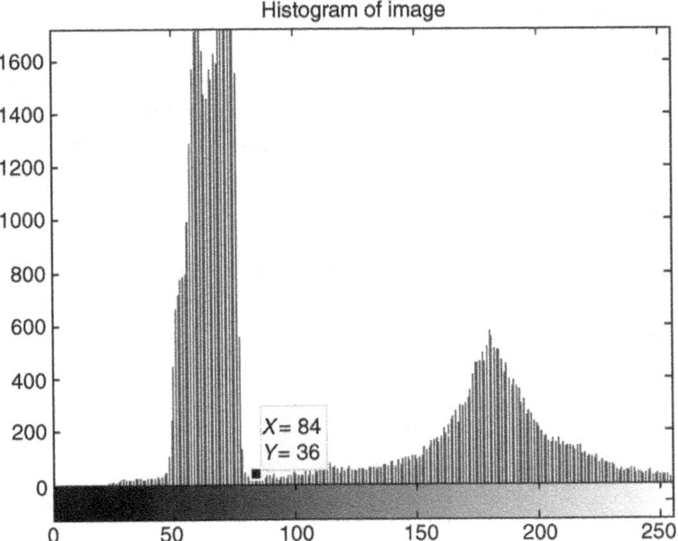

FIGURE 15.11 Histogram plot with data cursor selection.

We can inspect the X and Y values of the histogram plot by clicking the "inspect" icon and then selecting a particular bar on the graph.

3. Inspect the histogram near the right of the background pixels by activating the data cursor. To do so, click on the "inspect" icon.

Figure 15.11 illustrates what the selection should look like. The data cursor tool suggests that values between 80 and 85 could possibly be used as a threshold, since they fall immediately to the right of the leftmost peak in the histogram. Let us see what happens if we use a threshold value of 85.

4. Set the threshold value to 85 and generate the new image.

```
T = 85; I_thresh = im2bw(I,( T / 255));
figure, imshow(I_thresh), title('Threshold Image (heuristic)');
```

Question 2 What is the purpose of the `im2bw` function?

Question 3 Why do we divide the threshold value by 255 in the `im2bw` function call?

You may have noticed that several pixels—some white pixels in the background and a few black pixels where coins are located—do not belong in the resulting image. This small amount of noise can be cleaned up using the noise removal techniques discussed in chapter 12.

Question 4 Write one or more lines of MATLAB code to remove the noise pixels in the thresholded image.

The thresholding process we just explored is known as the *heuristic* approach. Although it did work, it cannot be extended to automated processes. Imagine taking on the job of thresholding a thousand images using the heuristic approach! MATLAB's IPT function `graythresh` uses Otsu's method [Ots79] for automatically finding the best threshold value.

5. Use the `graythresh` function to generate the threshold value automatically.

```
T2 = graythresh(I);
I_thresh2 = im2bw(I,T2);
figure, imshow(I_thresh2), title('Threshold Image (graythresh)');
```

Question 5 How did the `graythresh` function compare with the heuristic approach?

Adaptive Thresholding

Bimodal images are fairly easy to separate using basic thresholding techniques discussed thus far. Some images, however, are not as well behaved and require a more advanced thresholding technique such as *adaptive thresholding*. Take, for example, one of the images we used back in Tutorial 6.1: a scanned text document with a nonuniform gradient background.

6. Close all open figures and clear all workspace variables.
7. Load the `gradient_with_text` image and prepare a subplot.

```
I = imread('gradient_with_text.tif');
figure, imshow(I), title('Original Image');
```

Let us see what happens when we attempt to threshold this image using the techniques we have learned so far.

8. Globally threshold the image.

```
I_gthresh = im2bw(I,graythresh(I));
figure, imshow(I_gthresh), title('Global Thresholding');
figure, imhist(I), title('Histogram of Original');
```

As you may have noticed, we cannot pick one particular value to set as the threshold value because the image is clearly not bimodal. Adaptive thresholding may help us in this instance. To properly implement adaptive thresholding, we must use the `blkproc` function to perform an operation on small blocks of pixels one at a time.

In order to use the function, we must specify what is to be done on each block of pixels. This can be specified within a function that we will create manually. Let us first set up this function.

9. Close all open figures.
10. Start a new M-File in the MATLAB Editor.
11. Define the function as well as its input and output parameters in the first line.

```
function y = adapt_thresh(x)
```

This function will be used to define each new block of pixels in our image. Basically all we want to do is perform thresholding on each block individually, so the code to do so will be similar to the code we previously used for thresholding.

12. Add this line of code under the function definition.

```
y = im2bw(x,graythresh(x));
```

When the function is called, it will be passed a small portion of the image, and will be stored in the variable `x`. We define our output variable `y` as a black and white image calculated by thresholding the input.

13. Save your function as `adapt_thresh.m` in the current directory.

We can now perform the operation using the `blkproc` function. We will adaptively threshold the image, 10×10 pixel blocks at a time.

14. Perform adaptive thresholding by entering the following command in the command window. Note that it may take a moment to perform the calculation, so be patient.

```
I_thresh = blkproc(I,[10 10],@adapt_thresh);
```

15. Display the original and new image.

```
figure
subplot(1,2,1), imshow(I), title('Original Image');
subplot(1,2,2), imshow(I_thresh), title('Adaptive Thresholding');
```

The output is not quite what we expected. If you look closely, however, the operation was successful near the text, but everywhere else it was a disaster. This suggests that we need to add an extra step to our function to compensate for this unwanted effect. Over the next few steps, let us examine the standard deviation of the original image where there *is* text, and where there is *not*.

16. Calculate the standard deviation of two 10×10 blocks of pixels; one where there is text and another where there is not.

```
std_without_text = std2(I(1:10, 1:10))
std_with_text = std2(I(100:110, 100:110))
```

Question 6 What is the difference between the standard deviation of the two blocks of pixels? Explain.

Since there is such a difference between a block with and without text, we can use this information to improve our function. Let us replace the one line of code we previously wrote with the new code that will include an `if` statement: if the standard deviation of the block of pixels is low, then simply label it as background; otherwise, perform thresholding on it. This change should cause the function to only perform thresholding where text exists. Everything else will be labeled as background.

17. Replace the last line of our function with the following code. Save the function after the alteration.

```
if std2(x) < 1
    y = ones(size(x,1),size(x,2));
else
    y = im2bw(x,graythresh(x));
end
```

Question 7 How does our function label a block of pixels as background?

18. Now rerun the block process (in the command window) to see the result.

```
I_thresh2 = blkproc(I,[10 10],@adapt_thresh);
figure, subplot(1,2,1), imshow(I), title('Original Image');
subplot(1,2,2), imshow(I_thresh2), title('Adaptive Thresholding');
```

Question 8 How does the output of the new function compare with the old?

Question 9 What is the main limitation of the adaptive thresholding function developed in this tutorial?

WHAT HAVE WE LEARNED?

- Image segmentation is the process of grouping image pixels into meaningful, usually connected, regions. It is an important (and often required) step in many image processing solutions. It establishes the transition between treating the image as a whole to processing individual relevant regions.

- Image segmentation is a hard image processing problem: the quality of the results will depend on the algorithm, careful selection of algorithm's parameters, and the input image.
- Thresholding is an image processing technique by which an input (grayscale) image is requantized to two gray levels, that is, converted to a binary image. Each pixel in the original image is compared with a threshold; the result of such comparison will determine whether the pixel will be converted to black or white. The simplest thresholding algorithm (*global thresholding*, `im2bw` in MATLAB) employs one value for the entire image.
- Image segmentation techniques can be classified in three main groups: *intensity-based methods* (e.g., thresholding), *region-based methods* (e.g., region growing and split and merge), and *other methods* (e.g., segmentation based on texture, edges, and motion).

LEARN MORE ABOUT IT

- Entire books have been written on the topic of image segmentation, for example, [Wan08] and [Zha06a].
- Many surveys on image segmentation have been published during the past 30 years, among them (in reverse chronological order): [Zha06b], [FMR+02], [CJSW01], [PP93], [HS85], and [FM81].
- Chapters 6 and 7 of [SHB08] provide an extensive and very readable discussion of image segmentation algorithms. Several of these algorithms have been implemented in MATLAB [SKH08].
- Chapter 4 of [Dav04] is entirely devoted to thresholding techniques.
- Section 6.1.2 of [SHB08] discusses optimal thresholding techniques.
- The concept of thresholding can be extended to cases where the original image is to be partitioned in more than two regions, in what is known as *multiple thresholding*. Refer to Section 10.3.6 of [GW08] for a discussion on this topic.
- Chapters 4.7–4.9 of [Bov00a] discuss statistical, texture-based, and motion-based segmentation strategies, respectively.
- The discussion on watershed segmentation is expanded to include the use of gradients and markers in Section 10.5 of [GWE04].
- Comparing and evaluating different segmentation approaches can be a challenging task, for which no universally accepted benchmarks exist. This issue is discussed in [Zha96], [Zha01], and, more recently, [UPH05], expanded in [UPH07].

ON THE WEB

- Color-based segmentation using K-means clustering (IPT image segmentation demo)
  http://tinyurl.com/matlab-k-means

- Color-based segmentation using the L*a*b* color space (IPT image segmentation demo)
  http://tinyurl.com/matlab-lab
- Detecting a cell using image segmentation (IPT image segmentation demo)
  http://tinyurl.com/cell-seg
- Marker-controlled watershed segmentation (IPT image segmentation demo)
  http://tinyurl.com/watershed-seg
- Texture segmentation Using Texture Filters (IPT image segmentation demo)
  http://tinyurl.com/texture-seg
- University of Washington image segmentation demo
  http://www.cs.washington.edu/research/imagedatabase/demo/seg/

15.6 PROBLEMS

15.1 Explain in your own words why the image on the top right of Figure 15.1 is significantly harder to segment than the one on the bottom left of the same figure.

15.2 Modify the MATLAB code to perform iterative threshold selection on an input gray-level image (section 15.2.2) to include a variable that counts the number of iterations and an array that stores the values of T for each iteration.

15.3 Write a MATLAB script to demonstrate that thresholding techniques can be used to subtract the background of an image.

CHAPTER 16

COLOR IMAGE PROCESSING

WHAT WILL WE LEARN?

- What are the most important concepts and terms related to color perception?
- What are the main color models used to represent and quantify color?
- How are color images represented in MATLAB?
- What is pseudocolor image processing and how does it differ from full-color image processing?
- How can monochrome image processing techniques be extended to color images?

16.1 THE PSYCHOPHYSICS OF COLOR

Color perception is a psychophysical phenomenon that combines two main components:

1. The physical properties of light sources (usually expressed by their spectral power distribution (SPD)) and surfaces (e.g., their absorption and reflectance capabilities).
2. The physiological and psychological aspects of the human visual system (HVS).

Practical Image and Video Processing Using MATLAB®. By Oge Marques.

In this section, we expand on the discussion started in Section 5.2.4 and present the main concepts involved in color perception and representation.

16.1.1 Basic Concepts

The perception of color starts with a chromatic light source, capable of emitting electromagnetic radiation with wavelengths between approximately 400 and 700 nm. Part of that radiation reflects on the surfaces of the objects in a scene and the resulting reflected light reaches the human eye, giving rise to the sensation of color. An object that reflects light almost equally in all wavelengths within the visible spectrum is perceived as white, whereas an object that absorbs most of the incoming light, regardless of the wavelength, is seen as black. The perception of several shades of gray between pure white and pure black is usually referred to as *achromatic*. Objects that have more selective properties are considered *chromatic*, and the range of the spectrum that they reflect is often associated with a color name. For example, an object that absorbs most of the energy within the 565–590 nm wavelength range is considered yellow.

A chromatic light source can be described by three basic quantities:

- *Intensity (or Radiance)*: the total amount of energy that flows from the light source, measured in watts (W).
- *Luminance*: a measure of the amount of information an observer *perceives* from a light source, measured in lumen (lm). It corresponds to the radiant power of a light source weighted by a spectral sensitivity function (characteristic of the HVS).
- *Brightness*: the *subjective* perception of (achromatic) luminous intensity.

The human retina (the surface at the back of the eye where images are projected) is coated with photosensitive receptors of two different types: *cones* and *rods*. Rods cannot encode color but respond to lower luminance levels and enable vision under darker conditions. Cones are primarily responsible for color perception and operate only under bright conditions. There are three types of cone cells (L cones, M cones, and S cones, corresponding to long ($\approx$ 610 nm), medium ($\approx$ 560 nm), and short ($\approx$ 430 nm) wavelengths, respectively) whose spectral responses are shown in Figure 16.1.[1]

The existence of three specialized types of cones in the human eye was hypothesized more than a century before it could be confirmed experimentally by Thomas Young and his *trichromatic* theory of vision in 1802. Young's theory explains only part of the color vision process, though. It does not explain, for instance, why it is possible to speak of 'bluish green' colors, but not 'bluish yellow' ones. Such understanding came with the *opponent-process* theory of color vision, brought forth by Edward Herring in 1872. The colors to which the cones respond more strongly are known as the *primary colors of light* and have been standardized by the CIE

[1]The figure also shows the spectral absorption curve for rods, responsible for achromatic vision.

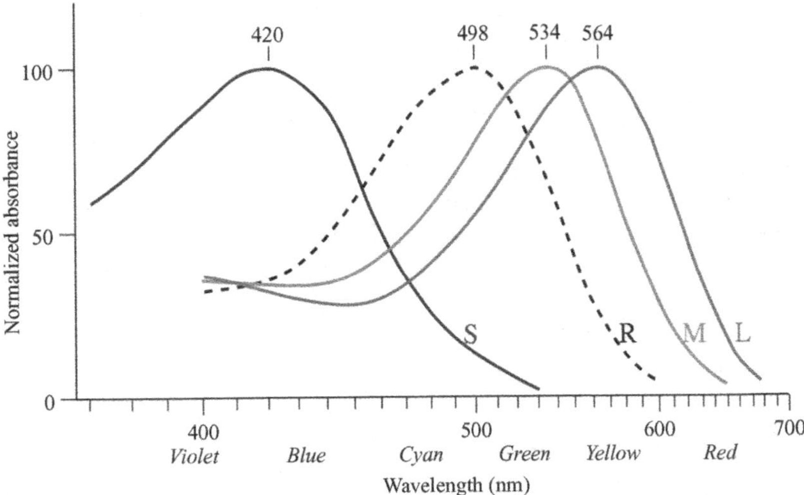

FIGURE 16.1 Spectral absorption curves of the short (S), medium (M), and long (L) wavelength pigments in human cone and rod (R) cells. Courtesy of Wikimedia Commons.

(*Commission Internationale de L'Éclairage*—International Commission on Illumination, an organization responsible for color standards) as red (700 nm), green (546.1 nm), and blue (435.8 nm).

The secondary colors of light, obtained by additive mixtures of the primaries, two colors at a time, are *magenta* (or *purple*) = red + blue, *cyan* (or *turquoise*) = blue + green, and *yellow* = green + red (Figure 16.2a).

For color mixtures using pigments (or paints), the primary colors are magenta, cyan, and yellow and the secondary colors are red, green, and blue (Figure 16.2b). It is important to note that for pigments a color is named after the portion of the spectrum

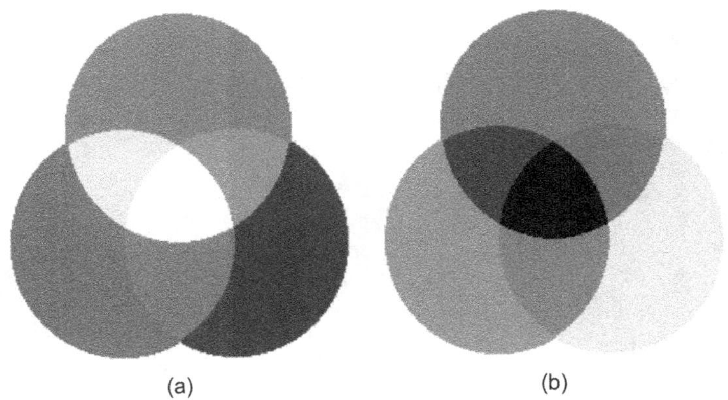

FIGURE 16.2 Additive (a) and subtractive (b) color mixtures.

that it absorbs, whereas for light a color is defined based on the portion of the spectrum that it emits. Consequently, mixing all three primary colors of light results in white (i.e., the entire spectrum of visible light), whereas mixing all three primary colors of paints results in black (i.e., all colors have been absorbed, and nothing remains to reflect the incoming light).

The use of the expression *primary colors* to refer to red, green, and blue may lead to a common misinterpretation: that *all* visible colors can be obtained by mixing different amounts of each primary color, which is not true. A related phenomenon of color perception, the existence of *color metamers*, may have contributed to this confusion. Color metamers are combinations of primary colors (e.g., red and green) perceived by the HVS as another color (in this case, yellow) that could have been produced by a spectral color of fixed wavelength (of ≈ 580 nm).

16.1.2 The CIE *XYZ* Chromaticity Diagram

In color matching experiments performed in the late 1920s, subjects were asked to adjust the amount of red, green, and blue on one patch that were needed to match a color on a second patch. The results of such experiments are summarized in Figure 16.3. The existence of negative values of red and green in this figure means that the second patch should be made brighter (i.e., equal amounts of red, green, and blue has to be added to the color) for the subjects to report a perfect match. Since adding amounts of primary colors on the second patch corresponds to subtracting them in the first, negative values can occur. The amounts of three primary colors in a three-component additive color model (on the first patch) needed to match a test color (on the second patch) are called *tristimulus values*.

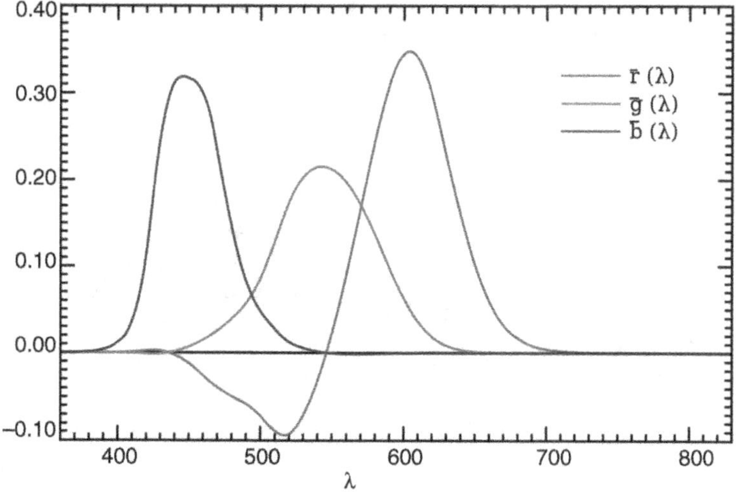

FIGURE 16.3 RGB color matching function (CIE 1931). Courtesy of Wikimedia Commons.

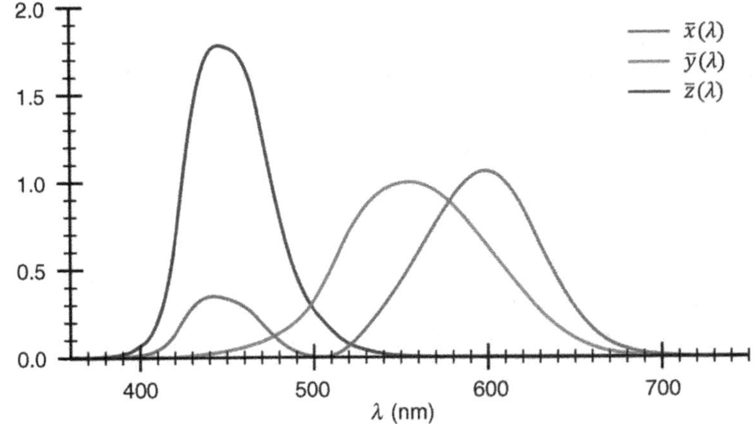

FIGURE 16.4 *XYZ* color matching function (CIE 1931). Courtesy of Wikimedia Commons.

To remove the inconvenience of having to deal with (physically impossible) negative values to represent observable colors, in 1931 the CIE adopted standard curves for a hypothetical *standard (colorimetric) observer*, considered to be the chromatic response of the average human viewing through a 2° angle, due to the belief at that time that the cones resided within a 2° arc of the fovea.[2] These curves specify how a SPD corresponding to the physical power (or *radiance*) of the light source can be transformed into a set of three numbers that specifies a color. These curves are not based on the values of R, G, and B, but on a new set of tristimulus values: X, Y, and Z.

This model, whose color matching functions are shown in Figure 16.4, is known as the *CIE XYZ* (or CIE 1931) model. The tristimulus values of X, Y, and Z in Figure 16.4 are related to the values of R, G, and B in Figure 16.3 by the following linear transformations:

$$\begin{bmatrix} X \\ Y \\ Z \end{bmatrix} = \begin{bmatrix} 0.431 & 0.342 & 0.178 \\ 0.222 & 0.707 & 0.071 \\ 0.020 & 0.130 & 0.939 \end{bmatrix} \begin{bmatrix} R \\ G \\ B \end{bmatrix} \tag{16.1}$$

and

$$\begin{bmatrix} R \\ G \\ B \end{bmatrix} = \begin{bmatrix} 3.063 & -1.393 & -0.476 \\ -0.969 & 1.876 & 0.042 \\ 0.068 & -0.229 & 1.069 \end{bmatrix} \begin{bmatrix} X \\ Y \\ Z \end{bmatrix} \tag{16.2}$$

[2]This understanding was eventually revised and a new model, CIE 1964, for a 10° standard observer, was produced. Interestingly enough, the CIE 1931 model is more popular than the CIE 1964 alternative even today.

The CIE XYZ color space was designed so that the Y parameter corresponds to a measure of the *brightness* of a color. The *chromaticity* of a color is specified by two other parameters, x and y, known as *chromaticity coordinates* and calculated as

$$x = \frac{X}{X + Y + Z} \tag{16.3}$$

$$y = \frac{Y}{X + Y + Z} \tag{16.4}$$

where x and y are also called *normalized tristimulus values*. The third normalized tristimulus value, z, can just as easily be calculated as

$$z = \frac{Z}{X + Y + Z} \tag{16.5}$$

Clearly, a combination of the three normalized tristimulus values results in

$$x + y + z = 1 \tag{16.6}$$

The resulting CIE XYZ chromaticity diagram (Figure 16.5) allows the mapping of a color to a point of coordinates (x, y) corresponding to the color's chromaticity. The complete specification of a color (chromaticity and luminance) takes the form of an xyY triple.[3]

To recover X and Z from x, y, and Y, we can use

$$X = \frac{x}{y} Y \tag{16.7}$$

$$Z = \frac{1 - x - y}{y} Y \tag{16.8}$$

The resulting CIE XYZ chromaticity diagram shows a horseshoe-shaped outer curved boundary, representing the *spectral locus* of wavelengths (in nm) along the visible light portion of the electromagnetic spectrum. The *line of purples* on a chromaticity diagram joins the two extreme points of the spectrum, suggesting that the sensation of purple cannot be produced by a single wavelength: it requires a mixture of shortwave and longwave light and for this reason purple is referred to as a *non-spectral color*. All colors of light are contained in the area in (x, y) bounded by the line of purples and the spectral locus, with pure white at its center.

The inner triangle in Figure 16.6a represents a *color gamut*, that is, a range of colors that can be produced by a physical device, in this case a CRT monitor. Different color image display and printing devices and technologies exhibit gamuts of different shape and size, as shown in Figure 16.6. As a rule of thumb, the larger the gamut, the better the device's color reproduction capabilities.

[3] This explains why this color space is also known as *CIExyY* color space.

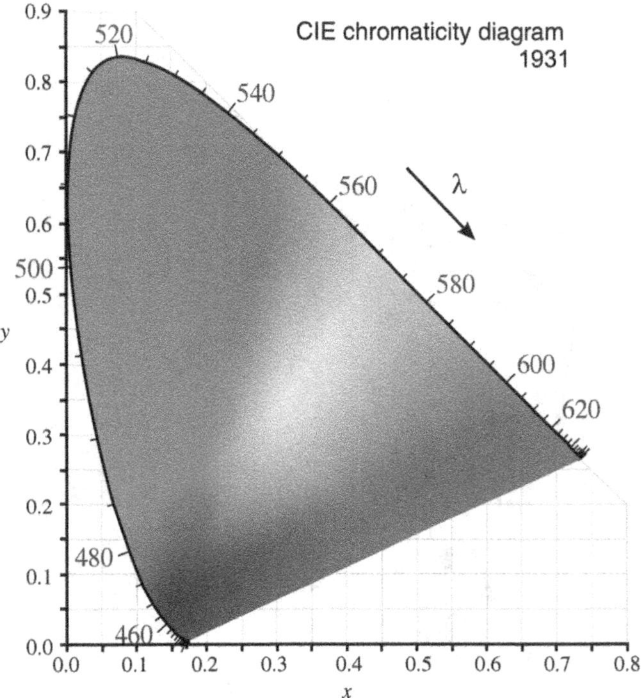

FIGURE 16.5 CIE *XYZ* color model.

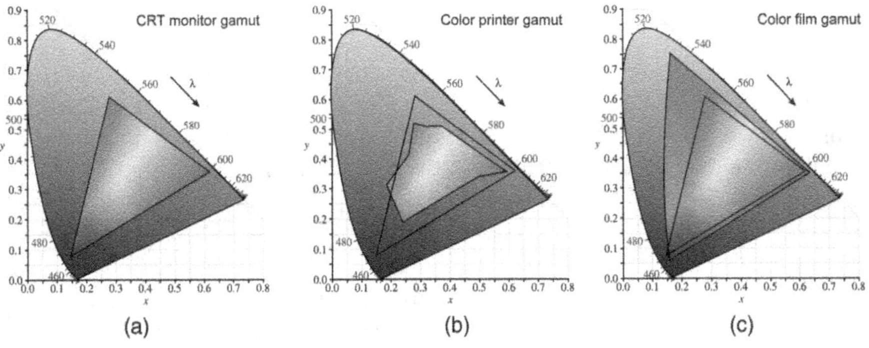

FIGURE 16.6 Color gamut for three different devices: (a) CRT monitor; (b) printer; (c) film. The RGB triangle is the same in all figures to serve as a reference for comparison.

16.1.3 Perceptually Uniform Color Spaces

One of the main limitations of the CIE *XYZ* chromaticity diagram lies in the fact that a distance on the *xy* plane does not correspond to the degree of difference between two colors. This was demonstrated in the early 1940s by David MacAdam, who conducted

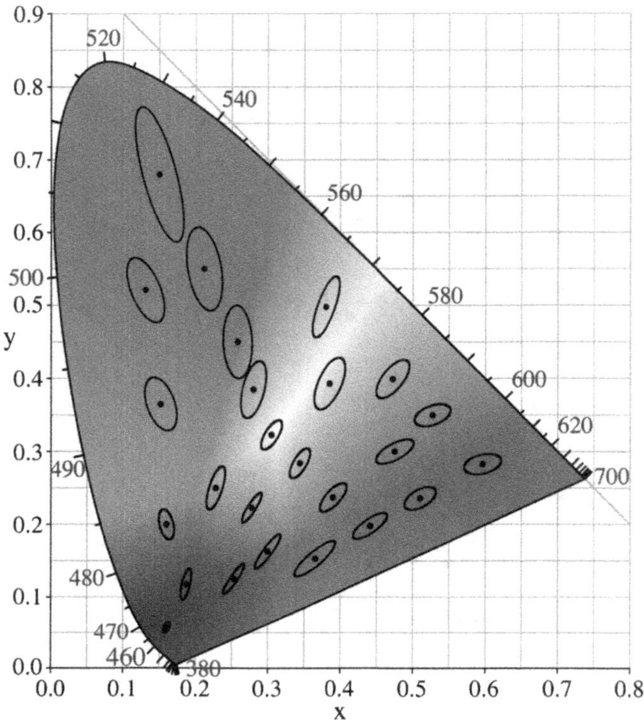

FIGURE 16.7 MacAdam ellipses overlapped on the CIE 1931 chromaticity diagram. Courtesy of Wikimedia Commons.

experiments asking subjects to report noticeable changes in color (relative to a starting color stimulus). The result of that study is illustrated in Figure 16.7, showing the resulting *MacAdam ellipses*: regions on the chromaticity diagram corresponding to all colors that are indistinguishable, to the average human eye, from the color at the center of the ellipse. The contour of the ellipse represents the just noticeable differences (JNDs) of chromaticity. Based on the work of MacAdam, several CIE color spaces—most notably the CIE L*u*v* (also known as CIELUV) and the CIE L*a*b* (also known as CIELAB)—were developed, with the goal of achieving *perceptual uniformity*, that is, have an equal distance in the color space corresponding to equal differences in color.

In MATLAB

The IPT has an extensive support for conversion among CIE color spaces. Converting from one color space to another is usually accomplished by using function `makecform` (to create a color transformation structure that defines the desired color space conversion) followed by `applycform`, which takes the color transformation structure as a parameter.

TABLE 16.1 IPT Functions for CIE *XYZ* and CIELAB Color Spaces

Function	Description
`xyz2double`	Converts an $M \times 3$ or $M \times N \times 3$ array of XYZ color values to `double`
`xyz2uint16`	Converts an $M \times 3$ or $M \times N \times 3$ array of XYZ color values to `uint16`
`lab2double`	Converts an $M \times 3$ or $M \times N \times 3$ array of L*a*b* color values to `double`
`lab2uint16`	Converts an $M \times 3$ or $M \times N \times 3$ array of L*a*b* color values to `uint16`
`lab2uint8`	Converts an $M \times 3$ or $M \times N \times 3$ array of L*a*b* color values to `uint8`
`whitepoint`	Returns a 3×1 vector of XYZ values scaled so that $Y = 1$

Other functions for manipulation of CIE XYZ and CIELAB values are listed in Table 16.1.

16.1.4 ICC Profiles

An ICC (International Color Consortium) profile is a standardized description of a color input or output device, or a color space, according to standards established by the ICC. Profiles are used to define a mapping between the device source or target color space and a *profile connection space* (PCS), which is either CIELAB (L*a*b*) or CIEXYZ.

In MATLAB

The IPT has several functions to support ICC profile operations. They are listed in Table 16.2.

TABLE 16.2 IPT Functions for *ICC* Profile Manipulation

Function	Description
`iccread`	Reads an ICC profile into the MATLAB workspace
`iccfind`	Finds ICC color profiles on a system, or a particular ICC color profile whose description contains a certain text string
`iccroot`	Returns the name of the directory that is the default system repository for ICC profiles
`iccwrite`	Writes an ICC color profile to disk file

16.2 COLOR MODELS

A *color model* (also called *color space* or *color system*) is a specification of a coordinate system and a subspace within that system where each color is represented by a single point.

There have been many different color models proposed over the last 400 years. Contemporary color models have also evolved to specify colors for different purposes (e.g., photography, physical measurements of light, color mixtures, etc.). In this section, we discuss the most popular color models used in image processing.

16.2.1 The RGB Color Model

The RGB color model is based on a Cartesian coordinate system whose axes represent the three primary colors of light (R, G, and B), usually normalized to the range [0, 1] (Figure 16.8). The eight vertices of the resulting cube correspond to the three primary colors of light, the three secondary colors, pure white, and pure black. Table 16.3 shows the R, G, and B values for each of these eight vertices.

RGB color coordinates are often represented in hexadecimal notation, with individual components varying from `00` (decimal 0) to `FF` (decimal 255). For example, a pure (100% saturated) red would be denoted `FF0000`, whereas a slightly desaturated yellow could be written as `CCCC33`.

The number of discrete values of R, G, and B is a function of the *pixel depth*, defined as the number of bits used to represent each pixel: a typical value is

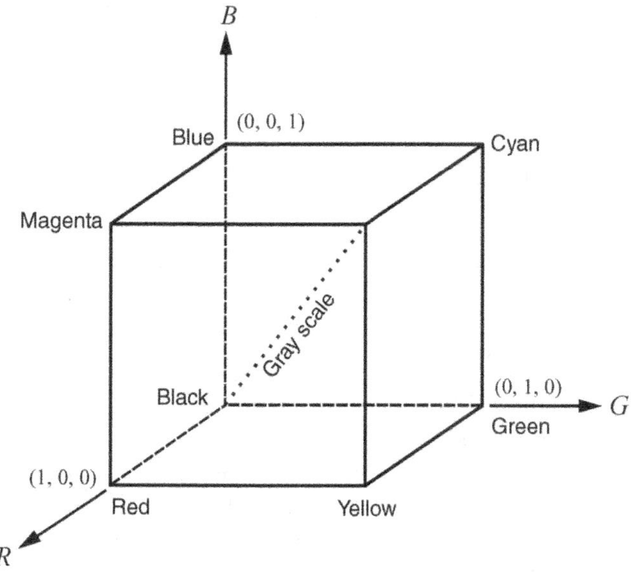

FIGURE 16.8 RGB color model.

TABLE 16.3 *R*, *G*, and *B* Values for Eight Representative Colors Corresponding to the Vertices of the *RGB* Cube

Color Name	*R*	*G*	*B*
Black	0	0	0
Blue	0	0	1
Green	0	1	0
Cyan	0	1	1
Red	1	0	0
Magenta	1	0	1
Yellow	1	1	0
White	1	1	1

24 bits = 3 image planes × 8 bits per plane. The resulting cube—with more than 16 million possible color combinations—is shown in Figure 16.9.

In MATLAB

The RGB cube in Figure 16.9 was generated using `patch`, a graphics function for creating *patch graphics objects*, made up of one or more polygons, which can be specified by passing the coordinates of their vertices and the coloring and lighting of the patch as parameters.

FIGURE 16.9 RGB color cube.

16.2.2 The CMY and CMYK Color Models

The CMY model is based on the three primary colors of pigments ($c$yan, $m$agenta, and $y$ellow). It is used for color printers, where each primary color usually corresponds to an ink (or toner) cartridge. Since the addition of equal amounts of each primary to produce black usually produces unacceptable, muddy looking black, in practice, a fourth color, $blacK$, is added, and the resulting model is called $CMYK$.

The conversion from RGB to CMY is straightforward:

$$\begin{bmatrix} C \\ M \\ Y \end{bmatrix} = \begin{bmatrix} 1 \\ 1 \\ 1 \end{bmatrix} - \begin{bmatrix} R \\ G \\ B \end{bmatrix} \tag{16.9}$$

The inverse operation (conversion from CMY to RGB), although it is equally easy from a mathematical standpoint, is of little practical use.

In MATLAB

Conversion between RGB and CMY in MATLAB can also be accomplished using the `imcomplement` function.

16.2.3 The HSV Color Model

Color models such as the RGB and $CMYK$ described previously are very convenient to specify color coordinates for display or printing, respectively. They are not, however, useful to capture a typical human description of color. After all, none of us goes to a store looking for a `FFFFCC` shirt to go with the `FFCC33` jacket we got for our birthday. Rather, the human perception of color is best described in terms of hue, saturation, and lightness. *Hue* describes the color type, or tone, of the color (and very often is expressed by the "color name"), *saturation* provides a measure of its purity (or how much it has been diluted in white), and *lightness* refers to the intensity of light reflected from objects.

For representing colors in a way that is closer to the human description, a family of color models have been proposed. The common aspect among these models is their ability to dissociate the dimension of *intensity* (also called *brightness* or *value*) from the *chromaticity*—expressed as a combination of *hue* and *saturation*—of a color.

We will look at a representative example from this family: the HSV (hue–saturation–value) color model.[4]

The HSV (sometimes called HSB) color model can be obtained by looking at the RGB color cube along its main diagonal (or *gray axis*), which results in a hexagon-shaped color palette. As we move along the main axis in the pyramid in Figure 16.10,

[4]The terminology for color models based on hue and saturation is not universal, which is unfortunate. What we call HSV in this book may appear under different names and acronyms elsewhere. Moreover, the distinctions among these models are very subtle and different acronyms might be used to represent slight variations among them.

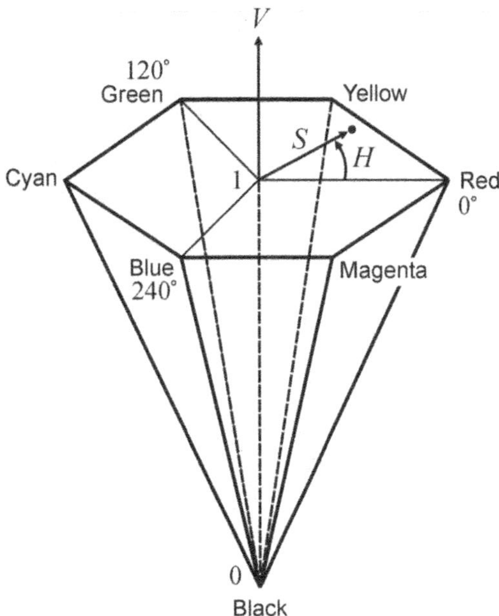

FIGURE 16.10 The HSV color model as a hexagonal cone.

the hexagon gets smaller, corresponding to decreasing values of V, from 1 (white) to 0 (black). For any hexagon, the three primary and the three secondary colors of light are represented in its vertices. Hue, therefore, is specified as an angle relative to the origin (the red axis by convention). Finally, saturation is specified by the distance to the axis: the longer the distance, the more saturated the color.

Figure 16.11 shows an alternative representation of the *HSV* color model in which the hexcone is replaced by a cylinder. Figure 16.12 shows yet another equivalent three-dimensional representation for the *HSV* color model, as a cone with circular-shaped base.

In summary, the main advantages of the *HSV* color model (and its closely related alternatives) are its ability to match the human way of describing colors and to allow for independent control over hue, saturation, and intensity (value). The ability to isolate the intensity component from the other two—which are often collectively called *chromaticity* components—is a requirement in many color image processing algorithms, as we shall see in Section 16.5. Its main disadvantages include the discontinuity in numeric values of hue around red, the computationally expensive conversion to/from RGB, and the fact that hue is undefined for a saturation of 0.

In MATLAB

Converting between *HSV* and *RGB* in MATLAB can be accomplished by the functions `rgb2hsv` and `hsv2rgb`. Tutorial 16.2 explores these functions in detail.

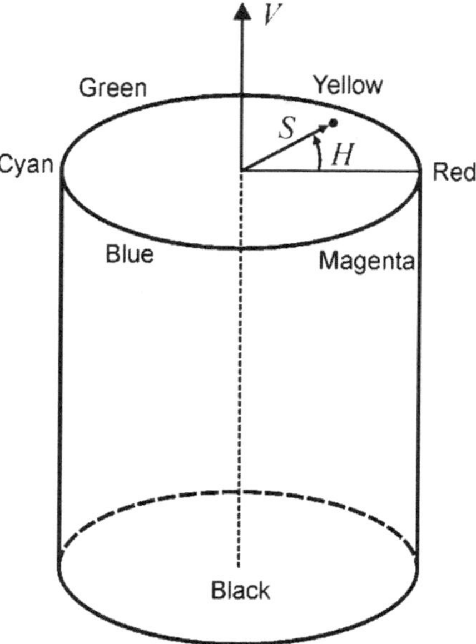

FIGURE 16.11 The *HSV* color model as a cylinder.

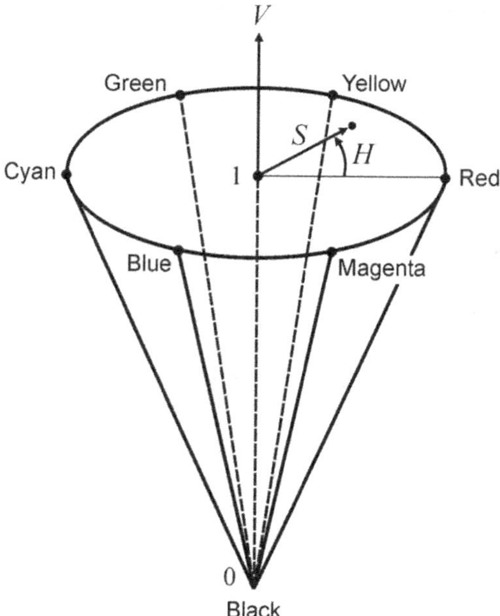

FIGURE 16.12 The HSV color model as a cone.

16.2.4 The YIQ (NTSC) Color Model

The NTSC color model is used in the American standard for analog television, which will be described in more detail in Chapter 20. One of the main advantages of this model is the ability to separate grayscale contents from color data, a major design requirement at a time when emerging color TV sets and transmission equipment had to be backward compatible with their B&W predecessors. In the NTSC color model, the three components are luminance (Y) and two color-difference signals hue (I) and saturation (Q).[5]

Conversion from RGB to YIQ can be performed using the transformation

$$\begin{bmatrix} Y \\ I \\ Q \end{bmatrix} = \begin{bmatrix} 0.299 & 0.587 & 0.114 \\ 0.596 & -0.274 & -0.322 \\ 0.211 & -0.523 & 0.312 \end{bmatrix} \begin{bmatrix} R \\ G \\ B \end{bmatrix} \tag{16.10}$$

In MATLAB

Converting between RGB and YIQ (NTSC) in MATLAB is accomplished using functions `rgb2ntsc` and `ntsc2rgb`.

16.2.5 The YCbCr Color Model

The $YCbCr$ color model is the most popular color representation for digital video.[6] In this format, one component represents luminance (Y), while the other two are color-difference signals: Cb (the difference between the blue component and a reference value) and Cr (the difference between the red component and a reference value).

Conversion from RGB to $YCbCr$ is possible using the transformation

$$\begin{bmatrix} Y \\ Cb \\ Cr \end{bmatrix} = \begin{bmatrix} 0.299 & 0.587 & 0.114 \\ -0.169 & -0.331 & 0.500 \\ 0.500 & -0.419 & -0.081 \end{bmatrix} \begin{bmatrix} R \\ G \\ B \end{bmatrix} \tag{16.11}$$

In MATLAB

Converting between $YCbCr$ and RGB in MATLAB can be accomplished by the functions `rgb2ycbcr` and `ycbcr2rgb`.

16.3 REPRESENTATION OF COLOR IMAGES IN MATLAB

As we have seen in Chapter 2, color images are usually represented as RGB (24 bits per pixel) or indexed with a palette (color map), usually of size 256. These representation

[5]The choice for letters I and Q stems from the fact that one of the components is *in* phase, whereas the other is off by 90°, that is, in *quadrature*.

[6]More details in Chapter 20.

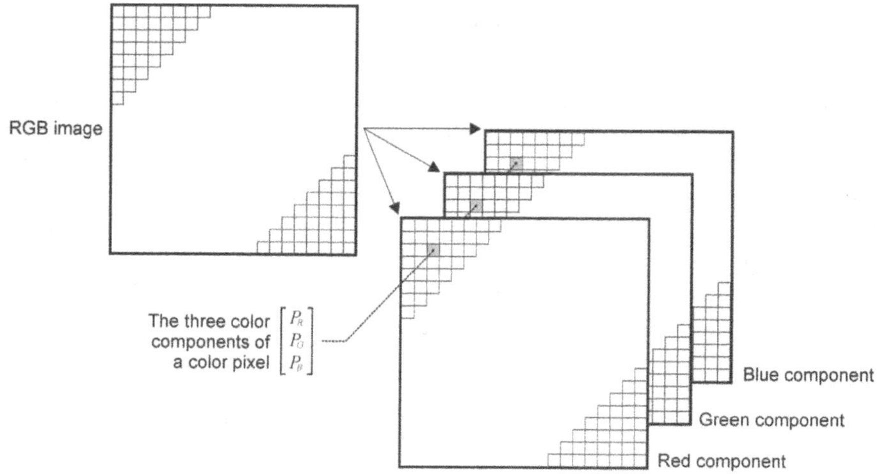

FIGURE 16.13 RGB color image representation.

modes are independent of file format (although GIF images are usually indexed and JPEG images typically are not). In this section, we provide a more detailed analysis of color image representation in MATLAB.

16.3.1 RGB Images

An RGB color image in MATLAB corresponds to a 3D array of dimensions $M \times N \times 3$, where M and N are the image's height and width (respectively) and 3 is the number of color planes (channels). Each color pixel is represented as a triple containing the values of its R, G, and B components (Figure 16.13). Each individual array of size $M \times N$ is called a *component image* and corresponds to one of the color channels: red, green, or blue. The data class of the component images determines their range of values. For RGB images of class `double`, the range of values is [0.0, 1.0], whereas for classes `uint8` or `uint16`, the ranges are [0, 255] and [0, 65535], respectively. RGB images typically have a *bit depth* of 24 bits per pixel (8 bits per pixel per component image), resulting in a total number of $(2^8)^3 = 16,777,216$ colors.

■ EXAMPLE 16.1

The following MATLAB sequence can be used to open, verify the size (in this case, $384 \times 512 \times 3$) and data class (in this case, `uint8`), and display an RGB color image (Figure 16.14).

```
I = imread('peppers.png');
size(I)
class(I)
subplot(2,2,1), imshow(I), title('Color image (RGB)')
```

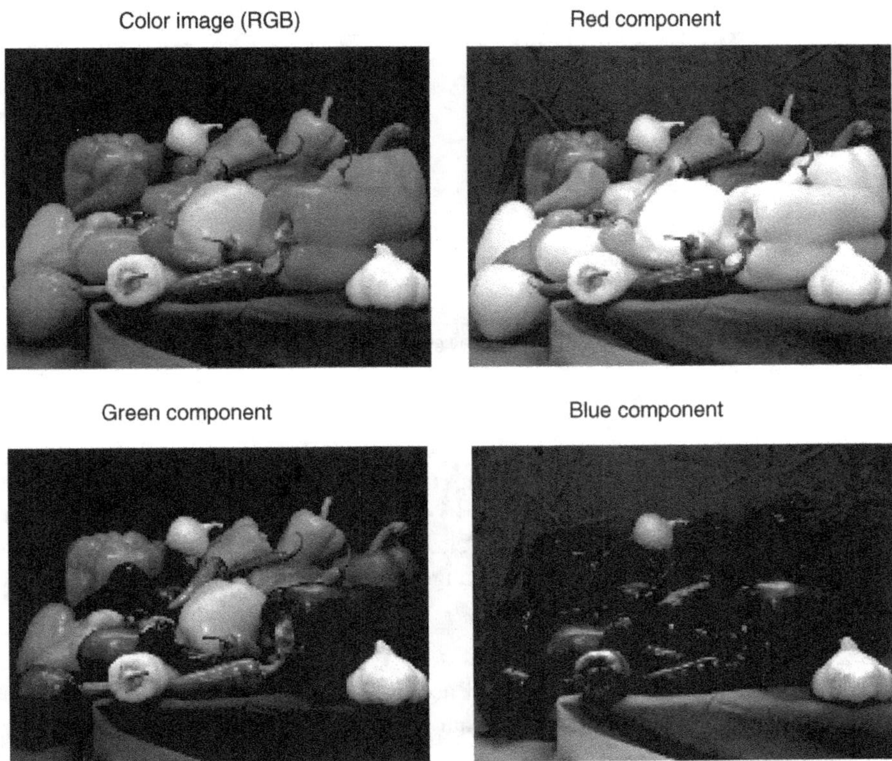

FIGURE 16.14 *RGB* image and its three color components (or *channels*). Original image: courtesy of MathWorks.

```
subplot(2,2,2), imshow(I(:,:,1)), title('Red component')
subplot(2,2,3), imshow(I(:,:,2)), title('Green component')
subplot(2,2,4), imshow(I(:,:,3)), title('Blue component')
```

16.3.2 Indexed Images

An indexed image is a matrix of integers (`X`), where each integer refers to a particular row of RGB values in a secondary matrix (`map`) known as a *color map*. The image can be represented by an array of class `uint8`, `uint16`, or `double`. The color map array is an $M \times 3$ matrix of class `double`, where each element's value is within the range [0.0, 1.0]. Each row in the color map represents R (red), G (green), and B (blue) values, in that order. The indexing mechanism works as follows (Figure 16.15): if `X` is of class `uint8` or `uint16`, all components with value 0 point to the first row in `map`, all components with value 1 point to the second row, and so on. If `X` is of class `double`, all components with value less than or equal to 1.0 point to the first row and so on.

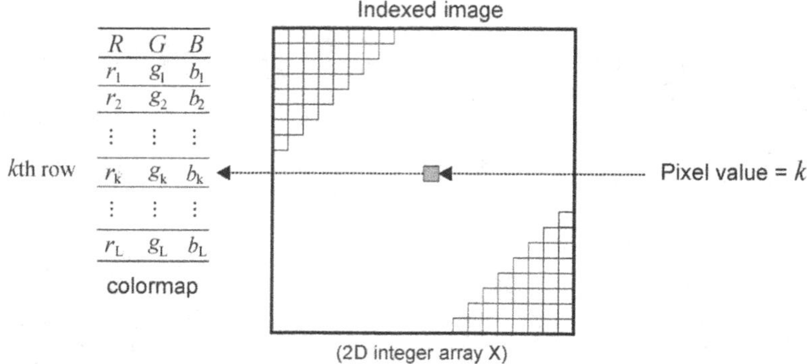

FIGURE 16.15 Indexed color image representation.

MATLAB has many built-in color maps (which can be accessed using the function `colormap`), briefly described in Table 16.4. In addition, you can easily create your own color map by defining an array of class `double` and size $M \times 3$, where each element is a floating-point value in the range [0.0, 1.0].

■ EXAMPLE 16.2

The following MATLAB sequence can be used to load a built-in indexed image, verify its size (in this case, 200×300) and data class (in this case, `double`), verify the

TABLE 16.4 Color Maps in MATLAB

Name	Description
`hsv`	Hue–saturation–value color map
`hot`	Black–red–yellow–white color map
`gray`	Linear gray-scale color map
`bone`	Gray scale with tinge of blue color map
`copper`	Linear copper-tone color map
`pink`	Pastel shades of pink color map
`white`	All white color map
`flag`	Alternating red, white, blue, and black color map
`lines`	Color map with the line colors
`colorcube`	Enhanced color-cube color map
`vga`	Windows color map for 16 colors
`jet`	Variant of HSV
`prism`	Prism color map
`cool`	Shades of cyan and magenta color map
`autumn`	Shades of red and yellow color map.
`spring`	Shades of magenta and yellow color map
`winter`	Shades of blue and green color map
`summer`	Shades of green and yellow color map

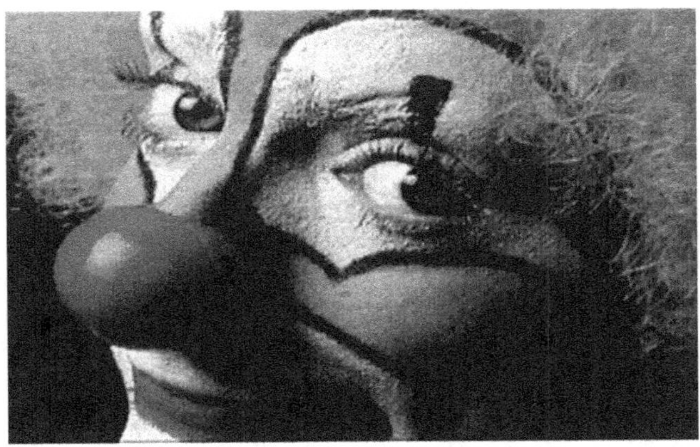

FIGURE 16.16 A built-in indexed image. Original image: courtesy of MathWorks.

color map's size (in this case, 81×3) and data class (in this case, `double`), and display the image (Figure 16.16).

```
load clown
size(X)
class(X)
size(map)
class(map)
imshow(X,map), title('Color (Indexed)')
```

MATLAB has many useful functions for manipulating indexed color images:

- If we need to approximate an indexed image by one with fewer colors, we can use MATLAB's `imapprox` function.
- Conversion between RGB and indexed color images is straightforward, thanks to functions `rgb2ind` and `ind2rgb`.
- Conversion from either color format to their grayscale equivalent is equally easy, using functions `rgb2gray` and `ind2gray`.
- We can also create an index image from an RGB image by dithering the original image using function `dither`.
- Function `grayslice` creates an indexed image from an intensity (grayscale) image by thresholding and can be used in pseudocolor image processing (Section 16.4).
- Function `gray2ind` converts an intensity (grayscale) image into its indexed image equivalent. It is different from `grayslice`. In this case, the resulting

image is monochrome, just as the original one;[7] only the internal data representation has changed.

16.4 PSEUDOCOLOR IMAGE PROCESSING

The purpose of pseudocolor image processing techniques is to enhance a monochrome image for human viewing purposes. Their rationale is that subtle variations of gray levels may very often mask or hide regions of interest within an image. This can be particularly damaging if the masked region is relevant to the application domain (e.g., the presence of a tumor in a medical image). Since the human eye is capable of discerning thousands of color hues and intensities, compared to only less than 100 shades of gray, replacing gray levels with colors leads to better visualization and enhanced capability for detecting relevant details within the image.

The typical solution consists of using a color lookup table (LUT) designed to map the entire range of (typically 256) gray levels to a (usually much smaller) number of colors. For better results, contrasting colors should appear in consecutive rows in the LUT. The term *pseudo*color is used to emphasize the fact that the assigned colors usually have no correspondence whatsoever with the truecolors that might have been present in the original image.

16.4.1 Intensity Slicing

The technique of *intensity* (or *density*) *slicing* is the simplest and best-known pseudocoloring technique. If we look at a monochrome image as if it were a 3D plot of gray levels versus spatial coordinates, where the most prominent peaks correspond to the brightest pixels, the technique corresponds to placing several planes parallel to the coordinate plane of the image (also known as the xy plane). Each plane "slices" the 3D function in the area of intersection, resulting in several gray-level intervals. Each side of the plane is then assigned a different color. Figure 16.17 shows an example of intensity slicing using only one slicing plane at $f(x, y) = l_i$ and Figure 16.18 shows an alternative representation, where the chosen colors are indicated as c_1, c_2, c_3, and c_4. The idea can easily be extended to M planes and $M + 1$ intervals.

In MATLAB

Intensity slicing can be accomplished using the `grayslice` function. You will learn how to use this function in Tutorial 16.1.

[7]The only possible differences would have resulted from the need to requantize the number of gray levels if the specified color map is smaller than the original number of gray levels.

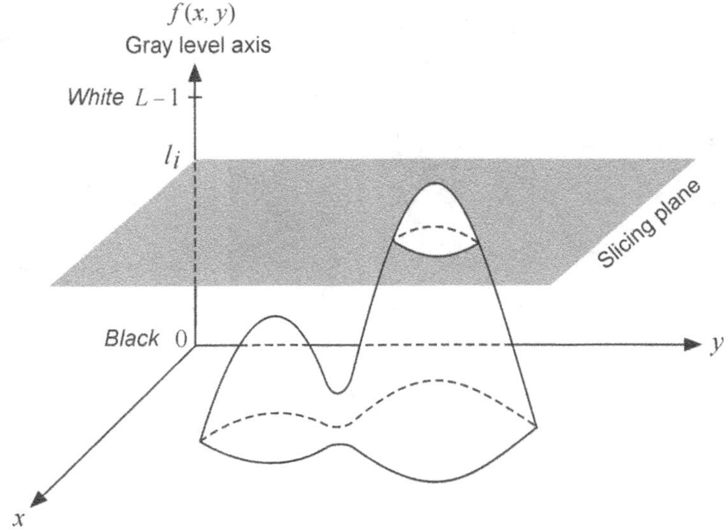

FIGURE 16.17 Pseudocoloring with intensity slicing.

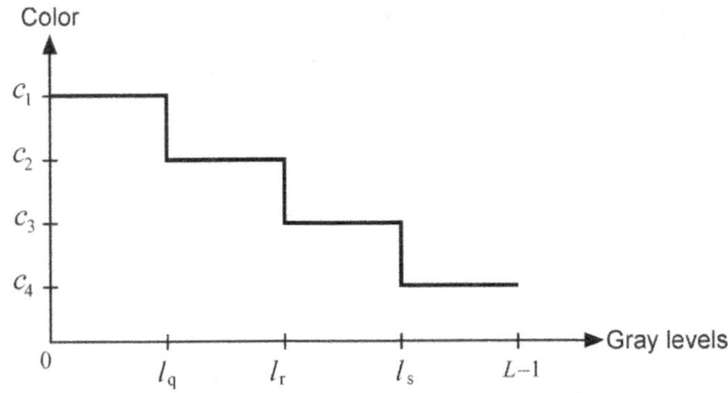

FIGURE 16.18 An alternative representation of the intensity slicing technique for an image with L gray levels pseudocolored using four colors.

■ EXAMPLE 16.3

Figure 16.19 shows an example of pseudocoloring with intensity slicing using 16 levels, the same input image (a), and three different color maps `summer` (b), `hsv` (c), and `jet` (d).

16.4.2 Gray Level to Color Transformations

An alternative approach to pseudocoloring consists of using three independent transformation functions on each pixel of the input image and assigning the results of each

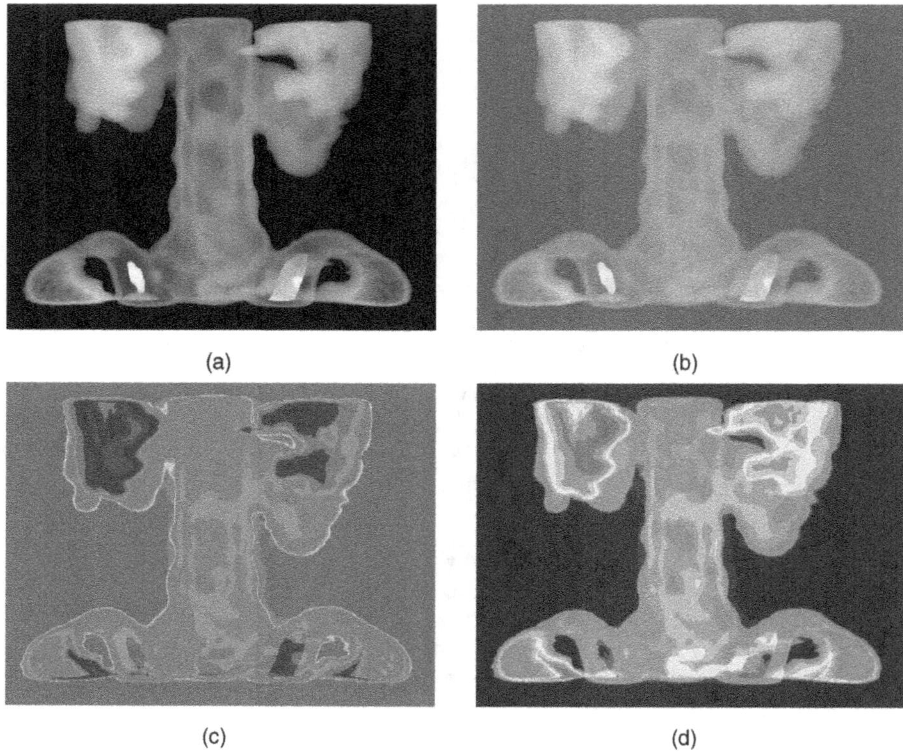

FIGURE 16.19 Pseudocoloring using intensity slicing: original image (a) and results of pseudocoloring using different color maps (b–d). Original image: courtesy of MathWorks.

function to a color channel (Figure 16.20). This method provides additional flexibility, since it allows creation of a composite color image whose contents can be modulated by each individual transformation function. Recall from our discussion in Chapter 8 that these are point functions, that is, the resulting value for each pixel does not depend on its spatial location or the gray level of its neighbors.

The intensity slicing method described in Section 16.4.1 is a particular case of gray level to color transformation, in which all transformation functions are identical and shaped like a staircase.

16.4.3 Pseudocoloring in the Frequency Domain

Pseudocoloring can also performed in the frequency domain[8] by applying a Fourier transform (FT) to the original image and then applying a low-pass, bandpass, and high-pass filters to the transformed data. The three individual filter outputs are then inverse transformed and used as the R, G, and B components of the resulting image.

[8]Frequency-domain techniques were discussed in Chapter 11.

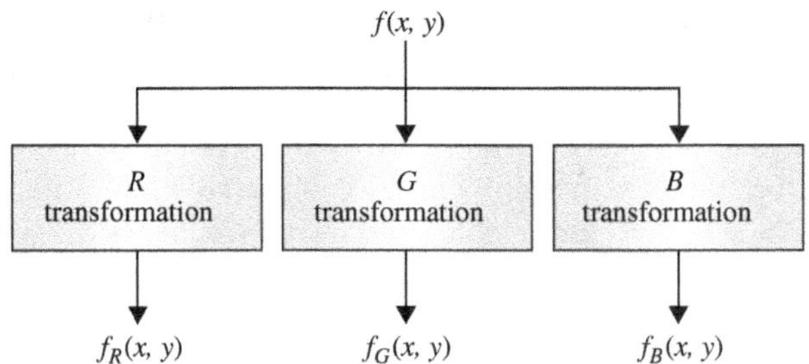

FIGURE 16.20 Block diagram for pseudocoloring using color transformation functions.

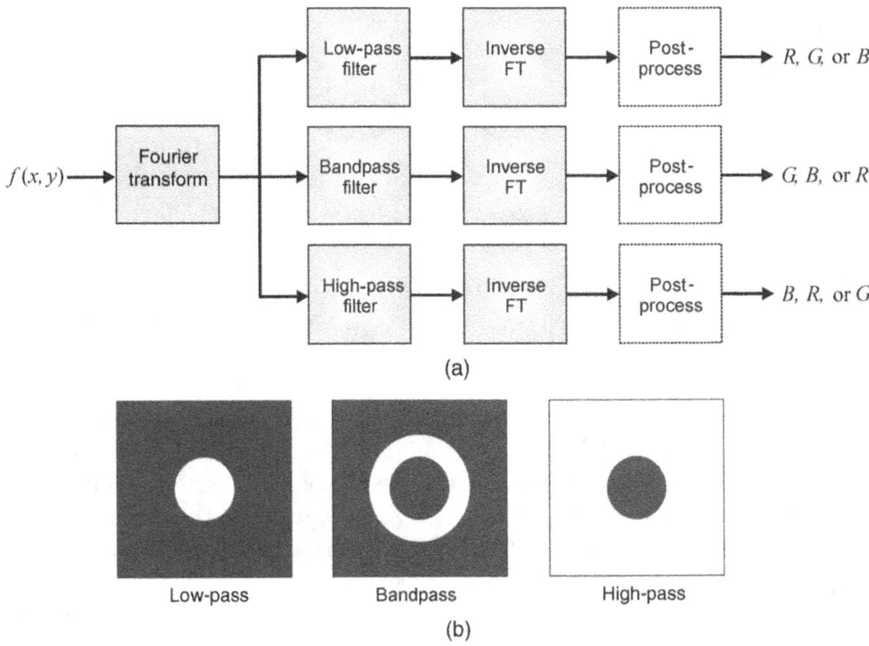

FIGURE 16.21 (a) Block diagram for pseudocoloring in the frequency domain; (b) frequency response of the filters. Redrawn from [Umb05].

Figure 16.21 shows a block diagram of the method as well as representative examples of the frequency response of the filters used in the process. The *postprocessing* stage is optional and application dependent.

16.5 FULL-COLOR IMAGE PROCESSING

This section examines techniques that process the full contents of a digital color image. Full-color image processing is a relatively young branch of digital image processing that has become increasingly popular and relevant in recent years, thanks to the wide availability of inexpensive hardware for capturing, storing, displaying, and printing color images.

There are several technical challenges involved in extrapolating monochrome image processing techniques to their color image equivalent, one of which is particularly important: the choice of the appropriate color model for the task. The impact of this choice will become evident as we look at specific examples in this section.

There are two ways to achieve color image processing:

- *Componentwise*: Having selected an appropriate color model, each component image (e.g., R, G, and B) is processed individually and then forms a composite processed image.
- *Vector Methods*: Color pixels are treated as vectors:

$$\mathbf{c}(x, y) = \begin{bmatrix} c_R(x, y) \\ c_G(x, y) \\ c_B(x, y) \end{bmatrix} = \begin{bmatrix} R(x, y) \\ G(x, y) \\ B(x, y) \end{bmatrix} \tag{16.12}$$

The two methods are equivalent if (and only if)

- The process is applicable to both vectors and scalars.
- The operation on each component of a vector is independent of the other components.

Vector methods for color image processing are mathematically intensive and beyond the scope of this text.[9] For the remaining of this discussion, we shall focus on componentwise color image processing and the role of the chosen color model.

Several color image processing techniques can be performed on the individual R, G, and B channels of the original image, whereas other techniques require access to a component that is equivalent to the monochrome version of the input image (e.g., the Y component in the YIQ color model or the V component in the HSV color model). The former can be called *RGB processing* (Figure 16.22), whereas the latter can be referred to as *intensity processing* (Figure 16.23).

16.5.1 Color Transformations

It is possible to extend the concept of grayscale transformations to color images. The original formulation (see Chapter 8)

$$g(x, y) = T[f(x, y)] \tag{16.13}$$

can be adapted for the case where the input and output images ($f(x, y)$ and $g(x, y)$) are color images, that is, where each individual pixel value is no longer an unsigned

[9]The interested reader will find a few useful references at the end of the chapter.

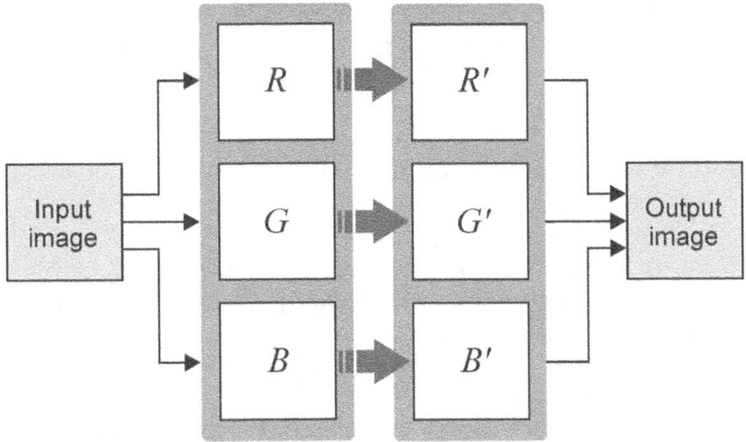

FIGURE 16.22 RGB processing.

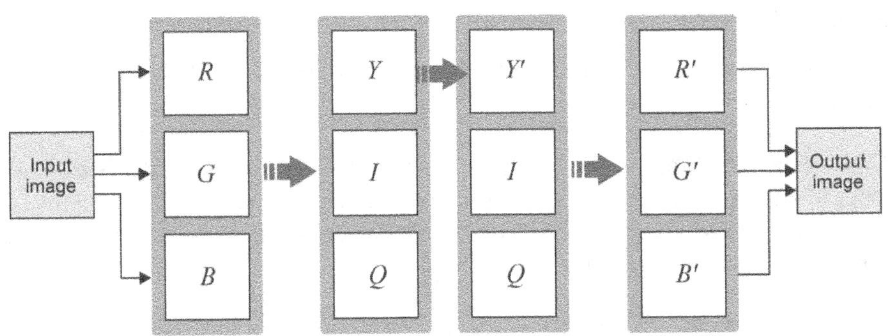

FIGURE 16.23 Intensity processing using RGB to YIQ color space conversions.

integer or double, but a triple of values (e.g., corresponding to the R, G, and B values for that pixel).

Since we know that such transformation functions are point processing functions (independent of the pixel's location or neighbors' values), we can adopt a modified version of the simplified notation introduced in Chapter 8:

$$s_i = T_i(r_1, r_2, ..., r_n), \quad i = 1, 2, ..., n \tag{16.14}$$

where r_i and s_i are the color components of the original ($f(x, y)$) and the processed ($g(x, y)$) image, respectively, n is the number of color components, and $T_1, T_2, ..., T_n$ is a set of color transformation (or *mapping*) functions that operate on r_i to produce s_i. For RGB images, $n = 3$, and r_1, r_2, and r_3 correspond to the R, G, and B values for each pixel in the input image.

Intensity Modification A simple example of color mapping function is the intensity modification function described by

$$g(x, y) = kf(x, y) \tag{16.15}$$

Clearly, if $k > 1$, the resulting image will be brighter than the original, whereas for $k < 1$, the output image will be darker than the input image.

Color Complements The color complement operation is the color equivalent of the grayscale negative transformation introduced in Chapter 8. It replaces each hue by its complement (sometimes called *opponent color*).

If the input image is represented using the RGB color model, the operation can be performed by applying a trivial transfer function to each individual color channel. If the input image is represented using the HSV color space, we must apply a trivial transfer function to V, a nontrivial transfer function—that takes care of the discontinuities in hue around $0°$—to H and leave S unchanged.

In MATLAB

Color complement for RGB images can be accomplished using the `imcomplement` function.

Color Slicing Color slicing is a mapping process by which all colors outside a range of interest are mapped to a "neutral" color (e.g., gray), while all colors of interest remain unchanged. The color range of interest can be specified as a cube or a sphere centered at a prototypical reference color.

16.5.2 Histogram Processing

The concept of histogram can be extended to color images, in which case each image can be represented using three histograms with N (typically, $4 \leq N \leq 256$) bins each. Histogram techniques (e.g., histogram equalization) can be applied to an image represented using a color model that allows separation between luminance and chrominance components (e.g., HSI): the luminance (intensity, I) component of a color image is processed, while the chromaticity components (H and S) are left unchanged. Figure 16.24 shows an example that uses the YIQ (NTSC) color model (and equalizes only its Y component). The resulting image is a modified version of the original one, in which background details become more noticeable. Notice that although the colors become somewhat washed out, they are faithful to their original hue.

16.5.3 Color Image Smoothing and Sharpening

Linear neighborhood-oriented smoothing and sharpening techniques can be extended to color images under the componentwise paradigm. The original 3×3 kernel typically used in monochrome operations (see Chapters 4 and 10) becomes an array of vectors (Figure 16.25).

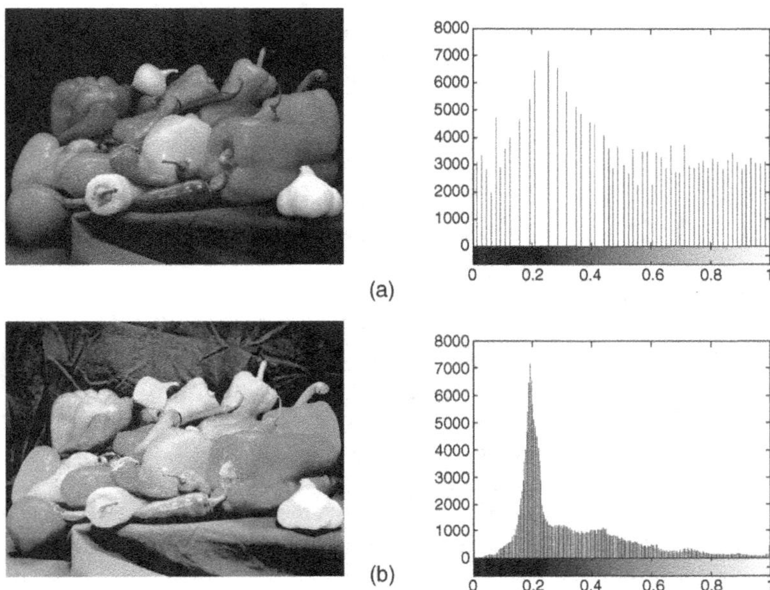

FIGURE 16.24 Example of color histogram equalization. (a) Original image and its Y channel histogram; (b) output image and its equalized Y channel histogram. Original image: courtesy of MathWorks.

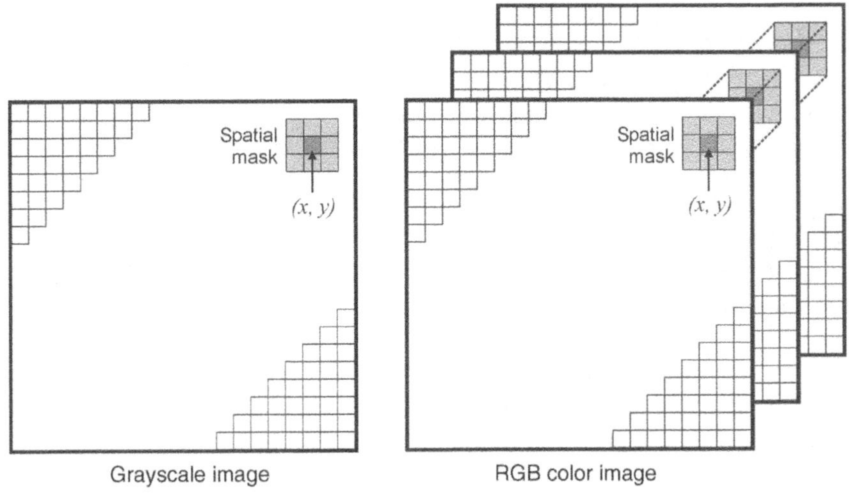

FIGURE 16.25 Spatial convolution masks for grayscale and RGB color images.

For example, the vector formulation of the averaging filter for an RGB image using a neighborhood S_{xy} centered around coordinates (x, y) becomes

$$\bar{\mathbf{c}}(x, y) = \begin{bmatrix} \frac{1}{K} \sum_{(s,t)\in S_{xy}} R(s, t) \\ \frac{1}{K} \sum_{(s,t)\in S_{xy}} G(s, t) \\ \frac{1}{K} \sum_{(s,t)\in S_{xy}} B(s, t) \end{bmatrix} \tag{16.16}$$

Equation (16.16) indicates that the result can be obtained by performing neighborhood averaging on each individual color channel using standard grayscale neighborhood processing. Similarly, sharpening a color image encoded using the RGB color model can be accomplished by applying the sharpening operator (e.g., Laplacian) to each component image individually and combining the results.

Smoothing and sharpening operations can also be performed by processing the intensity (luminance) component of an image encoded with the proper color model (e.g., YIQ or HSI) and combining the result with the original chrominance channels. Tutorial 16.2 explores smoothing and sharpening of color images.

16.5.4 Color Noise Reduction

The impact of noise on color images strongly depends on the color model used. Even when only one of the R, G, or B channels is affected by noise, conversion to another color model such as HSI or YIQ will spread the noise to all components. Linear noise reduction techniques (such as the mean filter) can be applied on each R, G, and B component separately with good results.

16.5.5 Color-Based Image Segmentation

Color Image Segmentation by Thresholding There are several possible ways to extend the image thresholding ideas commonly used for monochrome images (Chapter 15) to their color equivalent. The basic idea is to partition the color space into a few regions (which hopefully should correspond to meaningful objects and regions in the image) using appropriately chosen thresholds.

A simple option is to define one (or more) threshold(s) for each color component (e.g., R, G, and B), which results in a partitioning of the RGB cube from which the color range of interest (a smaller cube) can be isolated (Figure 16.26).

Color Image Segmentation in RGB Vector Space It is also possible to specify a threshold relative to a distance between any color and a reference color in the RGB space. If we call the reference color (R_0, G_0, B_0), the thresholding rule can be expressed as

$$g(x, y) = \begin{cases} 1 & d(x, y) \leq d_{\max} \\ 0 & d(x, y) > d_{\max} \end{cases} \tag{16.17}$$

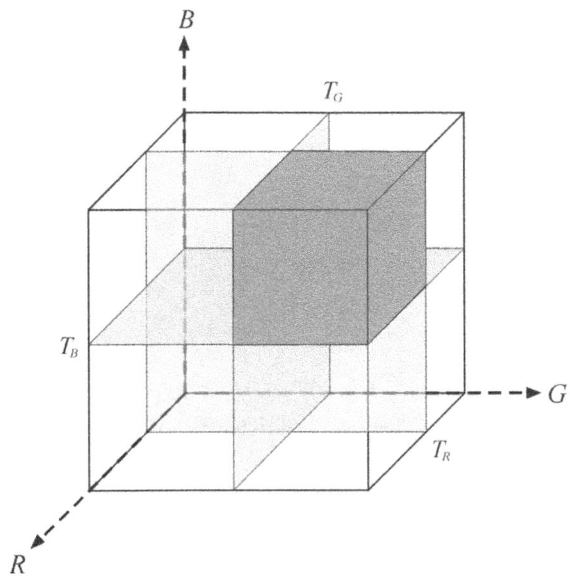

FIGURE 16.26 Thresholding in RGB space.

where

$$d(x, y) = \sqrt{[f_R(x, y) - R_0]^2 + [f_G(x, y) - G_0]^2 + [f_B(x, y) - B_0]^2} \qquad (16.18)$$

Thresholding according to equations (16.17) and (16.18) in fact specifies a sphere in the RGB space, whose center is the reference color. Any pixel whose color lies inside the sphere (or on its surface) will be set to 1; all other pixels will be assigned a value of 0.

Equations (16.17) and (16.18) can be generalized by specifying different threshold values for each primary color, which will result in an ellipsoid (rather than a sphere) being defined in RGB space (Figure 16.27).

In MATLAB

A simple way to segment color images in MATLAB is by using the `rgb2ind` function. The primary use of this function is to generate an indexed image based on an input truecolor image, but with less number of colors. When a specific number of colors are specified, MATLAB quantizes the image and produces an indexed image and a color map at the output. The resulting indexed image is essentially what we want from a segmentation process: a labeled image. If there are n regions in the image, there will be $n + 1$ labels, where the additional label is for the background.

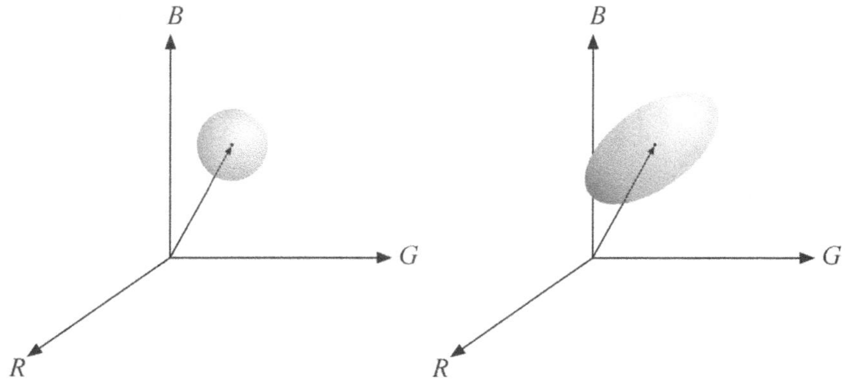

FIGURE 16.27 Defining spherical (ellipsoidal) regions in RGB space.

■ EXAMPLE 16.4

In this example, we use `rgb2ind` to segment a simple image containing three predominant colors: black (background), red (marker), and desaturated yellow (marker cap). Figure 16.28 shows the input image (a), the result of segmentation using `rgb2ind` and the default color map (b), and the result of segmentation displayed with a different color map (c). With few exceptions (caused by light reflections on the red marker), the algorithm did a good job, segmenting the input image into three main parts.

```
I = imread('marker.png');
n = 3;
[I2,map2] = rgb2ind(I,n,'nodither');
imshow(I)
figure, imshow(I2,map2)
figure, imshow(I2,hsv(3))
```

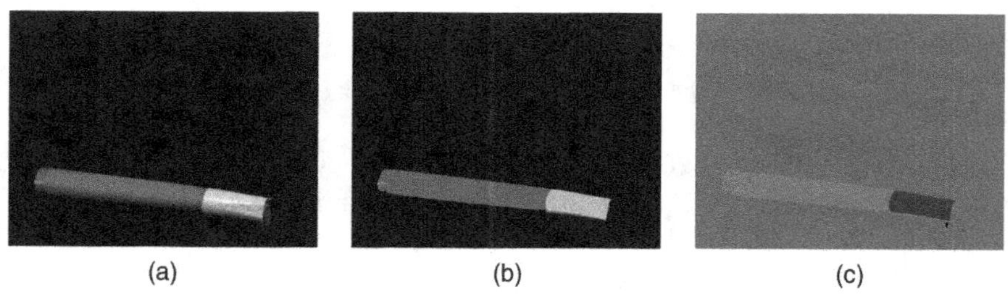

FIGURE 16.28 Example of color segmentation using requantization.

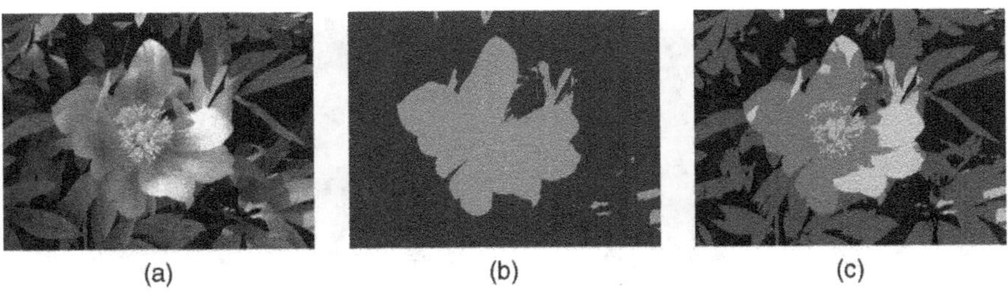

FIGURE 16.29 Another example of color segmentation using requantization: (a) original image; (b) requantized image with two color levels; (c) requantized image with five color levels.

In Figure 16.29, we replace the input image with a more difficult scene and show results for different values of n. The results are good: $n = 2$ provides good rough separation between "flower" and "leaves," whereas $n = 5$ segments the yellow filaments as well. Note that the uneven lighting pattern causes most of the imperfect results.

16.5.6 Color Edge Detection

In Chapter 14, we defined an edge as a boundary between two image regions having distinct characteristics. We then focused on edges in grayscale 2D images, which are relatively easy to define as "sharp variations of the intensity function across a portion of the image." In this section, we briefly look at the problem of color edge detection.

Several definitions of a color edge have been proposed in the literature [Pra07]:

1. An edge in a color image can be said to exist if and only if there is such an edge in the luminance channel. This definition ignores discontinuities in the other color channels (e.g., hue and saturation for *HSV* images).
2. A color edge is present if an edge exists in any of its three component images.
3. An edge exists if the sum of the results of edge detection operators applied to individual color channels exceeds a certain threshold. This definition is convenient and straightforward and commonly used when the emphasis is not on accuracy.

In general, extending the gradient-based methods for edge detection in monochrome images described in Chapter 14 to color images by simply computing the gradients for each component image and combining the results (e.g., by applying a logical OR operator) will lead to erroneous—but acceptable for most cases—results. If accuracy is important, we must resort to a new definition of the gradient that is applicable to vector quantities (e.g., the method proposed by Di Zenzo [DZ86]), which is beyond the scope of this book.

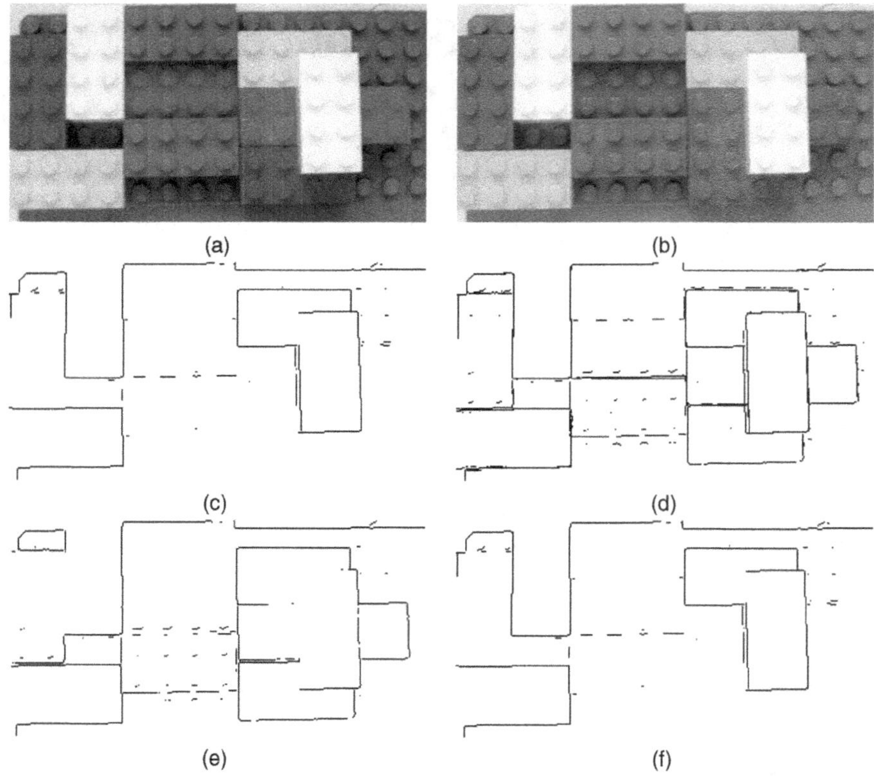

FIGURE 16.30 Color edge detection example: (a) original image; (b) grayscale equivalent; (c) edge detection on (b); (d) edge detection on individual *RGB* components; (e) edge detection on *Y* component only; (f) edge detection on *V* component only.

■ EXAMPLE 16.5

In this example, we show the results of using the `edge` function for different color models. More specifically, we read an *RGB* color image, convert it to gray, *HSV*, and *YIQ* representations, and compare the results (Figure 16.30) obtained when using

- Edge detection on the grayscale equivalent of the input image (c).
- The logical OR of edge detection applied to each individual channel in the original image (*R*, *G*, and *B*) (d).
- Edge detection on the *Y* component of the input image represented in the *YIQ* model (e).
- Edge detection on the *V* component of the input image represented in the *HSV* model (f).

16.6 TUTORIAL 16.1: PSEUDOCOLOR IMAGE PROCESSING

Goal

The goal of this tutorial is to learn how to display grayscale images using pseudocolors in MATLAB.

Objectives

- Learn how to use the `grayslice` function to perform intensity slicing.
- Learn how to specify color maps with a custom number of colors.

What You Will Need

- `grad.jpg`
- `mri.jpg`

Procedure

We will start by exploring the `grayslice` function on a gradient image.

1. Create and display a gradient image.

```
I = repmat(uint8([0:255]),256,1);
figure, subplot(1,2,1), subimage(I), title('Original Image');
```

2. Slice the image and display the results.

```
I2 = grayslice(I,16);
subplot(1,2,2), subimage(I2,colormap(winter(16))), ...
    title('Pseudo-colored with "winter" colormap')
```

Question 1 Why did we use the `subimage` function to display the images (instead of the familiar `imshow`)?

Question 2 What does the value 16 represent in the function call for `grayslice`?

Question 3 In the statement `subimage(I2,colormap(winter(16)))`, what does the value 16 represent?

In the above procedure, we sliced the image into equal partitions—this is the default for the `grayslice` function. We will now learn how to slice the range of grayscale values into unequal partitions.

3. Slice the image into unequal partitions and display the result.

```
levels = [0.25*255, 0.75*255, 0.9*255];
I3 = grayslice(I,levels);
figure, imshow(I3,spring(4))
```

Question 4 The original image consists of values in the range [0, 255]. If our original image values ranged [0.0, 1.0], how would the above code change?

Now that we have seen how pseudocoloring works, let us apply it to an image where this visual information might be useful.

4. Clear all variables and close any open figures.
5. Load and display the `mri.jpg`.

```
I = imread('mri.jpg');
figure, subplot(1,2,1), subimage(I), title('Original Image');
```

6. Pseudocolor the image.

```
I2 = grayslice(I,16);
subplot(1,2,2), subimage(I2, colormap(jet(16))), ...
    title('Pseudo-colored with "jet" colormap');
```

Question 5 In the previous steps, we have specified how many colors we want in our color map. If we do not specify this number, how does MATLAB determine how many colors to return in the color map?

16.7 TUTORIAL 16.2: FULL-COLOR IMAGE PROCESSING

Goal

The goal of this tutorial is to learn how to convert between color spaces and perform filtering on color images in MATLAB.

Objectives

- Learn how to convert from RGB to HSV color space using the `rgb2hsv` function.
- Learn how to convert from HSV to RGB color space using the `hsv2rgb` function.
- Explore smoothing and sharpening in the RGB and HSV color spaces.

Procedure

We will start by exploring the `rgb2hsv` function.

1. Load the `onions.png` image and display its RGB components.

```
I = imread('onion.png');
figure, subplot(2,4,1), imshow(I), title('Original Image');
subplot(2,4,2), imshow(I(:,:,1)), title('R component');
subplot(2,4,3), imshow(I(:,:,2)), title('G component');
subplot(2,4,4), imshow(I(:,:,3)), title('B component');
```

2. Convert the image to HSV and display its components.

```
Ihsv = rgb2hsv(I);
subplot(2,4,6), imshow(Ihsv(:,:,1)), title('Hue')
subplot(2,4,7), imshow(Ihsv(:,:,2)), title('Saturation');
subplot(2,4,8), imshow(Ihsv(:,:,3)), title('Value');
```

Question 1 Why do we not display the HSV equivalent of the image?

When viewing the components of an RGB image, the grayscale visualization of each component is intuitive because the intensity within that component corresponds to how much of the component is being used to generate the final color. Visualization of the components of an HSV image is not as intuitive. You may have noticed that when displaying the hue, saturation, and value components, hue and saturation do not give you much insight as to what the actual color is. The value component, on the other hand, appears to be a grayscale version of the image.

3. Convert the original image to grayscale and compare it with the value component of the HSV image.

```
Igray = rgb2gray(I);
figure, subplot(1,2,1), imshow(Igray), title('Grayscale');
subplot(1,2,2), imshow(Ihsv(:,:,3)), title('Value component');
```

Question 2 How does the grayscale version of the original image and the value component of the HSV image compare?

Procedures for filtering a color image will vary depending on the color space being used. Let us first learn how to apply a smoothing filter on an RGB image.

4. Apply a smoothing filter to each component and then reconstruct the image.

```
fn = fspecial('average');
I2r = imfilter(I(:,:,1), fn);
I2g = imfilter(I(:,:,2), fn);
I2b = imfilter(I(:,:,3), fn);
I2(:,:,1) = I2r;
I2(:,:,2) = I2g;
```

```
I2(:,:,3) = I2b;
figure, subplot(1,2,1), imshow(I), title('Original Image');
subplot(1,2,2), imshow(I2), title('Averaged Image');
```

Question 3 Do the results confirm that the RGB equivalent of averaging a grayscale image is to average each component of the RGB image individually?

Now let us see what happens when we perform the same operations on an HSV image. Note that in these steps, we will use the `hsv2rgb` function to convert the HSV image back to RGB so that it can be displayed.

5. Filter all components of the *HSV* image.

```
Ihsv2h = imfilter(Ihsv(:,:,1), fn);
Ihsv2s = imfilter(Ihsv(:,:,2), fn);
Ihsv2v = imfilter(Ihsv(:,:,3), fn);
Ihsv2(:,:,1) = Ihsv2h;
Ihsv2(:,:,2) = Ihsv2s;
Ihsv2(:,:,3) = Ihsv2v;
```

6. Display the results.

```
figure, subplot(2,3,1), imshow(Ihsv(:,:,1)), ...
    title('Original Hue');
subplot(2,3,2), imshow(Ihsv(:,:,2)), ...
    title('Original Saturation');
subplot(2,3,3), imshow(Ihsv(:,:,3)), ...
    title('Original Value');
subplot(2,3,4), imshow(Ihsv2(:,:,1)), ...
    title('Filtered Hue');
subplot(2,3,5), imshow(Ihsv2(:,:,2)), ...
    title('Filtered Saturation');
subplot(2,3,6), imshow(Ihsv2(:,:,3)), ...
    title('Filtered Value');
figure, subplot(1,2,1), imshow(I), title('Original Image');
subplot(1,2,2), imshow(hsv2rgb(Ihsv2)), ...
    title('HSV with all components filtered');
```

Question 4 Based on the results, does it make sense to say that the HSV equivalent of averaging a grayscale image is to average each component of the HSV image individually?

7. Filter only the value component and display the results.

```
Ihsv3(:,:,[1 2]) = Ihsv(:,:,[1 2]);
Ihsv3(:,:,3) = Ihsv2v;
figure, subplot(1,2,1), imshow(I), title('Original Image');
subplot(1,2,2), imshow(hsv2rgb(Ihsv3)), ...
    title('HSV with only value component filtered');
```

Question 5 How does this result compare with the previous one?

We can sharpen an HSV image following a similar sequence of steps.

8. Sharpen the HSV image and display the result.

```
fn2 = fspecial('laplacian',0);
Ihsv4v = imfilter(Ihsv(:,:,3), fn2);
Ihsv4(:,:,[1 2]) = Ihsv(:,:,[1 2]);
Ihsv4(:,:,3) = imsubtract(Ihsv(:,:,3),Ihsv4v);
figure, subplot(1,2,1), imshow(I), title('Original Image');
subplot(1,2,2), imshow(hsv2rgb(Ihsv4)), ...
    title('HSV sharpened');
```

Question 6 How would we perform the same sharpening technique on an *RGB* image?

WHAT HAVE WE LEARNED?

- This chapter introduced the most important concepts and terms related to color perception, representation, and processing. Understand the meaning of the main terms in colorimetry is a required step toward the understanding of color image processing.
- There are many color models used to represent and quantify color information. Some of the most popular (and their use) models are as follows:
  - RGB, CMY(K): display and printing devices
  - YIQ, YCbCr: television and video
  - XYZ: color standardization
  - CIELAB, CIELUV: perceptual uniformity
  - HSV, HSI, HSL: intuitive description of color properties
- Color images can be represented in MATLAB either as an $M \times N \times 3$ array (one per color channel) or as an $M \times N$ array of indices (pointers) to a secondary (usually 256×3) color palette. The former representation is called an *RGB image*, whereas the latter is called an *indexed image*.
- Pseudocolor image processing techniques assign color to pixels based on an interpretation of the data rather than the original scene color (which may not even be known). Full-color image processing methods, on the other hand, process

the pixel values of images whose colors usually correspond to the color of the original scene.

- Several monochrome image processing techniques, from edge detection to histogram equalization, can be extended to color images. The success of applying such techniques to color images depends on the choice of color model used to represent the images.

LEARN MORE ABOUT IT

The topics of color vision, colorimetry, color science, and color image processing have been covered at great length elsewhere. A modest list of useful references is as follows.

- For a deeper understanding of color vision, we recommend Chapter 5 of [Pal99].
- The book by Wyszecki and Styles [WS82] is considered a reference in the field of color science.
- The book by Westland and Ripamonti [WR04] combines color science and colorimetry principles with MATLAB.
- Chapter 3 of [Pra07] contains an in-depth analysis of colorimetry concepts and mathematical formulation.
- There are several books devoted entirely to color image processing, such as [LP06].
- Special issues of leading journals in the field devoted to this topic include [TSV05], [Luk07], and [TTP08].
- Chapter 12 of [BB08] covers color spaces and color space conversion in great detail.
- Chapter 6 of [GWE04] includes a GUI-based MATLAB tool for color image processing, transformations, and manipulations.

ON THE WEB

- The Color Model Museum (an exhibit of color representation models and diagrams dating back to the sixteenth century).
  http://www.colorcube.com/articles/models/model.htm
- The *Colour and Vision Research Laboratories* at the Institute of Ophthalmology (London, England): a rich repository of standard data sets relevant to color and vision research.
  http://www.cvrl.org/
- The Color FAQ document by Charles Poynton: an excellent reference on colorimetry and its implications in the design of image and video systems.
  http://poynton.com/ColorFAQ.html

- International Color Consortium.
  http://www.color.org

16.8 PROBLEMS

16.1 Use the MATLAB function `patch` to display the *RGB* cube in Figure 16.9.

16.2 Write MATLAB code to add two RGB color images. Test it with two test images (of the same size) of your choice. Are the results what you had expected?

16.3 What is wrong with the following MATLAB code to add an indexed color image to a constant for brightening purposes? Fix the code to achieve the desired goal.

```
[X, map] = imread('canoe.tif');
X = X + 20;
```

16.4 In our discussion of pseudocoloring, we stated that the intensity slicing method described in Section 16.4.1 is a particular case of the more general method (using transformation functions) described in Section 16.4.2.

Assuming that a 256-level monochrome image has been "sliced" into four colors, red, green, blue, and yellow, and that each range of gray levels has the same width (64 gray levels), plot the (staircase-shaped) transformation functions for each color channel (R, G, and B).

16.5 In our discussion of color histogram equalization (Section 16.5.2), we showed an example in which the equalization technique was applied to the Y channel of an image represented using the YUQ color model. Explain (using MATLAB) what happens if the original image were represented using the RGB color model and each color channel underwent histogram equalization individually.

16.6 Use the `edge` function in MATLAB and write a script to compute and display the edges of a color image for the following cases:

- **(a)** RGB image, combining the edges from each color channel by adding them up.
- **(b)** RGB image, combining the edges from each color channel with a logical OR operation.
- **(c)** YIQ image, combining the edges from each color channel by adding them up.
- **(d)** YIQ image, combining the edges from each color channel with a logical OR operation.

16.7 Repeat Problem 16.6 for noisy versions of the input color images.

CHAPTER 17

IMAGE COMPRESSION AND CODING

WHAT WILL WE LEARN?

- What is the meaning of compressing an image?
- Which types of redundancy can be exploited when compressing images and how is this usually done?
- What are the main (lossy and lossless) image compression techniques in use today?
- How can we evaluate the (subjective) quality of an image that has undergone some type of processing (e.g., lossy compression) in an objective way?
- What are the most popular contemporary image compression standards?

17.1 INTRODUCTION

Images can be represented in digital format in many different ways, the simplest of which consists in simply storing all the pixel values in a file following a certain convention (e.g., row first, starting from the top left corner). The result of using such raw image representations is the need for large amounts of storage space (and proportionally long transmission times in the case of file uploads/downloads). The need to save storage space and shorten transmission time and the human visual system

Practical Image and Video Processing Using MATLAB®. By Oge Marques.

tolerance to a modest amount of loss have been the driving factors behind image compression techniques.

Image compression techniques are behind the most popular image and video applications and devices in use today, such as DVD players, digital cameras, and web-based video streaming. Contemporary image compression algorithms are capable of encoding images with high compression efficiency and minimal noticeable degradation of visual quality.

Compression methods can be *lossy*—when a tolerable degree of deterioration in the visual quality of the resulting image is acceptable—or *lossless*—when the image is encoded in its full quality and the original image can be fully recovered at the decoding stage. The overall results of the compression process in terms of both storage savings—usually expressed with figures of merit such as compression ratio (CR) or bits per pixel (bpp)—and resulting quality loss (for the case of lossy techniques) may vary depending on the technique, format, options (such as the "quality" setting for JPEG), and image contents. As a general guideline, lossy compression should be used for general-purpose photographic images, whereas lossless compression should be preferred when dealing with line art, technical drawings, or images in which no loss of detail may be tolerable (most notably, space images and medical images).

In the remainder of this chapter, we provide an overview of image compression and coding. This is a vast and ever-growing field. Keeping up with the main goals of this book, we emphasize key concepts, provide a brief overview of the most representative contemporary image compression techniques and standards, and show how to evaluate image quality when lossy image compression is used. Readers interested in a more detailed analysis of specific techniques and standards will find many useful pointers to further their studies in the "Learn More About It" section at the end of the chapter.

17.2 BASIC CONCEPTS

The general problem of image compression is to reduce the amount of *data* (i.e., bits) required to represent a digital image. The underlying basis of the reduction process is the removal of redundant data. Mathematically, visual data compression typically involves transforming (encoding) a 2D pixel array into a statistically uncorrelated data set. This transformation is applied prior to storage or transmission. At some later time, the compressed image is decompressed to reconstruct the original image information (if lossless techniques are used) or an approximation of it (when lossy techniques are employed).

17.2.1 Redundancy

Image compression is a particular case of *data compression*: the process of reducing the amount of *data* required to represent a given quantity of *information*. If the same information can be represented using different amounts of data, it is reasonable to believe that the representation that requires more data contains what is technically called *data redundancy*.

(a) (b)

FIGURE 17.1 Two ways to represent the same information using different amounts of data. See text for details.

■ EXAMPLE 17.1

To understand the difference between *data* and *information*, these are five different ways of representing the same information (the quantity *four*) using different representations (in increasing order of size in bits):

- The binary equivalent of the number 4 (`100`): 3 bits
- The Roman numeral `IV` encoded using the ASCII characters for `I` and `V`: 16 bits
- The word "quatro," meaning "four" in Portuguese, encoded as a string of ASCII characters: 48 bits
- An uncompressed 64×64 binary image with four white squares against a black background (Figure 17.1a): 32,768 bits
- A compressed 143×231 grayscale image (Figure 17.1b): 21,7632 bits

The mathematical definition of data redundancy is given in equation (17.1). It is assumed that b_1 and b_2 represent the number of information carrying units (or simply *bits*) in two data sets that represent the same information. In this case, the relative redundancy (R) of the first data set (represented by b_1) can be defined as

$$R = 1 - \frac{1}{\text{CR}} \tag{17.1}$$

where the parameter CR, commonly known as *compression ratio*, is given by

$$\text{CR} = \frac{b_1}{b_2} \tag{17.2}$$

For the case $b_2 = b_1$, $\text{CR} = 1$, and $R = 0$, it can be concluded that the first data set contains no redundant data. When $b_2 \ll b_1$, $\text{CR} \to \infty$, and $R \to 1$, indicating high redundancy, and, consequently, high compression. Finally, when $b_2 \gg b_1$, $\text{CR} \to 0$

and $R \to -\infty$, so it can be concluded that the second data set contains much more data than the first, suggesting the undesirable case of data expansion. In general, CR and R lie within the $[0, \infty]$ and $[-\infty, 1]$, respectively. A compression ratio such as 100 (or 100:1) basically means that the first data set has 100 information units (e.g., bits) for each unit in the second (compressed) data set. The corresponding redundancy (0.99 in this case) indicates that 99% of the data in the first data set are redundant.

Image compression and coding techniques exploit three types of redundancies that are commonly present in 2D image arrays: coding redundancy, interpixel (spatial) redundancy, and psychovisual redundancy. The way each of them is exploited is briefly described next.

Coding Redundancy The 8 bits per pixel (per color) conventionally used to represent pixels in an image contain an amount of redundancy. From the perspective of coding theory, each 8-bit value is a fixed-length *codeword* used to represent a piece of information (in this case, the intensity or color values of each pixel). Entropy-based coding techniques (e.g., Huffman coding, Golomb coding, and arithmetic coding) replace those fixed-length codewords with a variable-length equivalent in such a way that the total number of bits needed to represent the same information (b_2) is less than the original number of bits (b_1).[1] This type of coding is always reversible and usually implemented using lookup tables (LUTs).

Interpixel Redundancy This type of redundancy—sometimes called *spatial* redundancy, *interframe* redundancy, or *geometric* redundancy—exploits the fact that an image very often contains strongly correlated pixels—in other words, large regions whose pixel values are the same or almost the same. This redundancy can be exploited in several ways, one of which is by predicting a pixel value based on the values of its neighboring pixels. To do so, the original 2D array of pixels is usually mapped into a different format, for example, an array of differences between adjacent pixels. If the original image pixels can be reconstructed from the transformed data set, the mapping is said to be *reversible*. Examples of compression techniques that exploit the interpixel redundancy include constant area coding (CAC), Lempel–Ziv–Welch (LZW), (1D or 2D) run-length encoding (RLE) techniques, and many predictive coding algorithms such as differential pulse code modulation (DPCM).

Psychovisual Redundancy Many experiments on the psychophysical aspects of human vision have proven that the human eye does not respond with equal sensitivity to all incoming visual information; some pieces of information are more important than others. The knowledge of which particular types of information are more or less relevant to the human visual system has led to image compression techniques that aim at eliminating or reducing any amount of data that is psychovisually redundant. The end result of applying these techniques is a compressed image file, whose size

[1] You might remember from Chapter 2 that the total number of bits occupied by an image in uncompressed form, b_1, can be easily calculated by multiplying the width, height, number of bits per color channel (usually 8), and number of color channels (1 for monochrome, 3 for color images) in the input image.

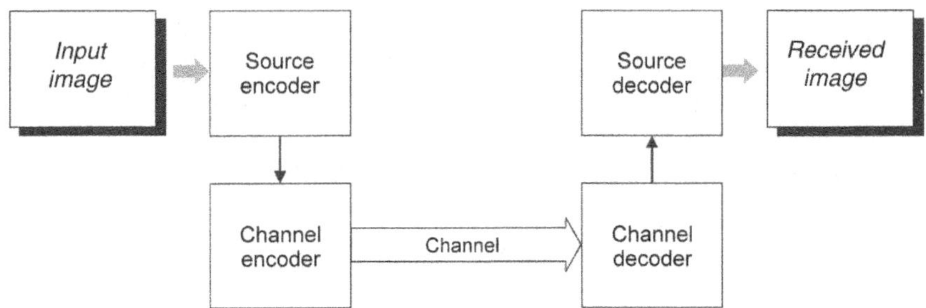

FIGURE 17.2 A general image encoding and decoding model.

and quality are smaller than those of the original information, but whose resulting quality is still acceptable for the application at hand. The loss of quality that ensues as a by-product of such techniques is frequently called *quantization*, so as to indicate that a wider range of input values is normally mapped into a narrower range of output values through an irreversible process.

Video compression techniques also exploit a fourth type of redundancy, interframe (or *temporal*) redundancy, caused by the fact that consecutive frames in time are usually very similar.

17.2.2 Image Encoding and Decoding Model

Figure 17.2 shows a general image encoding and decoding model. It consists of a source encoder, a channel encoder, the storage or transmission media (also referred to as *channel*), a channel decoder, and a source decoder. The source encoder reduces or eliminates any redundancies in the input image, which usually leads to bit savings. Source encoding techniques are the primary focus of this chapter. The channel encoder increases noise immunity of source encoder's output, usually adding extra bits to achieve its goals. If the channel is noise-free, the channel encoder and decoder may be omitted. At the receiver's side, the channel and the source decoder perform the opposite functions and ultimately recover (an approximation of) the original image.

Figure 17.3 shows the source encoder in further detail. Its main components are as follows:

- *Mapper*: It transforms the input data into a (usually nonvisual) format designed to reduce interpixel redundancies in the input image. This operation is generally

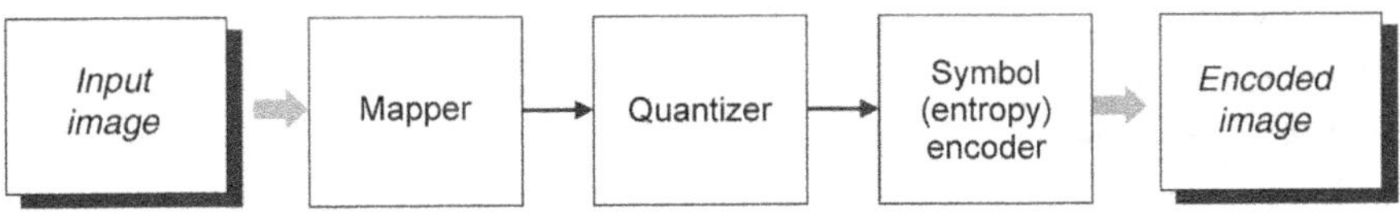

FIGURE 17.3 Source encoder.

reversible and may or may not directly reduce the amount of data required to represent the image.

- *Quantizer*: It reduces the accuracy of the mapper's output in accordance with some pre-established fidelity criterion. This block exploits the psychovisual redundancies of the input image. This operation is not reversible and must be omitted if lossless compression is desired.
- *Symbol (Entropy) Encoder*: It creates a fixed- or variable-length code to represent the quantizer's output and maps the output in accordance with the code. In most cases, a variable-length code is used. This block exploits the coding redundancies of the data produced at the output of the quantizer. This operation is reversible.

17.3 LOSSLESS AND LOSSY COMPRESSION TECHNIQUES

In this section, we present a broad overview of the techniques that lie at the core of the most common lossless and lossy compression algorithms and standards in use today.

17.3.1 Lossless Compression Techniques

Error-Free Compression Using Variable Length Coding (VLC) Error-free compression techniques usually rely on entropy-based encoding algorithms. The concept of entropy provides an upper bound on how much compression can be achieved, given the probability distribution of the source. In other words, it establishes a theoretical limit on the amount of lossless compression that can be achieved using entropy encoding techniques alone. Most entropy-based encoding techniques rely on assigning variable-length codewords to each symbol, whereas the most likely symbols are assigned shorter codewords. In the case of image coding, the symbols may be raw pixel values or the numerical values obtained at the output of the mapper stage (e.g., differences between consecutive pixels, run lengths, etc.). The most popular entropy-based encoding technique is the Huffman code, which is used in the JPEG standard (among others).

Run-Length Encoding RLE is one of the simplest data compression techniques. It consists of replacing a sequence (run) of identical symbols by a pair containing the symbol and the run length. It is used as the primary compression technique in the 1D CCITT Group 3 fax standard and in conjunction with other techniques in the JPEG image compression standard.

Differential Coding Differential coding techniques exploit the interpixel redundancy in digital images. The basic idea consists of applying a simple difference operator to neighboring pixels to calculate a difference image, whose values are likely to follow within a much narrower range than the original gray-level range.

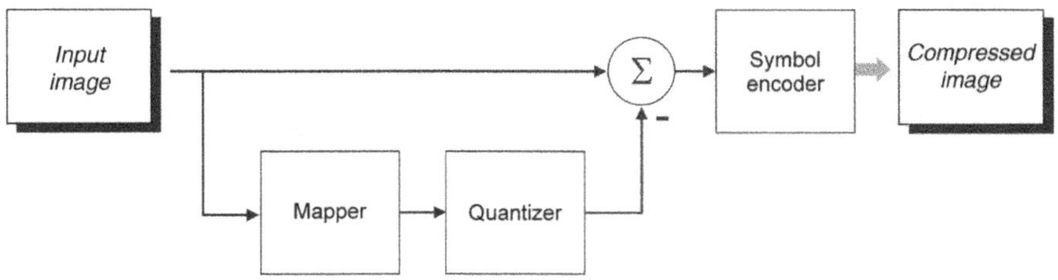

FIGURE 17.4 Lossless predictive encoder.

As a consequence of this narrower distribution, and consequently reduced entropy, Huffman coding or other VLC schemes will produce shorter codewords for the difference image.

Predictive Coding Predictive coding techniques also exploit interpixel redundancy. The basic idea, in this case, is to encode only the new information in each pixel, which is usually defined as the difference between the actual and the predicted value of that pixel. Figure 17.4 shows the main blocks of a lossless predictive encoder. The key component is the *predictor*, whose function is to generate an estimated (predicted) value for each pixel from the input image based on previous pixel values. The predictor's output is rounded to the nearest integer and compared with the actual pixel value: the difference between the two (called *prediction error*) is then encoded by a VLC encoder. Since prediction errors are likely to be smaller than the original pixel values, the VLC encoder will likely generate shorter codewords.

There are several local, global, and adaptive prediction algorithms in the literature. In most cases, the predicted pixel value is a linear combination of previous pixels.

Dictionary-Based Coding Dictionary-based coding techniques are based on the idea of incrementally building a dictionary (a special type of *table*) while receiving the data. Unlike VLC techniques, dictionary-based techniques use fixed-length codewords to represent variable-length strings of symbols that commonly occur together. Consequently, there is no need to calculate, store, or transmit the probability distribution of the source, which makes these algorithms extremely convenient and popular. The best-known variant of dictionary-based coding algorithms is the LZW (Lempel–Ziv–Welch) encoding scheme, used in popular multimedia file formats such as GIF, TIFF, and PDF.

17.3.2 Lossy Compression Techniques

Lossy compression techniques deliberately introduce a certain amount of distortion to the encoded image, exploring the psychovisual redundancies of the original image. These techniques must find an appropriate balance between the amount of error (loss) and the resulting bit savings.

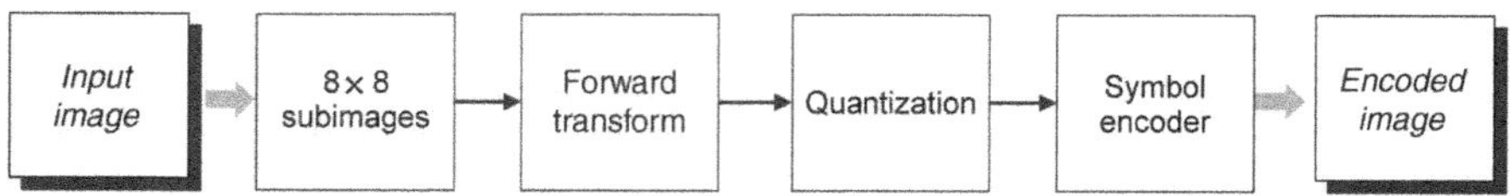

FIGURE 17.5 Transform coding diagram.

Quantization The quantization stage is at the core of any lossy image encoding algorithm. Quantization, in this context, is the process of partitioning of the input data range into a smaller set of values. There are two main types of quantizers: *scalar* quantizers and *vector* quantizers. A scalar quantizer partitions the domain of input values into a smaller number of intervals. If the output intervals are equally spaced, the process is called *uniform scalar quantization* or otherwise, for reasons usually related to minimization of total distortion, it is called *nonuniform scalar quantization*. Vector quantization (VQ) techniques extend the basic principles of scalar quantization to multiple dimensions. VQ-based coding schemes are computationally efficient because the decoding can be done using lookup tables.

Transform Coding The techniques discussed so far work directly on the pixel values and are usually called *spatial-domain techniques*. Transform coding techniques use a reversible, linear mathematical transform to map the pixel values onto a set of coefficients that are then quantized and encoded. The key factor behind the success of transform-based coding schemes lies in the ability of the selected mathematical transform to encode a significant amount of the visual information originally present in the image using the few first coefficients. Consequently, most of the resulting coefficients have small magnitudes and can be quantized (or discarded altogether) without causing significant distortion in the decoded image. The discrete cosine transform (DCT) has become the most widely used transform coding technique. DCT-based coding is an essential component of the JPEG and JPEG-LS compression standards.

Transform coding algorithms (Figure 17.5) usually work on a block basis. They start by partitioning the original image into subimages (blocks) of small size (usually 8×8).[2] For each block, the transform coefficients are calculated, effectively converting the original 8×8 array of pixel values into an array of coefficients within which the coefficients closer to the top left corner usually contain most of the information needed to quantize and encode (and eventually perform the reverse process at the decoder's side) the image with little perceptual distortion. The resulting coefficients are then quantized and the output of the quantizer is used by a (combination of) symbol encoding technique(s) to produce the output bitstream representing the encoded image. At the decoder's side, the reverse process takes place, with the obvious difference that the "dequantization" stage will generate only an approximated version of the original coefficient values; in other words, whatever loss was introduced by the quantizer in the encoder stage is not reversible at the decoder.

[2]An undesired side effect of this division is the introduction of noticeable artifacts (known as *blockiness*) in a compressed image encoded using low quality (see Tutorial 17.1).

Wavelet Coding Wavelet coding techniques are also based on the idea of using a transform that decorrelates the pixels of the input image, converting them into a set of coefficients that can be coded more efficiently than the original pixel values themselves. Contrary to DCT-based coding, the subdivision of the input image into smaller subimages is not necessary in wavelet coding. Wavelet coding is a core technique in the JPEG 2000 compression standard.

17.4 IMAGE COMPRESSION STANDARDS

Work on international standards for image compression started in the late 1970s with a United Nations organization called the *Consultative Committee of the International Telephone and Telegraph* (CCITT) (currently *International Telecommunications Union* (ITU-T)) and its need to standardize binary image compression algorithms for Group 3 facsimile communications. Since then, many other committees and standards have been formed to produce *de jure* standards (such as JPEG or JPEG 2000), while several commercially successful initiatives have effectively become *de facto* standards (such as the Adobe Systems' PDF format for document representation). Image compression standards bring many benefits, such as easier exchange of image files between different devices and applications, reuse of existing hardware and software for a wider array of products, and availability of benchmarks and reference data sets for new and alternative developments.

17.4.1 Binary Image Compression Standards

Work on binary image compression standards was initially motivated by CCITT Group 3 and 4 facsimile standards. The Group 3 standard uses Huffman coding and a nonadaptive, 1D RLE technique in which the last $K - 1$ lines of each group of K lines (for $K = 2$ or 4) are optionally coded in a 2D manner, using the Modified Relative Element Address Designate (MREAD) algorithm. The Group 4 standard uses only the MREAD coding algorithm. Both classes of algorithms are nonadaptive and were optimized for a set of eight test images, containing a mix of representative documents, which sometimes resulted in data expansion when applied to different types of documents (e.g., halftone images). The Joint Bilevel Image Group (JBIG)—a joint committee of the ITU-T and the International Standards Organization (ISO)—has addressed these limitations and proposed two new standards (JBIG—also known as JBIG1—and JBIG2) that can be used to compress binary and grayscale images of up to 6 gray-coded bits/pixel.

17.4.2 Continuous Tone Still Image Compression Standards

For photograph quality images (both grayscale and color), different standards have been proposed, mostly based on lossy compression techniques. The most popular standard in this category is the JPEG standard, a lossy, DCT-based coding algorithm.

Despite its great popularity and adoption, ranging from digital cameras to the World Wide Web, certain limitations of the original JPEG algorithm have motivated the recent development of two alternative standards, JPEG 2000 and JPEG-LS (lossless). JPEG, JPEG 2000, and JPEG-LS are described next.

17.4.3 JPEG

The JPEG format was originally published as a standard (ISO IS 10918-1) by the Joint Photographic Experts Group in 1994. It has become the most widely used format for storing digital photographs ever since. The JPEG specification defines how an image is transformed into a stream of bytes, but not how those bytes are encapsulated in any particular storage medium. Another standard, created by the independent JPEG group, called JFIF (JPEG File Interchange Format) specifies how to produce a file suitable for computer storage and transmission from a JPEG stream.

Even though the original JPEG specification defined four compression modes (sequential, hierarchical, progressive, and lossless), most JPEG files used today employ the sequential mode. The baseline JPEG encoder (Figure 17.6, top) consists of the following main steps:

1. The original RGB color image is converted to an alternative color model (YCbCr) and the color information is subsampled.
2. The image is divided into 8×8 blocks.
3. The 2D DCT is applied to each block image; the resulting 64 values are referred to as *DCT coefficients*.
4. DCT coefficients are quantized according to a quantization table; this is the step where acceptable loss is introduced.

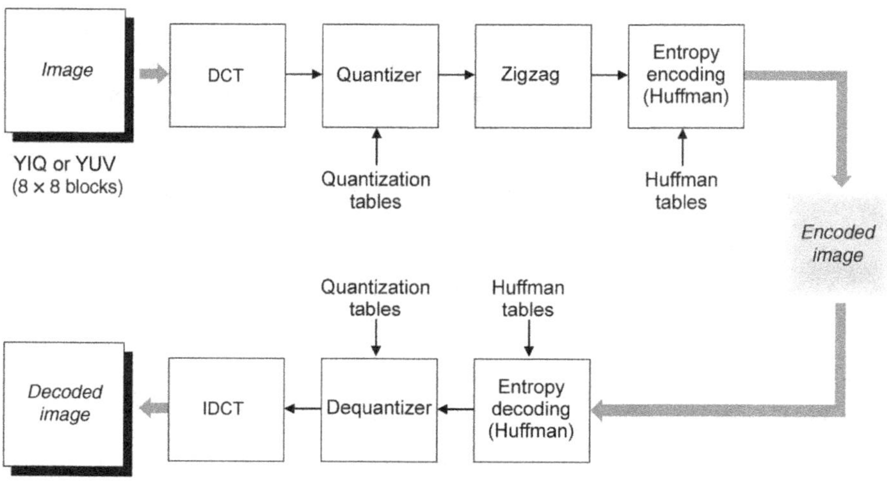

FIGURE 17.6 JPEG encoder and decoder.

5. Quantized DCT coefficients are scanned in a zigzag fashion (from top left to bottom right). The resulting sequence is run-length encoded, in preparation for the entropy encoding step.
6. The run-length encoded sequences are converted to variable-length binary codewords using Huffman encoding.

At the decoder side (Figure 17.6, bottom), the process is reversed; it should be noted that the loss introduced at the quantizer stage in the encoder cannot be canceled at the "dequantizer" stage in the decoder; that is, the values recovered at the output of the "dequantizer" block are approximations of the values produced at the output of the DCT block in the encoder.

17.4.4 JPEG 2000

The JPEG 2000 format [TM01,AT04] is a wavelet-based image compression standard (ISO/IEC 15444-1:2000), created by the Joint Photographic Experts Group committee with the intention of superseding their original DCT-based JPEG standard. The usual file extension is .jp2. It addresses several well-known limitations of the original JPEG algorithm and prepares the way for next-generation imagery applications. Some of its advertised advantages are as follows:

1. Low bit-rate compression
2. Superior lossless and lossy compression in a single bitstream
3. Ability to handle very large images without need for tiling
4. Single decompression architecture
5. Error resilience for transmission in noisy environments, such as wireless communication networks
6. Region of interest (ROI) coding
7. Metadata mechanisms for incorporating additional nonimage data as part of the file

JPEG 2000 is not yet widely supported in web browsers, and hence it is not generally used on the World Wide Web.

17.4.5 JPEG-LS

JPEG-LS is an image compression standard (ISO/IEC-14495-1/ITU-T Rec. T.87) designed to provide effective lossless and near-lossless compression of continuous-tone, grayscale, and color still images. *Near lossless*, in this context, means a guaranteed maximum error between the original image data and the reconstructed image data. One of its main potential driving forces for this encoding mode is the compression of medical images without any quality loss. The core algorithm behind JPEG-LS is called low complexity lossless compression for images (LOCO-I), proposed by

Hewlett-Packard. LOCO-I builds upon a concept known as *context modeling*. The main idea behind context modeling is to take advantage of the structure in the input source, modeled in terms of conditional probabilities of what pixel values follow from each other in the image.

17.5 IMAGE QUALITY MEASURES

The human visual system is the final link in the perception of images and the ultimate judge of image quality. Therefore, in the design and specification of image processing systems, it is necessary to establish methods for measuring the quality of the images displayed to the viewer. Such need can be justified by looking at two extreme cases: (1) there is no reason for designing a system whose image quality exceeds the ability of the human eye to perceive it; (2) on the other hand, since different processing and encoding techniques are likely to cause noticeable changes to the resulting images, it becomes necessary to evaluate the amount and impact of the resulting loss in visual quality.

Measuring image quality is a difficult and often imprecise task, in part because of the vast number of variables that may impact the final result (ranging from display technology to ambient lighting to the subject's mood, among many others). Assessment of visual quality is inherently a subjective process that has traditionally been evaluated with objective quality measures and criteria because such objective measures provide accurate, repeatable results and usually rely on fewer, controllable factors. Despite their mathematical convenience, however, most quantitative objective quality measures do not correlate well with the subjective experience of a human viewer watching the image. Consequently, the design of objective quality measurement systems that match the subjective human judgments is an open research topic.

17.5.1 Subjective Quality Measurement

The subjective quality of an image or video sequence can be established by asking human observers to report, choose, or rank their visual experiences within a controlled experiment. These measures can be *absolute* or *relative*.

Absolute techniques are those in which an image is classified on an isolated basis and without regard to another image. An example is the classification of an image as *excellent*, *fine*, *passable*, *marginal*, *inferior*, or *unusable* according to the criteria of the Television Allocations Study Organization [FB60].

Relative measurements ask subjects to compare an image against another and decide which is best. A popular protocol for these tests is the *double stimulus continuous quality scale* (DSCQS) method defined in ITU-R Recommendation BT.500-10 [ITU00]. In this method, the subject is presented with a pair of images or short video sequences A and B, one after another, and is asked to provide a score recorded on a continuous line with five gradings: *excellent*, *good*, *fair*, *poor*, and *bad*. Experiments usually consist of several tests in which either A or B is (randomly) assigned to the "original" image or video, whereas the other is the video after some processing, whose impact on perceptual quality is being evaluated. The DSCQS test is often used as a measure of subjective visual quality, but it suffers from several shortcomings, such as the dependency on prior knowledge among (*experts* and *nonexperts*) subjects.

17.5.2 Objective Quality Measurement

Image and video developers often rely on objective measures of visual quality for two main reasons:

- They provide an alternative to subjective measurements and their limitations.
- They are easy to compute.

The most common measures of image quality are the root mean square (RMS) error and the peak signal to noise ratio (PSNR). Let $f(x, y)$ be the original (or reference) image and $f'(x, y)$ the modified (or reconstructed) image. For every value of x and y, the error $e(x, y)$ between $f(x, y)$ and $f'(x, y)$ can be defined as

$$e(x, y) = f'(x, y) - f(x, y) \tag{17.3}$$

and the total error between the two images (whose sizes are $M \times N$) is

$$E = \sum_{x=0}^{M-1} \sum_{y=0}^{N-1} |f'(x, y) - f(x, y)| \tag{17.4}$$

The RMS error, e_{rms}, between $f(x, y)$ and $f'(x, y)$ can be calculated by

$$e_{\text{rms}} = \sqrt{\frac{1}{MN} \sum_{x=0}^{M-1} \sum_{y=0}^{N-1} \left[f'(x, y) - f(x, y)\right]^2} \tag{17.5}$$

If we consider the resulting image, $f'(x, y)$, as the "signal" and the error as "noise," we can define the RMS signal to noise ratio (SNR_{rms}) between modified and original images as

$$\text{SNR}_{\text{rms}} = \sqrt{\frac{\sum_{x=0}^{M-1} \sum_{y=0}^{N-1} f'(x, y)^2}{\sum_{x=0}^{M-1} \sum_{y=0}^{N-1} \left[f'(x, y) - f(x, y)\right]^2}} \tag{17.6}$$

If we express the SNR as a function of the peak value of the original image and the RMS error (equation (17.5)), we obtain another metric, the PSNR, defined as

$$\text{PSNR} = 10 \log_{10} \frac{(L-1)^2}{(e_{\text{rms}})^2} \tag{17.7}$$

where L is the number of gray levels (for 8 bits/pixel monochrome images, $L = 256$).

TABLE 17.1 Objective Quality Measures for Three JPEG Images with Different Quality Factors

Quality Factor	File Size (B)	RMS Error	PSNR (dB)
90	10,8792	2.1647	41.4230
20	20,781	6.3553	32.0681
5	5992	7.6188	30.4931

■ EXAMPLE 17.2

In this example, we use the `imwrite` function in MATLAB to generate three JPEG files (using different quality factors) from the same input image. The result of image degradation caused by lossy compression is measured by the RMS error (equation (17.5)) and the PSNR (equation (17.7)). Clearly, the use of lower quality factors leads to a degradation in subjective quality: the image in Figure 17.7c is significantly worse than the image in parts (a) or (b).

Table 17.1 summarizes the quality factors, file size (in B), RMS error, and PSNR for the images in Figure 17.7a–c.

Objective quality measures suffer from a number of limitations:

- They usually require an "original" image (whose quality is assumed to be perfect). This image may not always be available and the assumption that it is "perfect" may be too strong in many cases.
- They do not correlate well with human subjective evaluation of quality. For example, in Figure 17.8 most subjects would say that the rightmost image (in which an important portion of the image got severely blurred) has significantly worse subjective quality than the one on the center (where the entire image was subject to a mild blurring), even though their RMS errors (measured against the reference image on the left) are comparable.

Finding objective measures of image and video quality that match human quality evaluation is an ongoing field of research and the subject of recent standardization efforts by the ITU-T Video Quality Experts Group (VQEG), among other groups.

17.6 TUTORIAL 17.1: IMAGE COMPRESSION

Goal

The goal of this tutorial is to show how we can perform image compression using MATLAB.

Objectives

- Experiment with image compression concepts, such as compression ratio.
- Calculate objective quality measures on compressed images.

FIGURE 17.7 Measuring objective image quality after compression: (a) original; (b) compressed version of (a) (using quality factor = 90), $e_{rms} = 2.1647$, PSNR = 41.4230 dB; (c) compressed version of (a) (using quality factor = 5), $e_{rms} = 7.6188$, PSNR = 30.4931 dB.

(a)

(b)

(c)

FIGURE 17.8 The problem of poor correlation between objective and subjective measures of image quality: (a) original; (b) blurred version of (a) (using a 5 × 5 average filter), $e_{\mathrm{rms}} = 0.0689$, PSNR = 71.3623 dB; (c) partially blurred version of (a) (after applying a severe blurring filter only to a small part of the image), $e_{\mathrm{rms}} = 0.0629$, PSNR = 72.1583 dB.

What You Will Need

- coat_of_arms.jpg
- espresso_color.jpg
- statue.png

Procedure

In this tutorial, you will have a chance to adjust the quality factor in a JPEG encoder and make quantitative observations on the trade-off between the resulting compression ratio and the associated quality loss.

1. Use the sequence below to calculate the compression ratio in the JPEG test file coat_of_arms.jpg.

```
filename = 'coat_of_arms.jpg';
I = imread(filename);
fileInfo = dir(filename)
imageInfo = whos('I')
fileSize = fileInfo.bytes
imageSize = imageInfo.bytes
CR = imageSize / fileSize
```

2. Repeat the process for the JPEG test file espresso_color.jpg.

Question 1 Are the compression ratio values calculated for the two test images comparable? Explain.

3. Load file statue.png, convert it to JPEG using MATLAB's imwrite function, and choose five different values for quality, 50, 25, 15, 5, and 0, following the example below. Store these results in files named statue50.jpg, ..., statue00.jpg.

```
org = imread('statue.png');
imwrite(org, 'statue50.jpg', 'quality', 50);
```

4. Display the original image as well as the five compressed images (in decreasing degree of quality).
5. Calculate the compression ratio for each file following the example below:

```
K = imfinfo('statue50.jpg');
image_bytes = K.Width*K.Height*K.BitDepth/8;
compressed_bytes = K.FileSize;
CR = image_bytes / compressed_bytes
```

TABLE 17.2 Compression Ratio for Five JPEG Images with Different Quality Factors

File Name	File Size (B)	CR
`statue.png`	243825	N/A
`statue50.jpg`		
`statue25.jpg`		
`statue15.jpg`		
`statue05.jpg`		
`statue00.jpg`		

TABLE 17.3 Objective Quality Measures for Five JPEG Images with Different Quality Factors

File Name	RMS Error	PSNR (dB)	Subjective Quality
`statue.png`	N/A	N/A	Very good
`statue50.jpg`			
`statue25.jpg`			
`statue15.jpg`			
`statue05.jpg`			
`statue00.jpg`			

Question 2 How do the results for CR obtained with this method compare with the ones obtained with the method suggested in step 1?

6. Log the results to Table 17.2.
7. Write code to calculate and display the RMS error and the PSNR (dB) between the original image and each of the lower quality resulting images.
8. Show the results in Table 17.3 and compare them with your subjective evaluation of image quality.

WHAT HAVE WE LEARNED?

- Image compression is a process by which the total size of a digital image representation (in bytes) is reduced without significant loss in the visual quality of the resulting image. The compressed image will occupy less storage space on a computer and take less time to be transmitted from one computer (e.g., Web server) to another (e.g., Web client browser) over a computer network.
- Three types of redundancy are usually exploited when compressing images: coding, interpixel (spatial), and psychovisual. Video compression techniques also exploit interframe (temporal) redundancy to achieve higher compression ratios (fewer bits per pixel).

- Image compression techniques often employ general data compression techniques capable of exploiting coding redundancy (e.g., Huffman and arithmetic coding) and spatial redundancy (e.g., predictive coding).
- The main difference between image compression methods and general data compression techniques is that most image compression techniques allow a certain amount of loss to be introduced during the compression process. This loss, which usually can be controlled by some parameters in the encoding algorithm, is the consequence of exploiting psychovisual redundancies within the image. Image compression techniques are classified as *lossy* or *lossless* depending on whether such quality loss is present.
- The field of image compression has experienced intense activity over the past 30 years. As a result, many (lossy and lossless) image compression techniques have been proposed, a fraction of which have been adopted as part of widely used image standards and file formats.
- Subjective evaluation of image compression algorithms can be done in absolute or relative ways. Objective evaluation consists of straightforward methods and calculations (e.g., PSNR and RMS errors). The design of objective image and video quality measures with strong correlation to subjective perception of quality is an ongoing research topic.
- The most popular image compression standards in use today are JPEG, JPEG 2000, and JPEG-LS. Other image file formats that use some form of compression are GIF, TIFF, and PNG.

LEARN MORE ABOUT IT

- There are entire books dedicated to classic data compression, such as [NG95], [Say05], and [Sal06].
- Many books on image processing and multimedia have one or more chapters devoted to image coding and compression, for example, Chapter 8 of [GW08] and Chapter 10 of [Umb05].
- The JPEG standard is described in detail in [PM92].
- The JPEG 2000 standard is presented in detail in [TM01,AT04].

ON THE WEB

- The JPEG standard
  http://www.w3.org/Graphics/JPEG/itu-t81.pdf
- The JPEG committee home page
  http://www.jpeg.org/
- ITU-T Video Quality Experts Group
  http://www.its.bldrdoc.gov/vqeg/

CHAPTER 18

FEATURE EXTRACTION AND REPRESENTATION

WHAT WILL WE LEARN?

- What is feature extraction and why is it a critical step in most computer vision and image processing solutions?
- Which types of features can be extracted from an image and how is this usually done?
- How are the extracted features usually represented for further processing?

18.1 INTRODUCTION

This chapter discusses methods and techniques for representing and describing an image and its objects or regions of interest. Most techniques presented in this chapter assume that an image has undergone segmentation.[1] The common goal of feature extraction and representation techniques is to convert the segmented objects into representations that better describe their main features and attributes. The type and complexity of the resulting representation depend on many factors, such as the type of image (e.g., binary, grayscale, or color), the level of granularity (entire image or

[1]Image segmentation was the topic of Chapter 16.

Practical Image and Video Processing Using MATLAB®. By Oge Marques.

individual regions) desired, and the context of the application that uses the results (e.g., a two-class pattern classifier that tells circular objects from noncircular ones or an image retrieval system that retrieves images judged to be similar to an example image).

Feature extraction is the process by which certain features of interest within an image are detected and represented for further processing. It is a critical step in most computer vision and image processing solutions because it marks the transition from pictorial to nonpictorial (alphanumerical, usually quantitative) data representation. The resulting representation can be subsequently used as an input to a number of pattern recognition and classification techniques,[2] which will then label, classify, or recognize the semantic contents of the image or its objects.

There are many ways an image (and its objects) can be represented for image analysis purposes. In this chapter, we present several representative techniques for feature extraction using a broad range of image properties.

18.2 FEATURE VECTORS AND VECTOR SPACES

A *feature vector* is a $n \times 1$ array that encodes the n features (or measurements) of an image or object. The array contents may be symbolic (e.g., a string containing the name of the predominant color in the image), numerical (e.g., an integer expressing the area of an object, in pixels), or both. In the remaining part of our discussion, we shall focus exclusively on numerical feature vectors.

Mathematically, a numerical feature vector $\mathbf{x}$ is given by

$$\mathbf{x} = (x_1, x_2, \ldots, x_n)^T \qquad (18.1)$$

where n is the total number of features and T indicates the *transpose* operation.

The feature vector is a compact representation of an image (or object within the image), which can be associated with the notion of a *feature space*, an n-dimensional *hyperspace* that allows the visualization (for $n < 4$) and interpretation of the feature vectors' contents, their relative distances, and so on.[3]

■ EXAMPLE 18.1

Suppose that the objects in Figure 18.1a have been represented by their area and perimeter, which have been computed as follows:

Object	Area	Perimeter
Square (Sq)	1024	124
Large circle (LC)	3209	211
Small circle (SC)	797	105

[2] Pattern classification techniques will be discussed in Chapter 19.

[3] In Chapter 19, we shall discuss how feature vectors are used by pattern classifiers.

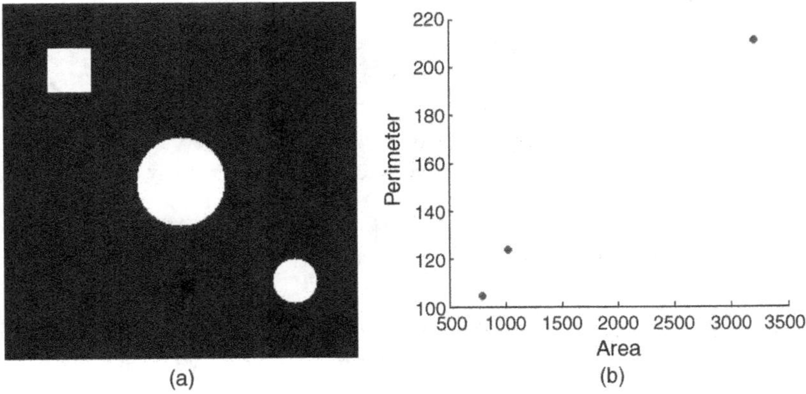

FIGURE 18.1 Test image (a) and resulting 2D feature vectors (b).

The resulting feature vectors will be as follows:
$\mathbf{Sq} = (1024, 124)^T$
$\mathbf{LC} = (3209, 211)^T$
$\mathbf{SC} = (797, 105)^T$

Figure 18.1b shows the three feature vectors plotted in a 2D graph whose axes are the selected features, namely, *area* and *perimeter*.

In MATLAB

In MATLAB, feature vectors are usually represented using cell arrays and structures.[4]

18.2.1 Invariance and Robustness

A common requirement for feature extraction and representation techniques is that the features used to represent an image be invariant to rotation, scaling, and translation, collectively known as *RST*. RST invariance ensures that a machine vision system will still be able to recognize objects even when they appear at different size, position within the image, and angle (relative to a horizontal reference). Clearly, this requirement is application dependent. For example, if the goal of a machine vision system is to ensure that an integrated circuit (IC) is present at the correct position and orientation on a printed circuit board, rotation invariance is no longer a requirement; on the contrary, being able to tell that the IC is upside down is an integral part of the system's functionality.

As you may recall from our discussion in Chapter 1, the goal of many machine vision systems is to emulate—to the highest possible degree—the human visual system's ability to recognize objects and scenes under a variety of circumstances. In order

[4]Refer to Chapter 3—particularly Tutorial 3.2—if you need to refresh these concepts.

to achieve such ability, the feature extraction and representation stage of a machine vision system should ideally provide a representation that is not only RST invariant but also *robust* to other aspects, such as poor spatial resolution, nonuniform lighting, geometric distortions (caused by different viewing angles), and noise. This is a big challenge for machine vision systems designers, which may require careful feature selection as well as pre- and postprocessing techniques to be fully accomplished.

18.3 BINARY OBJECT FEATURES

In this section, we present several useful features for binary objects. A binary object, in this case, is a connected region within a binary image $f(x, y)$, which will be denoted as O_i, $i > 0$.

Mathematically, we can define a function $O_i(x, y)$ as follows:

$$O_i(x, y) = \begin{cases} 1 & \text{if } f(x, y) \in O_i \\ 0 & \text{otherwise} \end{cases} \tag{18.2}$$

In MATLAB

The IPT function `bwlabel` – introduced in Chapter 2 and also discussed in Chapter 13 – implements the operation $i \times O_i(x, y)$ by labeling regions of connected pixels in a binary image: pixels labeled 0 correspond to the background; pixels labeled 1 and higher correspond to the connected components in the image.

18.3.1 Area

The area of the ith object O_i, measured in pixels, is given by

$$A_i = \sum_{x=0}^{M-1} \sum_{y=0}^{N-1} O_i(x, y) \tag{18.3}$$

18.3.2 Centroid

The coordinates of the centroid (also known as *center of area*) of object O_i, denoted $(\bar{x}_i, \bar{y}_i)$, are given by

$$\bar{x}_i = \frac{1}{A_i} \sum_{x=0}^{M-1} \sum_{y=0}^{N-1} x O_i(x, y) \tag{18.4}$$

and

$$\bar{y}_i = \frac{1}{A_i} \sum_{x=0}^{M-1} \sum_{y=0}^{N-1} y O_i(x, y) \tag{18.5}$$

where A_i is the area of object O_i, as defined in equation (18.3).

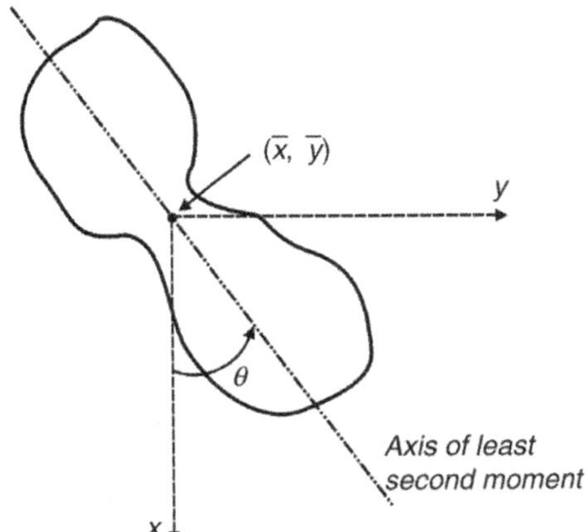

FIGURE 18.2 Axis of least second moment.

18.3.3 Axis of Least Second Moment

The axis of least second moment is used to provide information about the object's orientation relative to the coordinate plan of the image. It can be described as the axis of least inertia, that is, the line about which it takes the least amount of energy to rotate the object.

By convention, the angle θ represents the angle between the vertical axis and the axis of least second moment, measured counterclockwise. Figure 18.2 illustrates this concept. Note that the origin of the coordinate system has been moved to the center of area of the object.

Mathematically, θ can be calculated using equation (18.6):

$$\tan(2\theta_i) = 2 \times \frac{\sum_{x=0}^{M-1}\sum_{y=0}^{N-1} x O_i(x, y)}{\sum_{x=0}^{M-1}\sum_{y=0}^{N-1} x^2 O_i(x, y) - \sum_{x=0}^{M-1}\sum_{y=0}^{N-1} y^2 O_i(x, y)} \tag{18.6}$$

18.3.4 Projections

The horizontal and vertical projections of a binary object—$h_i(x)$ and $v_i(y)$, respectively—are obtained by equations (18.7) and (18.8):

$$h_i(x) = \sum_{x=0}^{M-1} O_i(x, y) \tag{18.7}$$

and

$$v_i(y) = \sum_{y=0}^{N-1} O_i(x, y) \tag{18.8}$$

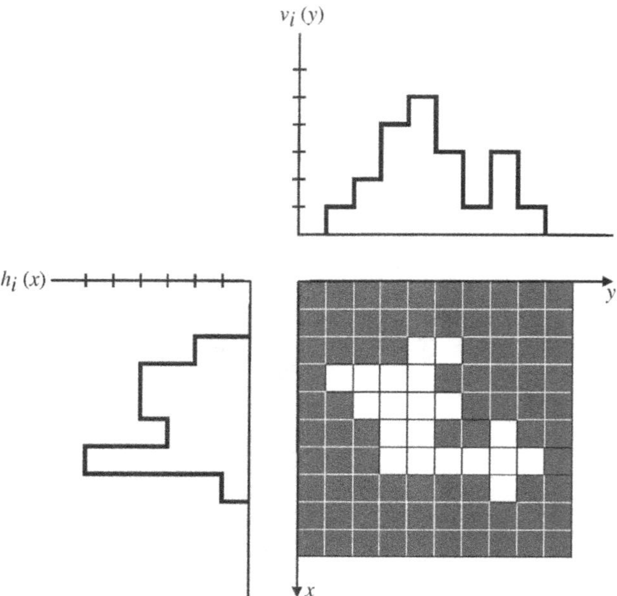

FIGURE 18.3 Horizontal and vertical projections.

Projections are very useful and compact shape descriptors. For example, the height and width of an object without holes can be computed as the maximum value of the object's vertical and horizontal projections, respectively, as illustrated in Figure 18.3.

The equations for the coordinates of the center of area of the object (equations (18.4) and (18.5)) can be rewritten as a function of the horizontal and vertical projections as follows:

$$\bar{x}_i = \frac{1}{A_i} \sum_{x=0}^{M-1} x h_i(x) \tag{18.9}$$

and

$$\bar{y}_i = \frac{1}{A_i} \sum_{y=0}^{N-1} y v_i(y) \tag{18.10}$$

18.3.5 Euler Number

The Euler number of an object or image (E) is defined as the number of connected components (C) minus the number of holes (H):

$$E = C - H \tag{18.11}$$

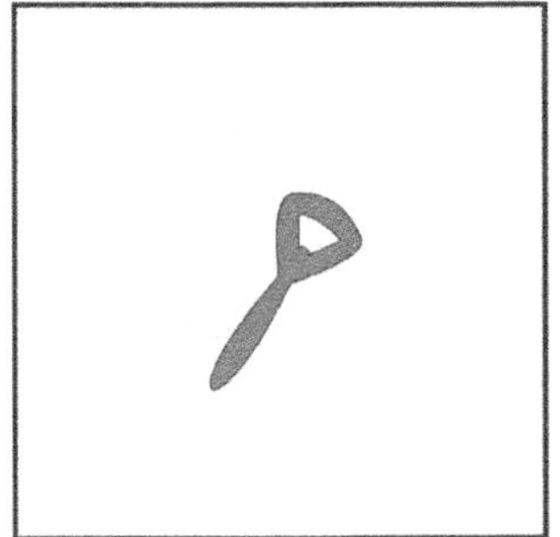

FIGURE 18.4 Examples of two regions with Euler numbers equal to 0 and −1, respectively.

Alternatively, the Euler number can be expressed as the difference between the number of *convexities* and the number of *concavities* in the image or region. The Euler number is considered a *topological descriptor*, because it is unaffected by deformations (such as *rubber sheet* distortions), provided that they do not fill holes or break connected components apart. Figure 18.4 shows examples of two different objects and their Euler numbers.

18.3.6 Perimeter

The perimeter of a binary object O_i can be calculated by counting the number of object pixels (whose value is 1) that have one or more background pixels (whose value is 0) as their neighbors. An alternative method consists in first extracting the edge (contour) of the object and then counting the number of pixels in the resulting border. Due to some inevitable imperfections in the digitization process (e.g., jagged curve outlines and serrated edges), the value of perimeter computed using either method is not 100% accurate; it has been suggested [Umb05] that those values should be multiplied by $\pi/4$ for better accuracy.

In MATLAB

The IPT function `bwperim`—introduced in Chapter 13—computes the perimeter of objects in a binary image using either 4- (which is the default) or 8-connectivity criteria.

18.3.7 Thinness Ratio

The thinness ratio T_i of a binary object O_i is a figure of merit that relates the object's area and its perimeter by equation (8.12):

$$T_i = \frac{4\pi A_i}{P_i^2} \tag{18.12}$$

where A_i is the area and P_i is the perimeter.

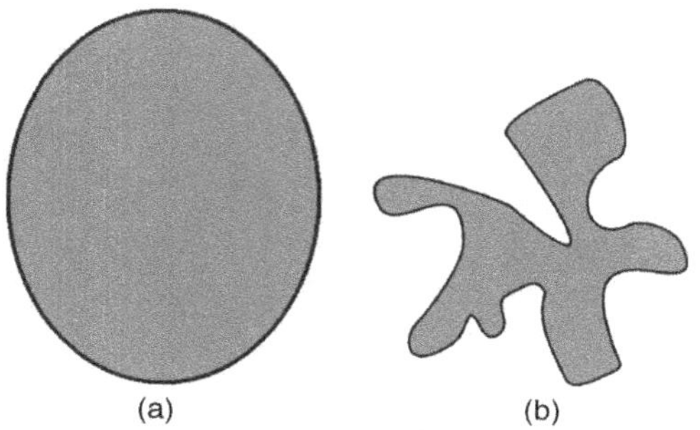

FIGURE 18.5 Examples of a compact (a) and a noncompact (b) regions.

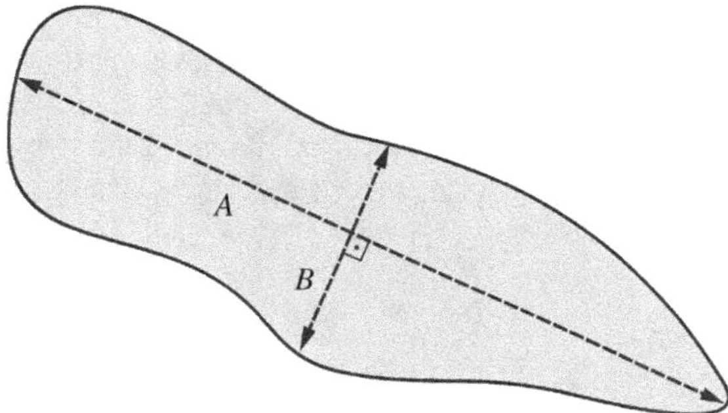

FIGURE 18.6 Eccentricity (A/B) of a region.

The thinness ratio is often used as a measure of roundness. Since the maximum value for T_i is 1 (which corresponds to a perfect circle), for a generic object the higher its thinness ratio, the more round it is. This figure of merit can also be used as a measure of regularity and its inverse, $1/T_i$ is sometimes called *irregularity* or *compactness* ratio. Figure 18.5 shows an example of a compact and a noncompact region.

18.3.8 Eccentricity

The eccentricity of an object is defined as the ratio of the major and minor axes of the object (Figure 18.6).

18.3.9 Aspect Ratio

The aspect ratio (AR) is a measure of the relationship between the dimensions of the bounding box of an object. It is given by

$$\mathrm{AR} = \frac{x_{\max} - x_{\min} + 1}{y_{\max} - y_{\min} + 1} \tag{18.13}$$

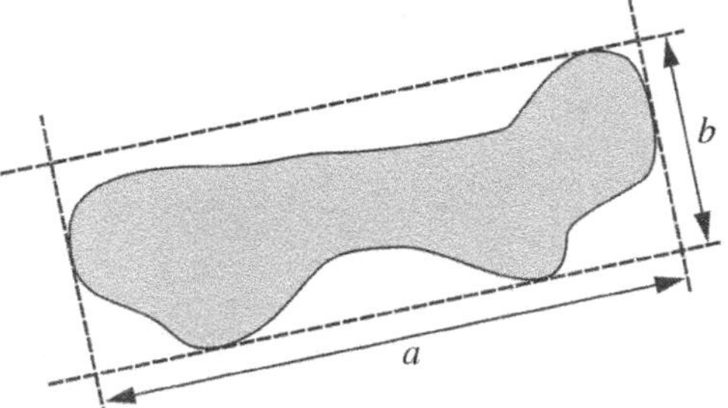

FIGURE 18.7 Elongatedness (a/b) of a region.

where $(x_{\min}, y_{\min})$ and $(x_{\max}, y_{\max})$ are the coordinates of the top left and bottom right corners of the bounding box surrounding an object, respectively.

It should be noted that the aspect ratio is not rotation invariant and cannot be used to compare rotated objects against one another unless it is normalized somehow, for example, by computing the AR after aligning the axis of least second moment to the horizontal direction. Such normalized representations of the aspect ratio are also referred to as the *elongatedness* of the object (Figure 18.7).

18.3.10 Moments

The 2D *moment of order $(p + q)$* of a digital image $f(x, y)$ is defined as

$$m_{pq} = \sum_{x=0}^{M-1} \sum_{y=0}^{N-1} x^p y^q f(x, y) \tag{18.14}$$

where M and N are the image height and width, respectively, and p and q are positive nonzero integers.

Central moments are the translation-invariant equivalent of moments. They are defined as

$$\mu_{pq} = \sum_{x=0}^{M-1} \sum_{y=0}^{N-1} (x - \bar{x})^p (y - \bar{y})^q f(x, y) \tag{18.15}$$

where

$$\bar{x} = \frac{m_{10}}{m_{00}} \quad \text{and} \quad \bar{y} = \frac{m_{01}}{m_{00}} \tag{18.16}$$

The *normalized central moments* are defined as

$$\eta_{pq} = \frac{\mu_{pq}}{\mu_{00}^{\gamma}} \tag{18.17}$$

TABLE 18.1 RST-Invariant Moments

$$\phi_1 = \eta_{20} + \eta_{02}$$
$$\phi_2 = (\eta_{20} - \eta_{02})^2 + 4\eta_{11}^2$$
$$\phi_3 = (\eta_{30} - 3\eta_{12})^2 + (3\eta_{21} - \eta_{03})^2$$
$$\phi_4 = (\eta_{30} + \eta_{12})^2 + (\eta_{21} + \eta_{03})^2$$
$$\phi_5 = (\eta_{30} - 3\eta_{12})(\eta_{30} + \eta_{12})\left[(\eta_{30} + \eta_{12})^2 - 3(\eta_{21} + \eta_{03})^2\right] + (3\eta_{21} - \eta_{03})(\eta_{21} + \eta_{03})\left[3(\eta_{30} + \eta_{12})^2 - (\eta_{21} + \eta_{03})^2\right]$$
$$\phi_6 = (\eta_{30} - \eta_{02})\left[(\eta_{30} + \eta_{12})^2 - (\eta_{21} + \eta_{03})^2\right] + 4\eta_{11}(\eta_{30} + \eta_{12})(\eta_{21} + \eta_{03})$$
$$\phi_7 = (3\eta_{21} - \eta_{03})(\eta_{30} + \eta_{12})\left[(\eta_{30} + \eta_{12})^2 - 3(\eta_{21} + \eta_{03})^2\right] - (\eta_{30} - 3\eta_{12})(\eta_{21} + \eta_{03})\left[3(\eta_{30} + \eta_{12})^2 - (\eta_{21} + \eta_{03})^2\right]$$

where

$$\gamma = \frac{p+q}{2} + 1 \tag{18.18}$$

for $(p+q) > 1$.

A set of seven *RST-invariant moments*, ϕ_1–ϕ_7, originally proposed by Hu in 1962 [Hu62], can be derived from second- and third-order normalized central moments. They are listed in Table 18.1.

In MATLAB

The IPT function `regionprops` measures a set of properties for each labeled region `L`. One or more properties from Table 18.2 can be specified as parameters. You will explore `regionprops` in Tutorial 18.1.

18.4 BOUNDARY DESCRIPTORS

The descriptors described in Section 18.3 are collectively referred to as *region based*. In this section, we will look at *contour-based* representation and description techniques. These techniques assume that the contour (or *boundary*) of an object can be represented in a convenient coordinate system (Cartesian—the most common, polar, or tangential) and rely exclusively on boundary pixels to describe the region or object. Object boundaries can be represented by different techniques, ranging from simple polygonal approximation methods to more elaborated techniques involving piecewise polynomial interpolations such as B-spline curves.

The techniques described in this section assume that the pixels belonging to the boundary of the object (or region) can be traced, starting from any background pixel, using an algorithm known as *bug tracing* that works as follows: as soon as the conceptual bug crosses into a boundary pixel, it makes a left turn and moves to the next pixel; if that pixel is a boundary pixel, the bug makes another left turn, otherwise it turns right; the process is repeated until the bug is back to the starting point. As the

TABLE 18.2 Properties of Labeled Regions

Property	Description
'Area'	The actual number of pixels in the region
'BoundingBox'	The smallest rectangle containing the region, specified by the coordinates of the top left corner and the length along each dimension
'Centroid'	A vector that specifies the center of mass of the region
'ConvexHull'	The smallest convex polygon that can contain the region
'ConvexImage'	Binary image that specifies the convex hull, that is, the image for which with all pixels within the hull are set to *true*
'ConvexArea'	The number of pixels in 'ConvexImage'
'Eccentricity'	The eccentricity of the ellipse that has the same second moments as the region. The eccentricity is a value in the [0,1] range (0 for a circle, 1 for a line segment). It is defined as the ratio of the distance between the foci of the ellipse and its major axis length
'EquivDiameter'	The diameter of a circle with the same area as the region, calculated as $\sqrt{(4 * \text{Area}/\pi)}$
'EulerNumber'	The difference between the number of objects in the region and the number of holes in these objects
'Extent'	The proportion of the pixels in the bounding box that are also in the region, that is, the ratio between the area of region and the area of the bounding box
'Extrema'	An 8×2 matrix containing the coordinates of the extrema points in the region
'FilledArea'	The number of pixels in FilledImage
'FilledImage'	A binary image of the same size as the bounding box of the region, where all *true* pixels correspond to the region and all holes have been filled in
'Image'	A binary image of the same size as the bounding box of the region, where all *true* pixels correspond to the region, and all other pixels are set to *false*
'MajorAxisLength'	The length (in pixels) of the major axis of the ellipse that has the same normalized second central moments as the region
'MaxIntensity'	The value of the pixel with the greatest intensity in the region
'MeanIntensity'	The mean of all the intensity values in the region
'MinIntensity'	The value of the pixel with the lowest intensity in the region
'MinorAxisLength'	The length (in pixels) of the minor axis of the ellipse that has the same normalized second central moments as the region

(*Continued*)

TABLE 18.2 *(Continued)*

Property	Description
`'Orientation'`	The angle (in degrees ranging from -90° to 90°) between the x-axis and the major axis of the ellipse that has the same second-moments as the region
`'Perimeter'`	Vector containing the perimeter of each region in the image
`'PixelIdxList'`	Vector containing the linear indices of the pixels in the region
`'PixelList'`	Matrix specifying the locations of pixels in the region
`'PixelValues'`	Vector containing the values of all pixels in a region
`'Solidity'`	The proportion of the pixels in the convex hull that are also in the region. Computed as `Area/ConvexArea`
`'SubarrayIdx'`	Vector of linear indices of nonzero elements returned in `Image`
`'WeightedCentroid'`	Vector of coordinates specifying the center of the region based on location and intensity value

conceptual bug follows the contour, it builds a list of coordinates of the boundary pixels being visited.

In MATLAB

The IPT function `bwtraceboundary` traces objects in a binary image. The coordinates of the starting point and the initial search direction (N, NE, E, SE, S, SW, W, or NW) must be passed as parameters. Optionally, the type of connectivity (4 or 8) and the direction of the tracing (clockwise or counterclockwise) can also be passed as arguments. The `bwtraceboundary` function returns an array containing the row and column coordinates of the boundary pixels.

■ EXAMPLE 18.2

Figure 18.8a shows an example of using `bwtraceboundary` to trace the first 50 pixels of the boundary of the leftmost object in the input image, starting from the top left corner and heading south, searching in counterclockwise direction. Figure 18.8b repeats the process, this time tracing the first 40 pixels of the boundary of *all* objects in the image, heading east, and searching in clockwise direction. The boundaries are shown in red.

In MATLAB

The IPT function `bwboundaries` traces region boundaries in binary image. It can trace the exterior boundaries of objects, as well as boundaries of holes inside these

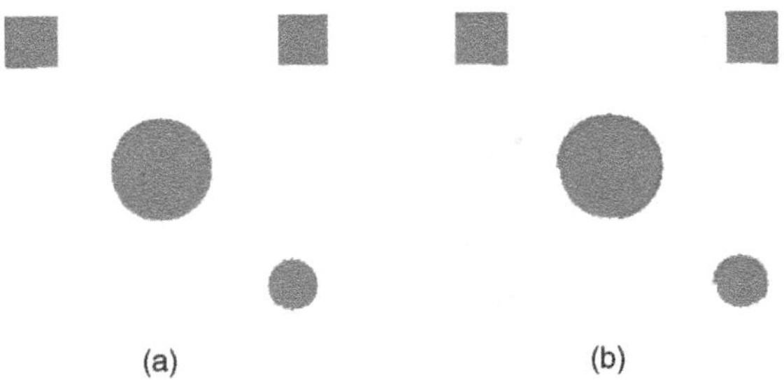

FIGURE 18.8 Tracing boundaries of objects.

objects. It can also trace the boundaries of children of parent objects in an image. The `bwboundaries` function returns a cell array (`B`) in which each cell contains the row and column coordinates of a boundary pixel. Optionally, it also returns a label matrix `L` where each object and hole is assigned a unique label, the total number of objects found (`N`), and an adjacency matrix used to represent parent–child–hole dependencies.

■ EXAMPLE 18.3

Figure 18.9a shows an example of using `bwboundaries` to trace the objects (represented in red) and holes (represented in green) in the input image using the syntax `[B,L,N] = bwboundaries(BW);`. Figure 18.9b repeats the process, this time using 4-connectivity and searching only for object boundaries, that is, no holes, in the input image using the syntax `[B,L,N] = bwboundaries(BW, 4, 'noholes');`.

18.4.1 Chain Code, Freeman Code, and Shape Number

Chain codes are alternative methods for tracing and describing a contour. A chain code is a boundary representation technique by which a contour is represented as a sequence of straight line segments of specified length (usually 1) and direction. The simplest chain code mechanism, also known as *crack code*, consists of assigning a

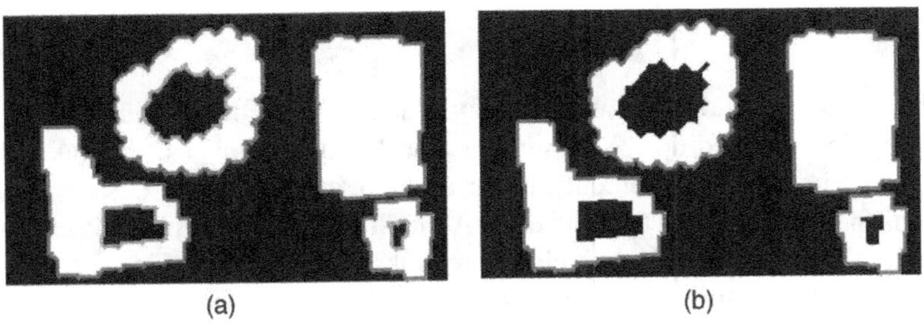

FIGURE 18.9 Tracing boundaries of objects and holes.

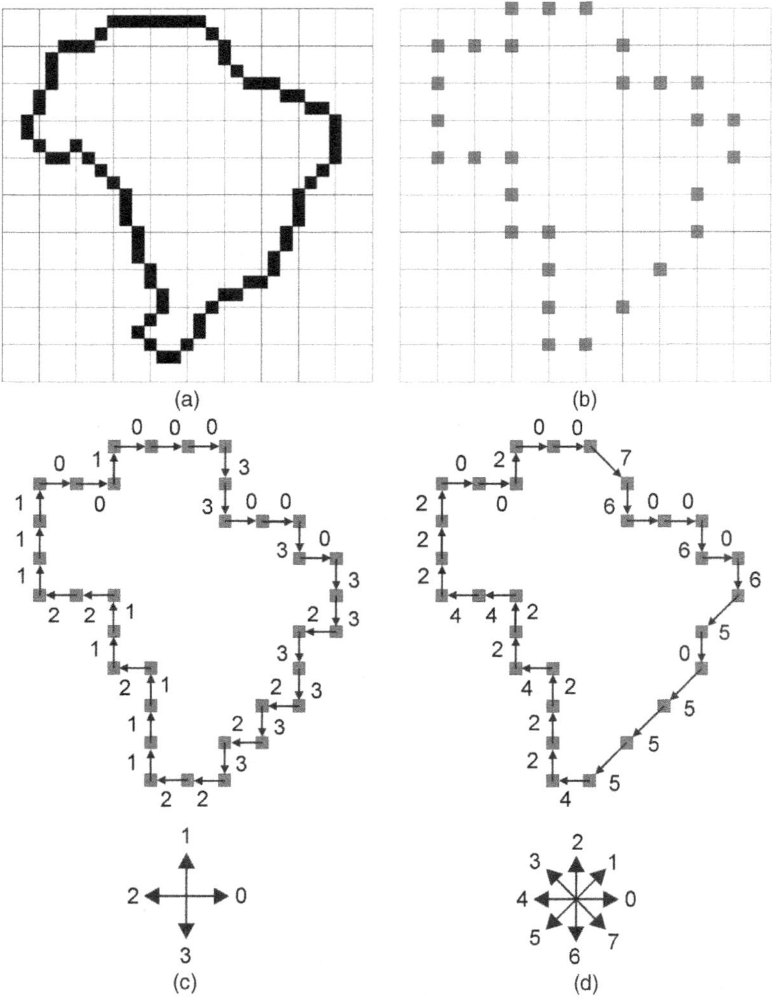

FIGURE 18.10 Chain code and Freeman code for a contour: (a) original contour; (b) subsampled version of the contour; (c) chain code representation; (d) Freeman code representation.

number to the direction followed by a bug tracking algorithm as follows: right (0), down (1), left (2), and up (3). Assuming that the total number of boundary points is p (the perimeter of the contour), the array C (of size p), where $C(p) = 0, 1, 2, 3$, contains the chain code of the boundary. A modified version of the basic chain code, known as the *Freeman code*, uses eight directions instead of four. Figure 18.10 shows an example of a contour, its chain code, and its Freeman code.

Once the chain code for a boundary has been computed, it is possible to convert the resulting array into a rotation-invariant equivalent, known as the *first difference*,

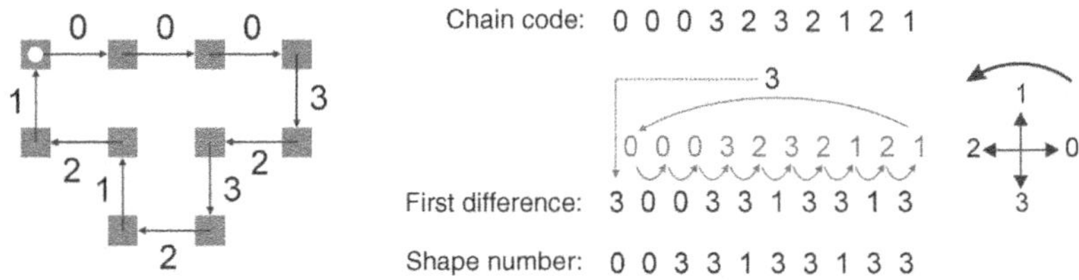

FIGURE 18.11 Chain code, first differences, and shape number.

which is obtained by encoding the number of direction changes, expressed in multiples of 90° (according to a predefined convention, for example, counterclockwise), between two consecutive elements of the Freeman code. The first difference of smallest magnitude—obtained by treating the resulting array as a circular array and rotating it cyclically until the resulting numerical pattern results in the smallest possible number—is known as the *shape number* of the contour. The shape number is rotation invariant and insensitive to the starting point used to compute the original sequence. Figure 18.11 shows an example of a contour, its chain code, first differences, and shape number.

In MATLAB

The IPT[5] does not have built-in functions for computing the chain code, Freeman code, and shape number of a contour. Refer to Chapter 12 of [McA04] or Chapter 11 of [GWE04] for alternative implementations.

18.4.2 Signatures

A signature is a 1D representation of a boundary, usually obtained by representing the boundary in a polar coordinate system and computing the distance r between each pixel along the boundary and the centroid of the region, and the angle θ subtended between a straight line connecting the boundary pixel to the centroid and a horizontal reference (Figure 18.12, top). The resulting plot of all computed values for $0 \leq \theta \leq 2\pi$ (Figure 18.12, bottom) provides a concise representation of the boundary that is translation invariant can be made rotation invariant (if the same starting point is always selected), but is *not* scaling invariant.

Figure 18.13 illustrates the effects of noise on the signature of a contour.

In MATLAB

The IPT does not have a built-in function for computing the signature of a boundary. Refer to Chapter 11 of [GWE04] for an alternative implementation.

[5] At the time of this writing, the latest version of the IPT is Version 7.2 (R2011a).

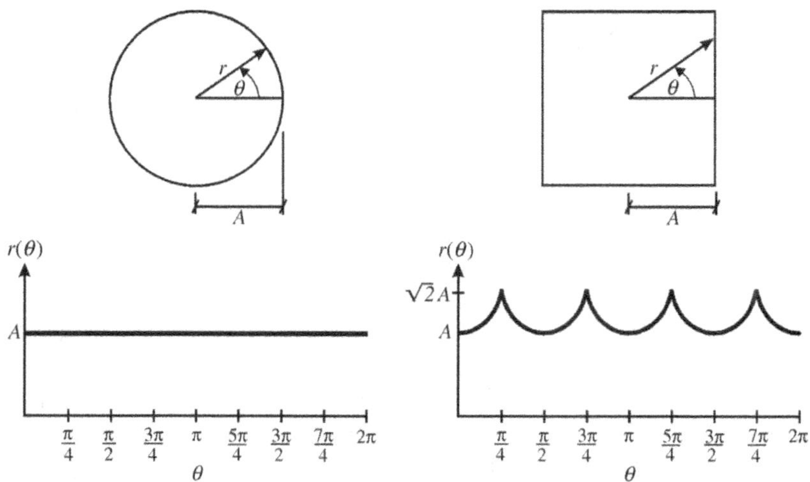

FIGURE 18.12 Distance × angle signatures for two different objects. Redrawn from [GW08].

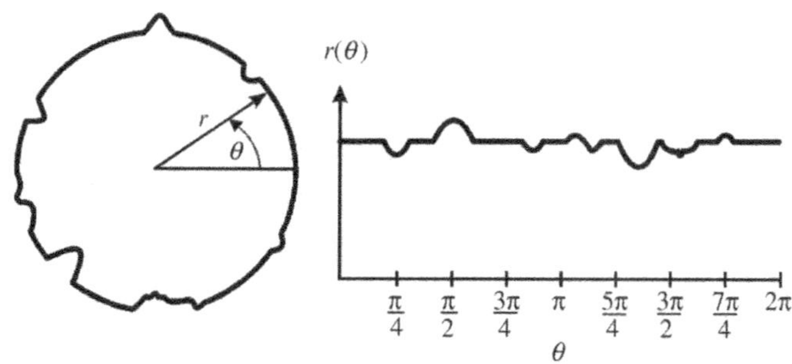

FIGURE 18.13 Effect of noise on signatures for two different objects. Redrawn from [GW08].

18.4.3 Fourier Descriptors

The idea behind Fourier descriptors is to traverse the pixels belonging to a boundary, starting from an arbitrary point, and record their coordinates. Each value in the resulting list of coordinate pairs $(x_0, y_0), (x_1, y_1), \ldots, (x_{K-1}, y_{K-1})$ is then interpreted as a complex number $x_k + jy_k$, for $k = 0, 1, \ldots, K - 1$. The discrete Fourier transform (DFT)[6] of this list of complex numbers is the *Fourier descriptor* of the boundary. The inverse DFT restores the original boundary. Figure 18.14 shows a K-point digital boundary in the xy plane and the first two coordinate pairs, (x_0, y_0) and (x_1, y_1).

[6]The discrete Fourier transform was introduced in Chapter 11.

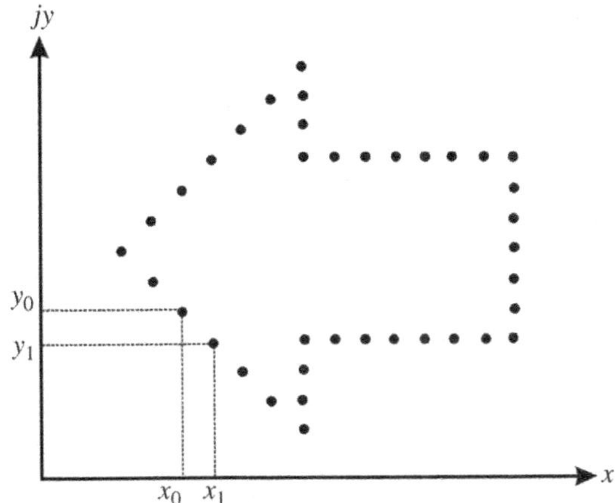

FIGURE 18.14 Fourier descriptor of a boundary.

In MATLAB

The IPT does not have a built-in function for computing the Fourier descriptor of a boundary, but it is not too difficult to create your own function. The basic procedure is to compute the boundaries of the region, convert the resulting coordinates into complex numbers (e.g., using the `complex` function), and apply the `fft` function to the resulting 1D array of complex numbers. Refer to Chapter 12 of [McA04] or Chapter 11 of [GWE04] for possible implementations.

One of the chief advantages of using Fourier descriptors is their ability to represent the essence of the corresponding boundary using very few coefficients. This property is directly related to the ability of the low-order coefficients of the DFT to preserve the main aspects of the boundary, while the high-order coefficients encode the fine details.

■ EXAMPLE 18.4

Figure 18.15 shows an example of a boundary containing 1467 points (part (a)) and the results of reconstructing it (applying the DFT followed by IDFT) using a variable number of terms. Part (b) shows the results of reconstructing with the same number of points for the sake of verification. Parts (c)–(f) show the results for reconstruction with progressively fewer points: 734, 366, 36, and 15, respectively. The chosen values for parts (c)–(f) correspond to approximately 50%, 25%, 2.5%, and 1% of the total number of points, respectively. A close inspection of the results shows that the image reconstructed using 25% of the points is virtually identical to the one obtained with 50% or 100% of the points. The boundary reconstructed using 2.5% of the points still preserves much of its overall shape, and the results using only 1% of the total number of points can be deemed unacceptable.

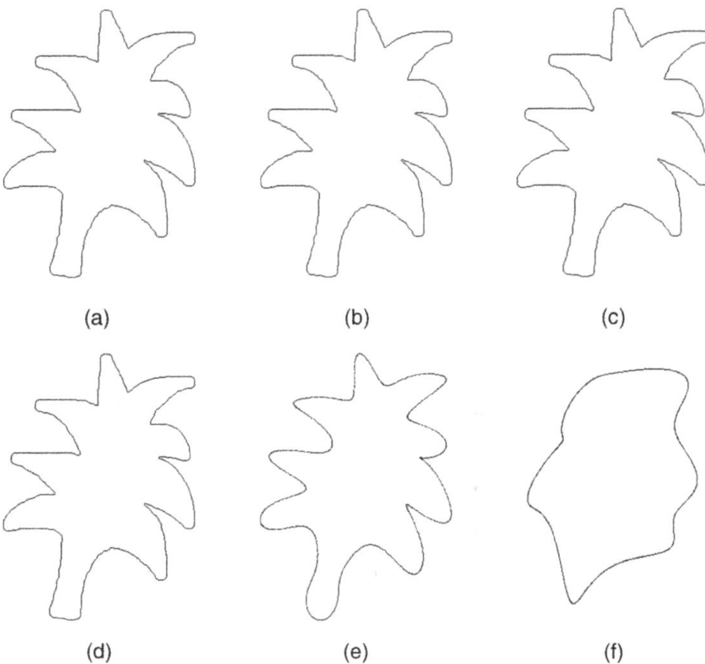

FIGURE 18.15 Example of boundary reconstruction using Fourier descriptors: (a) original image; (b–f) reconstructed image using 100%, 50%, 25%, 2.5%, and 1% of the total number of points, respectively.

18.5 HISTOGRAM-BASED (STATISTICAL) FEATURES

Histograms provide a concise and useful representation of the intensity levels in a grayscale image. In Chapter 9, we learned how to compute and display an image's histogram and we used histogram-based techniques for image enhancement. In Chapter 16, we extended that discussion to color images. In this section, we are interested in using histograms (or numerical descriptors that can be derived from them) as features[7] that describe an image (or its objects).

The simplest histogram-based descriptor is the *mean* gray value of an image, representing its average intensity m and given by

$$m = \sum_{j=0}^{L-1} r_j p(r_j) \tag{18.19}$$

where r_j is the jth gray level (out of a total of L possible values), whose probability of occurrence is $p(r_j)$.

[7]Histogram-based features are also referred to as *amplitude features* in the literature.

The mean gray value can also be computed directly from the pixel values from the original image $f(x, y)$ of size $M \times N$ as follows:

$$m = \frac{1}{MN} \sum_{x=0}^{M-1} \sum_{y=0}^{N-1} f(x, y) \tag{18.20}$$

The mean is a very compact descriptor (one floating-point value per image or object) that provides a measure of the overall brightness of the corresponding image or object. It is also RST invariant. On the negative side, it has very limited expressiveness and discriminative power.

The *standard deviation* σ of an image is given by

$$\sigma = \sqrt{\sum_{j=0}^{L-1} (r_j - m)^2 p(r_j)} \tag{18.21}$$

where m is given by equation (18.19) or equation (18.20).

The square of the standard deviation is the *variance*, which is also known as the *normalized second-order moment* of the image.

The standard deviation provides a concise representation of the overall contrast. Similar to the mean, it is compact and RST invariant, but has limited expressiveness and discriminative power.

The *skew* of a histogram is a measure of its asymmetry about the mean level. It is defined as

$$\text{skew} = \frac{1}{\sigma^3} \sum_{j=0}^{L-1} (r_j - m)^3 p(r_j) \tag{18.22}$$

where σ is the standard deviation given by equation (18.21).

The sign of the skew indicates whether the histogram's tail spreads to the right (positive skew) or to the left (negative skew). The skew is also known as the *normalized third-order moment* of the image.

If the image's mean value (m), standard deviation (σ), and *mode*—defined as the histogram's highest peak—are known, the skew can be calculated as follows:

$$\text{skew} = \frac{m - \text{mode}}{\sigma} \tag{18.23}$$

The *energy* descriptor provides another measure of how the pixel values are distributed along the gray-level range: images with a single constant value have maximum energy (i.e., energy $= 1$); images with few gray levels will have higher energy than the ones with many gray levels. The energy descriptor can be calculated as

$$\text{energy} = \sum_{j=0}^{L-1} [p(r_j)]^2 \tag{18.24}$$

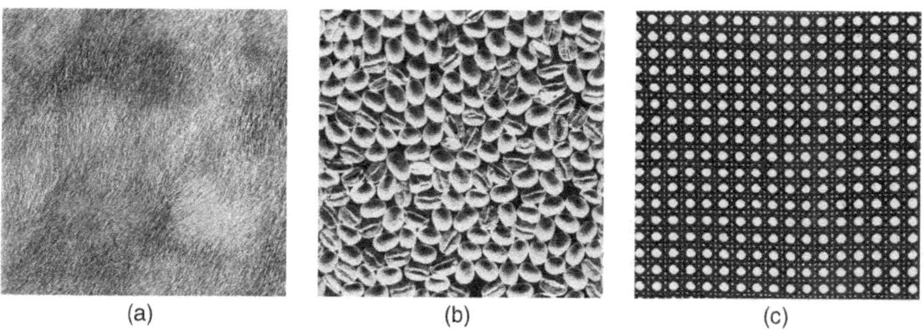

FIGURE 18.16 Example of images with smooth (a), coarse (b), and regular (c) texture. Images from the Brodatz textures data set. Courtesy of http://tinyurl.com/brodatz.

Histograms also provide information about the complexity of the image, in the form of *entropy* descriptor. The higher the entropy, the more complex the image.[8] Entropy and energy tend to vary inversely with one another. The mathematical formulation for entropy is

$$\text{entropy} = -\sum_{j=0}^{L-1} p(r_j)\log_2[p(r_j)] \tag{18.25}$$

Histogram-based features and their variants are usually employed as texture descriptors, as we shall see in Section 18.6.

18.6 TEXTURE FEATURES

Texture can be a powerful descriptor of an image (or one of its regions). Although there is not a universally agreed upon definition of texture, image processing techniques usually associate the notion of texture with image (or region) properties such as *smoothness* (or its opposite, *roughness*), *coarseness*, and *regularity*. Figure 18.16 shows one example of each and Figure 18.17 shows their histograms.

There are three main approaches to describe texture properties in image processing: structural, spectral, and statistical. In this book, we will focus on the statistical approaches, due to their popularity, usefulness, and ease of computing.[9]

One of the simplest set of statistical features for texture description consists of the following histogram-based descriptors of the image (or region): mean, variance (or its square root, the standard deviation), skew, energy (used as a measure of *uniformity*),

[8]This information also has implications for image coding and compression schemes (Chapter 17): the amount of coding redundancy that can be exploited by such schemes is inversely proportional to the entropy.

[9]Refer to "Learn More About It" section at the end of the chapter for references on structural and spectral methods for describing texture properties.

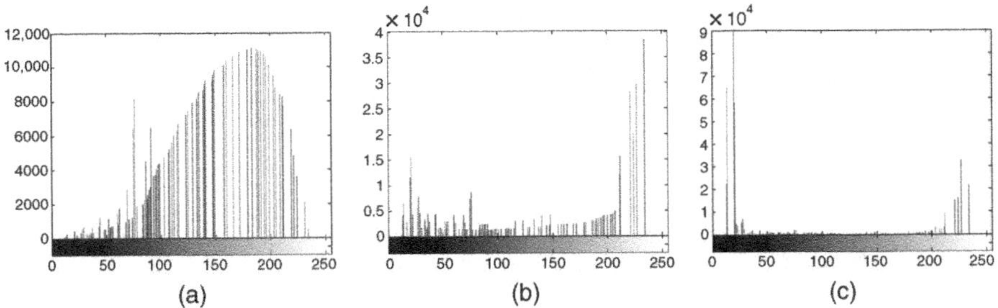

FIGURE 18.17 Histograms of images in Figure 18.16.

and entropy, all of which were introduced in Section 18.5. The variance is sometimes used as a normalized descriptor of roughness (R), defined as

$$R = 1 - \frac{1}{1 + \sigma^2} \tag{18.26}$$

where σ^2 is the normalized (to a [0, 1] interval) variance. $R = 0$ for areas of constant intensity, that is, smooth texture.

■ EXAMPLE 18.5

Table 18.3 shows the values of representative statistical texture descriptors for the three images in Figure 18.16, whose histograms are shown in Figure 18.17.

As expected, the regular texture has the highest uniformity (and lowest entropy) of all three. Moreover, the coarse texture shows a higher value for normalized roughness than the smooth texture, which is also consistent with our expectations.

Histogram-based texture descriptors are limited by the fact that the histogram does not carry any information about the spatial relationships among pixels. One way to circumvent this limitation consists in using an alternative representation for the pixel values that encodes their relative position with respect to one another. One such representation is the *gray-level cooccurrence matrix* **G**, defined as a matrix whose element $g(i, j)$ represents the number of times that pixel pairs with intensities z_i and z_j occur in image $f(x, y)$ in the position specified by an operator **d**. The vector **d** is known as *displacement vector*:

$$\mathbf{d} = (d_x, d_y) \tag{18.27}$$

TABLE 18.3 Statistical Texture Descriptors for the Three Images in Figure 18.16

Texture	Mean	Standard deviation	Roughness R	Skew	Uniformity	Entropy
Smooth	147.1459	47.9172	0.0341	−0.4999	0.0190	5.9223
Coarse	138.8249	81.1479	0.0920	−1.9095	0.0306	5.8405
Regular	79.9275	89.7844	0.1103	10.0278	0.1100	4.1181

0	1	5	5	2	0
3	6	3	0	7	6
7	7	5	7	0	1
3	2	6	3	1	7
6	3	6	3	5	1
4	7	5	3	5	4

(a)

$i \downarrow$ / $j \rightarrow$	0	1	2	3	4	5	6	7
0	0	2	0	0	0	0	0	1
1	0	0	0	0	0	1	0	1
2	1	0	0	0	0	0	1	0
3	1	1	1	0	0	2	2	0
4	0	0	0	0	0	0	0	1
5	0	1	1	1	1	2	0	0
6	0	0	0	4	0	0	0	0
7	1	0	0	0	0	2	1	1

(b)

FIGURE 18.18 An image (a) and its cooccurrence matrix for $\mathbf{d} = (0, 1)$ (b).

where d_x and d_y are the displacements, in pixels, along the rows and columns, of the image, respectively.

Figure 18.18 shows an example of gray-level cooccurrence matrix for $\mathbf{d} = (0, 1)$. The array on the left is an image $F(x, y)$ of size 4×4 and $L = 8$ (L is the total number of gray levels). The array on the right is the gray-level cooccurrence matrix $\mathbf{G}$, using a convention $0 \leq i, j < L$. Each element in $\mathbf{G}$ corresponds to the number of occurrences of a pixel of gray-level i occurs to the left of a pixel of gray-level j. For example, since the value 6 appears to the left of the value 3 in the original image four times, the value of $g(6, 3)$ is equal to 4.

The contents of $\mathbf{G}$ clearly depend on the choice of $\mathbf{d}$. If we had chosen $\mathbf{d} = (1, 0)$ (which can be interpreted as "one pixel below"), the resulting gray-level cooccurrence matrix for the same image in Figure 18.18a would be the one in Figure 18.19.

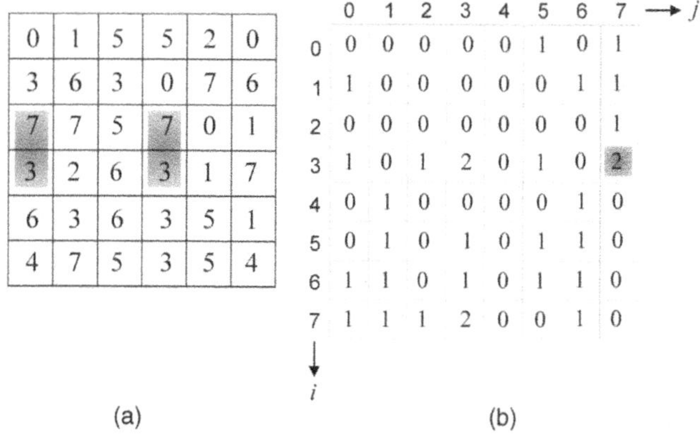

0	1	5	5	2	0
3	6	3	0	7	6
7	7	5	7	0	1
3	2	6	3	1	7
6	3	6	3	5	1
4	7	5	3	5	4

(a)

$i \downarrow$ / $j \rightarrow$	0	1	2	3	4	5	6	7
0	0	0	0	0	0	1	0	1
1	1	0	0	0	0	0	1	1
2	0	0	0	0	0	0	0	1
3	1	0	1	2	0	1	0	2
4	0	1	0	0	0	0	1	0
5	0	1	0	1	0	1	1	0
6	1	1	0	1	0	1	1	0
7	1	1	1	2	0	0	1	0

(b)

FIGURE 18.19 An image (a) and its cooccurrence matrix for $\mathbf{d} = (1, 0)$ (b).

The gray-level cooccurrence matrix can be normalized as follows:

$$N_g(i, j) = \frac{g(i, j)}{\sum_i \sum_j g(i, j)} \tag{18.28}$$

where $N_g(i, j)$ is the normalized gray-level cooccurrence matrix. Since all values of $N_g(i, j)$ lie between 0 and 1, they can be thought of as the probability that a pair of points satisfying **d** will have values (z_i, z_j).

Cooccurrence matrices can be used to represent the texture properties of an image. Instead of using the entire matrix, more compact descriptors are preferred. These are the most popular texture-based features that can be computed from a normalized gray-level cooccurrence matrix $N_g(i, j)$:

$$\text{Maximum probability} = \max_{i,j} N_g(i, j) \tag{18.29}$$

$$\text{Energy} = \sum_i \sum_j N_g^2(i, j) \tag{18.30}$$

$$\text{Entropy} = -\sum_i \sum_j N_g(i, j) \log_2 N_g(i, j) \tag{18.31}$$

$$\text{Contrast} = \sum_i \sum_j (i - j)^2 N_g(i, j) \tag{18.32}$$

$$\text{Homogeneity} = \sum_i \sum_j \frac{N_g(i, j)}{1 + |i - j|} \tag{18.33}$$

$$\text{Correlation} = \frac{\sum_i \sum_j (i-\mu_i)(j-\mu_j) N_g(i, j)}{\sigma_i \sigma_j} \tag{18.34}$$

where μ_i, μ_j are the means and σ_i, σ_j are the standard deviations of the row and column sums $N_g(i)$ and $N_g(j)$, defined as

$$N_g(i) = \sum_i N_g(i, j) \tag{18.35}$$

$$N_g(j) = \sum_j N_g(i, j) \tag{18.36}$$

■ EXAMPLE 18.6

Table 18.4 shows the values of representative statistical texture descriptors for the three images in Figure 18.16, whose histograms are shown in Figure 18.17. These descriptors were obtained by computing the normalized gray-level cooccurrence matrices with $\mathbf{d} = (0, 1)$.

As expected, the regular texture has the highest uniformity, highest homogeneity, and lowest entropy of all three. Although distinguishing between the coarse and smooth textures using these descriptors is possible (e.g., the image with coarse texture shows significantly higher correlation, uniformity, and homogeneity than the image with smooth texture), it is not as intuitive.

TABLE 18.4 Statistical Texture Descriptors for the Three Images in Figure 18.16

Texture	Max Probability	Correlation	Contrast	Uniformity (Energy)	Homogeneity	Entropy
Smooth	0.0013	0.5859	33.4779	0.0005	0.0982	7.9731
Coarse	0.0645	0.9420	14.5181	0.0088	0.3279	6.8345
Regular	0.1136	0.9267	13.1013	0.0380	0.5226	4.7150

18.7 TUTORIAL 18.1: FEATURE EXTRACTION AND REPRESENTATION

Goal

The goal of this tutorial is to learn how to use MATLAB to extract features from binary images and use these features to recognize objects within the image.

Objectives

- Learn how to use the `regionprops` function to extract features from binary objects.
- Learn how to perform feature selection and use the selected features to implement a simple, application-specific, heuristic classifier.

What You Will Need

- Test images `TPTest1.png`, `shapes23.png`, and `Test3.png`.

Procedure

1. Load test image `TPTest1.png` (Figure 18.20a) and display its contents.

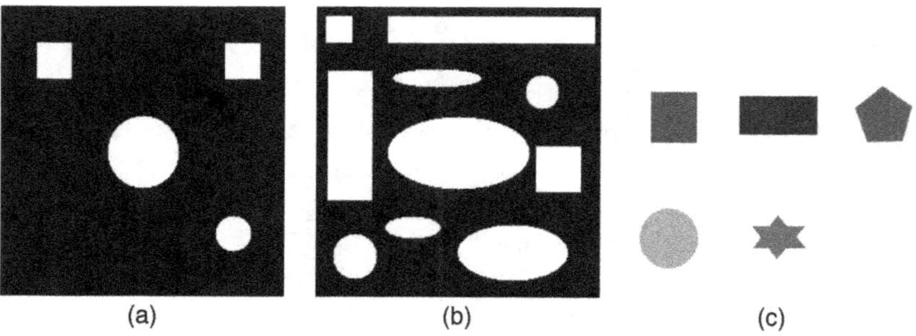

FIGURE 18.20 Test images for this tutorial: (a) steps 1–6; (b) step 7; (c) step 11.

```
J = imread('TPTest1.png');
imshow(J)
```

2. Use `bwboundaries` to display the boundaries of the objects in the test image.

```
[B,L] = bwboundaries(J);
figure; imshow(J); hold on;
for k=1:length(B),
    boundary = B{k};
    plot(boundary(:,2),boundary(:,1),'g','LineWidth',2);
end
```

3. Use `bwlabel` to label the connected regions (i.e., objects) in the test image, pseudocolor them, and display each of them with an associated numerical label.

```
[L, N] = bwlabel(J);
RGB = label2rgb(L,'hsv',[.5 .5 .5], 'shuffle');

figure; imshow(RGB); hold on;
for k=1:length(B),
    boundary = B{k};
    plot(boundary(:,2),boundary(:,1),'w','LineWidth',2);
    text(boundary(1,2)-11,boundary(1,1)+11,num2str(k),'Color','y',...
       'FontSize',14,'FontWeight','bold');
end
```

Question 1 What is the value of `N` returned by `bwlabel`? Does it make sense to you?

4. Use `regionprops` to extract the following binary features for each object in the image (top left square, top right square, small circle, big circle): area, centroid, orientation, Euler number, eccentricity, aspect ratio, perimeter, and thinness ratio.
5. Organize the feature values and object names in a table (see Table 18.5), for easier comparative analysis.

TABLE 18.5 Table for Feature Extraction Results

Object	Area	Centroid (row, col)	Orientation (degrees)	Euler number	Eccentricity	Aspect ratio	Perimeter	Thiness ratio
Top left square								
Big circle								
Small circle								
Top right square								

```
stats = regionprops(L,'all');
temp = zeros(1,N);
for k = 1:N
    % Compute thinness ratio
    temp(k) = 4*pi*stats(k,1).Area / (stats(k,1).Perimeter)^2;
    stats(k,1).ThinnessRatio = temp(k);

    % Compute aspect ratio
    temp(k) = (stats(k,1).BoundingBox(3))/(stats(k,1).BoundingBox(4));
    stats(k,1).AspectRatio = temp(k);
end
```

Question 2 Do the results obtained for the extracted features correspond to your expectations? Explain.

Question 3 Which of the extracted features have the **best** discriminative power to help tell squares from circles? Explain.

Question 4 Which of the extracted features have the **worst** discriminative power to help tell squares from circles? Explain.

Question 5 Which of the extracted features are ST invariant, that is, robust to changes in size and translation? Explain.

Question 6 If you had to use only one feature to distinguish squares from circles, in a ST-invariant way, which feature would you use? Why?

6. Plot the 2D feature vectors obtained using the area and thinness ratio of each object.

```
areas = zeros(1,N);
for k = 1:N
    areas(k) = stats(k).Area;
end

TR = zeros(1,N);
for k = 1:N
    TR(k) = stats(k).ThinnessRatio;
end

cmap = colormap(lines(16))
for k = 1:N
    scatter(areas(k), TR(k), [], cmap(k,:), 'filled'), ...
        ylabel('Thinness Ratio'), xlabel('Area')
    hold on
end
```

7. Repeat steps 1–6 for a different test image, `Test3.png` (Figure 18.20b).
8. Write MATLAB code to implement a heuristic three-class classifier capable of discriminating *squares* from *circles* from *unknown* shapes. *Hints*: Use a subset of features with enough discriminative power and encode your solution using `if-else-if` statements. Use the code snippet below to get started.[10]

```
name = cell(1,N);
for k = 1:N
    if (TR(k) > 0.9)
        name{1,k}='circle';
    else if (TR (k) > 0.8)
            name{1,k} = 'square';
        else
            name{1,k} = 'other';
        end
    end
end
```

9. Test your solution using the `TPTest1.png` and `Test3.png` test images.
10. Test your solution using different test images.
11. Extend your classifier to be able to process color images, for example, `shapes23.png` (Figure 18.20c).

WHAT HAVE WE LEARNED?

- Feature extraction is the process by which certain features of interest within an image are detected and represented for further processing. It is a critical step in most computer vision and image processing solutions because it marks the transition from pictorial to nonpictorial (alphanumerical, usually quantitative) data representation. From a pragmatic standpoint, having extracted meaningful features from an image enables the use of a vast array of pattern recognition and classification techniques to be applied to the resulting data.
- The types of features that can be extracted from an image depend on the type of image (e.g., binary, gray-level, or color), the level of granularity (entire image or individual regions) desired, and the context of the application. Several representative techniques for feature extraction using pixel intensity, texture, and relative positioning have been described in this chapter.
- Once the features have been extracted, they are usually represented in an alphanumerical way for further processing. The actual representation depends on the technique used: chain codes are normally treated as arrays of numbers,

[10]This solution, although naive and inelegant, works for both test images used so far. In Chapter 19, you will learn how to build better classifiers.

whereas quantitative features (e.g., measures of an object's area, perimeter, and number of holes) are usually encoded into a *feature vector*.

LEARN MORE ABOUT IT

- Polygonal approximation techniques are described in Chapter 11 of [GW08], Chapter 11 of [GWE04], Section 5.3 of [SOS00], and Section 8.2 of [SHB08].
- Section 11.1 of [GW08] and Section 17.6 of [Pra07] discuss boundary following algorithms in greater detail.
- Many structural, spectral, and statistical texture descriptors have been proposed in the literature and entire books have been written on the topic of texture analysis, among them are [MXS08], [Pet06], and [TT90].

18.8 PROBLEMS

18.1 Compute the first difference for the chain code: 0103212111021103.

18.2 Sketch the signature plots for the following geometrical figures:

(a) Rectangle

(b) Isosceles triangle

(c) Ellipse

(d) 6-point star

18.3 Write a MATLAB function to compute the gray-level cooccurrence matrix for an image $f(x, y)$ and a displacement vector **d**, which should be passed as parameters.

CHAPTER 19

VISUAL PATTERN RECOGNITION

WHAT WILL WE LEARN?

- What is visual pattern recognition and how does it relate to general pattern recognition?
- What are patterns and pattern classes?
- What is a pattern classifier?
- Which steps are normally needed to design, build, and test a visual pattern classifier?
- How can the performance of visual pattern classifiers be evaluated?

19.1 INTRODUCTION

This chapter presents the basic concepts of pattern recognition (also known as *pattern classification*) and introduces a few representative techniques used in computer vision. These techniques assume that an image has been acquired and processed and its contents have been represented using one or more of the techniques described in Chapter 18. The goal of pattern classification techniques is to assign a class to each image (or object within an image) based on a numerical representation of the image's (or object's) properties that is most suitable for the problem at hand.

Practical Image and Video Processing Using MATLAB®. By Oge Marques.

Pattern classification techniques are usually classified into two main groups: statistical and structural (or syntactic). In this chapter, we exclusively focus on statistical pattern recognition techniques, which assume that each object or class can be represented as a *feature vector* and make decisions on which class to assign to a certain pattern based on distance calculations or probabilistic models. Since the techniques presented in this chapter work with numerical feature vectors, regardless of the meaning of their contents (and their relationship to the original images), they can also be applied to many other classes of problems outside image processing and computer vision. Nonetheless, since our focus is on *visual* pattern recognition, all examples will refer to images and objects within images, keeping this discussion consistent with the "big picture" presented in Chapter 1 and the feature extraction methods introduced in Chapter 18.

19.2 FUNDAMENTALS

In this section, we introduce some of the most important concepts and terminologies associated with pattern classification. The common goal of pattern classification techniques is to assign a *class* to an unknown *pattern* based on previously acquired knowledge about objects and the classes to which they belong. Figure 19.1 shows a schematic diagram indicating how a statistical pattern classifier processes numerical information from the feature vectors, computes a series of distances or probabilities, and uses those results to make decisions regarding which class label $C(\mathbf{x})$ should be assigned to each input pattern $\mathbf{x}$.

19.2.1 Design and Implementation of a Visual Pattern Classifier

The design and implementation of a visual pattern recognition system is usually an interactive process that also involves the selection and computation of features

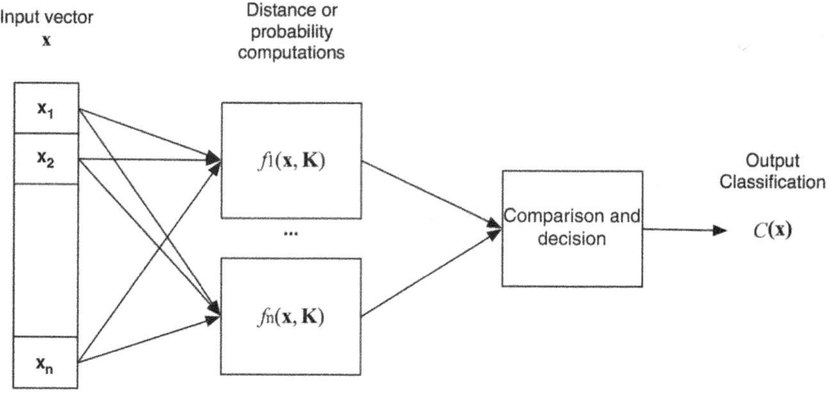

FIGURE 19.1 Diagram of a statistical pattern classifier. Redrawn from [SS01].

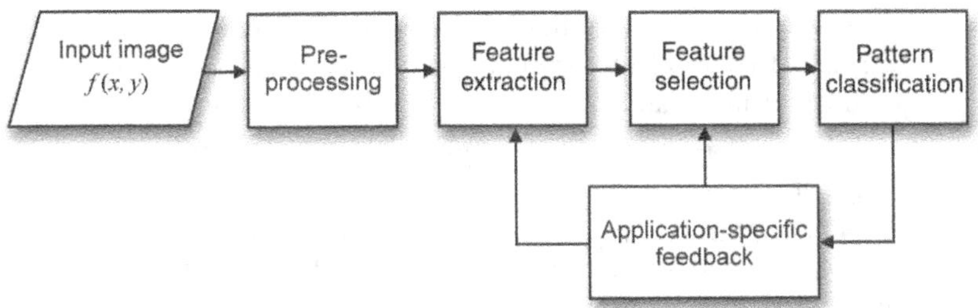

FIGURE 19.2 The interplay between feature extraction, feature selection, and pattern classification as a function of the application at hand. Adapted and redrawn from [Umb05].

from the images (or objects) that we want to classify. Moreover, as with most tasks in computer vision, the decisions are often application dependent. Figure 19.2 illustrates the process.

The design of a statistical visual pattern classifier usually consists of the following steps:

1. Define the problem and determine the number of classes involved.

   This is where it all begins. Legitimate questions to ask at this stage include the following: How many classes are there? How well do these classes describe the objects or images? Are certain classes subcategories of others? Is there a *reject class*[1] in this case?

2. Extract features that are most suitable to describe the images and allow the classifier to label them accordingly.

   This is the step where the interactive and application-dependent process of *preliminary* feature extraction and selection takes place. It is also the step where the designer of a machine vision solution is faced with many options of *types* of features (e.g., color based, texture based, boundary oriented, etc.) and the specific *methods* to extract them (e.g., color histograms, Tamura descriptors, Fourier descriptors, etc.). Refer to Chapter 18 for detailed information.

3. Select a classification method or algorithm.

   At this point, we can benefit from the vast array of tools and techniques in the field of data mining and machine learning and choose one that best suits our needs, based on their complexity, computational cost, training capabilities, and other properties. The selected technique can be as simple as a minimum distance classifier or as complex as a support vector machine (SVM).

4. Select a data set.

   At this stage, we collect representative images that can be used to train and test the solution. If the problem domain (e.g., object recognition) has associated

[1] A *reject class* is a generic class for objects that could not be successfully labeled as belonging to any of the other classes.

data sets that are publicly available and widely used (e.g., the Caltech 101 and Caltech 256 for object recognition), we should consider using them. Using standardized datasets allows for benchmarking of our solution against others.

5. Select a subset of images and use them to train the classifier.

   Many pattern classification strategies require a *training stage*, where a small subset of images is used to "teach" the classifier about the classes it should be able to recognize, as well as adjust some of the classifier's parameters.

6. Test the classifier.

   At this step, we measure success and error rates, compute relevant figures of merit (e.g., precision and recall), and compile the main numerical results into representative plots (e.g., ROC curves).

7. Refine and improve the solution.

   After having analyzed the results computed in step 6, we might need to go back to an earlier step, process a few changes (e.g., select different features, modify a classifier's parameters, collect additional images, etc.), and test the modified solution.

19.2.2 Patterns and Pattern Classes

A *pattern* can be defined as an arrangement of *descriptors* (or *features*). Patterns are usually encoded in the form of feature vectors, strings, or trees.[2]

A feature vector—as seen in Chapter 18—is an $n \times 1$ array of numbers corresponding to the descriptors (or features) used to represent a certain pattern:

$$\mathbf{x} = (x_1, x_2, \ldots, x_n)^{\mathrm{T}} \tag{19.1}$$

where n is the total number of features and T indicates the *transpose* operation.

The total number of features and their meaning will depend on the selected properties of the objects and the representation techniques used to describe them. This number can vary from one (if a certain property is discriminative enough to enable a classification decision) to several thousand (e.g., the number of points for a boundary encoded using Fourier descriptors, as seen in Section 18.4.3).

A *class* is a set of patterns that share some common properties. An ideal class is one in which its members are very similar to one another (i.e., the class has high intraclass similarity) and yet significantly different from members of other classes (i.e., interclass differences are significant). Pattern classes will be represented as follows: $\omega_1, \omega_2, \ldots, \omega_K$, where K is the total number of classes.

Figure 19.3 shows a simple example of a 2D plot of feature vectors representing the height and weight of a group of table tennis players and a group of sumo wrestlers:

$$\mathbf{x} = (x_1, x_2)^{\mathrm{T}} \tag{19.2}$$

where x_1 = weight and x_2 = height.

[2] Strings and trees are used in structural pattern classification methods and will not be discussed any further.

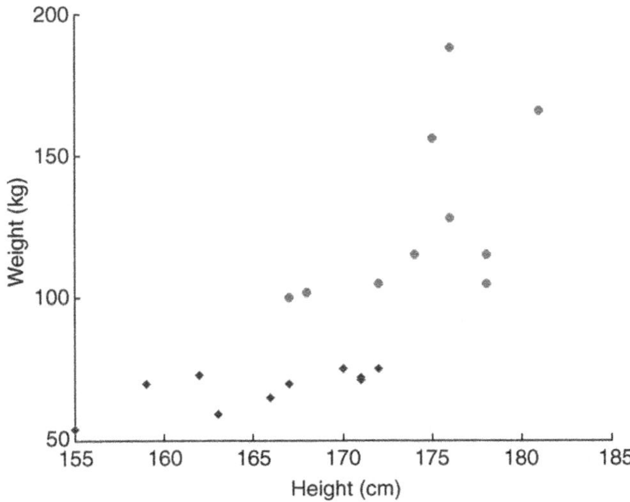

FIGURE 19.3 Example of two classes (*sumo wrestlers*—red circles—and *table tennis players*—blue diamonds) described by two measurements (*weight* and *height*).

In this example, ω_1 is the *sumo wrestlers* class, whereas ω_2 is the *table tennis players* class. A simple visual inspection allows us to see a clear separation between the two clusters of data points, which typically translates into a relatively easy task for a pattern classifier in charge of telling the two groups apart and classifying a new instance according to which group it most likely belongs to. Another observation that can be derived from the same figure is that the *weight* feature (x_1) is more discriminative than the *height* feature (x_2) in this particular problem.

19.2.3 Data Preprocessing

Before the numerical data (e.g., a collection of feature vectors) can be input to a pattern classifier, it is often necessary to perform an additional step known as *data preprocessing*. Common preprocessing techniques include the following:

- *Noise Removal* (also known as *Outlier Removal*): A preprocessing step where data samples that deviate too far from the average value for a class are removed, under the rationale that (a) there may have been a mistake while measuring (or extracting) that particular sample and (b) the sample is a poor example of the underlying structure of the class.
- *Normalization*: Feature vectors may need to be normalized before distance, similarity, and probability calculations take place. These are some representative normalization techniques [Umb05]:
  - *Unit Vector Normalization*: It enforces that all feature vectors have a magnitude of 1.

- *Standard Normal Density (SND)*: Here, each element of a feature vector $\mathbf{x}$ $(x_1, x_2, \ldots, x_n)$ is replaced by a normalized value $\tilde{x}_i$, given by

$$\tilde{x}_i = \frac{x_i - \mu}{\sigma} \tag{19.3}$$

  where μ and σ are the mean and the standard deviation of the elements in $\mathbf{x}$.
- *Other Linear and Nonlinear Techniques*: For example, *Min–Max Normalization* and *Softmax Scaling*: Their goal is to limit the feature values to a specific range, for example [0, 1].

• *Insertion of Missing Data*: In this (optional) last preprocessing step, additional data items—provided that they follow a similar probabilistic distribution and do not bias the results—are added to the data set.

19.2.4 Training and Test Sets

The process of development and testing of pattern classification algorithms usually requires that the data set be divided into two subgroups: the *training set*, used for algorithm development and fine-tuning, and the *test set*, used to evaluate the algorithm's performance. The training set contains a small (typically 20% or less) but representative subsample of the data set that can be selected manually or automatically (i.e., randomly). The size of the training set and the method used to build it are often dependent on the selected pattern classification technique. The goal of having two separate sets—one for designing, improving, and fine-tuning the algorithms, and the other for a systematic quantitative evaluation—is to avoid bias in reporting the success rates of the approach. After all, if the designer is allowed to work on the same data set the whole time, it is quite possible to tweak the solution enough to produce a nearly perfect performance on that particular collection of images. There would be no guarantee, however, that the same method and choice of parameters would work well for other images and data sets.

19.2.5 Confusion Matrix

The confusion matrix is a 2D array of size $K \times K$ (where K is the total number of classes) used to report raw results of classification experiments. The value in row i, column j indicates the number of times an object whose true class is i was labeled as belonging to class j. The main diagonal of the confusion matrix indicates the number of cases where the classifier was successful; a perfect classifier would show all off-diagonal elements equal to zero.

■ EXAMPLE 19.1

Figure 19.4 shows an example of a confusion matrix for four generic classes $\omega_1, \ldots,$ ω_4. A careful analysis of the confusion matrix shows that inputs labeled as class ω_3 were correctly classified all the time. Classification errors were highest (11%) for inputs labeled as class ω_2. The most common confusion incurred by the classifier was

	ω_1	ω_2	ω_3	ω_4
ω_1	97	0	2	1
ω_2	0	89	10	1
ω_3	0	0	100	0
ω_4	0	3	5	92

FIGURE 19.4 Example of 4 × 4 confusion matrix.

labeling an input of class ω_2 as class ω_3 (10% of the time). Moreover, the classifier's performance for class ω_1 is also worth commenting: although three inputs labeled as class ω_1 were incorrectly classified (two as class ω_3 and one as class ω_4), the classifier did not label any input from other classes as class ω_1 (i.e., the remaining values for the ω_1 column are all zeros).

19.2.6 System Errors

A quantitative performance analysis of pattern classifiers usually involves measuring *error rates*. Other measures, such as speed and computational complexity, may be important, of course, but error rates are essential. Error rates measure how many times a classifier incurred in a classification error, that is, classified an input object as class ω_p when the correct class is ω_q, $p \neq q$. The error rate of a classifier is usually determined empirically, that is, by measuring the number (percentage) of errors observed when testing the classifier against a test set.

■ **EXAMPLE 19.2**

Given the confusion matrix in Figure 19.4 and assuming that all classes had the same number of objects, the classifier's overall error rate would be

$$(3 + 11 + 0 + 8)/(4 \times 100) = 5.5\%.$$

19.2.7 Hit Rates, False Alarm Rates, and ROC Curves

Many visual pattern classification problems employ a two-class classifier. A classical computer vision example is the *object detection* task, where a computer vision algorithm is presented with an image and the question: "Is the object present in this image or not?" If the algorithm successfully answers *yes* (and points to where in the image the object is located) when the object is present, it is called a *true positive*. If the algorithm correctly answers *no* when the object is absent, it is called a *true negative*. There are two possible errors the algorithm can make: answering *yes* in the absence of an object (this is called a *false alarm* or *false positive*) or answering *no* when the object is present, that is, missing the object (this is called a *false negative*).

The cost of a false positive or a false negative is application dependent and can lead to quite different outcomes. In surveillance applications, for example, it will probably be best to have occasional false positives (e.g., alerting a human operator

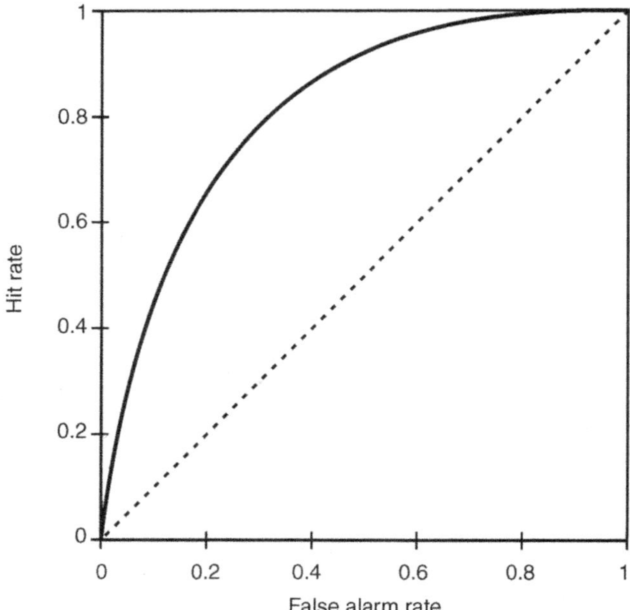

FIGURE 19.5 Example of ROC curve.

for the presence of a suspicious object on the screen where no such object exists) than miss real suspicious objects (or persons or events) altogether. In such case, a legitimate goal is to minimize the false negative rate, even if it can only be achieved by tolerating relatively high false positive rates.

The *receiver operating characteristic* (or simply *ROC*) curve is a plot that shows the relationship between the correct detection (true positive) rate (also known as *hit rate*) and the *false alarm* (false positive) *rate*. Figure 19.5 shows an example of generic ROC curve. It also shows a dashed straight line that corresponds to the performance of a classifier operating by chance (i.e., guessing *yes* or *no* at each time). The ideal ROC curve is one in which the "knee" of the curve is as close to the top left corner of the graph as possible, suggesting hit rate close to 100% with a false alarm rate close to zero.

19.2.8 Precision and Recall

Certain image processing applications, notably image retrieval, have the goal to retrieve *relevant* images while not retrieving *irrelevant* ones. The measures of performance used in image retrieval borrow from the field of (document) information retrieval and are based on two primary figures of merit: *precision* and *recall*. *Precision* is the number of relevant documents retrieved by the system divided by the total number of documents retrieved (i.e., true positives plus false alarms). *Recall* is the number of relevant documents retrieved by the system divided by the total number of relevant documents in the database (which should, therefore, have been retrieved).

Precision can be interpreted as a measure of exactness, whereas recall provides a measure of completeness. A perfect precision score of 1.0 means that every retrieved document (or image in our case) was relevant, but does not provide any insight as to whether all relevant documents were retrieved. A perfect recall score of 1.0 means that all relevant images were retrieved, but says nothing about how many irrelevant images might have also been retrieved.

Precision (P) and recall (R) measures can also be adapted to and used in classification tasks and expressed in terms of true positives (tp), false positives (fp), and false negatives (fn) as follows:

$$P = \frac{\text{tp}}{\text{tp} + \text{fp}} \tag{19.4}$$

and

$$R = \frac{\text{tp}}{\text{tp} + \text{fn}} \tag{19.5}$$

In this case, a precision score of 1.0 for a class ω_i means that every item labeled as belonging to class ω_i does indeed belong to class ω_i, but says nothing about the number of items from class ω_i that were not labeled correctly. A recall of 1.0 means that every item from class ω_i was labeled as belonging to class ω_i, but says nothing about how many other items were also incorrectly labeled as belonging to class ω_i.

■ EXAMPLE 19.3

Given the confusion matrix in Figure 19.4, the precision and recall per category can be calculated as follows:

$P_1 = 97/(97 + 0 + 0 + 0) = 100\%$
$P_2 = 89/(0 + 89 + 0 + 3) = 96.74\%$
$P_3 = 100/(2 + 10 + 100 + 5) = 85.47\%$
$P_4 = 92/(1 + 1 + 0 + 92) = 97.87\%$
$R_1 = 97/(97 + 0 + 2 + 1) = 97\%$
$R_2 = 89/(0 + 89 + 10 + 1) = 89\%$
$R_3 = 100/(0 + 0 + 100 + 0) = 100\%$
$R_4 = 92/(0 + 3 + 5 + 92) = 92\%$

In this case, the classifier shows perfect precision for class ω_1 and perfect recall for class ω_3.

Precision and recall are often interrelated and the cost of increasing one of them is an undesired decrease in the other. In the case of document (or image) retrieval systems, the choice of retrieving fewer documents boosts precision at the expense of a low recall, whereas the retrieval of too many documents improves recall at the expense of lower precision. This trade-off is often expressed in a plot, known as the *precision–recall* (or simply *PR*) *graph*. A PR graph is obtained by calculating the precision at various recall levels. The ideal PR graph shows perfect precision values at every recall level until the point where all relevant documents (and only those) have been retrieved; from this point on it falls monotonically until the point where recall reaches 1. Figure 19.6 shows an example of a generic PR graph.

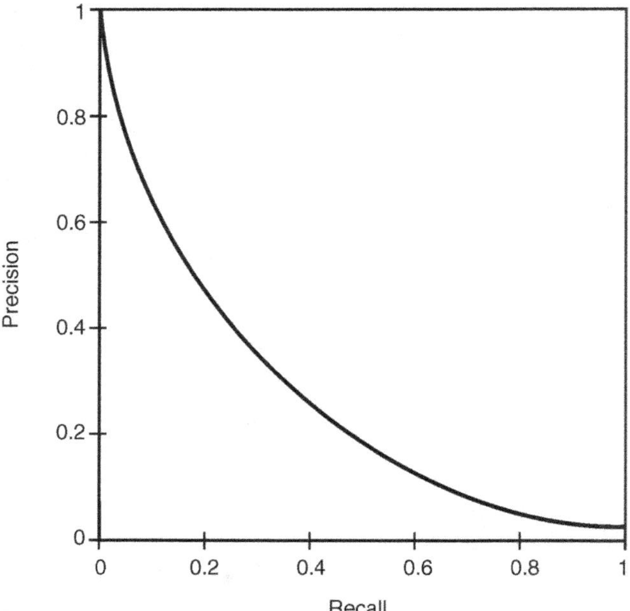

FIGURE 19.6 Example of precision–recall (PR) graph.

A more compact representation of the precision and recall properties of a system consists in combing those two values into a single figure of merit, such as the *F measure* (also known as *F1 measure*) that computes the weighted harmonic mean of precision and recall:

$$F1 = 2 \times \frac{\text{precision} \times \text{recall}}{\text{precision} + \text{recall}} \tag{19.6}$$

■ EXAMPLE 19.4

An image retrieval system produced the following 10 ranked results for a search operation against a database of 500 images, of which 5 are relevant to the query:

Rank	Result
1	*R*
2	*R*
3	*N*
4	*R*
5	*N*
6	*N*
7	*N*
8	*R*
9	*N*
10	*R*

where *R* means *relevant* and *N* means *not relevant*.

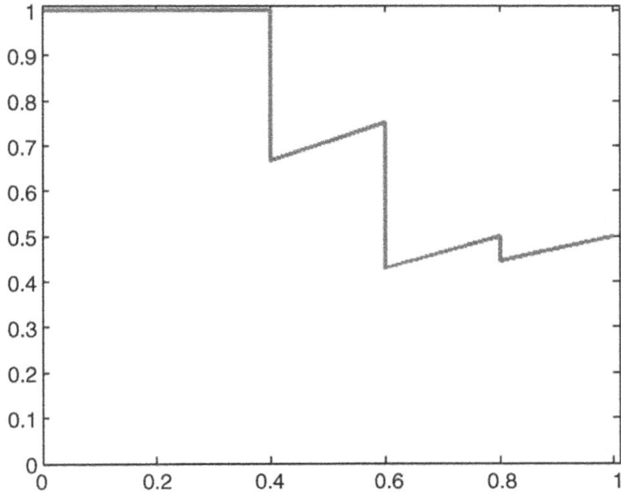

FIGURE 19.7 Precision–recall graph for Example 19.4.

If we calculate the precision at all recall levels, we will have the following results:

Recall	Precision
0.2	1.0000
0.4	1.0000
0.4	0.6667
0.6	0.7500
0.6	0.6000
0.6	0.5000
0.6	0.4286
0.8	0.5000
0.8	0.4444
1.0	0.5000

The corresponding PR graph is shown in Figure 19.7.

19.2.9 Distance and Similarity Measures

Two feature vectors can be compared with each other by calculating (i.e., measuring) the distance between them or, conversely, establishing their degree of similarity.

There are many distance measures in use in visual pattern classification. Given two feature vectors $\mathbf{a} = (a_1, a_2, \ldots, a_n)^T$ and $\mathbf{b} = (b_1, b_2, \ldots, b_n)^T$, the following are the equations for the most widely used distance measures:[3]

- Euclidean distance:

$$d_E = \sqrt{\sum_{i=1}^{n} (a_i - b_i)^2} \quad (19.7)$$

- Manhattan (or city block) distance:

$$d_M = \sum_{i=1}^{n} |a_i - b_i| \quad (19.8)$$

- Minkowski distance:

$$d_M = \left[\sum_{i=1}^{n} |a_i - b_i|^r \right]^{1/r} \quad (19.9)$$

  where r is a positive integer. Clearly, the Minkowski distances for $r = 1$ and $r = 2$ are the same as the Manhattan and Euclidean distances, respectively.

In MATLAB

Computing distances in MATLAB is straightforward, thanks to MATLAB's matrix handling abilities. For the Euclidean distance between two vectors `x` and `y`, we can use the `norm` function as an elegant alternative: `d_E = norm(x - y)`.

Although distance measures are *inversely* related to the notion (and measure) of similarity, there are other similarity measures in the literature, such as:

- The vector inner product:

$$\sum_{i=1}^{n} a_i b_i = a_1 b_1 + a_2 b_2 + \cdots + a_n b_n \quad (19.10)$$

- The Tanimoto metric, which establishes a percentage(%) of similarity and is expressed as

$$\frac{\sum_{i=1}^{n} a_i b_i}{\sum_{i=1}^{n} a_i^2 + \sum_{i=1}^{n} b_i^2 - \sum_{i=1}^{n} a_i b_i} \quad (19.11)$$

[3]Some of these equations may seem familiar to you, since they appeared in a slightly different format back in Chapter 2. However, in Chapter 2 we were measuring distances between pixels, whereas in this chapter we are measuring distances between feature vectors in a feature space.

19.3 STATISTICAL PATTERN CLASSIFICATION TECHNIQUES

In this section, we present the basics of three statistical pattern classification techniques: the minimum distance classifier, the k-nearest neighbors (KNN) classifier, and the maximum likelihood (or Bayesian) classifier. The goal of this section is to present a few alternatives for building a visual pattern recognition using the knowledge from the previous chapters, particularly the feature extraction and representation techniques learned in Chapter 18. In Tutorial 19.1, you will have an opportunity for putting this knowledge into practice.

We have learned so far that objects' properties can be represented using *feature vectors* that are projected onto a *feature space*. If the features used to represent the objects (and the classes to which they belong) are properly chosen, the resulting points in the n-dimensional feature space will be distributed in a way that correlates proximity in the feature space with similarity among the actual objects. In other words, feature vectors associated with objects from the same class will appear close together as *clusters* in the feature space.

The job of a statistical pattern classification technique is to find a discrimination curve (or a *hypersurface*, in the case of an n-dimensional feature space) that can tell the clusters (and the classes to which they correspond) apart. Figure 19.8 illustrates the concept for a three-class classifier in a 2D feature space.

A statistical pattern classifier has n inputs (the features of the object to be classified, encoded into feature vector $\mathbf{x} = (x_1, x_2, \ldots, x_n)^T$) and one output (the class to which

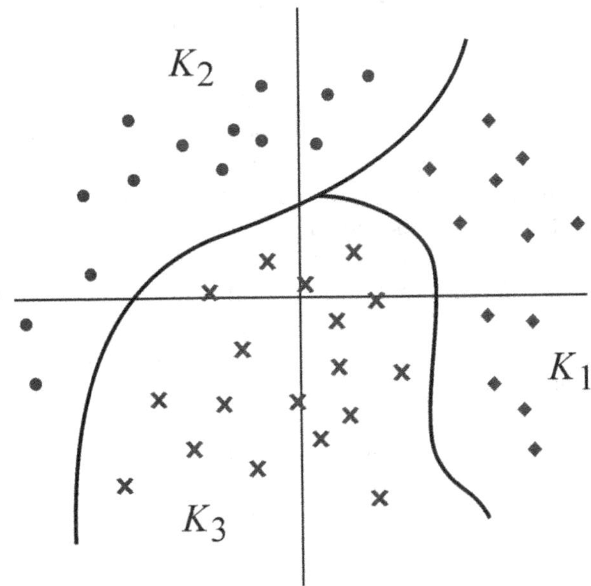

FIGURE 19.8 Discrimination functions for a three-class classifier in a 2D feature space.

the object belongs, $C(\mathbf{x})$, represented by one of the symbols $\omega_1, \omega_2, \ldots, \omega_W$, where W is the total number of classes). The symbols ω_w are called *class identifiers*.

Classifiers make comparisons among the representation of the unknown object and the known classes. These comparisons provide information to make decision about which class to assign to the unknown pattern. The decision of assigning an input pattern to class ω_i rather than another class ω_j is based on which side of the discrimination hypersurfaces among classes the unknown object sits on.

Mathematically, the job of the classifier is to apply a series of *decision rules* that divide the feature space into W disjoint subsets K_w, $w = 1, 2, \ldots, W$, each of which includes the feature vectors $\mathbf{x}$ for which $d(\mathbf{x}) = \omega_w$, where $d(\cdot)$ is the decision rule.

19.3.1 Minimum Distance Classifier

The minimum distance classifier (also known as the *nearest-class mean* classifier) works by computing a distance metric between an unknown feature vector and the centroids (i.e., mean vectors) of each class:

$$d_j(\mathbf{x}) = \|\mathbf{x} - \mathbf{m}_j\| \tag{19.12}$$

where d_j is a distance metric (between class j and the unknown feature vector $\mathbf{x}$) and $\mathbf{m}_j$ is the mean vector for class j, defined as

$$\mathbf{m}_j = \frac{1}{N_j} \sum_{\mathbf{x} \in \omega_j} \mathbf{x}_j \tag{19.13}$$

where N_j is the number of pattern vectors from class ω_j.

Figure 19.9 shows an example of two classes and their mean vectors.

The minimum distance classifier works well for few compact classes, but cannot handle more complex cases, such as the one depicted in Figure 19.10. In this case, there are two problems worth mentioning: (i) class A (clusters K_1 and K_4) is multimodal, that is, its samples lie in two disjoint (although they are compact) clusters, and the mean vector lies in a point in the feature space that is actually *outside* both clusters; (ii) classes B (cluster K_2) and C (cluster K_3) have quite irregular shapes that would potentially lead to different classification decisions for two unknown patterns situated at the same distance from their mean vectors. The latter problem can be alleviated by using a modified distance metric, known as *scaled Euclidean distance* (equation (19.14)), whereas the former problem usually calls for more complex classifier schemes.

$$\tilde{d}_E = \|\mathbf{x} - \mathbf{x}_j\| = \sqrt{\sum_{i=1}^{n} [(x[i] - x_j[i])/\sigma_i]^2} \tag{19.14}$$

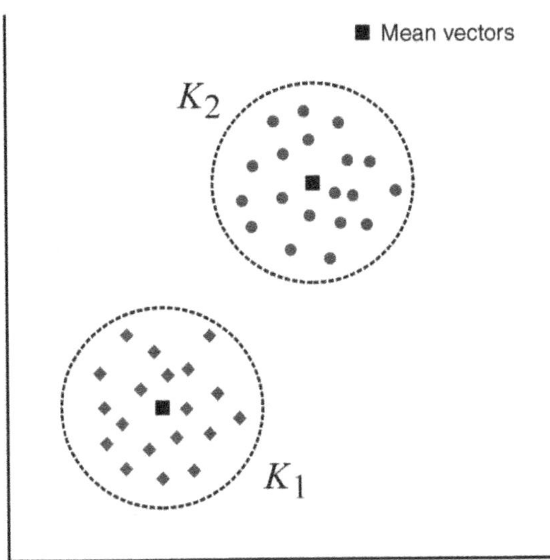

FIGURE 19.9 Example of two classes and their mean vectors.

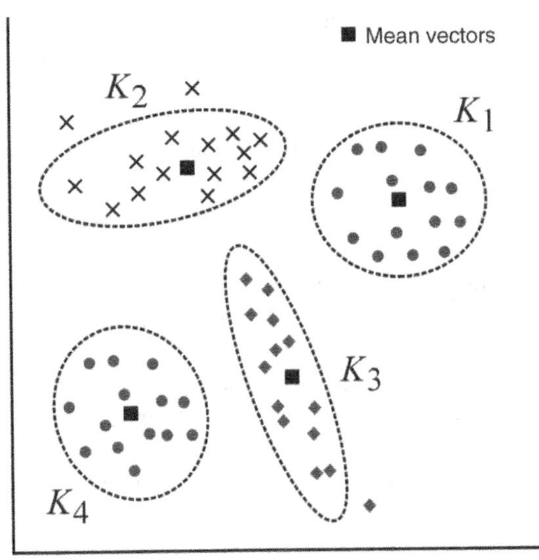

FIGURE 19.10 Example of three classes with relatively complex structure.

where $\tilde{d}_E$ is the scaled Euclidean distance between an unknown pattern and a class j, $\mathbf{x}$ is the feature vector of the unknown pattern, $\mathbf{x}_j$ is the mean vector of class j, σ_i is the standard deviation of class j along dimension i, and n is the dimensionality of the feature space.

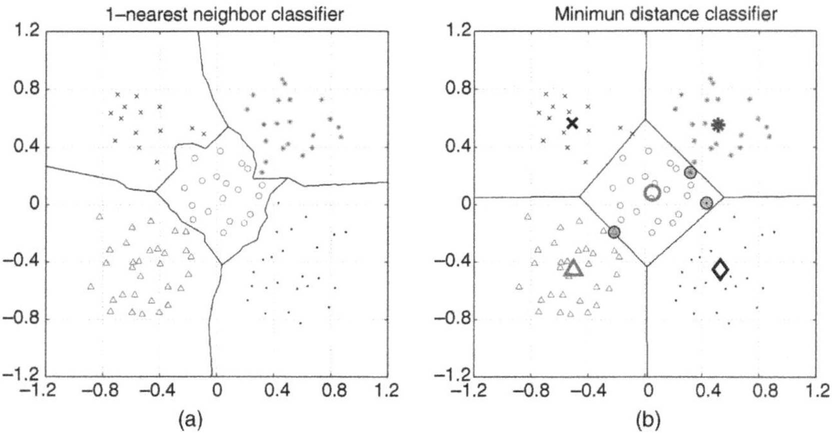

FIGURE 19.11 (a) Example of a KNN classifier ($k = 1$) for a five-class classifier in a 2D feature space (obtained using the STPRTool toolbox). (b) Minimum distance classifier results for the same data set.

19.3.2 *k*-Nearest Neighbors Classifier

A k-nearest neighbors (KNN) classifier works by computing the distance between an unknown pattern's feature vector $\mathbf{x}$ and the k closest *points*[4] to it in the feature space, and then assigning the unknown pattern to the class to which the majority of the k sampled points belong. The main advantages of this approach are its simplicity (e.g., no assumptions need to be made about the probability distributions of each class) and versatility (e.g., it handles overlapping classes or classes with complex structure well). Its main disadvantage is the computational cost involved in computing distances between the unknown sample and many (potentially *all*, if a brute force approach is used) stored points in the feature space.

Figure 19.11a illustrates the concept for a five-class classifier in a 2D feature space, where $k = 1$. It clearly shows that the KNN classifier is able to derive irregularly shaped discrimination functions among classes. This is in contrast to the minimum distance classifier, which would be constrained to using only straight lines as discrimination functions, as shown in Figure 19.11b, which also highlights the mean vectors for each class and the three data points that would be left out of their classes.

19.3.3 Bayesian Classifier

The rationale behind Bayesian classifiers is that a classification decision can be made based on the probability distributions of the training samples for each class; that is, an unknown object is assigned to the class to which it is *more likely* to belong based on the observed features.

[4]Note that we are referring to the k-nearest points, *not* the total number of classes, denoted as K.

The mathematical calculations performed by a Bayesian classifier require three probability distributions:

- The *a priori* (or *prior*) probability for each class ω_k, denoted by $P(\omega_k)$.
- The unconditional distribution of the feature vector representing the measured pattern $\mathbf{x}$, denoted by $p(\mathbf{x})$.
- The class conditional distribution, that is, the probability of $\mathbf{x}$ given class ω_k, denoted by $p(\mathbf{x}|\omega_k)$.

These three distributions are then used, applying Bayes' rule, to compute the *a posteriori* probability that a pattern $\mathbf{x}$ comes from class ω_k, represented as $p(\omega_k|\mathbf{x})$, as follows:

$$p(\omega_k|\mathbf{x}) = \frac{p(\mathbf{x}|\omega_k)P(\omega_k)}{p(\mathbf{x})} = \frac{p(\mathbf{x}|\omega_k)P(\omega_k)}{\sum_{k=1}^{W} p(\mathbf{x}|\omega_k)P(\omega_k)} \tag{19.15}$$

The design of a Bayes classifier requires that the prior probability for each class ($P(\omega_k)$) and the class conditional distribution ($p(\mathbf{x}|\omega_k)$) be known. The prior probability is easy to compute, based on the number of samples per class and the total number of samples. Estimating the class conditional distribution is a much harder problem, though, and it is often handled by modeling probability density function (PDF) of each class as a Gaussian (normal) distribution.

In MATLAB

There are many excellent MATLAB toolboxes for statistical pattern classification available on the Web (the "On the Web" section at the end of the chapter lists a few of them).

19.4 TUTORIAL 19.1: PATTERN CLASSIFICATION

Goal

The goal of this tutorial is to learn how to use MATLAB and publicly available toolboxes and data sets to build a visual pattern classifier.

Objectives

- Learn how to build a small optical character recognition (OCR) solution to classify digits (between 0 and 9) using `regionprops` and a KNN classifier.[5]
- Learn how to prepare and process the training and test data sets.

[5]The KNN classifier—and some of the supporting functions—used in this tutorial comes from the *Statistical Pattern Recognition Toolbox (STPRtool)*, available at http://cmp.felk.cvut.cz/cmp/software/stprtool/ .

- Learn how to perform feature selection.
- Learn how to present the classification results using a confusion matrix.

What You Will Need

- Scripts and test images available from the book web site (ocr_example.zip).

Procedure

1. Download the ocr_example.zip file and unzip its contents, keeping the folder structure. This file contain all the scripts and data you need, organized into meaningful folders.
2. Run script load_data.m. This script will load the 1000 training images and 1000 test images (10 images per character between 0 and 9) to the proper folders.

```
load_data
```

3. Examine the contents of script load_data.m to better understand *how* it does what it does.
4. Run script preprocess_images.m. This script will preprocess the training and test images and perform outlier removal.

```
preprocess_images
```

Question 1 What type of preprocessing operations are applied to all training and test images?

Question 2 How are outliers identified and handled?

5. Extract features from preprocessed training images using regionprops and selecting two properties with potential discriminative power: *eccentricity* and *Euler number*.
6. Store the results into a 2×1000 array corresponding to the 1000 feature vectors (of size 2×1 each).

```
fv = zeros(2,Ntrain);
for k = 1:Ntrain
    class = trn_data.y(k);
    I = imread([out_dir, sprintf('train_image_%02d_%02d_bw.png',...
        class,rem(k,100))]);
    [L, N] = bwlabel(I);
    % compute features and  build feature vector using 2 properties
    stats = regionprops(L,'all');
```

```
    fv(1,k) = stats.Eccentricity;
    fv(2,k) = stats.EulerNumber;
end
```

7. Save feature vector to a MAT file.

```
save([out_dir,'train_fv.mat'], 'fv');
```

8. Format feature vectors in a way that the KNN classifier will understand.

```
trn_data_binary.X = fv;
trn_data_binary.y = trn_data.y;
```

9. Create a KNN classifier and organize data for classification.

```
model = knnrule(trn_data_binary, 1);
```

10. Plot the 2D feature space and inspect it carefully. You should see a plot identical to Figure 19.12.

```
plot_feature_space
```

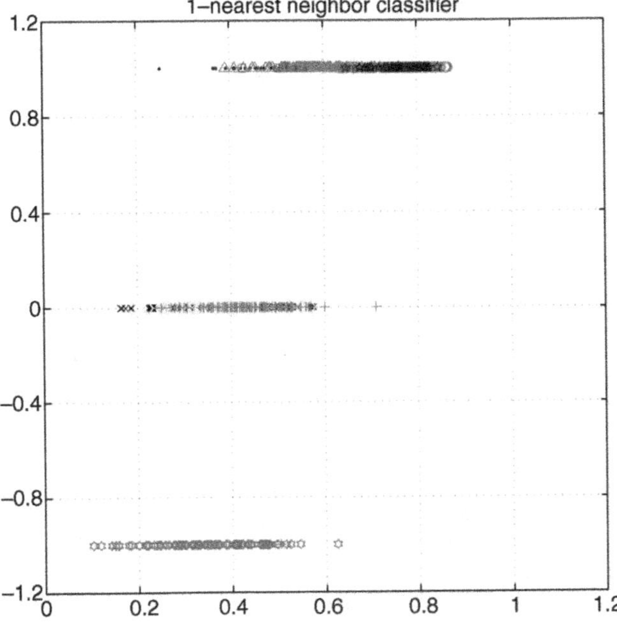

FIGURE 19.12 Feature space for training set. Obtained using the *Statistical Pattern Recognition Toolbox (STPRtool)*, available at http://cmp.felk.cvut.cz/cmp/software/stprtool/.

Question 3 Which feature is being represented on each axis of the plot?

Question 4 After inspecting the plot (and comparing it with that shown in Figure 19.11), what can you conclude so far?

11. Extract features from preprocessed test images using `regionprops` and selecting the same two properties used for the training set, namely, eccentricity and Euler number.
12. Store the results into a 2×1000 array corresponding to the 1000 feature vectors (of size 2×1 each).

```
fv_test = zeros(2,Ntest);
for k = 1:Ntest
    class = tst_data.y(k);
    I = imread([out_dir, sprintf('test_image_%02d_%02d_bw.png',...
        class,rem(k,100))]);
    [L, N] = bwlabel(I);
    stats = regionprops(L,'all');
    fv_test(1,k) = stats.Eccentricity;
    fv_test(2,k) = stats.EulerNumber;
end
```

13. Save feature vector to a MAT file.

```
save([out_dir,'test_fv.mat'], 'fv_test');
```

14. Format feature vectors in a way that the KNN classifier will understand.

```
tst_data_binary.X = fv_test;
tst_data_binary.y = tst_data.y;
```

15. Run the KNN classifier to assign labels to the test images.

```
labels = knnclass(tst_data_binary.X,model);
display_kNN_results;
```

The results of running the kNN classifier on the test data set will show the following:

```
classification_result on test set data: 583 out of 1000 missclassified
class "0"  missclassified   66 times
class "1"  missclassified   46 times
class "2"  missclassified   61 times
class "3"  missclassified   68 times
class "4"  missclassified   77 times
class "5"  missclassified   71 times
class "6"  missclassified   67 times
```

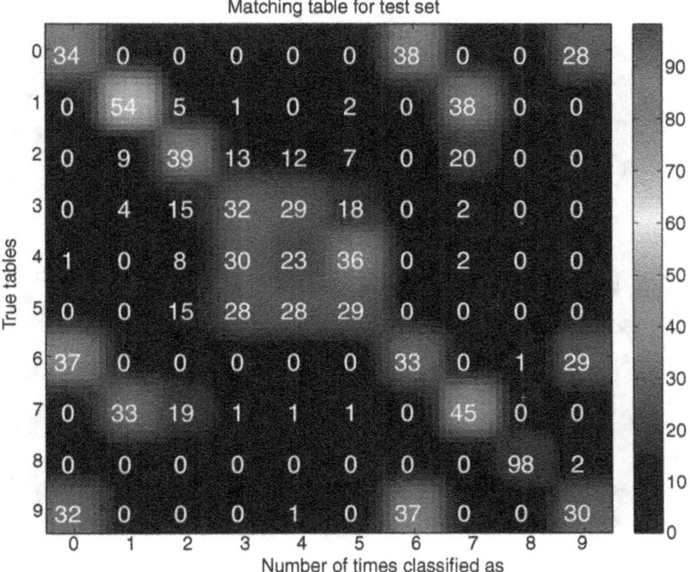

FIGURE 19.13 Confusion matrix with results of KNN classifier for the selected features. Obtained using the *Statistical Pattern Recognition Toolbox (STPRtool)*, available at http://cmp.felk.cvut.cz/cmp/software/stprtool/.

```
class "7"  missclassified   55 times
class "8"  missclassified    2 times
class "9"  missclassified   70 times
```

Figure 19.13 shows the resulting confusion matrix.

Question 5 Why are the results so poor (less than 42% success rate) overall?

Question 6 Why is the performance for one class (the digit 8) so much better than any other class?

Reflection

Our classifier did not perform as well as expected. It is time to look back at *each and every step* of our procedure and consider what needs to be changed or improved. At this point, all the evidence points in the direction of better, that is, more descriptive and discriminative, features.

In case you are wondering about the quality of the classifier, Figure 19.14 shows the resulting confusion matrix if we use the exact same classifier and test images, but different feature vectors. In this case, the actual gray values of the original 13×13 images were used as "features" that is, each image was represented using a 169×1 feature vector, without incurring in any of the preprocessing and feature extraction stages used in our solution. The results are remarkably better (98.2% overall success

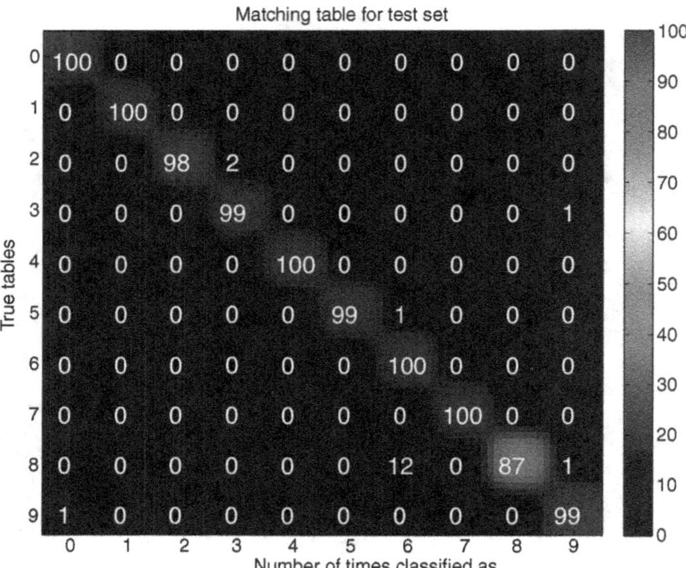

FIGURE 19.14 Number of Confusion matrix with results of KNN classifier for the case where the images' gray values are used as "features." Obtained using the *Statistical Pattern Recognition Toolbox (STPRtool)*, available at http://cmp.felk.cvut.cz/cmp/software/stprtool/.

rate) than the ones we obtained with our choice of features (Figure 19.13). Another interesting remark: the worst-performing class (the number 8) for the modified design was the one that performed best in our original method.

Question 7 Does the alternative method scale well for larger (and more realistic) image sizes? Explain.

Question 8 Which changes would you make to our design (assuming the same dataset and classifier) and in which sequence would you experiment with them? Be precise.

WHAT HAVE WE LEARNED?

- Visual pattern recognition is the collection of methods used to recognize and classify patterns within images. It is a subset of the large research area of pattern recognition, in which the data used for the recognition and classification tasks correspond to the visual contents of a scene (and have been extracted from such scenes using one or more of the techniques described in Chapter 18).
- A *pattern* is an arrangement of *descriptors* (or *features*). The two most common arrangements are feature vectors (for quantitative descriptions) and strings (for

structural descriptions). A *pattern class* is a family of patterns that share a set of common properties.

- A *pattern classifier* is a (mathematical) method by which each sample in a data set is assigned a pattern class to which it is most likely to belong. Decision-theoretic pattern classifiers perform such assignment decisions based on mathematical properties such as distance measures between feature vectors or probability distributions of the pattern vectors within a class.

LEARN MORE ABOUT IT

- Pattern classification is a vast and complex field and entire books have been written on the topic. Two recent examples (both of which come with their own companion MATLAB toolbox) are [DHS01] and [vdHDdRT04].
- For information on structural pattern recognition techniques, including MATLAB code, refer to Chapter 8 of [DHS01] (and its companion manual) or Chapter 12 of [GWE04].

ON THE WEB

- Statistical Pattern Recognition Toolbox (STPRtool)
http://cmp.felk.cvut.cz/cmp/software/stprtool/manual/
- PRTools: The Matlab Toolbox for Pattern Recognition
http://prtools.org/
- DCPR (Data Clustering and Pattern Recognition) Toolbox
http://neural.cs.nthu.edu.tw/jang/matlab/toolbox/DCPR/

19.5 PROBLEMS

19.1 Given the confusion matrix in Figure 19.15, use MATLAB to calculate the precision and recall per category.

19.2 Design and implement (in MATLAB) a machine vision solution to count the number of pennies, nickels, dimes, and quarters in an image containing U.S. coins.

	ω_1	ω_2	ω_3	ω_4
ω_1	97	0	2	1
ω_2	0	94	5	1
ω_3	0	0	100	0
ω_4	3	0	5	92

FIGURE 19.15 Confusion matrix for Problem 19.1.

PART II

VIDEO PROCESSING

CHAPTER 20

VIDEO FUNDAMENTALS

WHAT WILL WE LEARN?

- What is an analog video raster and what are its main components and parameters?
- What are the most popular analog TV and video standards?
- What is digital video and how it differs from analog video?
- What are the most popular digital video standards?
- How is color information encoded in analog and digital video systems?
- How can we read, manipulate, and play digital video files in MATLAB?

20.1 BASIC CONCEPTS AND TERMINOLOGY

In this section, we present a list of technical concepts and terms used in analog and digital TV and video systems.[1] Similar to what we have done earlier (Section 1.2), this section is structured in a question-and-answer format in a sequence that starts with relatively simple concepts and builds up to more elaborated ones.

[1]Since many of these concepts are interconnected, there is no perfect sequence by which they should be presented. The reader is encouraged to read the whole chapter even if certain parts of the text refer to concepts that have not been officially introduced yet. I believe and hope that at the end of the chapter, the entire picture will be clear in the reader's mind.

Practical Image and Video Processing Using MATLAB®. By Oge Marques.

What is a Video Signal?

A *video signal* is a one-dimensional (1D) analog or digital signal varying over time whose spatiotemporal contents represent a sequence of images (or *frames*) according to a predefined scanning convention. Mathematically, a continuous (analog) video signal will be denoted by $f(x, y, t)$, where t is the temporal variable.

An *analog* video signal refers to a 1D electrical signal $f(t)$ obtained by sampling $f(x, y, t)$ in the vertical and temporal dimensions. A *digital* video signal is also sampled along the horizontal axis of each frame.

What is Scanning?

Scanning is a method used by all video systems as part of the process of converting optical images into electrical signals. Figure 20.1 shows the basic scanning operation in a television camera. During the scanning process, an electronic sensing spot moves across the image in a pattern known as a *raster*. The sensing spot converts differences in brightness into differences in instantaneous voltages. Starting at the upper left corner of the image, the spot moves in a horizontal direction across the frame to produce a scanning line. It then quickly returns to the left edge of the frame (a process called *horizontal retrace*) and begins scanning another line. These lines are slightly tilted downward, so that after the retrace the spot is just below the previously scanned line and ready to scan a new line. After the last line is scanned (i.e., when the sensing spot reaches the bottom of the image), both horizontal and vertical retraces occur, bringing the spot back to the upper left corner of the image. A complete scan of the image is called a *frame* [LI99].

Scanning also occurs at the time of reproducing the frames on a display device. The main difference, of course, is the replacement of the sensing spot by a spot of

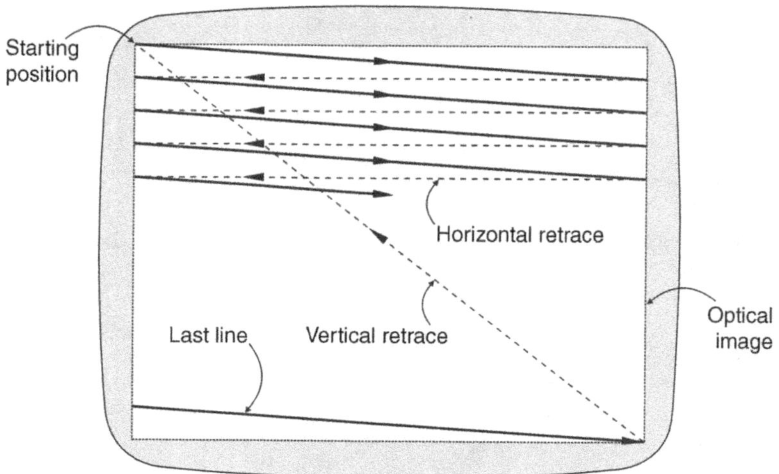

FIGURE 20.1 Scanning raster. Redrawn from [LI99].

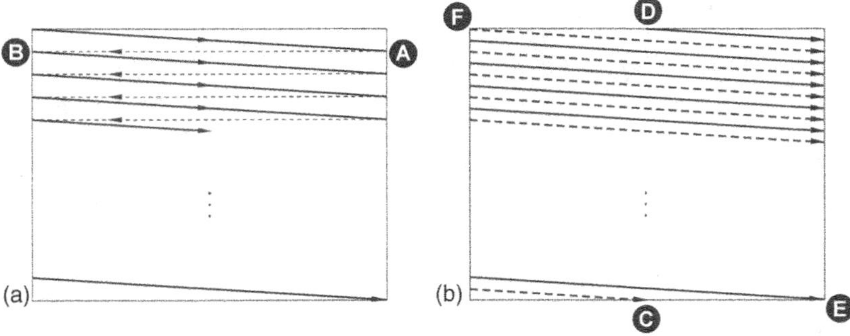

FIGURE 20.2 Scan and retrace: (a) progressive scan (dashed lines indicate horizontal retrace); (b) interlaced scan (solid and dashed lines represent even and odd fields, respectively). Adapted and redrawn from [WOZ02].

light whose intensity is controlled by the electrical signal originally produced at the output of the camera.

What are the Differences Between Interlaced and Progressive Scanning?

Progressive scanning is the process by which each image is scanned in one single pass, called *frame*, at every Δt (Figure 20.2a). It is used, for example, in computer displays, where a representative value for Δt is 1/72 s. Progressive scanning is a natural way to create an analog video raster. However, due to technological limitations at the time the early analog TV systems were being developed, it was not possible to implement it. As a result, the concept of *interlaced scanning* was proposed, which trades off spatial resolution for temporal resolution.[2]

With interlaced scanning (Figure 20.2b), each frame is scanned in two successive vertical passes, first the odd numbered lines and then the even numbered ones. Each pass is called a *field*. Since each field contains half of the lines in a frame, it lasts only one-half of the duration of an entire frame. Consequently, fields flash at a rate twice as fast as entire frames would, creating a better illusion of motion. If the field flash rate is greater than the *critical flicker frequency* (CFF) for the human eye, the result is successful and the motion is perceived as smooth. Perhaps more importantly, even though each field contains only half of the lines (in other words, half of the horizontal resolution that one would expect), viewers tolerate this drop in spatial resolution and are usually content with the overall quality of the resulting video.

[2]Appendix A has a more detailed explanation of spatial and temporal resolution, the relationship between them, and the implications of human visual perception experiments in the design of analog and digital video systems.

What is Blanking Interval?

It is the time interval at the end of each line (*horizontal retrace*, A to B in Figure 20.2a) or field (*vertical retrace*, C to D and E to F in Figure 20.2b) during which the video signal must be blanked before a new line or field is scanned.

What is Refresh Rate?

Most displays for moving images involve a period when the reproduced image is absent from the display, that is, a fraction of the frame time during which the display is black. To avoid objectionable flicker, it is necessary to flash the image at a rate higher than the rate necessary to portray motion. The rate at which images are flashed is called *flash* (or *refresh*) rate.

Typical refresh rates are 48 (cinema), 60 (conventional TV), and 75 Hz (computer monitors). The refresh rate highly depends on the ambient illumination: the brighter the environment, the higher the flash rate must be in order to avoid flicker. Refresh rates also depend on the display technology. In conventional movie theater projection systems, a refresh rate of 48 Hz is obtained by displaying each negative of a 24 frames/s movie twice. For progressive scan systems, the refresh rate is the same as the *frame rate*, whereas in the case of interlaced scan systems, the refresh rate is equivalent to the *field rate* (twice the frame rate).

What is the Meaning of Notation Such As 525/60/2:1, 480i29.97, or 720p?

There is no universally adopted scanning notation for video systems, which may be a potential source of confusion, due to the lack of consistency among different sources.[3]

Analog monochrome video scanning systems can be denoted by

- Total number of lines (including sync and blanking overhead)
- Refresh rate (in Hz) (which will be equal to the field rate for interlaced scan or the frame rate for progressive scan)
- Indication of interlaced (2:1) or progressive (1:1) scan

According to this notation, the analog TV system used in North America and Japan would be represented by 525/60/2:1 (or, more accurately, 525/59.94/2:1) and the European SDTV (standard definition TV) would be 625/50/2:1.

A more compact way of indicating the same information is by concatenating the number of visible lines per frame with the type of scanning (*p*rogressive or *i*nterlaced)

[3]For example, the use of NTSC to refer to 525 lines/60 Hz/interlaced monochrome video systems is common, but not accurate (after all, NTSC is a color TV standard). Worse yet is the use of PAL to refer to the 625 lines/50 Hz/interlaced monochrome video system used in great part of Europe, since PAL not only is a color encoding standard, but also contains a great number of variants. For a concrete example of how confusing this may get, in Brazil, the analog color TV standard adopted for TV broadcast (PAL-M) uses the PAL color encoding on a 525 lines/60 Hz/interlaced monochrome system.

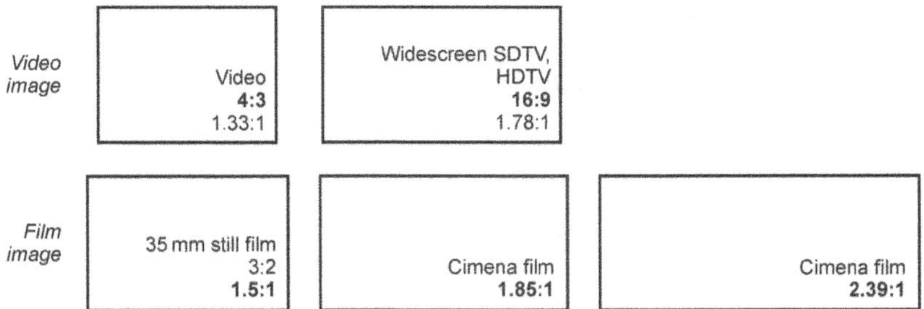

FIGURE 20.3 Aspect ratios of SDTV, HDTV, and film. Redrawn from [Poy03].

and the vertical frequency (roughly equal to the frame rate), for example, 480i29.97 and 576i25.

HDTV (high-definition TV) standards are usually represented in an even more compact notation that includes the number of lines and the type of scanning. Contemporary examples include the 720p and the 1080i standards.

What is Aspect Ratio?

Aspect ratio (AR) is the ratio of frame width to height. In the case of digital video, both dimensions can be specified in terms of numbers of pixels (e.g., 640 × 480 or 1920 × 1200). For analog video, however, only the number of lines can be counted and expressed as an integer value.

The most common values for TV are 4:3 (1.33:1) for SDTV and 16:9 (1.78:1) for HDTV. For movies, it usually varies between 1.66:1 and 2.39:1. Figure 20.3 shows representative examples of aspect ratios.

What is Gamma Correction?

Video acquisition and display devices are inherently nonlinear: the intensity of light sensed at the camera input or reproduced at the display output is a nonlinear function of the voltage levels.

For the cameras, this relationship is usually expressed as

$$v_c = B_c{}^{-\gamma_c} \tag{20.1}$$

where B_c represents the actual luminance (light intensity), v_c is the resulting voltage at the output of the camera, and γ_c is a value usually between 1.0 and 1.9.

Similarly, for display devices, the nonlinear relationship between the input voltage v_d and the displayed color intensity B_d is expressed as

$$B_d = v_d{}^{\gamma_d} \tag{20.2}$$

where γ_d is a value usually between 2.2 and 3.0.

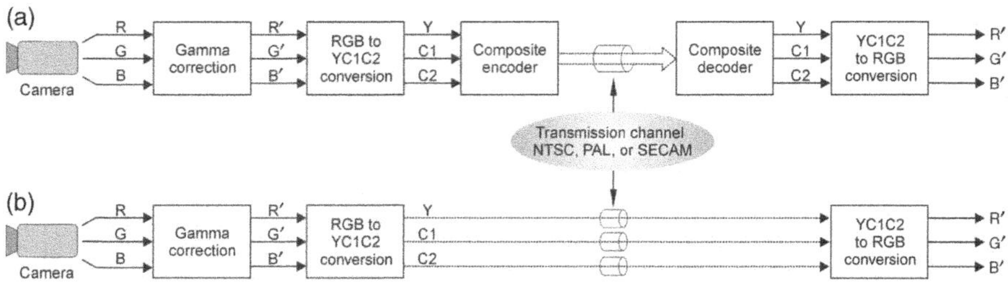

FIGURE 20.4 Gamma correction in video and TV systems: (a) composite video; (b) component video.

The process of compensating for this nonlinearity is known as *gamma correction* and is typically performed at the transmitter side. By gamma correcting the signal before displaying it, the intensity output of the display is roughly linear.

In addition to precompensating for the nonlinearity of the CRT display, gamma correction also codes the luminance information into a perceptually uniform space, thus compensating for the nonlinear characteristics of the human visual system in a way that Poynton calls an "amazing coincidence" [Poy03]. Moreover, the gamma-corrected signal also becomes less sensitive to noise.

Figure 20.4 shows how gamma correction is performed in a typical analog video system. First, a nonlinear transfer function is applied to each of the linear color components: R, G, and B right at the output of the color camera, resulting in the *gamma-corrected primaries* R', G', and B'. Then, a weighted sum of the nonlinear components is computed to form two generic chrominance signals (C1 and C2) and a luma signal, Y', representative of brightness, given by:[4]

$$Y' = 0.30R' + 0.59G' + 0.11B' \tag{20.3}$$

The resulting signals (Y', C1, and C2) are then transmitted—with or without being combined into a single composite signal—to the receiver, where eventually they are converted back into the gamma-corrected primaries (R', G', and B'), which will be used to drive the display device.

What are Component, Composite, and S-Video?

From our discussion on color (Chapter 16), we know that color images can be described by assigning three values (usually the R, G, and B color components) to each pixel. An extension of this representation scheme to color video requires using three independent one-dimensional color component signals (e.g., $R'G'B'$ or $Y'C_BC_R$), free from mutual interference, in what is known as *component analog video* (CAV).

Component video representation is convenient and feasible for certain applications (e.g., connecting a DVD player to a TV set) but could not be adopted for color TV

[4]The values of the coefficients in equation (20.3) are approximate. The exact values will vary among different analog video standards.

broadcast systems. In that case, primarily for backward compatibility reasons, a video encoding system was designed that combines the intensity and color information into a composite signal. This is known as *composite video*. A composite video signal relies on the fact that the chrominance components can be encoded using a significantly smaller bandwidth than the luminance component. Composite video systems combine brightness and color into one signal, at the expense of introducing a certain degree of mutual interference. For more details, refer to Section 20.3.

The *S-video* (also known as *Y/C video*) standard is an intermediate solution consisting of two components, luminance and a multiplexed chrominance component. S-video systems require less bandwidth (or data rate) than component video and produce better image quality than composite video.

20.2 MONOCHROME ANALOG VIDEO

In this section, we expand the discussion of the fundamental concepts and the most important aspects of monochrome (black-and-white) analog video.

20.2.1 Analog Video Raster

An analog video raster can be defined by two parameters: the *frame rate* (*FR*) (in frames/s or fps or Hz) that defines the temporal sampling rate and the *line number* (N_L) (in lines/frame or lines/picture height) that corresponds to the vertical sampling rate. These two parameters can be used to define other useful terms, such as

- Line rate (in lines/s): the product of FR and N_L.
- Temporal sampling interval (or frame interval): $\Delta_t = 1/\text{FR}$.
- Vertical sampling interval (or line spacing): $\Delta_y = \text{PH}/N_L$, where PH is the picture height.
- Line interval: $T_l = \Delta_t/N_L$, which is the total time required to scan a line (including the horizontal retrace, T_h).
- Actual scanning time for a line: $T_l' = T_l - T_h$.
- Number of active lines: $N_L' = (\Delta_t - T_v)/T_L = N_L - T_v/T_l$, where T_v is the *vertical retrace* and is usually chosen to be a multiple of T_l .

A typical waveform for an interlaced analog video raster is shown in Figure 20.5.

The maximum rate at which an analog video signal can change from one amplitude level to another is determined by the bandwidth of the video system. The bandwidth requirements of an analog TV system can be calculated by

$$\text{BW} = \frac{K\ \text{AR}\ N_L'}{2T_l'} \tag{20.4}$$

where BW is the system bandwidth (in hertz), AR is the aspect ratio, N_L' is the number of active TV scanning lines per frame, T_l' is the duration of a line, and K is

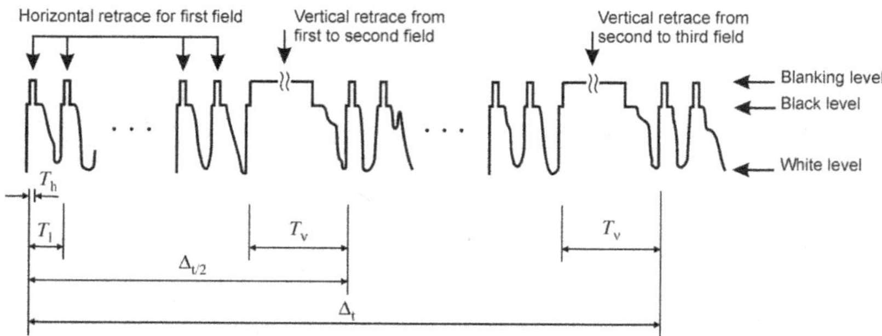

FIGURE 20.5 Typical interlaced video raster. Redrawn from [WOZ02].

the *Kell factor*: an attenuation factor (whose value is usually 0.7) that accounts for the differences between the *actual* vertical resolution of a frame and the *perceived* resolution when the human visual acuity is taken into account.

■ EXAMPLE 20.1

Calculate the required bandwidth for the luminance signal of an analog TV system with the following characteristics:

- 2:1 interlaced scanning
- 525 horizontal lines/frame, 42 of which are blanked for vertical retrace
- 30 frames per second
- 4:3 aspect ratio
- 10 μs of horizontal blanking
- Kell factor = 0.7

Solution

$T_l = 1/(525 \times 30) = 63.5\ \mu s$
$T_l' = 63.5 - 10 = 53.5\ \mu s$
$N_L' = 525 - 42 = 483$ lines
$BW = (0.7 \times 4/3 \times 483)/(2 \times 53.5 \times 10^{-6}) = 4.21$ MHz

20.2.2 Blanking Intervals

Most analog TV and video systems use a composite signal that includes all the information needed to convey a picture (brightness, color, synchronization) encoded into a single one-dimensional time-varying signal. As discussed previously, part of the time occupied by the signal is used to bring the scanning starting point back to the beginning of a new line or frame, a process known as horizontal and vertical retraces, respectively. These retrace periods are also known as *blanking intervals*, during which the amplitude of the signal is such that it cannot be seen on the screen (indicated in Figure 20.5 as "blanking level").

The duration of the *vertical blanking interval* (VBI) (also known as *vertical retrace*, T_v) is approximately 7.5% of the frame interval (Δ_t). The *horizontal blanking interval* (also known as *horizontal retrace*, T_h) usually lasts 14–18% of the total line interval (T_l).

20.2.3 Synchronization Signals

In analog TV and video systems, it is necessary to establish a way by which the scanning process at the display device is synchronized with the scanning process that took place in the imager (camera). This is done by adding synchronizing (or simply *sync*) pulses to the horizontal and vertical blanking intervals. To ensure that these pulses do not interfere with the purpose of the blanking intervals, their amplitude is such as to correspond to "blacker than black" luminance levels. This also allows them to be easily separated from the rest of the video by simply clipping the composite video signal, in a process known as *sync separation*.

20.2.4 Spectral Content of Composite Monochrome Analog Video

A monochrome analog video signal usually covers the entire frequency spectrum from DC to the maximum frequency determined by the desired picture resolution (see equation (20.4)). However, the frequency spectrum is not continuous—it contains a fine-grained structure determined by the scanning frequencies (Figure 20.6).

The spectrum shown in Figure 20.6 reflects the fact that video signals contain three periodic components: one repeating at the line rate, one at the field rate, and one at the

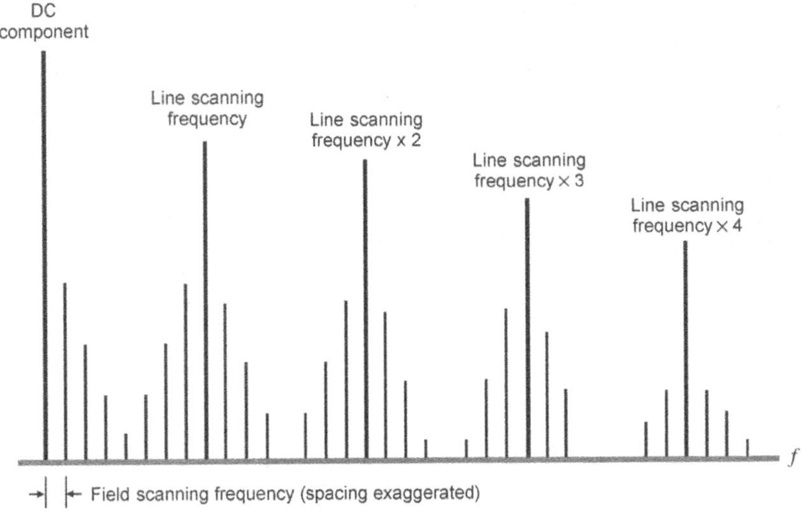

FIGURE 20.6 Fine-grained frequency spectrum of a monochrome analog video signal. Redrawn from [LI99].

frame rate (at 15,750, 60, and 30 Hz, respectively, for the NTSC standard adopted in the United States). Each component can be expressed as a Fourier series—a sequence of sinusoidal terms with frequencies that are multiple (harmonics) of the repetition rate and amplitudes determined by the waveform of the component. The summation of the amplitudes of the terms for the three periodic components over the entire video spectrum is the frequency content of the signal.

Figure 20.6 also shows that the spectrum of a monochrome TV signal consists of a series of harmonics of the line frequency, each surrounded by a cluster of frequency components separated by the field frequency. The amplitude of the line frequency components is determined by the horizontal variations in brightness, whereas the amplitude of the field frequency components is determined by the vertical brightness variations.

The gaps in the frequency spectrum of a monochrome TV signal have been utilized for two main purposes:

1. Interleave frequency components of the *color subcarrier* and its sidebands, making color TV systems backward compatible with their monochrome predecessors (see Section 20.3).
2. Improve the high-frequency signal to noise ratio using *comb filters*.

20.3 COLOR IN VIDEO

In this section, we discuss the most relevant issues associated with representing, encoding, transmitting, and displaying color information in analog TV and video systems. It builds upon the general knowledge of color in image and video (Chapter 16) and focuses on specific aspects of color modulation and encoding in analog TV and video systems.

Historically, the development of color TV came about at a time where monochrome TV already enjoyed widespread popularity. As a result of the existing equipment and technologies, and frequency spectrum regulations already in place, the development of color TV needed to be an extension of the existing system and was required to maintain *backward compatibility* with its predecessor. This requirement will be the underlying theme of the discussion that follows.

Analog color TV and video systems utilize red, green, and blue as primary colors that are then gamma corrected (resulting in R', G', B') and used to calculate a luma value:

$$Y' = 0.299R' + 0.587G' + 0.114B' \tag{20.5}$$

The R', G', B', and Y' signals are then combined into *color-difference* signals $(B' - Y')$ and $(R' - Y')$, in a process known as *matrixing*, according to the following equations:

$$B' - Y' = -0.299R' - 0.587G' + 0.889B' \tag{20.6}$$

$$R' - Y' = 0.701R' - 0.587G' - 0.114B' \tag{20.7}$$

The remaining color-difference signal $(G' - Y')$ does not need to be transmitted. It can be re-created at the receiver's side by a simple arithmetic combination of the other two color-difference signals.

The color-difference signals are then multiplied by a scaling factor (e.g., 0.493 for the $(B' - Y')$ and 0.877 for the $(R' - Y')$) and in the case of NTSC rotated by a certain angle (33°). Those two signals are then usually modulated in quadrature (90° phase difference) and the result is used to modulate a color subcarrier whose bandwidth is usually between 0.5 and 1.5 MHz (significantly narrower than the 4.5 MHz typically used for the luma component), resulting in a single-wire composite video signal with a total bandwidth suited to the specific transmission standard.

At the receiver's side, first a decoder separates the composite signal into luma and chroma. The chroma portion is then bandpass filtered and demodulated to recover the color-difference signals that are then added to a delayed (to account for the processing time through filters and demodulators) version of the luma signal to a matrix circuit in charge of regenerating the gamma-corrected primaries needed as input signals for the monitor.

The details of the resulting composite video varies from one standard to the next (see Section 20.4) but have the following characteristics in common [RP00]:

- *Monochrome Compatibility*: A monochrome receiver must reproduce the brightness content of a color TV signal correctly and without noticeable interference from the color information.
- *Reverse Compatibility*: A color receiver must reproduce a monochrome signal correctly in shades of gray without spurious color components.
- *Scanning Compatibility*: The scanning system used for color systems must be identical to the one used by the existing monochrome standard.
- *Channel Compatibility*: The color signal must fit into the existing monochrome TV channel and use the same spacing (in Hz) between the luminance and audio carriers.
- *Frequency-Division Multiplexing*: All systems use two narrowband color-difference signals that modulate a color subcarrier and the chrominance and luminance signals are frequency-division multiplexed to obtain a single-wire composite video signal with a total bandwidth suited to the specific transmission standard.

Composite color encoding has three major disadvantages [Poy03]:

1. Some degree of mutual interference between luma and chroma is inevitably introduced during the encoding process.
2. It is impossible to perform certain video processing operations directly in the composite domain. Even something as simple as resizing a frame requires decoding.

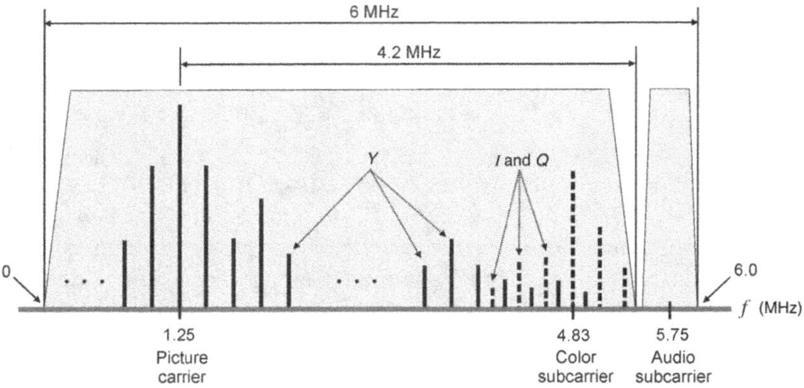

FIGURE 20.7 NTSC spectrum, showing how luminance (Y) and chrominance (I and Q) signals are interleaved. Redrawn from [LD04].

3. Digital compression techniques such as MPEG (Motion Pictures Expert Group) cannot be directly applied to composite video signals; moreover, the artifacts of NTSC or PAL encoding are destructive to MPEG encoding.

The spectral contents of composite color analog video differ from their monochrome equivalent because color information encoded on a high-frequency subcarrier is superimposed on the luminance signal (Figure 20.7). The frequency values have been carefully calculated so that the new spectral components due to color occupy gaps in the original monochrome spectrum shown earlier (Figure 20.6).

20.4 ANALOG VIDEO STANDARDS

In this section, we present a summary of some of the most relevant analog TV and video standards. There are two scanning standards used for conventional analog TV broadcast: the 480i29.97 used primarily in North America and Japan, and the 576i25 used in Europe, Asia, Australia, and Central America. The two systems share many features, such as 4:3 aspect ratio and interlaced scanning. Their main differences reside on the number of lines per frame and the frame rate.

Analog broadcast of 480i usually employs NTSC color coding with a color subcarrier of about 3.58 MHz; analog broadcast of 576i usually adopts PAL color encoding with a color subcarrier of approximately 4.43 MHz. Exceptions to these rules include the PAL-M system (480i scanning combined with PAL color coding) used in Brazil and the PAL-N system (576i scanning combined with a 3.58 MHz color subcarrier nearly identical to the NTSC's subcarrier) used in Argentina, among others.

20.4.1 NTSC

The NTSC (National Television System Committee) TV standard is an analog composite video standard that complies with the 480i component video standard. It uses two color-difference signals (U and V) that are scaled versions of the color-difference signals ($B' - Y'$) and ($R' - Y'$), respectively. The U and V color difference components are subject to low-pass filtering and combined into a single chroma signal, C:

$$C = U \sin \omega t + V \cos \omega t \tag{20.8}$$

where $\omega = 2\pi f_{sc}$ and f_{sc} is the frequency of color subcarrier (approximately 3.58 MHz).

In the past, to be compliant with FCC regulations for broadcast, an NTSC modulator was supposed to operate on I and Q components (scaled and rotated versions of U and V), where the Q component (600 kHz) was bandwidth limited more severely than the I component (1.3 MHz):

$$C = Q \sin(\omega t + 33^\circ) + I \cos(\omega t + 33^\circ) \tag{20.9}$$

Figure 20.7 shows details of the frequency interleaving between luma and chroma harmonics, as well as the overall spectral composition of the NTSC signal (including the audio component). In analog TV systems, audio is transmitted by a separate transmitter operating at a fixed frequency offset (in this case, 4.5 MHz) from the video transmitter. Contemporary NTSC equipment modulate equiband U and V color-difference signals.

20.4.2 PAL

The PAL (phase alternating line) TV standard is an analog composite video standard that is often used in connection with the 576i component video standard.

It uses two color-difference signals (U and V) that are scaled versions of the color-difference signals ($B' - Y'$) and ($R' - Y'$), respectively. The U and V color difference components are subject to low-pass filtering and combined into a single chroma signal, C:

$$C = U \sin \omega t \pm V \cos \omega t \tag{20.10}$$

where $\omega = 2\pi f_{sc}$ and f_{sc} is the frequency of color subcarrier (approximately 4.43 MHz).

The main difference between equations (20.10) and (20.8) reflects the fact that the V component switches phases on alternating lines, which is the origin of the acronym PAL and its most distinctive feature.

TABLE 20.1 Parameters of Analog Color TV Systems

Parameter	NTSC	PAL	SECAM
Field rate	59.94	50	50
Line number/frame	525	625	625
Line rate (lines/s)	15,750	15,625	15,625
Image aspect ratio (AR)	4:3	4:3	4:3
Color space	YIQ	YUV	YDbDr
Luminance bandwidth (MHz)	4.2	5.0, 5.5	6.0
Chrominance bandwidth (MHz)	1.5 (I), 0.5 (Q)	1.3 (U, V)	1.0 (U, V)
Color subcarrier (MHz)	3.58	4.43	4.25 (Db), 4.41 (Dr)
Color modulation	QAM	QAM	FM
Audio subcarrier (MHz)	4.5	5.5, 6.0	6.5
Composite signal bandwidth (MHz)	6.0	8.0, 8.5	8.0

Reproduced from [WOZ02].

20.4.3 SECAM

The SECAM (*Séquentiel couleur à mémoire*) standard is a color TV system with 576i25 scanning used in France, Russia, and a few other countries. Table 20.1 summarizes the main parameters of SECAM and their equivalent in the PAL (576i25) and NTSC (480i29.97) standards.

20.4.4 HDTV

HDTV is a name usually given to TV and video systems where each frame has 720 or more active lines. The most common variants of HDTV in use today are the 1280 × 720 and 1920 × 1080 image formats, also referred to as 720p60 and 1080i30 (or simply 720p and 1080i), respectively.

In addition to significantly higher spatial resolution, two salient differences between HDTV and SDTV are the aspect ratio of 16:9 instead of 4:3 and the use of progressive scanning.

20.5 DIGITAL VIDEO BASICS

A digital video may be obtained by sampling a raster scan (which requires analog-to-digital conversion (ADC), see Section 20.6) or directly using a digital video camera. Most of contemporary digital video cameras use CCD sensors. Similar to their analog counterpart, digital cameras sample the imaged scene over time, resulting in discrete frames. Each frame consists of output values from a CCD array, which is by nature discrete in both horizontal and vertical dimensions.

The result of the video acquisition stage—whether the digitization takes place within the camera or is performed by an external ADC embedded, for example, in a video capture card—is a collection of *samples*. These samples are numerical representation of pixel values along a line, for all the lines within a frame. Similar to

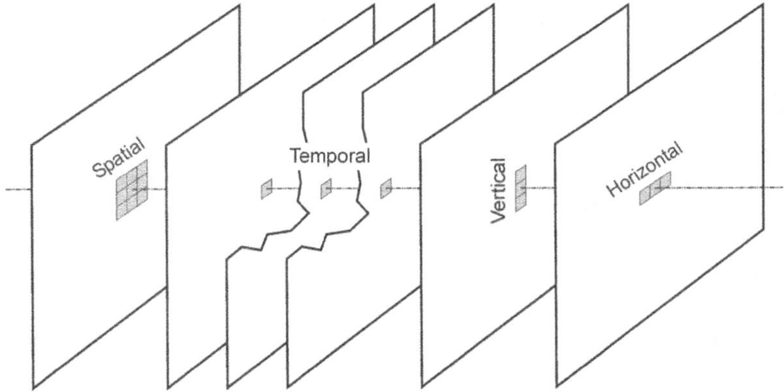

FIGURE 20.8 Sampling in the horizontal, vertical, and temporal dimensions. Redrawn from [Poy03].

their analog counterpart, to portray a smooth motion, digital video frames are captured and displayed at a specified frame rate.

Digital video can be understood as an alternative means of carrying a video waveform, in which the limitations of analog signals (any analog signal received at the destination is a valid one, regardless of any distortion, noise, or attenuation introduced on the original analog signal) are overcome by modulation techniques (such as PCM (pulse code modulation)) that enforce that each sample is encoded with a certain amplitude (from a finite set of values, known as *quantization levels*).

In analog video systems, the time axis is sampled into frames and the vertical axis is sampled into lines. Digital video simply adds a third sampling process along the lines (horizontal axis) (Figure 20.8).

20.5.1 Advantages of Digital Video

Digital representation of signals in general, and video in particular, have a number of well-known advantages over their analog counterpart [LI99]:

- Robustness to signal degradation. Digital signals are inherently more robust to signal degradation by attenuation, distortion, or noise than their analog counterpart. Moreover, error correction techniques can be applied so that distortion does not accumulate over consecutive stages in a digital video system.
- Smaller, more reliable, less expensive, and easier to design hardware implementation.
- Certain processes, such as signal delay or video special effects, are easier to accomplish in the digital domain.
- The possibility of encapsulating video at multiple spatial and temporal resolutions in a single scalable bitstream.
- Relatively easy software conversion from one format to another.

20.5.2 Parameters of a Digital Video Sequence

A digital video signal can be characterized by

- The frame rate ($f_{s,t}$)
- The line number ($f_{s,y}$)
- The number of samples per line ($f_{s,x}$)

From the above three quantities, we can find

- The temporal sampling interval or *frame interval* $\Delta_t = 1/f_{s,t}$
- The vertical sampling interval $\Delta_y = \text{PH}/f_{s,y}$, where PH is the picture height
- The horizontal sampling interval $\Delta_x = \text{PW}/f_{s,x}$, where PW is the picture width

In this book, we will use the notation $f(m, n, k)$ to represent a digital video, where m and n are the row and column indices and k is the frame number. The relationships between these integer indices and the actual spatial and temporal locations are $x = m\Delta_x$, $y = n\Delta_y$, and $t = k\Delta_t$.

Another important parameter of digital video is the number of bits used to represent a pixel value, N_b. For monochrome video, $N_b = 8$, whereas color videos require 8 bits per color component, that is, $N_b = 24$.

The data rate of the digital video, R, can be determined as

$$R = f_{s,t} \times f_{s,x} \times f_{s,y} \times N_b \quad \text{(in bps)} \tag{20.11}$$

Since the sampling rates for luma and chroma signals are usually different (see Section 20.7), N_b should reflect the equivalent number of bits used for each pixel in the sampling grid for the luma component. For example, if the horizontal and vertical sampling rates for each of the two chroma components are both half of that for the luma, then there are two chroma samples for every four luma samples. If each sample is represented by 8 bits, the equivalent number of bits per sample in the Y' (luma) resolution is $(4 \times 8 + 2 \times 8)/4 = 12$ bits.

When digital video is displayed on a monitor, each pixel is rendered as a rectangular region with constant color. The ratio of the width to the height of this rectangular area is the *pixel aspect ratio* (*PAR*). It is related to the aspect ratio of the entire frame (AR) by

$$\text{PAR} = \text{AR}\frac{f_{s,x}}{f_{s,y}} \tag{20.12}$$

Computer displays usually adopt a PAR of 1. In the TV industry, nonsquare pixels are used for historical reasons and PAR may vary from 8/9 (for 525/60 systems) to 16/15 (for 625/50 systems).

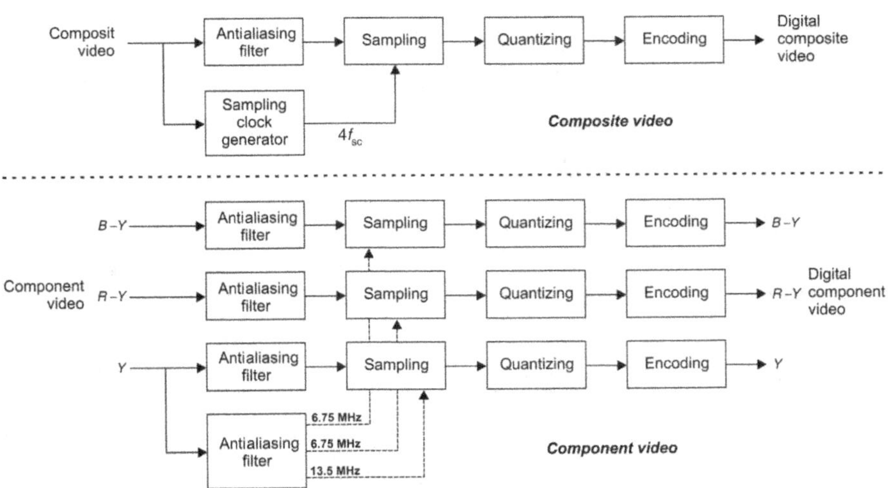

FIGURE 20.9 Analog-to-digital converters for composite (top) and component (bottom) video. Redrawn from [LI99].

20.5.3 The Audio Component

In analog TV systems, audio is transmitted by a separate transmitter operating at a fixed frequency offset from the video transmitter. In computer originated digital video, audio is typically encoded in a data stream that can be handled separately or interleaved with video data. In digital TV, audio is interleaved with video. In all cases, the audio part of the system involves processes equivalent to the ones used in video: creation, storage, transmission, and reproduction.

Digital audio formats are usually based on the PCM technique and its variants. The number of bits per sample and the sampling frequency are the determinant factors of the resulting data rate (and associated audio quality): from 12 kbps (or less) for speech to 176 kbps (or more) for CD quality stereo music.

20.6 ANALOG-TO-DIGITAL CONVERSION

Since digital video sequences are often the result of capturing and encoding a real-world analog scene, at some point an analog-to-digital conversion step is needed.[5] All digital cameras and camcorders perform ADC at the imager. Analog cameras output analog video that eventually needs to undergo ADC.

Figure 20.9 shows schematic block diagrams for analog-to-digital converters for composite and component video systems. The key components are described below [LI99].

[5]This statement implies that if the video sequence is generated from scratch by a computer application, no such conversion is necessary.

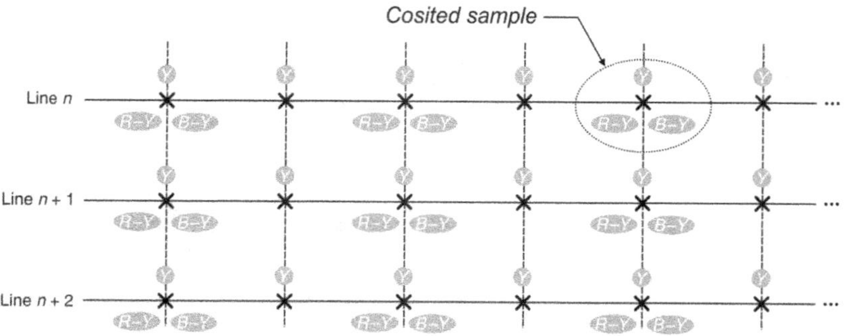

FIGURE 20.10 Location of sampling points in component video signals. Redrawn from [LI99].

Antialiasing Filter An optional low-pass filter with a cutoff frequency below the Nyquist limit (i.e., below one-half the sampling rate) whose primary function is to eliminate signal frequency components that could cause *aliasing*.[6]

Sampling The *sampling* block samples pixel values along a horizontal line at a sampling rate that is standardized for the main video formats as follows:

- NTSC $4f_{sc}$ (composite): 114.5 Mbps
- PAL $4f_{sc}$ (composite): 141.9 Mbps
- Rec. 601 (component) (luma): 108 Mbps
- Rec. 601 (component) (chroma): 54 Mbps

For component systems, the rates standardized by Rec. 601 were chosen to represent a sampling rate of 13.5 MHz, which is an integral multiple of both NTSC and PAL line rates. For composite signals, a sampling rate of $4f_{sc}$ (where f_{sc} is the frequency of the color subcarrier) exceeds the Nyquist limit by a comfortable margin and has become part of the SMPTE standard 244M.

Figure 20.10 shows the location of sampling points on individual lines for the Rec. 601 format. The luma and two color-difference sampling pulses are synchronized so that the color difference points are *cosited* with alternate luminance points. For composite signals, the sampling points follow a specified relationship with the phase of the color sync burst. For both NTSC and PAL at $4f_{sc}$, there are four sampling points for each cycle of the burst: for PAL, the sampling points are at 0°, 90°, 180°, and 270° points of the burst waveform. For NTSC, the first sample is located at 57° (the I axis), the second at 147° (the Q axis), and so on.

The sampling rate and phase must be synchronized with the line and subcarrier frequencies and phases to maintain these precise locations. This task is performed by the *sampling clock generator*.

[6]The concept of *aliasing* was introduced in Section 5.4.1.

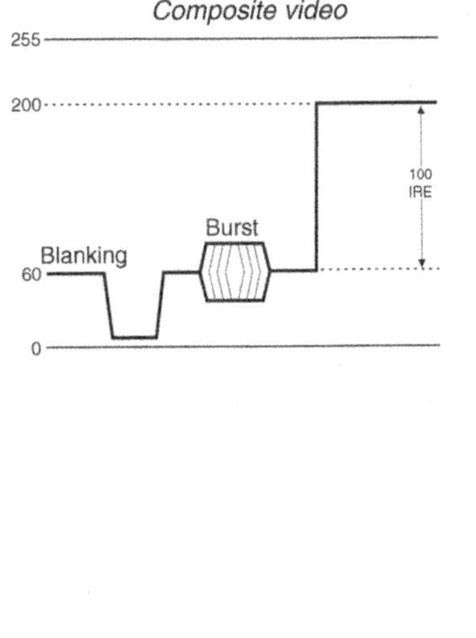

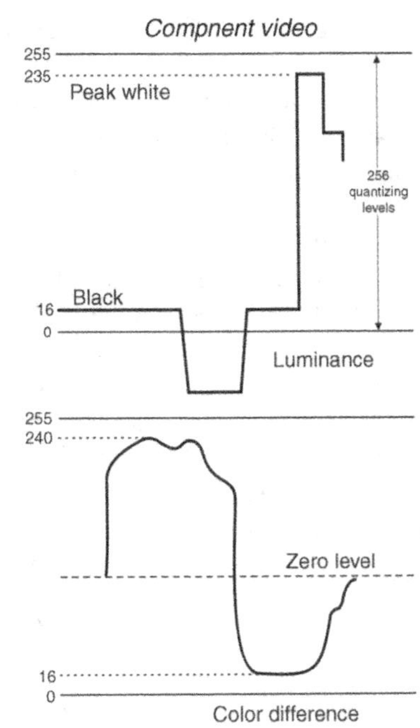

FIGURE 20.11 Assignment of quantization levels for component and composite video. Redrawn from [LI99].

Quantizing The next step in the ADC process is to quantize the samples using a finite number of bits per sample. This is achieved by dividing the amplitude range into discrete intervals and establishing a *quantum level* for each interval. The difference between the quantum level and the analog signal at the sampling point is called the *quantizing error.*

Most video quantization systems are based on 8-bit words, that is, 256 possible discrete values per sample (1024 levels are sometimes used for pregamma signals in cameras). Not all levels may be used for the video range because of the need to use some of the values for data or control signals.

Figure 20.11 shows how the resulting 256 quantized levels are typically used for composite and component video signals. In composite systems, the entire composite signal, including sync pulses, is transmitted within the quantized range of levels 4–200. Luma occupies levels 16–235. Since color-difference signals can be positive or negative, the zero level is placed at the center of the quantized range. In both cases, the values 0 and 255 are reserved for synchronization codes.

Encoding The final step in the ADC process is the encoding of the quantized levels of the signal samples. This is a topic that has been explored extensively, especially in connection with video compression (Section 20.9).

20.7 COLOR REPRESENTATION AND CHROMA SUBSAMPLING

A monochrome video frame can be represented using just one (typically 8-bit) value per spatiotemporal sample. This number usually indicates the gamma-corrected luminance information at that point, which we call *luma*: the larger the number, the brighter the pixel.

Color video requires multiple (usually three) values per sample. The meaning of each value depends on the adopted color model. The most common color representation for digital video adopts the $Y'CrCb$ color model. In this format, one component represents luma (Y'), while the other two are color-difference signals: Cb (the difference between the blue component and a reference value) and Cr (the difference between the red component and a reference value). $Y'CbCr$ is a scaled and offset version of the $Y'UV$ color space.

The exact equations used to calculate Y', Cr, and Cb vary among standards (see Chapter 3 of [Jac01] for details). For the sake of illustration, the following are the equations for the Rec. 601 SDTV standard:

$$Y'_{601} = 0.299R' + 0.587G' + 0.114B' \tag{20.13}$$

$$Cb = -0.172R' - 0.339G' + 0.511B' + 128 \tag{20.14}$$

$$Cr = 0.511R' - 0.428G' - 0.083B' + 128 \tag{20.15}$$

The key advantage of $Y'CrCb$ over RGB is that the Cr and Cb components may be represented with a lower spatial resolution than Y' because the HVS is less sensitive to variations in color than in luminance. This reduces the amount of data required to represent the chroma components, without significantly impacting the resulting visual quality. This process is known as *chroma subsampling*.

Figure 20.12 shows three of the most common patterns for chroma subsampling. The numbers indicate the relative sampling rate of each component in the *horizontal* direction. 4:4:4 means that the three components (Y', Cr, and Cb) have the same resolution; that is, there is a sample of each component at every pixel position. Another way of explaining it is to say that for every four luma samples, there are four Cr

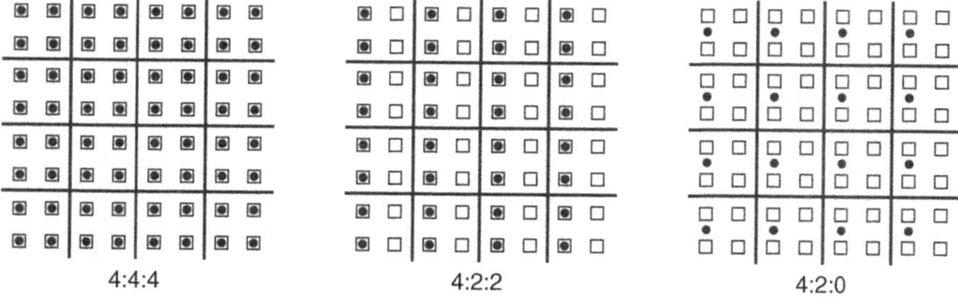

FIGURE 20.12 The most common chroma subsampling patterns: 4:4:4, 4:2:2, and 4:2:0.

and four *Cb* samples. The 4:4:4 sampling pattern preserves the full fidelity of the chroma components. In 4:2:2 sampling, the chroma components have the same vertical resolution but only half the horizontal resolution; that is, for every four luma samples in the horizontal direction, there are two *Cr* and two *Cb* samples. The 4:2:0 pattern corresponds to the case where both *Cr* and *Cb* have half the horizontal and the vertical resolution of Y'. The 4:2:0 pattern requires half as many samples—and, consequently, half as many bits—as the 4:4:4 video.

When chroma subsampling is used, the encoder discards selected color-difference samples after filtering. At the decoder side, these missing samples are approximated by interpolation. The most common interpolation scheme consists of using averaging filters whose size and shape vary according to the chroma subsampling pattern used at the encoder side.

■ EXAMPLE 20.2

Calculate the number of bits per frame required to encode a color Rec. 601 625/50 (720×480 pixels per frame) video, using the following chroma subsampling patterns: 4:4:4, 4:2:2, and 4:2:0.

Solution

The key to this problem is to calculate the equivalent number of bits per pixel (N_{b}) for each case. Once the value of N_{b} is obtained, it is a simple matter of multiplying that value by the frame height and width.

(a) $N_{\text{b}} = 8 \times 3 = 24$. The total number of bits is $720 \times 480 \times 24 = 8{,}294{,}400$.

(b) Since for every group of four pixels, eight samples are required, $N_b = 8 \times 8/4 = 16$. The total number of bits is $720 \times 480 \times 16 = 5{,}529{,}600$.

(c) Since for every group of four pixels, six samples are required, $N_{\text{b}} = 8 \times 6/4 = 12$. The total number of bits is $720 \times 480 \times 12 = 4{,}147{,}200$.

20.8 DIGITAL VIDEO FORMATS AND STANDARDS

The topic of digital video formats and standards is a broad and ever-changing one about which many entire books have been written. The enormous amount of formats and standards—and the myriad of technical details within each and every one of them—can easily overwhelm the reader. In this section, we present a brief summary of relevant digital video formats and standards. This list will be expanded in Section 20.9 when we incorporate (an even larger number of) standards for compressed video.

Digital video formats vary according to the application. For digital video with quality comparable to analog SDTV, the most important standard is the ITU-R (formerly CCIR) Recommendation BT.601-5, 4:2:2[7] (Section 20.8.1). For many video coding applications—particularly in video conferencing and video telephony—a family of

[7]For most of this chapter, we refer to this format simply as *Rec. 601*.

TABLE 20.2 Representative Digital Video Formats

Format (Application)	Y' size ($H \times V$)	Color Sampling	Frame Rate	Raw Data (Mbps)
QCIF (video telephony)	176 × 144	4:2:0	30p	9.1
CIF (videoconference)	352 × 288	4:2:0	30p	37
SIF (VCD, MPEG-1)	352 × 240 (288)	4:2:0	30p/25p	30
Rec. 601 (SDTV distribution)	720 × 480 (576)	4:2:0	60i/50i	124
Rec. 601 (video production)	720 × 480 (576)	4:2:2	60i/50i	166
SMPTE 296M (HDTV distribution)	1280 × 720	4:2:0	24p/30p/60p	265/332/664
SMPTE 274M (HDTV distribution)	1920 × 1080	4:2:0	24p/30p/60i	597/746/746

Reproduced from [WOZ02].

intermediate formats is used (Section 20.8.2). For digital television, different standards are adopted in different parts of the world. Table 20.2 provides a sample of relevant digital video formats, ranging from low bit rate QCIF for video telephony to high-definition SMPTE 296M and SMPTE 274M for HDTV broadcast.

20.8.1 The Rec. 601 Digital Video Format

The ITU-R Recommendation BT.601-5 is a digital video format widely used for television production. The luma component of the video signal is sampled at 13.5 MHz (which is an integer multiple of the line rate for both 50i and 60i standards) and the chroma at 6.75 MHz to produce a 4:2:2 $Y'CrCb$ component signal. The parameters of the sampled digital signal depend on the video frame rate (25 or 30 fps) and are shown in Table 20.3. It can be seen that the higher (30 fps) frame rate is compensated by a lower resolution, so the total bit rate is the same in both cases (216 Mbps). Figure 20.13 shows the resulting *active area*—720 × 480 and 720 × 576—for each case as well as the number of pixels left out to make up for horizontal and vertical blanking intervals (shaded portion of the figure).

The Rec. 601 formats are used in high-quality (standard definition) digital video applications. The 4:4:4 and 4:2:2 are typically used for video production and editing, whereas the 4:2:0 variant is used for video distribution, whether on DVDs, video on demand, or other format. The MPEG-2 compression standard was primarily developed for compression of Rec. 601 4:2:0 signals, although it has been made flexible enough to also handle video signals in lower and higher resolutions. The typical compression ratio achieved by MPEG-2-encoded Rec. 601 4:2:0 videos allows a reduction in data rate from 124 Mbps to about 4–8 Mbps.

TABLE 20.3 ITU-R Recommendation BT.601-5 Parameters

Parameters	525 lines, 30 (29.97) fps	625 lines, 25 fps
Fields Per Second	60 (59.94)	50
Luma channel		
Bandwidth (MHz)	5.5	5.5
Sampling frequency (MHz)	13.5	13.5
Number of samples per line	858	864
Number of samples per active line	720	720
Bits per sample	8	8
Bit rate (Mbps)	108	108
Chroma (color-difference) channels		
Bandwidth (MHz)	2.2	2.2
Sampling frequency (MHz)	6.75	6.75
Number of samples per line	429	432
Number of samples per active line	355	358
Bits per sample	8	8
Bit rate (Mbps)	54	54
Total bit rate (Mbps)	216	216

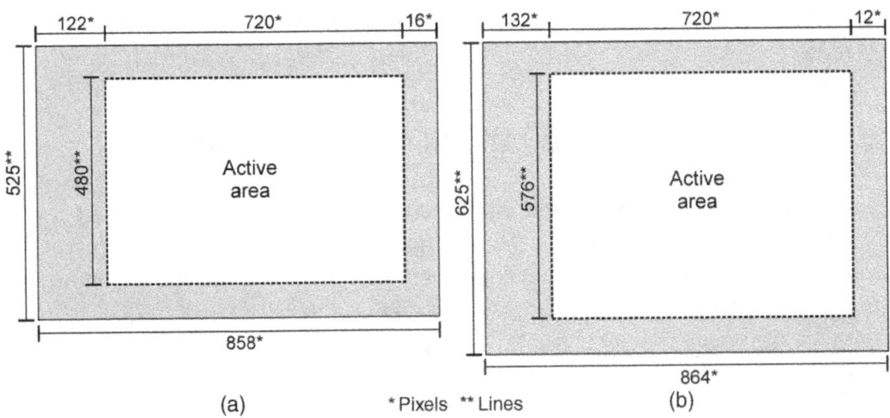

FIGURE 20.13 Rec. 601 format: screen dimensions and active area for the 525/59.94/2:1 (a) and 625/50/2:1 (b) variants. Redrawn from [WOZ02].

20.8.2 The Common Intermediate Format

For video coding applications, video is often converted into an "intermediate format" before compression and transmission. The names and frame sizes for the common intermediate format (CIF) family are presented in Table 20.4. The CIF format has about half of the horizontal and the vertical resolution of the Rec. 601 4:2:0. It was primarily developed for videoconference applications. The QCIF format, whose resolution is half of that of CIF in either dimension—hence the prefix *quarter*—is used for videophone and similar applications. Both are noninterlaced.

TABLE 20.4 Intermediate Formats

Format	Luma Resolution ($H \times V$)
Sub-QCIF	128×96
Quarter CIF (QCIF)	176×144
CIF	352×288
4 CIF	704×576

20.8.3 The Source Intermediate Format

The SIF source intermediate format is an ISO-MPEG format whose characteristics are about the same as CIF. It was targeted at video applications with medium quality such as Video CD (VCD) and video games. There are two variants of SIF: one with a frame rate of 30 fps and a line number of 240 and the other with a frame rate of 25 fps and a line number of 288. Both have 352 pixels per line. There is also a corresponding set of SIF-I (2:1 interlaced) formats. SIF files are often used in connection with the MPEG-1 compression algorithm that can reduce the necessary data rate to transmit them from 30 to approximately 1.1 Mbps with a visual quality comparable to a standard VHS VCR tape (which is lower than broadcast SDTV). MPEG-1-based VCDs have become all but obsolete with the popularization of MPEG-2-based DVDs. This story might take yet another turn with the popularization of *Blu-Ray* high-definition DVDs.

20.9 VIDEO COMPRESSION TECHNIQUES AND STANDARDS

In this section, we present a brief and broad overview of video compression principles, techniques, and standards. As anticipated in Chapter 17, video compression techniques exploit a type of redundancy not available in the case of image compression, but is quite intuitive for the case of video sequences, based on the fact that consecutive frames in time are usually very similar. This is often called interframe (or *temporal*) redundancy.

The simplest way to leverage the similarity between consecutive frames and save bits while encoding a video sequence is to use *predictive coding* techniques. The basic idea behind predictive coding is to predict the contents of frame $k + 1$ based on frame k, compute the difference between the predicted and the actual frame, and encode that difference. The simplest predictor would be one that claims that frame $k + 1$ is identical to frame k, and the simplest measure of difference would be the subtraction of one frame from the other. For certain video sequences (e.g., sequences with fixed background and little foreground activity), this naive approach may work surprisingly well. After all, regardless of the entropy-based technique used to convert the differences into bits, the frame differences are likely to be small (i.e., a significant entropy reduction will be obtained) and the bit savings will be significant.

In practice, predictive video coding techniques do better than that: they use block-based motion estimation (ME) techniques (described in Chapter 22) to estimate motion vectors containing the amount and direction of motion of blocks of pixels in the frame, compute a predicted frame based on the ME results, and encode the errors (between the actual pixel values and the predicted ones). Once again, since the differences are significantly smaller than the actual pixel motion vectors' values, the bit savings will be substantial. Moreover, the method will work for videos with virtually any amount or type of motion.

There is no universally accepted taxonomy for video compression and coding techniques. For the purposes of this book, we will classify them into four major groups:

- *Transform Based*: This group includes techniques that leverage the properties of mathematical transforms, particular the discrete cosine transform (DCT).
- *Predictive*: Techniques that encode the error terms obtained by comparing an actual frame with a predicted version of that frame based on one or more of its preceding (and succeeding) frames.
- *Block-Based Hybrid*: Combine transform-based techniques (applied to nonoverlapping blocks in each frame) with predictive methods (using motion compensated predictions of frame contents based on adjacent frames). Used in several influential video coding standards, such as H.261, H.263, MPEG-1, and MPEG-2.
- *Advanced*: In this group, we include techniques based on 2D shape and texture, as well as scalable, region-based, and object-based video coding techniques used in standards such as MPEG-4 and H.264.

20.9.1 Video Compression Standards, Codecs, and Containers

Numerous video compression standards have been proposed during the past 20 years. Some have been widely adopted and incorporated into consumer electronics products and services, while some others were little more than an academic exercise. Navigating the landscape of video coders and decoders (or *codecs* for short) can be a daunting task: there are too many standards, their scope and applications often overlap, their availability ranges from proprietary to open source, the terminology is not always consistent, and their relevance can be fleeting. To compound the problem a bit further, end users normally manipulate video files using *containers* (e.g., AVI or MOV) mapping to different subsets of codecs, which cannot simply be determined by the file extension and may not be widely supported by different video players.

To help you make sense of this picture and adopt a consistent terminology, we will refer to a *standard* as a collection of official documents usually issued by an international organization, a *codec* as a software implementation of one or more standards, and a *container* as a file format that acts as a wrapper around the encoded audiovisual data, additional information (e.g., subtitles), and associated metadata. Here is an overview of relevant video coding and compression standards, codecs, and containers at the time of this writing.

MPEG Standards The MPEG has overseen the development of video coding standards for consumer applications. The best-known representatives of the MPEG family of video compression standards are MPEG-1 (for VCD quality videos), MPEG-2 (which became a standard for digital broadcast TV), and MPEG-4 (which is significantly more complex than its predecessors and covers a broad range of applications).

ITU-T Standards The International Telecommunications Union (ITU-T) regulated the standardization of video formats for telecommunication applications, for example, video telephony and videoconferencing over conventional telephone lines. The most prominent results of those efforts were the H.261 standard for video telephony, the H.263 standard (and its variants) for videoconferencing, and the H.264 standard, which would become—in a joint effort with the MPEG-4 Part 10, or MPEG-4 AVC (for *Advanced Video Coding*) standard—the most successful video coding standard of the early twenty-first century.

Open Source Codecs There are many free software libraries containing open source implementations of popular video coding standards available online, such as x264 (for H.264/MPEG-4 AVC), Xvid (for MPEG-4), and FFmpeg (for many formats and standards).

Proprietary Codecs Proprietary codecs include DivX (for MPEG-4), Sorenson (used in Apple's QuickTime and Adobe Flash), Microsoft's Windows Media Video (WMV) (which is also the name of a container), and RealNetworks's RealVideo.

Popular Video Containers The most popular video containers in use today are the 3GP (for 3G mobile phones), Microsoft's Advanced Systems Format (.asf, .wma, .wmv), Microsoft's AVI, the DivX Media Format, Adobe Systems' Flash Video (whose notable users include YouTube, Google Video, and Yahoo! Video), MP4 (MPEG-4 Part 14), MPEG video file (.mpg, .mpeg), and Apple's Quicktime (.mov, .qt).

20.10 VIDEO PROCESSING IN MATLAB

MATLAB provides the necessary functionality for basic video processing using short video clips and a limited number of video formats. Not long ago, the only video container supported by built-in MATLAB functions was the AVI container, through functions such as `aviread`, `avifile`, `movie2avi`, and `aviinfo` (explained in more detail later). Moreover, the support was operating system dependent (on UNIX platforms, only uncompressed AVI files are supported) and limited only to a few codecs. Starting with MATLAB Version 7.5 (R2007b), a new library function (`mmreader`) was added to extend video support to formats such as AVI, MPEG, and WMV in a platform-dependent way: Windows machines can be used to read AVI (.avi), MPEG-1 (.mpg), Windows Media Video (.wmv, .asf, .asx), and any format supported by Microsoft DirectShow, whereas Mac users can employ it to read

AVI (.avi), MPEG-1 (.mpg), MPEG-4 (.mp4, .m4v), Apple QuickTime Movie (.mov), and any format supported by QuickTime.

MATLAB's ability to handle matrices makes it easy to create and manipulate 3D or 4D data structures to represent monochrome and color video, respectively, provided that the video sequences are short (no more than a few minutes worth of video). Moreover, once a frame needs to be processed individually, it can be converted to an image using the `frame2im` function, which can then be processed using any of the functions available in the Image Processing Toolbox (IPT).

Finally, it is worth mentioning that another MathWorks product, Simulink, contains a Video and Image Processing Blockset[8] that can be integrated with MATLAB and its closest toolboxes, particularly the IPT and the Image Acquisition Toolbox (IAT).

20.10.1 Reading Video Files

The MATLAB functions associated with reading video files are as follows:

- `aviread`: reads an AVI movie and store the frames into a MATLAB `movie` structure.
- `aviinfo`: returns a structure whose fields contain information (e.g., frame width and height, total number of frames, frame rate, file size, etc.) about the AVI file passed as a parameter.
- `mmreader`: constructs a multimedia reader object that can read video data from a variety of multimedia file formats.

You will learn how to use these functions in Tutorial 20.1.

20.10.2 Processing Video Files

Processing video files usually consist of the following steps (which can embedded in a `for` loop if the same type of processing is to be applied to all frames in a video):

1. Convert frame to an image using `frame2im`.
2. Process the image using any technique such as the ones described in Part I.
3. Convert the result back into a frame using `im2frame`.

20.10.3 Playing Video Files

The MATLAB functions associated with playing back video files are as follows:

- `movie`: primitive built-in video player
- `implay`: fully functional image and video player with VCR-like capabilities

You will learn how to use these functions in Tutorial 20.1. (page 528).

[8] A *blockset* in Simulink is equivalent to a *toolbox* in MATLAB.

20.10.4 Writing Video Files

The MATLAB functions associated with writing video files are as follows:

- `avifile`: creates a new AVI file that can then be populated with video frames in a variety of ways.
- `movie2avi`: creates an AVI file from a MATLAB `movie`.

You will learn how to use some of these functions in Tutorial 20.1.

20.11 TUTORIAL 20.1: BASIC DIGITAL VIDEO MANIPULATION IN MATLAB

Goal

The goal of this tutorial is to learn how to read and view video data in MATLAB, as well as extract and process individual frames.

Objectives

- Learn how to gather video file information using the `aviinfo` function.
- Learn how to read video data into a variable using the `aviread` function.
- Explore the `montage` function for viewing multiple frames simultaneously.
- Learn how to play a video using the `movie` function and the `implay` movie player.
- Learn how to convert from frame to image and vice versa using the `frame2im` and `im2frame` functions.
- Explore techniques for assembling images into video, including the `immovie` function.
- Learn how to write video data to a file using the `movie2avi` function.
- Learn how to read and play video files in different formats using the `mmreader` function.

What You Will Need

- Test files `original.avi` and `shopping_center.mpg`.

Procedure

We will start by learning how to use built-in functions to read information about video files and load them into the workspace. The `aviinfo` function takes a video file name as its parameter and returns information about the file, such as compression and number of frames.

Reading Information About Video Files

1. Read information about the `original.avi` file and save it in a local variable.

```
file_name = 'original.avi';
file_info = aviinfo(file_name);
```

2. View the video compression and the number of frames for this file.

```
file_info.VideoCompression
file_info.NumFrames
```

Question 1 What other information does the `aviinfo` function provide?

Question 2 Try viewing information for another AVI video file. What parameters are different for the new file?

Reading a Video File

The function `aviread` allows us to load an AVI file into the MATLAB workspace. The data are stored as a structure, where each field holds information for each frame.

3. Load the example.avi file using the `aviread` function.

```
my_movie = aviread(file_name);
```

Question 3 What is the size of the `my_movie` structure?

We can also load individual frames from a video file by specifying the frame numbers as a second parameter.

4. Load frames 5, 10, 15, and 20.

```
frame_nums = [5 10 15 20];
my_movie2 = aviread(file_name, frame_nums);
```

Question 4 What is the size of `my_movie2` structure?

Viewing Individual Frames

When we use `aviread` to load a movie file, each element of the structure holds information for that particular frame. This information is stored in two fields: `cdata`, which is the actual image data for that frame, and `colormap`, which stores the color map for the `cdata` field when the image type is indexed. If the image is truecolor, then the `colormap` field is left blank.

5. Inspect the first frame of the my_movie structure.

```
my_movie(1)
```

6. View the first frame as an image using the imshow function.

```
imshow(my_movie(1).cdata)
```

We can view all the frames simultaneously using the montage function. This function will display images in an array all at once in a grid-like fashion.

7. Preallocate a 4D array that will hold the image data for all frames.

```
image_data = uint8(zeros(file_info.Height, file_info.Width, 3, ...
    file_info.NumFrames));
```

8. Populate the image_data array with all the image data in my_movie.

```
for k = 1:file_info.NumFrames
    image_data(:,:,:,k) = my_movie(k).cdata;
end
```

9. Use the montage function to display all images in a grid.

```
montage(image_data)
```

Question 5 Explain how the data for each frame are stored in the image_data variable.

Playing a Video File

The function movie will play video data.

10. Play the video with default settings.

```
movie(my_movie)
```

Question 6 What is the default frame rate when playing a video?

11. Play the video five times with a frame rate of 30 fps.

```
movie(my_movie, 5, 30)
```

12. Play only frames 1–10.

```
frames = [5 1:10];
movie(my_movie, frames, 30)
```

Question 7 How many times will this movie play?

Question 8 At what frame rate will this movie play?

As you have probably noticed, the `movie` function has very limited functionality, with no support for simple operations such as pausing and stepping through frames. To make video analysis easier, we will use the `implay` function.

13. Play the movie with the `implay` function.

```
implay(my_movie)
```

The `implay` function opens a movie player with VCR-like controls and several familiar options available on other video players, for example, Apple QuickTime or Microsoft Windows Media Player.

Question 9 Explore the user interface of the movie player. How can we specify the frame rate of playback?

Question 10 How do we play back the movie in a continuous loop?

Processing Individual Frames

To perform image processing operations on individual frames, we can use the `frame2im` function, which will convert a specified frame to an image. Once we have this image, we can use familiar image processing tools such as the ones described in Part I.

14. Convert frame 10 to an image for further processing.

```
old_img = frame2im(my_movie(10));
```

15. Blur the image using an averaging filter and display the result.

```
fn = fspecial('average',7);
new_img = imfilter(old_img, fn);
figure
subplot(1,2,1), imshow(old_img), title('Original Frame');
subplot(1,2,2), imshow(new_img), title('Filtered Frame');
```

16. Using another frame, create a negative and display the result.

```
old_img2 = frame2im(my_movie(15));
image_neg = imadjust(old_img2, [0 1], [1 0]);
figure
subplot(1,2,1), imshow(old_img2), title('Original Frame');
subplot(1,2,2), imshow(image_neg), title('Filtered Frame');
```

Now that we have processed our image, we can convert it back to frame using the `im2frame` function.

17. Convert the images back to frames and save the new frames in the video structure.

```
my_movie2 = my_movie;
new_frame10 = im2frame(new_img);
new_frame15 = im2frame(image_neg);
my_movie2(10) = new_frame10;
my_movie2(15) = new_frame15;
```

Question 11 Use `implay` to view the new video with the two processed frames. Which of the two is more noticeable during playback?

A straightforward way to perform processing on all frames is to use a `for` loop.

18. Create a negative of all the frames and reconstruct the frames into a video.

```
movie_neg = my_movie;
for k = 1:file_info.NumFrames
    cur_img = frame2im(movie_neg(k));
    new_img = imadjust(cur_img, [0 1], [1 0]);
    movie_neg(k) = im2frame(new_img);
end
implay(movie_neg)
```

If you have an array of image data, another way to construct a video structure is through the `immovie` function. This function will take an array of image data and return a video structure.

19. Create an array of negative images from the original movie.

```
my_imgs = uint8(zeros(file_info.Height, file_info.Width,3, ...
    file_info.NumFrames));
for i = 1:file_info.NumFrames
    img_temp = frame2im(my_movie(i));
    my_imgs(:,:,:,i) = imadjust(img_temp, [0 1], [1 0]);
end
```

20. Now construct a video structuring using the `immovie` function.

```
new_movie = immovie(my_imgs);
implay(new_movie);
```

Question 12 How does this procedure differ from using the `im2frame` function in a loop? Is it beneficial?

Writing to a Video File

To write movies, that is, image sequences, to a video file, we use the `movie2avi` function. Note that in the following steps an AVI file will be created in your current directory, so make sure you have the permissions to do so.

21. Set the file name of the new movie to be created on disk.

```
file_name = 'new_video.avi';
```

22. Create the AVI file.

```
movie2avi(new_movie, file_name, 'compression', 'None');
```

23. Read and play the resulting AVI file.

```
my_movie3 = aviread(file_name);
implay(my_movie3);
```

Reading and Playing Video Files in Different Formats

24. Use the sequence below to read and play the first 100 frames of an MPEG movie.

```
obj = mmreader('shopping_center.mpg');
video = read(obj, [1 100]);
frameRate = get(obj,'FrameRate')
implay(video, frameRate)
```

Question 13 What are the frame rate, frame width, frame height, and total number of frames in file `shopping_center.mpg`?

FIGURE 20.14 Visual representation of a YUV file.

20.12 TUTORIAL 20.2: WORKING WITH YUV VIDEO DATA

Goal

The goal of this tutorial is to learn how to read, process, and display YUV video data[9] in MATLAB.

Objectives

- Explore the contents of a YUV video file.
- Explore the `readYUV` function used to read in YUV video data.

What You Will Need

- Test files `miss_am.yuv`, `foreman.yuv`, and `whale_show.yuv`
- The `readYUV.m` script

Procedure

A YUV video file contains raw video data stored as separate components: the luminance component Y is first, followed by the U and V chrominance components (Figure 20.14). Owing to chroma subsampling (Section 20.7), each U and V component contains only 1/4 of the data that Y does.

For the following steps, note that the `miss_am.yuv` video contains 30 frames, each formatted as `QCIF_PAL`, that is, 176×144 pixels.

1. Ensure that your current directory contains the file `miss_am.yuv`.

Question 1 Based on the format and number of frames in the video, calculate the size of the file (in kB).

Question 2 Does this value coincide with what MATLAB displays in the *Current Directory* pane?

YUV files do not use headers to store any information about the video itself, for example, frame size, frame rate, or color standard. Consequently, you must know in advance what those parameters are and use them whenever needed. Once we obtain this information, we can parse the video file, reading the proper data for each of

[9]YUV video files are very common in video processing research and will also be used in the following chapters.

the components and then converting it to a MATLAB video structure for further manipulation.

To read in a YUV file, we will use the `readYUV` function[10] and the test file `miss_am.yuv`.

2. Define all relevant data before calling the `readYUV` function.

```
file_name = 'miss_am.yuv';
file_format = 'QCIF_PAL';
num_of_frames = 30;
```

3. Read the video data and display the movie using `implay`.

```
[yuv_movie, yuv_array] = readYUV(file_name, num_of_frames, ...
    file_format);
implay(yuv_movie)
```

Question 3 Describe the contents and record the dimensions of the `yuv_movie` variable.

Question 4 Describe the contents and record the dimensions of the `yuv_array` variable. How is this variable different from `yuv_movie`?

4. Try to read and play the `foreman.yuv` and `whale_show.yuv` video files.

```
file_name = 'foreman.yuv';
file_format = 'QCIF_PAL';
num_of_frames = 25;
[yuv_movie, yuv_array] = readYUV(file_name, num_of_frames, ...
    file_format);
implay(yuv_movie)

file_name = 'whale_show.yuv';
file_format = 'NTSC';
num_of_frames = 25;
[yuv_movie, yuv_array] = readYUV(file_name, num_of_frames, ...
    file_format);
implay(yuv_movie)
```

Question 5 After reading both these files, show how do their sizes compare and why?

[10]This function was developed by Jeremy Jacob and can be downloaded from the book web site.

Question 6 When using the `readYUV` function, what happens if you specify more frames than there actually are in the file?

Question 7 What happens if you specify fewer frames than there are in the file?

WHAT HAVE WE LEARNED?

- Video is the electronic representation of visual information whose spatial distribution of intensity values varies over time.
- The video signal is a 1D analog or digital signal varying over time in such a way that the spatiotemporal information is ordered according to a predefined scanning convention that samples the signal in the vertical and temporal dimensions.
- An analog video raster is a fixed pattern of parallel scanning lines disposed across the image. A raster's main parameters include the line rate, the frame interval, the line spacing, the line interval, and the number of active lines.
- Some of the most important concepts and terminologies associated with (analog) video processing include
  - *Aspect Ratio*: the ratio of frame width to height.
  - *Vertical Resolution*: the number of horizontal black and white lines in the image (vertical detail) that can be distinguished or resolved in the picture height; it is a function of the number of scanning lines per frame.
  - *Horizontal Resolution*: the number of vertical lines in the image (horizontal detail) that can be distinguished in a dimension equal to the picture height; it is determined by the signal bandwidth in analog systems.
  - *Progressive Scanning*: the process by which each image is scanned in one single pass called *frame*.
  - *Interlaced Scanning*: the process by which each frame is scanned in two successive vertical passes, first the odd numbered lines and then the even numbered ones. Each pass is called a *field*.
  - *Blanking Interval*: the time interval at the end of each line (horizontal retrace) or field (vertical retrace) during which the video signal must be blanked before a new line or field is scanned.
  - *Component Video*: an analog video representation scheme that uses three 1D color component signals.
  - *Composite Video*: an analog video representation scheme that combines luminance and chrominance information into a single composite signal.
- Gamma correction is a process by which the nonlinearity of an image acquisition or display device is precompensated (on the transmitter's side) in such a way as to ensure the display of the correct dynamic range of color values within the signal.
- Color information is encoded in analog video systems in a backward-compatible way. The (gamma-corrected) luma signal and the three (gamma-corrected)

primary color channels are combined into color-difference signals that are then used to modulate a color subcarrier. The frequency of the color subcarrier is carefully chosen so that its spectral components fit within the gaps of the existing (monochrome) spectrum.

- The most popular analog SDTV and video standards used worldwide are NTSC and PAL (with SECAM in a distant third place). Although they share many common ideas (interlaced scanning, QAM color modulation, etc.), these standards use different values for most of the main raster parameters (such as lines per frame and frames per second).
- A digital video is a sampled two-dimensional (2D) version of a continuous three-dimensional (3D) scene. Digital video not only employs vertical and temporal sampling—similar to analog video—but also includes horizontal sampling, that is, sampling of pixel values along a line.
- The main parameters that characterize a digital video sequence are the frame rate, the line number, and the number of samples per line. The product of these three parameters times the average number of bits per pixel provides an estimate of the data rate needed to transmit that video in its uncompressed form.
- Video digitization, or analog-to-digital conversion, involves four main steps: (1) antialiasing filter—an optional low-pass filter used to eliminate signal frequency components that could cause aliasing; (2) sampling—pixel values are sampled along a horizontal line at a standardized rate; (3) quantizing—the samples are represented using a finite number of bits per sample; and (4) encoding—the quantized samples are converted to binary codewords.
- Chroma subsampling is the process of using a lower spatial resolution, and consequently fewer bits, to represent the color information of a digital video frame.
- Some of the most popular contemporary digital video formats are QCIF, CIF, SIF, Rec. 601, SMPTE 296M, and SMPTE 274M.
- MATLAB can be used to read, process, and play back digital video files in several different formats.

LEARN MORE ABOUT IT

- The following is a list of selected books on video processing and related fields:

  - Bovik, A. (ed.), *Handbook of Image and Video Processing*, San Diego, CA: Academic Press, 2000.
  - Grob, B. and Herndon, C. E., *Basic Television and Video Systems*, 6th ed., New York: McGraw-Hill, 1999.
  - Haskell, B. G., Puri, A., and Netravali, A. N., *Digital Video: an Introduction to MPEG-2*, Norwell, MA: Kluwer Academic Publishers, 1997.
  - Jack, K., *Video Demystified: A Handbook for the Digital Engineer*, 3rd ed., Eagle Rock, VA: LLH Technology Publishing, 1993.

  - Luther, A. C. & Inglis, A. F., *Video Engineering*, 3rd ed., New York: McGraw-Hill, 2000.
  - Poynton, C., *Digital Video and HDTV Algorithms and Interfaces*, San Francisco, CA: Morgan Kaufmann, 2003.
  - Poynton, C., *A Technical Introduction to Digital Video*, New York: Wiley, 1996.
  - Robin, M. and Poulin, M., *Digital Television Fundamentals: Design and Installation of Video and Audio Systems*, 2nd ed., New York: McGraw-Hill, 2000.
  - Tekalp, A. M., *Digital Video Processing*, Upper Saddle River, NJ: Prentice Hall, 1995.
  - Wang, Y., Ostermann, J., and Zhang, Y.-Q., *Video Processing and Communications*, Upper Saddle River, NJ: Prentice-Hall, 2002.
  - Watkinson, J., *The Art of Digital Video*, 3rd ed., Oxford: Focal Press, 2000.
  - Whitaker, J. C. and Benson, K. B. (Ed.), *Standard Handbook of Video and Television Engineering*, 3rd ed., New York: McGraw-Hill, 2000.
  - Woods, J. W., *Multidimensional Signal, Image, and Video Processing and Coding*, San Diego, CA: Academic Press, 2006.
- The book by Grob and Herndon [GH99] provides a broad and detailed coverage of analog TV and video systems.
- Chapter 2 of [RP00] and Chapters 8 and 9 of [Jac01] describe analog video standards in great detail.
- For more on gamma correction, please refer to Chapter 23 of [Poy03] and Section 5.7 of [BB08].
- Chapter 3 of [RP00] has a very detailed explanation of sampling, quantization, and (component and composite) digital video standards from a TV engineering perspective.
- Chapter 10 of [GH99] explains in detail the contents of VBI in analog TV systems.
- Chapter 13 of [WOZ02] provides a good overview of video compression standards.
- Chapter 2 of [Wat00] contains a broad overview of video principles.

ON THE WEB

- World TV Standards
  http://www.videouniversity.com/standard.htm
- Advanced Television Systems Committee (ATSC): American standards
  http://www.atsc.org/
- Digital Video Broadcasting Project (DVB): European standards
  http://www.dvb.org/

- Society of Motion Picture and Television Engineers (SMPTE)
  http://www.smpte.org/home
- The MPEG home page
  http://www.chiariglione.org/mpeg/
- MPEG Industry Forum
  http://www.mpegif.org/
- Test images and YUV videos—Stanford University
  http://scien.stanford.edu/pages/labsite/scien_test_images_videos.php
- YUV 4:2:0 video sequences—Arizona State University
  http://trace.eas.asu.edu/yuv/index.html
- Test video clips (with ground truth) from the CAVIAR project
  http://homepages.inf.ed.ac.uk/rbf/CAVIARDATA1/
- Charles Poynton's "Gamma FAQ"
  http://www.poynton.com/GammaFAQ.html
- Adobe digital audio and video primers
  http://www.adobe.com/motion/primers.html
- Keith Jack's blog
  http://keithjack.net/

20.13 PROBLEMS

20.1 Explain in your own words the differences between composite and component video systems in terms of bandwidth, historical context, and image quality.

20.2 The human visual system is much more sensitive to changes in the brightness of an image than to changes in color. In other words, our color spatial resolution is poor compared to the achromatic spatial resolution.

Knowledge of this property has been somehow embedded in the design of both analog and digital video systems. Answer the following questions:

(a) How did this property get exploited in analog TV systems?

(b) How did it get factored into the design of digital video formats?

20.3 Provide an objective explanation as to *why* the luminance signal in PAL/NTSC systems is calculated as $Y = 0.299R + 0.587G + 0.114B$, and *not* $Y = (R + G + B)/3$.

20.4 Calculate the raw data rate of a digital video signal with the following characteristics:

- 2:1 interlaced scanning
- 352×240 luminance samples per frame
- 30 frames per second
- 4:2:2 chroma subsampling

20.5 Regarding progressive and interlaced scanning methods,

(a) What are the pros and cons of each method?

(b) For the same line number per frame, what is the relation between the maximum temporal frequency that a progressive raster can have and that of an interlaced raster that divides each frame into two fields?

(c) What about the relation between the maximum vertical frequencies?

20.6 Which considerations would you use to determine the frame rate and line number when designing a video capture or display system?

20.7 Why does a computer monitor use a higher temporal refresh rate and line number than the one adopted by a typical TV monitor?

20.8 Regarding NTSC and PAL color TV systems,

(a) What do Y, I, and Q stand for in NTSC?

(b) What do Y, U, and V stand for in PAL?

(c) How are I and Q related to U and V?

(d) What is the relationship between NTSC's Y, PAL's Y, and CMY(K) color model's Y?

20.9 Write a MATLAB script to

(a) read an RGB color image and convert it to $Y'CrCb$;

(b) subsample this image into the 4:2:0 format;

(c) upsample the Cr and Cb components to full size (4:4:4 format) and convert the result back to RGB;

(d) Compute the difference between the original and processed RGB images.

20.10 Prove (using simple numerical calculations) that the raw data rate required to transmit an SMPTE 295M digital video signal (1920×1080 luminance samples per frame, 30 frames per second, progressive scanning, 4:2:0 chroma subsampling) is approximately 746 Mbps.

CHAPTER 21

VIDEO SAMPLING RATE AND STANDARDS CONVERSION

WHAT WILL WE LEARN?

- What is sampling rate conversion?
- What are the main practical aspects involved in converting a video sequence from one format to another?
- Which steps are involved in PAL to NTSC (and vice versa) standard conversion?

21.1 VIDEO SAMPLING

As discussed in Chapter 20, although video signals vary continuously in space and time, most TV cameras capture a video sequence by sampling it in the temporal and vertical dimensions. The resulting signal is stored in a 1D raster scan, which is a concatenation of the intensity (or color) variations along successive horizontal scan lines. A third type of sampling (along each line) takes place when an analog video is converted to a digital representation. In all three cases, the fact that selected intensity values are being sampled (while others are being left off) implies that there might be some loss in the process. To minimize such losses, the original video sequence should be sampled at the highest possible sampling rate.

Practical Image and Video Processing Using MATLAB®. By Oge Marques.

Based on Nyquist's sampling theorem, the sampling rate in each dimension (x, y, or t) should be at least twice the highest frequency along that direction. On the other hand, studies of the human visual system (HVS) (see Appendix A) have shown that our visual system cannot distinguish spatial and temporal variations beyond certain high frequencies. Consequently, the *visual cutoff frequencies*—which are the highest spatial and temporal frequencies that can be perceived by the HVS—should be the driving factor in determining the sampling rates for video. After all, there is no need to accommodate frequency components beyond those values. In addition to the frequency content of the underlying signal and the visual thresholds in terms of spatial and temporal cutoff frequencies imposed by the HVS, video sampling rates are also determined by the capture and display device characteristics and the associated processing, storage, and transmission costs.

Video sampling is a complex problem that can be mathematically modeled using *lattice theory*, which is beyond the scope of this book.[1] Lattice theory provides an elegant framework to model relationships between the engineering decisions that require adopting less than ideal sampling rates and the impact of those decisions on the frequency-domain representation of the sampled signals.

One of the consequences of using relatively low sampling rates is the problem of *aliasing*, which can manifest itself within a frame (*spatial aliasing*) or across multiple frames (*temporal aliasing*). Spatial aliasing, which is common to images and videos, particularly the appearance of *Moiré patterns*, was discussed in Chapter 5. Temporal aliasing is a related phenomenon, particularly noticeable in video sequences where wheels sometimes appear to move backward and slower, in what is known as the *wagon wheel effect* (also known as the *reverse rotation effect*).

21.2 SAMPLING RATE CONVERSION

The sampling rate conversion problem consists in converting a video sequence with a certain spatial and temporal resolution into another sequence in which one or more of those parameters have changed. Sampling rate conversion falls within one of the following two cases:

- When the original sequence contains *less* sampling points than the desired result, the problem is known as *up-conversion* (or *interpolation*). Up-conversion usually consists of filling all points that are in the desired video sequence but not in the original one with zeros (a process known as *zero padding*), and then estimating the values of these points, which is accomplished by an interpolation filter. Examples of up-conversion involving contemporary TV and video standards include
  - 480i59.94 to 720p59.94
  - 480i59.94 to 1080i59.94

[1] See Chapter 3 of [WOZ02] for a good explanation of video sampling using lattice theory.

  - 576i50 to 576p50
  - 576i50 to 1080i50

  In practice, up-conversion requires increase in resolution and calls for deinterlacing techniques (see Section 21.3.1).

- When the original sequence contains *more* sampling points than the desired result, the problem is known as *down-conversion* (or *decimation*). This is not as simple as just removing additional samples from the original sequence, which would cause aliasing in the down-sampled signal. Proper down-conversion requires applying a prefilter to the signal to limit its bandwidth, therefore avoiding aliasing. In Tutorial 21.1 (page 548), you will learn how to perform line down-conversion using MATLAB.

21.3 STANDARDS CONVERSION

Sampling rate conversion is also sometimes referred to as *standards conversion*. In the context of this book, a *standard* can be defined as a video format that can be specified by a combination of four main parameters: color encoding (composite/component), number of lines per field/frame, frame/field rate, and scanning method (interlaced/progressive). Hence, standard conversion will be understood as the process of converting one or more of those parameters into another format. Contemporary standard conversion examples include

- Frame and line rate conversion, for example, 1250i50 to 525i60 and 625i50 to 1250i50.
- Color standard conversion, for example, PAL to NTSC (see Section 21.3.2).
- Line and field doubling, for example, 625i50 to 1250i100.
- Film to video conversion, for example, 24 Hz film to 60 Hz video, a process known as *3:2 pull-down* (see Section 21.3.5).

21.3.1 Deinterlacing

The problem of deinterlacing is to fill in the skipped lines in each field, as shown in Figure 21.1.

Deinterlacing can be achieved with the use of relatively simple methods[2] (which can be treated as filters in the spatio-temporal domain) that fall into the following categories:

- Vertical interpolation within the same field
  - *Line Averaging*: a simple filter in which the missing line is estimated by averaging the lines above and below: $D = (C + E)/2$.

[2]In Tutorial 21.2 (page 550), you will learn how to perform basic deinterlacing methods using MATLAB.

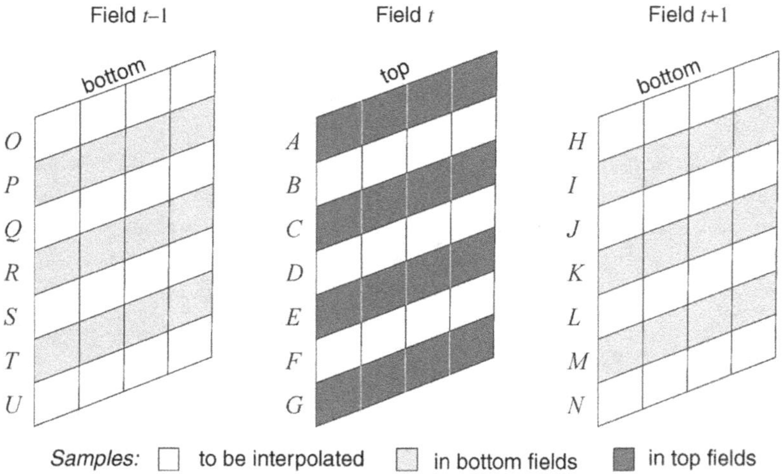

FIGURE 21.1 The deinterlacing process. Fields t and $t + 1$ form one interlaced frame.

  - *Line Averaging Variant*: improves the vertical frequency response upon the previous technique, bringing it closer to the desired (ideal) low-pass filter. In this technique, the missing line is calculated as $D = (A + 7C + 7E + G)/16$.
- Temporal interpolation
  - *Field Merging*: technique by which the deinterlaced frame is obtained by combining (merging) the lines from both fields: $D = K$ and $J = C$. The frequency response of the equivalent filter shows an all-pass behavior along the vertical axis and a low-pass behavior along the temporal axis.
  - *Field Averaging*: improved alternative in which $D = (K + R)/2$, which results in a filter with better (symmetric) frequency response along the vertical axis. The main disadvantage of this technique is that it involves three fields for interpolating any field, increasing storage and delay requirements.
- Temporal interpolation and vertical interpolation combined
  - *Line and Field Averaging*: a technique designed with the goal of achieving a compromise between spatial and temporal artifacts, in which the missing line is calculated as $D = (C + E + K + R)/4$. The frequency response of the equivalent filter is close to ideal, but the method requires storage of two frames.

When there is little or no motion between the two fields, the missing even lines in an odd field should be exactly the same as the corresponding even lines in the previous and the following even field. In these cases, temporal interpolation will yield perfect results. On the other hand, when there is motion in the scene, the corresponding lines in adjacent fields may not correspond to the same object location and temporal interpolation will result in unacceptable artifacts, such as double images. In these cases, although the results of vertical interpolation methods may be acceptable (at the

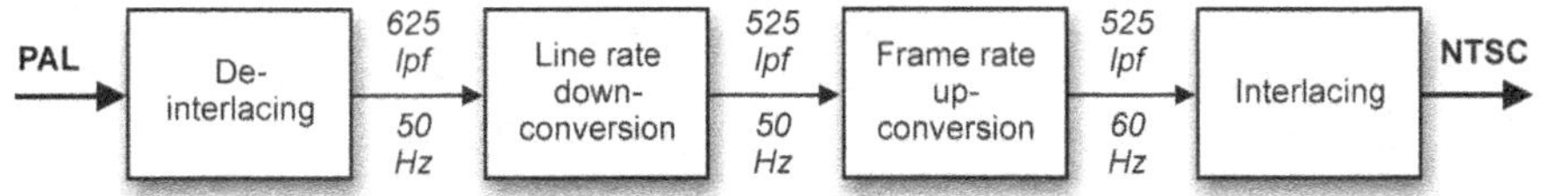

FIGURE 21.2 A practical method for converting PAL to NTSC formats.

cost of lower spatial resolution), the preferred alternative is to use motion-adaptive and motion-compensated interpolation methods.

21.3.2 Conversion between PAL and NTSC Signals

The conversion between PAL and NTSC signals (which in fact should be called *625i50 to 525i60 conversion*) provides a practical example of line number and frame rate conversions combined. This is a particularly challenging technical problem because the vertical and temporal sampling rates of the two video sequences are not integer multiples of each other.

In practice, the problem can be solved in four steps (Figure 21.2):

1. Deinterlacing each field in the PAL signal, resulting in a 625-line frame.
2. Line rate down-conversion from 625 to 525 within each deinterlaced frame.
3. Frame rate up-conversion from 50 to 60 Hz.
4. Splitting each frame into two interlacing fields.

The PAL to NTSC conversion process can be further optimized by

- Replacing the 625-to-525 lines conversion by an equivalent 25-to-21 lines conversion problem.
- Performing the temporal conversion from 50 to 60 Hz by converting every five frames to six frames.

The process of converting NTSC video to its PAL equivalent follows a similar sequence of steps, with obvious modifications: the line rate undergoes up-conversion (from 525 to 625), while the frame rate is down-converted (from 60 to 50 Hz). In Tutorial 21.3 (page 556), you will learn how to perform NTSC to PAL conversion using MATLAB.

21.3.3 Color Space Conversion

Color space conversion is the process of transforming the input color space to that of the output. As discussed previously, there are multiple color spaces in use by different video systems. To ensure the best possible color reproduction quality when converting from one standard to another, it might be necessary to adjust the RGB primaries and the white point to account for the differences in the standards.

In particular, the conversion from Rec. 601 SDTV images to SMPTE 274M HDTV images requires a manipulation of the color matrices to account for, among other things, the different phosphors used in HDTV monitors [Dea01].

The SMPTE 125M—Rec. 601 (Standard Definition) standard uses SMPTE RP145 for chromaticity coordinates:[3]

- Red ($x = 0.6400$, $y = 0.3300$)
- Green ($x = 0.3000$, $y = 0.6000$)
- Blue ($x = 0.1500$, $y = 0.0600$)
- White ($x = 0.3127$, $y = 0.3290$)
- $Y' = 0.2126R' + 0.7152G' + 0.0722B'$
- $U = -0.1146R' - 0.3854G' + 0.5000B'$
- $V = 0.5000R' - 0.4542G' - 0.0458B'$

The SMPTE 274M/296M—1080i/720p (High Definition) standard uses ITU-R BT.709 for all colorimetry measurements:

- Red ($x = 0.6300$, $y = 0.34004$)
- Green ($x = 0.3100$, $y = 0.5950$)
- Blue ($x = 0.1550$, $y = 0.0700$)
- White ($x = 0.3127$, $y = 0.3290$)
- $Y' = 0.2990R' + 0.5870G' + 0.1140B'$
- $U = -0.1690R' - 0.3310G' + 0.5000B'$
- $V = 0.5000R' - 0.4190G' - 0.0810B'$

Color space conversion can be performed by applying a linear transformation to the gamma-corrected components, which produces errors that are still acceptable for most applications. High-end format converters perform additional inverse gamma correction before the linear transformation stage and restore gamma correction after the colors have been linearly mapped to the desired color space, a process that is significantly more complex and expensive to implement [Dea01].

21.3.4 Aspect Ratio Conversion

Aspect ratio (AR) conversion is another aspect that must be addressed when performing certain types of conversion, for example, from SDTV (4:3 aspect ratio) to HDTV (16:9 aspect ratio) or from "widescreen" movies (AR greater or equal to 1.85) to a "full screen" SDTV equivalent (AR equal to 1.33). AR conversion techniques include cropping, stretching, or squeezing each frame, which are not very complex operations. These operations do, however, bring about a number of creative issues

[3]See Section 16.1.2 for an explanation of basic concepts related to chromaticity.

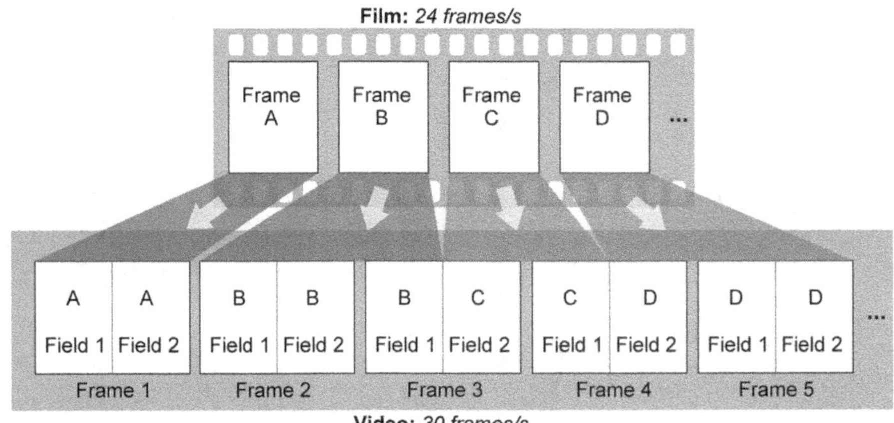

FIGURE 21.3 3:2 pull-down. Redrawn from [Ack01].

related to how frames can be resized and reshaped while keeping most of the contents as intended by the author of the original sequence.

21.3.5 3:2 Pull-Down

3:2 pull-down[4] refers to a technique used to convert material originated at 24 frames per second (fps), typically shot on film, to 30 or 60 fps video for broadcast.[5] In the case of 30 fps interlaced television, this means that each film of four frames is converted to 10 fields or 5 frames of video. In the case of 60 fps progressive scan television, each film of four frames is converted into 10 frames of video.

Figure 21.3 shows the 3:2 pull-down process for converting film material to interlaced video. The film frames are labeled A, B, C, and D; the video frames are numbered 1–5 and the corresponding video fields are called A1, A2, B1, B2, B3, C1, C2, D1, D2, and D3. Note that video frame 3 will contain video from film frames B and C and video frame 4 will contain video from film frames C and D. This is potentially problematic and may cause *mixed frames* to appear in the resulting video sequence when a scene change has occurred between a B and a C frame or between a C and a D frame in the original film sequence.

Another technical issue associated with the conversion from motion picture films to their standard video equivalent (a process known as *telecine*) is the appearance of *judder*. Judder is a motion artifact that causes smooth motion in the original film sequence to appear slightly jerky when telecined. Judder is particularly apparent during slow, steady camera movements. Figure 21.4 shows how the process of copying film frames into video fields using 3:2 pull-down causes the smooth optic flow in the

[4]3:2 pull-down is also known as *2:3 pull-down* in some references, for example, [Ack01].

[5]The process of 3:2 pull-down is necessary only for video formats based on 60 Hz. In 50 Hz systems, the film is simply transferred from 24 to 25 fps (50 fields/s) by running the telecine 4% faster than normal.

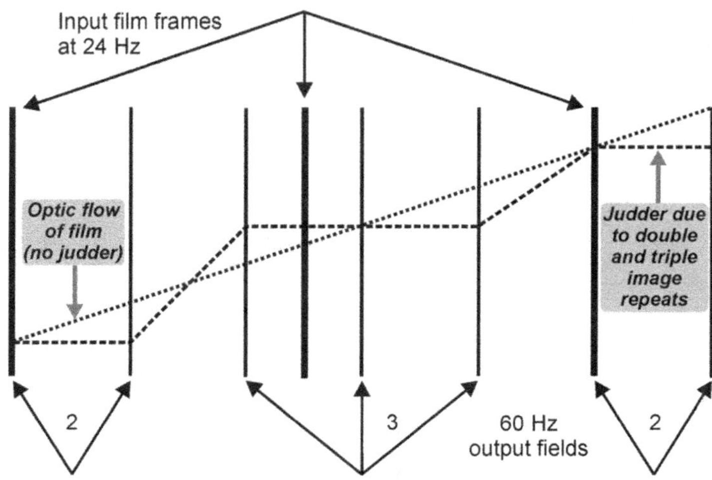

FIGURE 21.4 The problem of judder in telecine using 3:2 pull-down. Redrawn from [Wat94b].

original film sequence to become jerky, with no motion for two or three fields followed by sudden jumps in motion.

21.4 TUTORIAL 21.1: LINE DOWN-CONVERSION

Goal

The goal of this tutorial is to learn how to perform line down-conversion on a video sequence.

Objectives

- Learn how to extract a YUV frame from a video sequence using the `readYUV` function.
- Explore the MATLAB `resample` function to perform up- and down-conversion.
- Explore line down-conversion using the `downConversion` function.

What You Will Need

- MATLAB files `readYUV.m` and `downConversion.m`[6]
- Test file `miss_am.yuv`

[6]These functions were developed by Jeremy Jacob and can be downloaded from the book web site.

Procedure

We will use the readYUV function to read an individual frame from a YUV video file. Recall that when we use this function, both a MATLAB video structure and the YUV frame(s) are returned.

1. Read in the first frame of the miss_am.yuv file.

```
filename = 'miss_am.yuv';
fileformat = 'QCIF_PAL';
numframes = 1;
[movie, yuv] = readYUV(filename, numframes, fileformat);
```

The function resample will resample an image by a factor of p/q, where p and q are defined in the function call. Note that this function will resample only the columns of a matrix, so we must perform the operation twice: once on the transpose of the image and once on the transpose of the transpose, which will bring back the original orientation.

2. Resample the Y component of the frame by a factor of 2.

```
currY = yuv(:, :, 1, 1);
currY_res = double(currY);
currY_res = resample(currY_res', 2, 1);
currY_res = resample(currY_res', 2, 1);
```

Question 1 What is the size of the currY_res image? How does this relate to the original image (currY)?

To perform line down-conversion, we use the downConversion function.

3. Perform line down-conversion on the current frame.

```
currY_dwn = downConversion(currY_res);
figure, subplot(1,2,1), imshow(uint8(currY_res)), ...
    title('2X resampled frame');
subplot(1,2,2), imshow(uint8(currY_dwn)), ...
    title('line down-converted frame');
```

Question 2 How does the size of the converted image compare to the original resampled image?

4. Crop the original image at the bottom and compare it with the line down-converted image.

```
[r, c] = size(currY_dwn);
currY_crp = currY_res(1:r,:);
figure, subplot(1,2,1), imshow(uint8(currY_dwn)), ...
    title('Line down-converted');
subplot(1,2,2), imshow(uint8(currY_crp)), ...
    title('Cropped');
```

Question 3 How does line down-conversion compare to cropping an image?

21.5 TUTORIAL 21.2: DEINTERLACING

Goal

The goal of this tutorial is to learn several methods for deinterlacing a video sequence.

Objectives

- Learn how to obtain the two fields (top and bottom) of an interlaced movie frame.
- Explore deinterlacing by spatial interpolation.
- Explore deinterlacing by line averaging.
- Learn how to perform deinterlacing by temporal interpolation: field merging and field averaging methods.
- Explore deinterlacing by line and field averaging.
- Learn how to create a new *YUV* frame composed of the deinterlaced *Y* frame and the *U*, *V* color component frames.

What You Will Need

- `readYUV.m`
- Test file `foreman.yuv`

Procedure

To begin working with interlaced frames, we will read three frames of a *YUV* video sequence using the `readYUV` function.

1. Set the appropriate variables and read in the first three frames of the `foreman.yuv` video file.

```
filename = 'foreman.yuv';
numframes = 3;
fileformat = 'QCIF_PAL';
[mov,yuv] = readYUV(filename, numframes, fileformat);
```

For the first exercise, we will need only one frame, so we will use the second frame in the sequence. Later we will see that having a frame before and after it will be useful.

2. Extract the Y component of the second frame and store its size for later use.

```
currY = yuv(:, :, 1, 2);
[rows, cols] = size(currY);
```

Question 1 Explain how the statement `currY = yuv(:, :, 1, 2);` above extracts the Y component of the second frame.

Interlacing

To simulate interlaced frames, we can simply extract the even lines and store them in one variable and the odd lines in another.

3. Create interlaced frames and display them with the original Y component.

```
topField = currY(1:2:rows,:);
bottomField = currY(2:2:rows,:);
figure, subplot(2,2,1), imshow(uint8(currY));
subplot(2,2,3), imshow(uint8(topField));
subplot(2,2,4), imshow(uint8(bottomField));
```

4. Restore the interlaced frames to the original size of the frame by filling the blank lines with zeros.

```
topField2 = zeros(rows,cols);
bottomField2 = zeros(rows,cols);
topField2(1:2:rows,:) = topField;
bottomField2(2:2:rows,:) = bottomField;
figure, subplot(2,2,1), imshow(uint8(currY));
subplot(2,2,3), imshow(uint8(topField2));
subplot(2,2,4), imshow(uint8(bottomField2));
```

The lines in the interlaced frames may look distorted due to the resolution at which MATLAB displays the images in the figure. To see a particular frame without distortion, display the frame in its own figure without any subplots.

Line Averaging

We can use spatial interpolation to generate a full frame from an interlaced frame. To do this, we average two lines together and use this to fill in the missing lines.

5. Deinterlace the top and bottom frames using line averaging.

```
deIntField1 = topField2;
deIntField2 = bottomField2;
for m=1:2:rows-2
     for n=1:cols
        lineAverage1(m+1,n) = (deIntField1(m,n) + ...
            deIntField1(m+2,n)) / 2;
     end
end
lineAverage1(1:2:rows,:) = topField2(1:2:rows,:);
for m=2:2:rows-1
     for n=1:cols
        lineAverage2(m+1,n)=(bottomField2(m,n) + ...
            bottomField2(m+2,n)) / 2;
     end
end
lineAverage2(2:2:rows,:) = bottomField2(2:2:rows,:);
```

6. Display the results.

```
figure
subplot(2,2,1), imshow(uint8(currY)), ...
    title('Original frame');
subplot(2,2,3), imshow(uint8(lineAverage1)), ...
    title('Top field averaged');
subplot(2,2,4), imshow(uint8(lineAverage2)), ...
    title('Bottom field averaged');
```

The deinterlaced top frame, stored in `lineAverage1`, is missing one line, as you might have noticed by inspecting its dimensions compared to the original frame. This last line is skipped in the above procedure because there is no line available below (since it is the last line in the frame). Therefore, we will add a line of zeros just to keep the dimensions consistent.

7. Compensate for the missing line in the `lineAverage1` frame.

```
lineAverage1(rows,:) = zeros(1,cols);
```

8. Display the results.

```
figure
subplot(2,2,1), imshow(uint8(currY))
subplot(2,2,3), imshow(uint8(lineAverage1))
subplot(2,2,4), imshow(uint8(lineAverage2))
```

Question 2 This "missing line effect" is also noticeable in `lineAverage2` (the bottom field), but it is at the top of the frame. Why do you think we need to compensate for this in the top field but not the bottom?

Question 3 What are the visual differences between the original frame and the deinterlaced frame?

We can use the `imabsdiff` function to see how the interpolated frames compare to the original.

9. Display the difference between the top and bottom fields to the original frame.

```
dif1 = imabsdiff(double(currY), lineAverage1);
dif2 = imabsdiff(double(currY), lineAverage2);
figure
subplot(1,2,1), imshow(dif1,[]), title('Difference of top field');
subplot(1,2,2), imshow(dif2,[]), title('Difference of bottom field');
```

Because we already have the *U* and *V* components available, we can combine them with one of the new interpolated frames to see what it would look like as an *RGB* image.

10. Convert the interpolated top field to an *RGB* image using the original *U* and *V* components.

```
yuv_deint = yuv(:,:,:,2);
yuv_deint(:,:,1) = lineAverage1;
rgb_deint = ycbcr2rgb(uint8(yuv_deint));
rgb_org = ycbcr2rgb(uint8(yuv(:,:,:,2)));
figure
subplot(1,2,1), imshow(rgb_org), title('Original Frame')
subplot(1,2,2), imshow(rgb_deint), title('Interpolated top field')
```

Question 4 How does the interpolated *RGB* image compare to its original?

Of course, since we have both top and bottom fields, we can merge them, which will result in the original image.

11. Merge the two fields into a single frame.

```
mergeFields = zeros(rows, cols);
mergeFields(1:2:rows,:) = topField2(1:2:rows,:);
mergeFields(2:2:rows,:) = bottomField2(2:2:rows,:);
figure
subplot(1,2,1), imshow(uint8(currY)), ...
    title('Original frame');
```

```
subplot(1,2,2), imshow(uint8(mergeFields)), ...
    title('Merged fields');
```

Question 5 Show that this frame is equivalent to the original by using the `imabsdiff` function.

Field Averaging

Deinterlacing by field averaging requires information from both the preceding frame and the next frame in the sequence. To perform the averaging, we use the lines from the bottom field of the previous frame and the next frame to calculate the missing lines for the top field of the current frame. Similarly, we will use the lines from the top field in the previous and next frames to calculate the missing lines in the bottom field of the current frame.

12. Close any open figures.
13. Display the Y components of the previous, current, and next frame.

```
prevY = yuv(:, :, 1, 1);
nextY = yuv(:, :, 1, 3);
figure
subplot(1,3,1), imshow(uint8(prevY)), title('Previous Frame');
subplot(1,3,2), imshow(uint8(currY)), title('Current Frame');
subplot(1,3,3), imshow(uint8(nextY)), title('Next Frame');
```

14. Extract the top and bottom fields for both the previous and next frames.

```
topFieldPrev = zeros(rows,cols);
topFieldNext = zeros(rows,cols);
topFieldCurr = topField2;
bottomFieldPrev = zeros(rows,cols);
bottomFieldNext = zeros(rows,cols);
bottomFieldCurr = bottomField2;
topFieldPrev(1:2:rows,:) = prevY(1:2:rows,:);
topFieldNext(1:2:rows,:) = nextY(1:2:rows,:);
bottomFieldPrev(2:2:rows,:) = prevY(2:2:rows,:);
bottomFieldNext(2:2:rows,:) = nextY(2:2:rows,:);
```

15. Perform the averaging.

```
for m=2:2:rows
    for n=1:cols
        fieldAverage1(m,n)=(bottomFieldPrev(m,n) + ...
            bottomFieldNext(m,n)) / 2;
```

```
    end
end
fieldAverage1(1:2:rows,:) = topFieldCurr(1:2:rows,:);
for m=1:2:rows
    for n=1:cols
        fieldAverage2(m,n) = (topFieldPrev(m,n) + ...
            topFieldNext(m,n)) / 2;
    end
end
fieldAverage2(2:2:rows,:) = bottomFieldCurr(2:2:rows,:);
```

16. Display the results.

```
figure
subplot(1,2,1), imshow(uint8(fieldAverage1)), ...
    title('Top field');
subplot(1,2,2), imshow(uint8(fieldAverage2)), ...
    title('Bottom field');
```

Question 6 How does this technique compare to line averaging? More specifically, under what circumstances does this technique perform better/worse than the line averaging scheme?

Line and Field Averaging

Deinterlacing by line and field averaging also requires information from the previous and the next frame. This method is a combination of line averaging and field averaging. The missing line of the current top field is filled with the previous and next line of the current top field averaged together with the lines from the bottom field in the previous frame and lines from the bottom field in the next frame.

17. Perform line and field averaging using the frame fields previously defined.

```
for m=2:2:rows-2
    for n=1:cols
        lineFieldAverage1(m,n) = (topFieldCurr(m-1,n) + ...
            topFieldCurr(m+1,n)+ ...
            bottomFieldPrev(m,n) + ...
            bottomFieldNext(m,n)) / 4;
    end
end
lineFieldAverage1(1:2:rows,:) = topFieldCurr(1:2:rows,:);
for m=3:2:rows
    for n=1:cols
```

```
        lineFieldAverage2(m,n) = (bottomFieldCurr(m-1,n) + ...
            bottomFieldCurr(m+1,n) + ...
            topFieldPrev(m,n) + ...
            topFieldNext(m,n)) / 4;
    end
end
lineFieldAverage2(2:2:rows,:) = bottomFieldCurr(2:2:rows,:);
figure
subplot(1,2,1), imshow(uint8(lineFieldAverage1)), ...
    title('Top field');
subplot(1,2,2), imshow(uint8(lineFieldAverage2)), ...
    title('Bottom field');
```

Question 7 How does line and field averaging compare to just field averaging?

21.6 TUTORIAL 21.3: NTSC TO PAL CONVERSION

Goal

The goal of this tutorial is to learn how to convert an NTSC formatted video sequence to its equivalent in the PAL format.

Objectives

- Explore the steps in the function `ntsc2pal`, which converts NTSC video to PAL video.

What You Will Need

- MATLAB m-files `readYUV.m`, `ntsc2pal.m`, `downConversion.m`, and `deinterlace.m`
- Test file `whale_show.yuv`

Procedure

The MATLAB function `ntsc2pal.m`[7] performs the NTSC to PAL conversion.

1. Open `ntsc2pal.m` in the MATLAB Editor and answer the questions below.

Question 1 What deinterlacing method is used in this case?

[7]This function was developed by Jeremy Jacob and can be downloaded from the book web site.

Question 2 What other deinterlacing possibilities are available with the `deinterlace` function used in the above code? If we were to use other methods, would we need to change the code for it to work?

Let us now perform an NTSC to PAL conversion of the `whale_show.yuv` video sequence.[8]

2. Load the first 30 frames of the YUV video sequence `whale_show.yuv` using the `readYUV` function.

```
[mov_org, yuv_org] = readYUV('whale_show.yuv', 30, 'NTSC');
```

3. Convert the video sequence to PAL.

```
[mov_new, yuv_new] = ntsc2pal(yuv_org);
```

Question 3 Inspect the dimensions of the original *YUV* array and the new *YUV* array. How do they compare?

Question 4 Why is the number of frames in the new video sequence different from the original?

4. Use the `implay` function to view the original and new video sequences (one at a time, or simultaneously if enough system resources are available). Be sure to set the frame rate appropriately (30 fps for the original and 25 fps for the new video).

Question 5 Subjectively, how does the quality of the new video sequence compare to the original?

21.7 TUTORIAL 21.4: 3:2 PULL-DOWN

Goal

The goal of this tutorial is to learn how to convert from 24 fps film to 30 fps NTSC video using the 3:2 pull-down method.

Objectives

- Explore the steps in the function `pulldown_32`, which converts 24 fps film to 30 fps video.

[8]Note that when the function runs, it uses a lot of memory resources, and so it is recommended that you close any other programs you are not currently using in order to free system resources.

What You Will Need

- `readYUV.m`
- `pulldown_32.m`
- Test file `whale_show.yuv`

Procedure

This tutorial teaches the process of converting 24 fps film to standard 30 fps NTSC video using the 3:2 pull-down method, implemented in function `pulldown_32.m`.[9]

1. Load the `whale_show.yuv` video file, but load only the first 24 frames.

```
[mov_org, yuv_org] = readYUV('whale_show.yuv', 24, 'NTSC');
```

2. Perform a 3:2 pull-down conversion on the video sequence.

```
[mov_new, yuv_new] = pulldown_32(yuv_org);
```

Question 1 Use `implay` to compare the quality between the original and converted video sequences.

Question 2 Write a small script that will simply copy the first four frames and generate the fifth frame by making a copy of the fourth. How does this compare to the 3:2 pull-down method?

WHAT HAVE WE LEARNED?

- Sampling rate conversion is a common technical problem in video processing. Particular cases include up-conversion, down-conversion, deinterlacing, and conversion between PAL and NTSC signals.
- The choice of sampling frequencies to be used in the spatial and temporal directions depends on the frequency content of the underlying signal, the visual thresholds in terms of the spatial and temporal cutoff frequencies, the capture and display device characteristics, and the associated processing, storage, and transmission costs.
- The sampling artifact known as *aliasing* can be minimized—but not eliminated completely—by a proper selection of sampling rates. In specific video sequences (e.g., scenes containing fine striped patterns or moving wheels), some (spatial or temporal) aliasing may still be noticeable.

[9]This function was developed by Jeremy Jacob and can be downloaded from the book web site.

- Converting a video sequence from one format to another is a technologically challenging and economically expensive problem.
- PAL to NTSC conversion involves four main steps: (1) deinterlacing each field in the PAL signal, resulting in a 625-line frame, (2) line rate down-conversion from 625 to 525 within each deinterlaced frame, (3) frame rate up-conversion from 50 to 60 Hz, and (4) splitting each frame into two interlacing fields.

LEARN MORE ABOUT IT

- E. Dubois (in [Bov00a], Chapter 7.2) presents a good summary of video sampling and interpolation.
- Chapters 3 and 4 of [WOZ02] and Chapters 3 and 4 of [Tek95] discuss video sampling and sampling rate conversion using lattice theory as a framework.
- Chapters 16–18 of [Poy03] offer a detailed view of the processes of sampling, filtering, resampling, and reconstruction for 1D, 2D, and 3D signals.
- The booklet by Watkinson [Wat94b] explains standards conversion techniques from an engineering standpoint.
- Two articles by Ackerman [Ack01,Ack02] discuss standards conversion—particularly 3:2 pull-down, color space, and aspect ratio conversion—in more detail.
- Chapter 16 of [Tek95] provides a detailed coverage of standards conversion principles and techniques.
- Chapter 7 of [Jac01] and Chapters 36 and 37 of [Poy03] discuss standards conversion, deinterlacing, and 3:2 pull-down, among many other video processing techniques.

21.8 PROBLEMS

21.1 Explain in your own words the phenomenon known as *aliasing* in video processing. In which situations is aliasing more visible? How can it be prevented/reduced?

21.2 The notion of sampling is usually associated with digitization (analog-to-digital conversion). Give one example of sampling in the analog domain.

CHAPTER 22

DIGITAL VIDEO PROCESSING TECHNIQUES AND APPLICATIONS

WHAT WILL WE LEARN?

- What is motion estimation (ME) and why is it relevant?
- Which techniques and algorithms can be used to estimate motion within a video sequence?
- Which techniques can be used to filter a video sequence?
- What is the role of motion compensation (MC) in video filtering?

22.1 FUNDAMENTALS OF MOTION ESTIMATION AND MOTION COMPENSATION

Motion is an essential aspect of video sequences. The ability to estimate, analyze, and compensate for relative motion is a common requirement of many video processing, analysis and compression algorithms and techniques. In this section, we present the fundamental concepts associated with motion estimation and compensation.

Motion estimation is the process of analyzing successive frames in a video sequence to identify objects that are in motion. The motion of an object is usually described by a two-dimensional (2D) *motion vector* (MV), which encodes the length and direction of motion.

Practical Image and Video Processing Using MATLAB®. By Oge Marques.

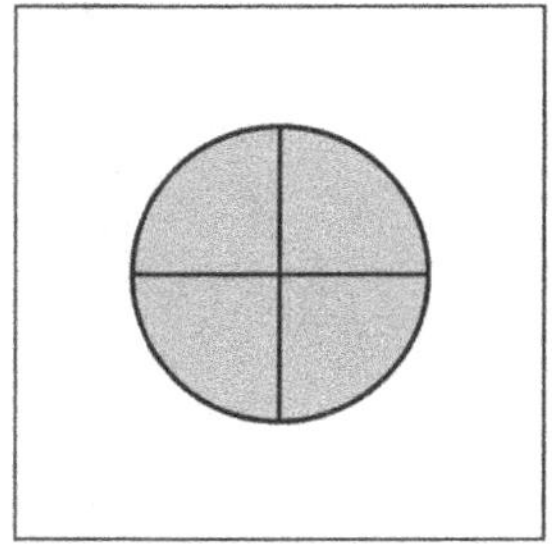
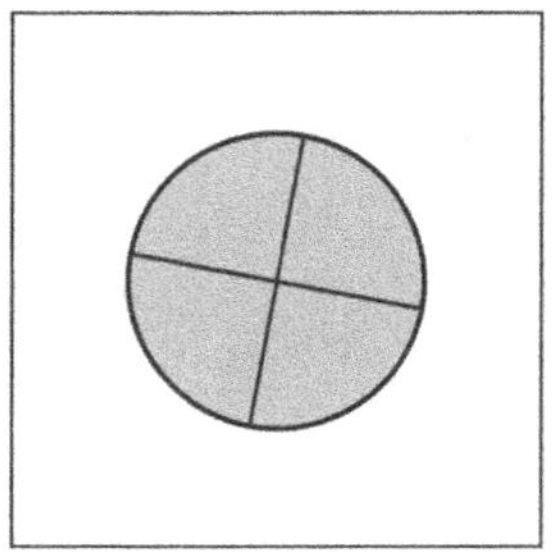
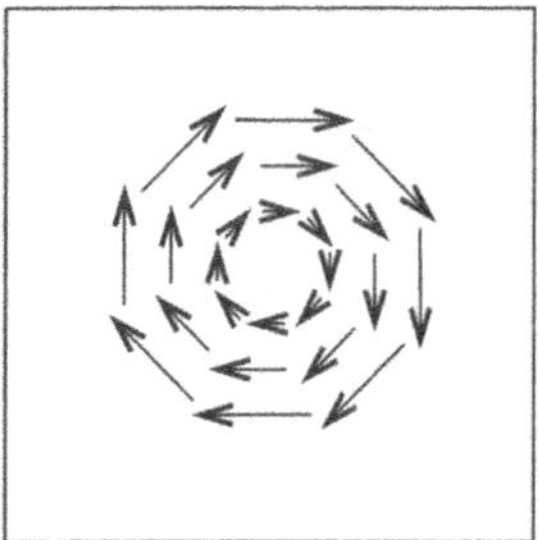

FIGURE 22.1 Two frames at different time instants (t_1 and $t_1 + \Delta_t$) and the resulting optical flow. Redrawn from [SHB08].

Many two-dimensional motion estimation algorithms are based on the concept of *optical flow*, which can be defined as the pattern of *apparent* motion of objects, surfaces, and edges in a visual scene. The optical flow—and consequently the apparent motion—can be computed by measuring variations (i.e., gradients) of intensity values over time. The resulting pattern is a vector field—known as the *optical flow field*—containing the motion vectors for each support region (e.g., pixel, block, or region) in the frame, representing the best approximation of the 3D motion of object points across the 2D frame. Figure 22.1 shows a schematic example of two consecutive frames and the resulting optical flow.

Optical flow computations[1] are based on two assumptions [SHB08]:

1. The observed brightness of an object is constant over time. This is referred to as the *constant intensity assumption*.
2. Points that are spatially close to one another in the image plane tend to move in a similar manner (the *velocity smoothness* constraint).

The two-dimensional motion estimation techniques aim at encoding motion information from intensity and color values obtained from the frames in a video sequence. This is a complex problem for many reasons, such as the following:

- The actual scene is three dimensional (and so are the objects in it), but 2D motion estimation algorithms have access only to a 2D representation of it. Consequently, these algorithms must rely on the *apparent motion* of the objects relative to the frame, which does not always correspond to true motion in the original scene.

  There are many cases in which actual object motion would go undetected or—conversely—a 2D motion estimation algorithm would estimate motion where none existed. An example of the former would be a sphere rotating on its own

[1]The complete mathematical formulation of optical flow methods is beyond the scope of this text. Refer to "Learn More About It" section at the end of the chapter for references.

axis. The latter problem can arise as a result of lighting variations, distracting objects, or many other scenarios.

- We can speak of motion only in *relative* terms, that is, by measuring how camera (or observer), objects of interest, and background move relatively to one another. There are six general cases to consider:
  1. Still camera, single moving object, constant background.
  2. Still camera, several moving objects, constant background.
  3. Still camera, single moving object, moving (cluttered) background.
  4. Still camera, several moving objects, moving (cluttered) background.
  5. Moving camera, fixed object(s), constant background.
  6. Moving camera, several moving objects, moving (cluttered) background.

  Most motion estimation algorithms in use today can handle the first two cases rather easily. There is a significant amount of ongoing work on cases where the background is complex and contains motion (e.g., wavering tree branches, light reflected on the ocean, etc.). The fifth case is usually treated as a separate problem, for which the goal is to determine the type and amount of camera movement. Finally, the sixth category is beyond the reach of contemporary computer vision solutions.
- There is an inherent ambiguity in the process of estimating motion based on edges of objects within the frame, which is known in the vision literature as the *aperture problem*, illustrated in Figure 22.2. If we view the image through a small circular aperture, we cannot tell exactly where a point (x_i, y_i) on an edge moves to. Optical flow-based ME algorithms usually rely on the motion vector perpendicular to the edge (indicated as $\vec{u}_n$), suggesting that the edge has moved up and to the right in this case. Clearly, it is possible that the actual motion was

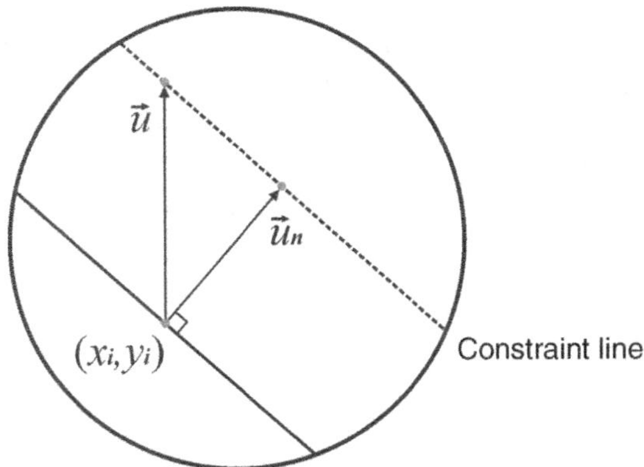

FIGURE 22.2 The aperture problem.

different—from purely up (indicated as $\vec{u}$) to purely to the right—but there is no way to know it for sure.

- ME algorithms must work at a level of granularity that is appropriate to the task at hand. Representing a MV for each pixel is usually not a good idea, and neither is using a single MV for the entire frame. Many algorithms find a compromise between these two extreme cases, by working with nonoverlapping blocks.
- In many cases (e.g., object tracking applications), we might be interested in computing the motion of *objects* in the scene, which presupposes that they can be accessed as individual entities. This increases the complexity of the overall solution, since it would require (accurate) object segmentation *before* the ME calculations.
- ME methods strongly depend on the intended application. For example, in video compression applications, the main goal of ME algorithms is to generate motion vectors that allow a decoder to re-create a frame based on the motion-compensated differences between the current frame and a previously decoded frame, while saving bits in the process. In motion-compensated standards conversion, on the other hand, the goal is to align the interpolation axis to the direction of motion of objects across multiple frames, and therefore reduce artifacts such as *judder*.

Motion compensation is the process by which the results of the motion estimation algorithms (usually encoded in the form of motion vectors) are employed to improve the performance of video processing algorithms. In its simplest case (e.g., a linear camera motion such as *panning*), motion compensation computes where an object of interest will be in an intermediate target field/frame and then shifts the object to that position in each of the source fields/frames.

In more general cases, the results of the motion estimation algorithm (i.e., the motion vectors) are used to create the equivalent of interpolation axes, aligned with the trajectory of each moving object (Figure 22.3). This causes a significant reduction in motion and temporal aliasing and has a dramatic positive effect on the output of motion-compensated algorithms for tasks such as interframe filtering (see Section 22.4.2), de-interlacing (see Section 21.3.1), and standards conversion (see Section 21.3), among many others. The overall performance of the motion-compensated video processing algorithms is determined primarily by the accuracy of the motion vectors, which underscores the importance of the motion estimation stage, which is described in more detail in the following sections.

22.2 GENERAL METHODOLOGIES IN MOTION ESTIMATION

The 2D motion estimation techniques described in this chapter address the problem of estimating the motion between two frames: $f(x, y, t_1)$, which will be called the *anchor frame*, and $f(x, y, t_2)$, which will be referred to as the *target frame*. The anchor frame may be either before or after the target frame in time. We will call it *forward*

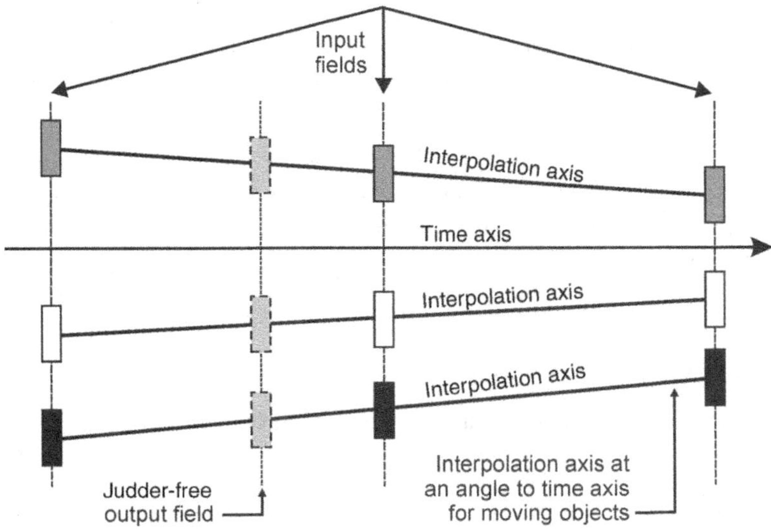

FIGURE 22.3 Motion compensation: interpolation axes are aligned with each moving object. Redrawn from [Wat94b].

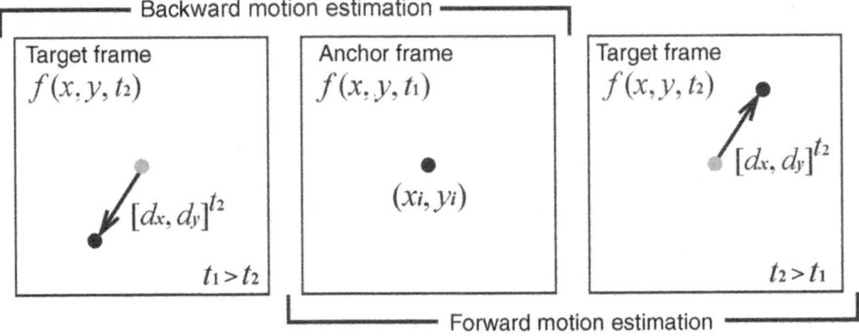

FIGURE 22.4 Anchor and target frames in forward and backward motion estimation.

motion estimation the case where $t_2 > t_1$ and *backward motion estimation* the case where $t_1 > t_2$. A motion vector $\mathbf{d}(\mathbf{x}) = [d_x d_y]^T$ will represent the displacement of a particular point $\mathbf{x} = (x_i, y_i)$ between time t_1 and t_2. Figure 22.4 illustrates these concepts and associated notation.

There are two main categories of ME approaches:

- *Feature-Based Methods*: Establish a correspondence between pairs of selected feature points in the two frames and attempt to fit them into a motion model using methods such as least-squares fitting. Feature-based methods are more often used in applications such as object tracking and 3D reconstruction from 2D video.

- *Intensity-Based Methods*: Rely on the constant intensity assumption described earlier and approach the motion estimation problem from the perspective of an optimization problem, for which three key questions must be answered:
  - How to represent the motion field?
  - Which criteria to use to estimate motion parameters?
  - How to search for optimal motion parameters?

  This section focuses exclusively on intensity-based methods. The answers to the three questions above will appear in Sections 22.2.1–22.2.3, respectively.

22.2.1 Motion Representation

Motion representation methods vary in terms of the size (and shape) of the region of support, which can vary among the following four types (illustrated in Figure 22.5):

1. *Global*: In this case, the region of support is the entire frame and a single motion model applies to all points within the frame. This is the most constrained motion model, but it is also the one that requires fewest parameters to estimate. Global motion representation is often used to estimate camera-induced motion, that is, apparent motion caused by camera operations such as pan, zoom, or tilt.

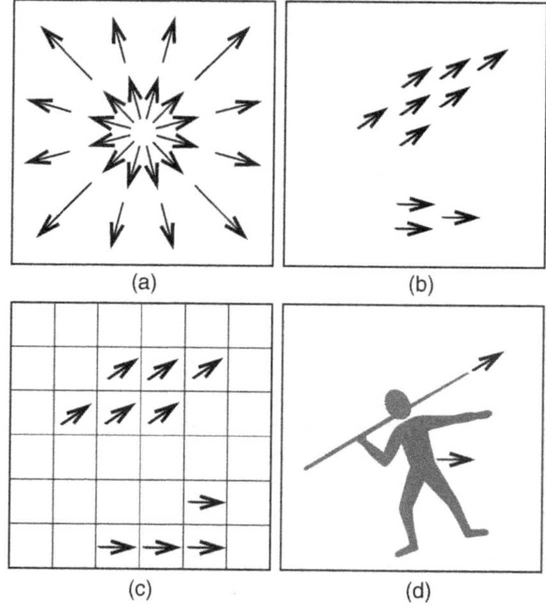

FIGURE 22.5 Motion estimation methods: (a) global; (b) pixel-based; (c) block-based; (d) object-based.

2. *Pixel-Based*: This model produces the densest possible motion field, with one MV per pixel, representing the 2D displacement of that point across successive frames. A smoothness constraint between adjacent MVs is often employed to ensure a more coherent motion representation. The computational complexity of pixel-based methods is high, due to the enormous number of values (twice the number of pixels in the frame) that need to be estimated.
3. *Block-Based*: In this case, the entire frame is divided into nonoverlapping blocks, and the problem is reduced to finding the MV that best represent the motion associated with each block. This model is widely used in video compression standards such as H.261, H.263, MPEG-1, and MPEG-2.
4. *Region-Based*: In this model, the entire frame is divided into irregularly shaped regions, with each region corresponding to an object or subobject with consistent motion, which can be represented by a few parameters. It is used in the MPEG-4 standard.

22.2.2 Motion Estimation Criteria

Once a model and region of support have been chosen, the problem shifts to the estimation of the model parameters. One of the most popular motion estimation criteria—and the only one described in this book—is based on the displaced frame difference (DFD). It consists in computing the differences in intensity between every point in the anchor frame and the corresponding point in the target frame. Mathematically, the objective function can be written as

$$E_{\text{DFD}}(\mathbf{d}) = \sum_{\mathbf{x}\in\mathcal{R}} (f_2(\mathbf{x}+\mathbf{d}) - f_1(\mathbf{x}))^p \tag{22.1}$$

where $f_1(\mathbf{x})$ is the pixel value at $\mathbf{x}$ in time t_1, $f_2(\mathbf{x}+d)$ is the pixel value at $(\mathbf{x}+\mathbf{d})$ in time t_2, $\mathbf{d}$ is the estimated displacement, and $\mathcal{R}$ is the region of support.

When $p = 1$, the error $E_{\text{DFD}}(\mathbf{d})$ is called the *mean absolute difference* (MAD), and when $p = 2$, it is called the *mean squared error* (MSE).

22.2.3 Optimization Methods

The error functions described by equation (22.1) can be minimized using different methods. The three most common optimization methods are as follows:

- *Exhaustive Search*: As the name suggests, this method searches through all possible parameter value combinations and is guaranteed to reach the global optimal at the expense of significant computational cost. It is typically used in conjunction with the MAD criterion described earlier.
- *Gradient-Based Search*: Uses numerical methods for gradient calculation and reaches the local optimal point closest to the initial solution. It is typically used in conjunction with the MSE criterion described earlier.

- *Multiresolution Search*: This method searches the motion parameters starting at a coarse resolution and propagating the partial solutions to finer resolutions, which can then be successively refined. It is a good trade-off between speed (i.e., it is faster than exhaustive search) and does not suffer from the problem of being trapped into a local minimum (as the gradient-based method does).

22.3 MOTION ESTIMATION ALGORITHMS

In this section, we present a selected subset of block-based ME algorithms. These algorithms assume that a frame has been divided into M nonoverlapping blocks that together cover the entire frame. Moreover, the motion in each block is assumed to be constant, that is, it is assumed that the entire block undergoes a translation that can be encoded in the associated motion vector. The problem of block-based ME algorithms is to find the best MV for each block. These algorithms are also called *block matching algorithms* (BMA).

22.3.1 Exhaustive Search Block Matching Algorithm

The exhaustive search block matching algorithm (EBMA) (Figure 22.6) works as follows: for each block $\mathcal{B}_{\mathrm{m}}$ in the anchor frame, it searches all candidate blocks (within a certain range) $\mathcal{B}'_{\mathrm{m}}$ in the target frame, and selects the one with the minimum error, that is, the one that optimizes a cost function such as MAD. The search range is indicated in Figure 22.6 by R_x and R_y.

The estimation accuracy of the EBMA method is determined by the step size for the block displacements. In the simplest case, the step size is one pixel, and the process is known as *integer-pixel accuracy search*. For more accurate motion

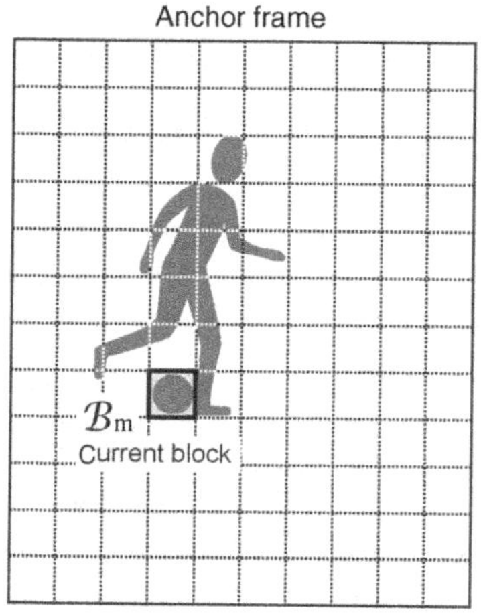

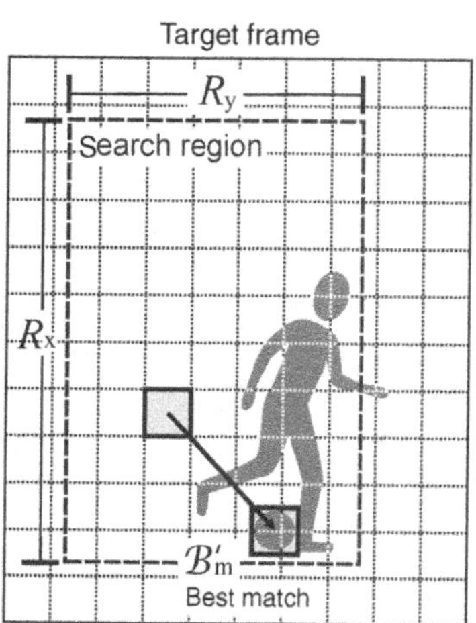

FIGURE 22.6 Exhaustive search block matching algorithm (EBMA).

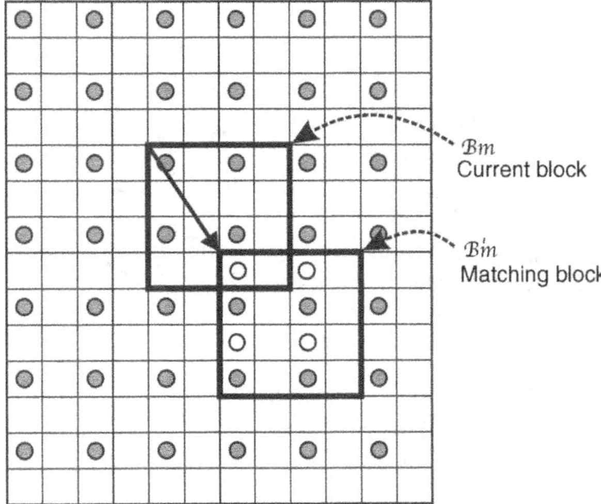

FIGURE 22.7 Block matching with half-pixel accuracy. The MV in this case is (1, 1.5). Redrawn from [WOZ02].

representation, *half-pixel accuracy search* is often used. In order to achieve subpixel accuracy, the target frame must be up-sampled (with the proper interpolation method, usually bilinear interpolation) by the proper factor. Figure 22.7 illustrates the process of block matching using half-pixel accuracy. In this figure, solid circles correspond to samples that exist in the original target frame, whereas hollow circles correspond to interpolated samples. In Tutorial 22.1 (page 579), you will learn how to perform motion estimation using the EBMA method (using both full-pixel and half-pixel accuracy) in MATLAB.

The EBMA algorithm has several weaknesses:

- It causes a blocking artifact (noticeable discontinuity across block boundaries) in the predicted frame.
- The resulting motion field can be somewhat chaotic because MVs are independently estimated from block to block.
- Flat regions are more likely to produce wrong MVs because motion is indeterminate when spatial gradient is near zero (a problem shared by all optical flow-based methods).

■ EXAMPLE 22.1

Figure 22.8 shows the full-pixel EBMA algorithm at work. Parts (a) and (b) show the target and anchor frame, respectively. Part (c) shows the motion field, which exemplifies the last two limitations described earlier. Finally, part (d) shows a reconstructed frame, that is, a frame obtained by computing the estimated motion applied to the target frame. Ideally, part (d) should result in a frame identical to the one in

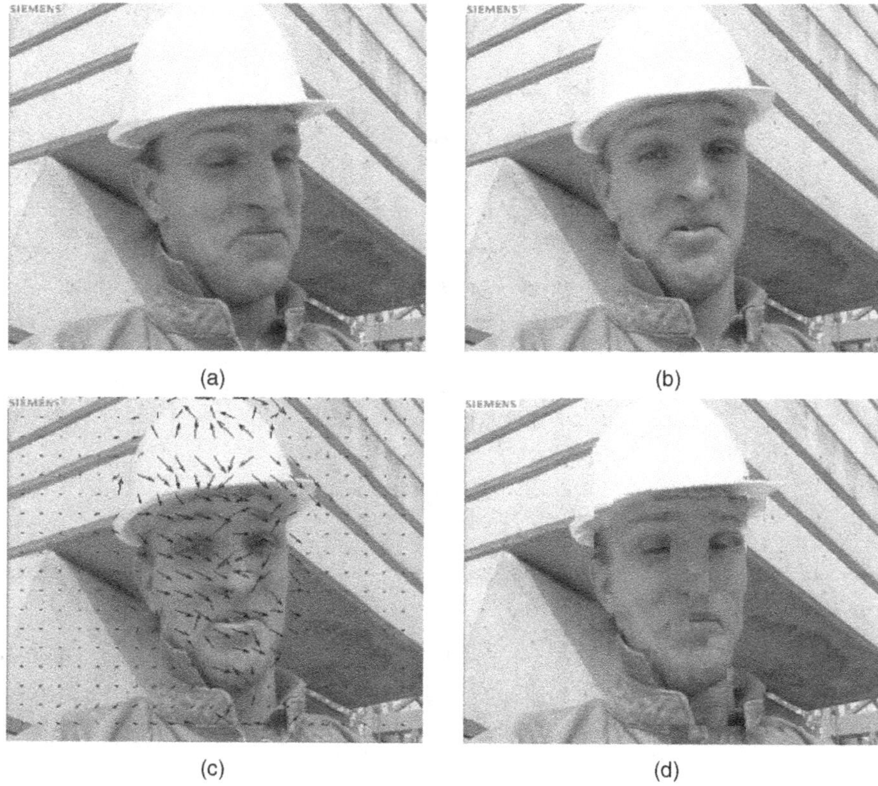

FIGURE 22.8 EBMA example: (a) target frame; (b) anchor frame; (c) motion field overlapped on anchor frame; (d) reconstructed frame.

part (b), but this is clearly not the case: the presence of blocking artifacts (besides other imperfections) in the reconstructed frame is clearly visible.

22.3.2 Fast Algorithms

The EBMA algorithm is computationally expensive (and significantly more so if fractional pixel accuracy is used), which motivated the development of several fast algorithms for block matching. The two most popular fast search algorithms are the 2D log and the three-step search methods. They are briefly described next.

2D Log Search Method The 2D log search method (Figure 22.9) starts from the position corresponding to zero displacement and tests five search points, arranged in a diamond shape. It selects the block that yields minimum error (usually MAD is the chosen figure of merit) and uses it to form a new search region around it. If the best matching point is the center point, it proceeds with a new search region that has an offset of half the amount of the previous offset; otherwise, the offset remains

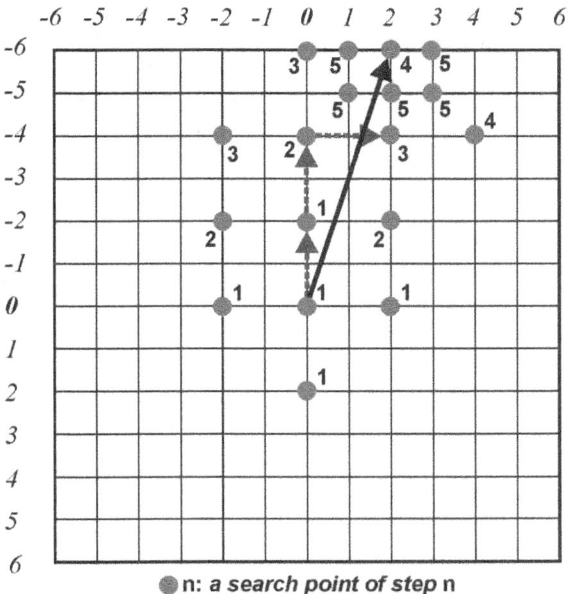

FIGURE 22.9 2D log search. In this case, the final MV is $(-6, 2)$. Redrawn from [MPG85].

the same. The process is repeated for successively smaller ranges, until the offset is equal to 1.

Three-Step Search Method The three-step search method (Figure 22.10) starts from the position corresponding to zero displacement and tests nine search points, with offset $p/2$, where p is half of the search range in each direction (horizontal or vertical). It selects the block that yields minimum error and uses it to form a new search region around it. The new search region has an offset of half the amount of the previous offset. The process is repeated for successively smaller ranges, until the offset is equal to 1.

22.3.3 Hierarchical Block Matching Algorithm

The hierarchical block matching algorithm (HBMA) is a multiresolution ME algorithm. It works by first estimating the motion vectors in a coarse resolution using a low-pass filtered, down-sampled frame pair. It then proceeds to modify and refine the initial solution using successively finer resolutions and proportionally smaller search ranges. The most common approach uses a pyramid structure (Figure 22.11) in which the resolution is reduced by half in both dimensions between successive levels. The HBMA algorithm strikes a compromise between execution time and quality of results: it has lower computational cost than the EBMA counterpart and produces better quality motion vectors than the fast algorithms described earlier. In Tutorial 22.1

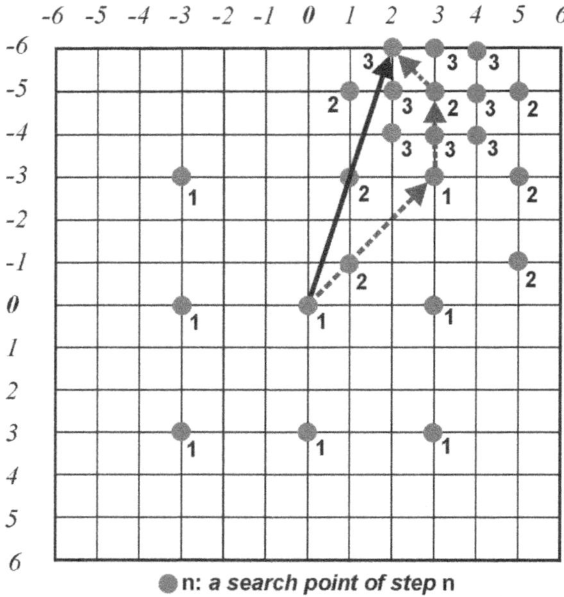

FIGURE 22.10 Three-step search. In this case, the final MV is (−6, 2). Redrawn from [MPG85].

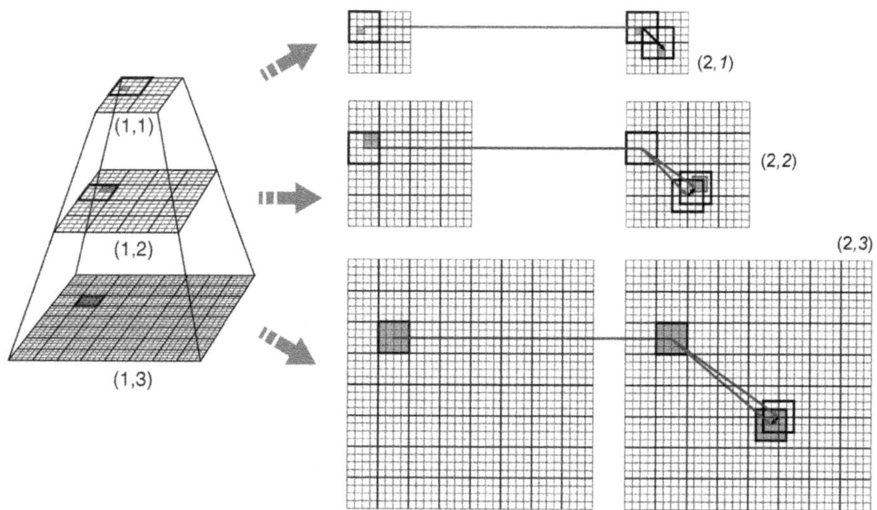

FIGURE 22.11 Hierarchical block matching algorithm (HBMA) using three levels. Redrawn from [WOZ02].

(page 579), you will learn how to perform motion estimation using the HBMA method in MATLAB.

22.3.4 Phase Correlation Method

The phase correlation method is a frequency-domain motion estimation technique that uses a normalized cross-correlation function computed in the frequency domain to estimate the relative shift between two blocks (one in the anchor frame, another in the target frame). The basic idea consists in performing spectral analysis on two successive fields and computing the phase differences. These differences undergo a reverse transform, which reveals peaks whose positions correspond to motions between fields.

1. Compute the discrete Fourier transform (DFT) of each block in the target and anchor frames (which we will denote frames k and $k+1$).
2. Calculate the normalized cross-power spectrum between the resulting DFTs according to equation (22.2):

$$C_{k,k+1} = \frac{\mathfrak{F}_{k+1}\mathfrak{F}_k^*}{|\mathfrak{F}_{k+1}\mathfrak{F}_k^*|} \tag{22.2}$$

   where $\mathfrak{F}_k$ and $\mathfrak{F}_{k+1}$ are the DFT of two blocks at the same spatial location in frames k and $k+1$ and * denotes the complex conjugate.
3. Compute the *phase correlation function*, PCF(**x**), defined as the inverse DFT of $C_{k,k+1}$, as follows:

$$\text{PCF}(\mathbf{x}) = \mathfrak{F}^{-1}(C_{k,k+1}) = \delta(\mathbf{x} + \mathbf{d}) \tag{22.3}$$

   where **d** is the displacement vector.
4. Locate the peaks of the PCF.

The basic algorithm described above, however, has a major limitation. It lacks the ability to tell which peak corresponds to which moving object in the frames. This problem can be resolved by adding a (selective) block matching stage, used to distinguish between valid motion vectors and false alarms. The modified phase correlation method has the following strengths: it is fast, subpixel accurate, and amplitude (brightness) independent. In Tutorial 22.1 (page 579), you will learn how to perform motion estimation using the phase correlation method in MATLAB.

22.4 VIDEO ENHANCEMENT AND NOISE REDUCTION

In this section, we provide a brief overview of how video sequences can be processed using filtering techniques aimed at enhancing the image quality or reducing noise and other artifacts. The simplest approach consists of applying (spatial- or frequency-domain) image processing algorithms to each individual frame: this will be called *intraframe filtering*. Many techniques described in Part I can be used to implement intraframe filtering, such as averaging filter and its variants, median and weighted median filters, or Wiener filters, to mention but a few.

Better results, however, can be obtained by exploiting the contents of adjacent frames with significant redundant information, using video-specific (spatiotemporal) algorithms, which will be referred to as *interframe filtering* (Section 22.4.2). In Tutorial 22.2 (page 585), you will experiment with inter- and intraframe video filtering techniques in MATLAB.

22.4.1 Noise Reduction in Video

Video acquisition, transmission, and recording systems are not perfect. Consequently, video sequences may be subject to different types of noise and distortion.

For analog video, the most common types are as follows:

- *Thermal (snow) Noise*: Caused by temperature-dependent random signal-level fluctuations in amplifiers or signal attenuation due to rain. The visible effect on the screen resembles a snowstorm.
- *Composite Triple Beats (CTB)*: Caused by cumulative effect of third-order intermodulation products and overdriven amplifiers. Produces graininess and tearing in the horizontal lines of the screen.
- *Composite Second-Order Beat (CSO)*: Caused by nonlinearities in the electronics. Appears on the screen as graininess texture or diagonal lines over the entire picture.
- *Single-Frequency Modulation Distortion*: Caused by intermodulation products. Its visible effect on the screen consists of horizontal or diagonal bands, broad bands of color variation across the screen.
- *Impulse Noise (Short Duration and High Energy)*: Caused by lightning, car starters, industrial machines, and other sources. Produces spots on the screen and sharp click sounds in the audio.
- *Cochannel Interference*: Caused when two pictures are received at the same time from two different TV transmitters at different locations, but which share the same TV channel. The visible effect occurs on the overlap of the unwanted service and is often accompanied by a "venetian blind" effect and distorted sound.
- *Ghosting*: Caused by multiple transmission paths, resulting from signal reflections. The visible effect consists in the appearance of a delayed version of the picture on the screen.

For digital video, most of the unwanted artifacts result from side effects of the lossy compression techniques used to store or transmit the video using fewer bits. These are the main types of artifacts that may be present in digital video sequences:

- *Block Edge Effect (or Simply Blockiness)*: Caused by coarse quantization of DCT coefficients in block-based compression schemes (e.g., M-JPEG and MPEG-1), which result in noticeable intensity discontinuities at the boundaries of adjacent blocks in the decoded frame.
- *False Edges (or False Contouring)*: Caused by coarse quantization of amplitude levels. Its visible effect in the frame is the appearance of edges at places where a smooth intensity transition should occur.
- *Ringing*: Caused by the use of ideal filters. Produces a rippling along high-contrast edges.
- *Blurring and Color Bleeding*: Caused by imperfections in compression algorithms. Results in loss of spatial details, running or smearing of color in areas with complex textures or edges.
- *Mosquito Noise*: Caused by imperfections in compression algorithms. Produces fluctuation of luminance/chrominance levels around high-contrast edges or moving video objects.

Video sequences, particularly old movies, can also be subject to other types of artifacts, such as the following:

- *Blotches*: Large, uncorrelated, bright or dark spots, typically present in film material, due to mishandling or aging.
- *Intensity Flicker*: Unnatural temporal fluctuation of frame intensities that do not originate from the original scene.

22.4.2 Interframe Filtering Techniques

Interframe filtering techniques employ a spatiotemporal support region of size $m \times n \times k$, where m and n correspond to the size of a neighborhood in a frame and k denotes the number of frames taken into account. For the special case where $m = n = 1$, the resulting filter is called a *temporal filter*, whereas the case where $k = 1$ corresponds to intraframe filtering. Interframe filtering techniques can be motion compensated, motion adaptive, or neither.

The simplest interframe filtering technique is the spatiotemporal averaging filter that consists of an extension of the basic averaging filter to the spatiotemporal domain. A purely temporal version of the algorithm is known as *frame averaging*. Frame averaging is based on the fact that averaging multiple observations of essentially the same pixel in different frames eliminates noise while resulting in no loss of spatial image resolution. The amount of noise reduction is proportional to the number of frames used, but so is the amount of blurring. In other words, besides the spatial blurring of 2D averaging filters, we now have a potential motion blur, due to movement of objects between consecutive frames.

There are two solutions to the blurring problem:

- Motion-adaptive, edge-preserving filters, where frame-to-frame motion is treated as "temporal edges."
- Motion-compensated filters, involving: low-pass filters that are applied along the motion trajectory of each pixel, as determined by a motion estimation algorithm.

Motion-compensated filters work under the assumption that the variation of pixel gray levels over any motion trajectory is mainly due to noise. Consequently, noise can be reduced by low-pass filtering over the respective motion trajectory at each pixel. Several motion-compensated filtering techniques have been proposed in the literature. They differ according to the following factors:

- The motion estimation method.
- The support of the filter, where *support* is defined as the union of predetermined spatial neighborhoods centered about the pixel (subpixel) locations along the motion trajectory.
- The filter structure (adaptive versus nonadaptive).

The effectiveness of motion-compensated filters is strongly related to the accuracy of the motion estimates. In the ideal case (perfect motion estimation), direct averaging of image intensities along a motion trajectory provides effective noise reduction. In practice, however, motion estimation is hardly ever perfect due to noise and sudden scene changes, as well as changes in illumination and camera views.

22.5 CASE STUDY: OBJECT SEGMENTATION AND TRACKING IN THE PRESENCE OF COMPLEX BACKGROUND

In this section, we present an example of object segmentation and tracking using MATLAB and the Image Processing Toolbox (IPT).[2] Object detection and tracking are essential steps in many practical applications, for example, video surveillance systems.

The solution presented in Figure 22.12 uses an object segmentation algorithm for video sequences with complex background, originally proposed by Socek et al. [SCM+05] and further refined and redesigned by Ćulibrk et al. [CMS+06,CMS+07]. The output of the segmentation algorithm is a series of binary images (frames), where background pixels are labeled as 0 and any foreground objects are labeled as 1. The

[2]The MATLAB code for this example was developed by Jeremy Jacob and can be downloaded from the book web site.

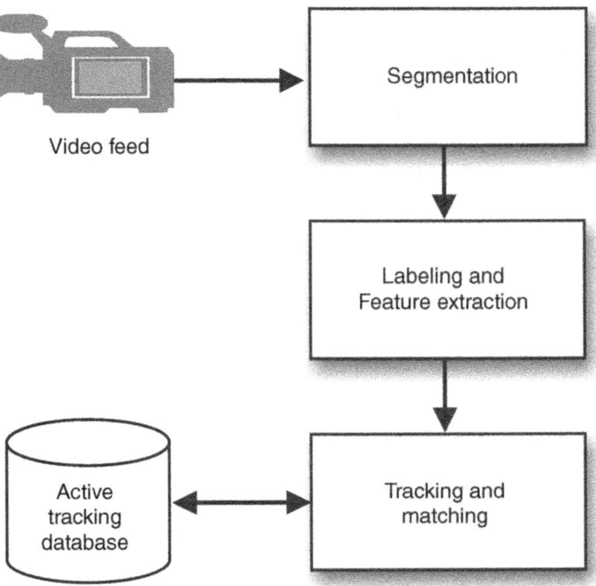

FIGURE 22.12 Object detection and tracking system.

"labeling and feature extraction" stage uses the IPT `bwlabel` function to label each connected component in the foreground of the frame and the IPT `regionprops` function to extract relevant features pertaining to each object, such as the coordinates of the centroid of the object, an index list of the pixels contained in the object, and its bounding box.

At the "tracking and matching" stage, the objects detected in the current frame are compared with the objects currently being tracked from previous frames in order to answer the following questions:

- Is any object new to the scene?
- Has any object left the scene?
- Where did the objects currently being tracked go?

When a new object is discovered (i.e., one for which no current objects match), this object is placed in a "three-frame buffer." The buffer is necessary to dismiss false alarms caused, for example, by noise artifacts introduced in the early stages of segmentation. If the same object is successfully tracked over three frames, it is then removed from the buffer and entered in the "active tracking list." Any object that is currently in this list is considered as an object officially being tracked, and any tracking information collected here will be reported in the final output. Figure 22.13 illustrates the process.

To determine the new position of an existing object A (which is contained in the active tracking list), we consider any object in the current frame whose centroid falls

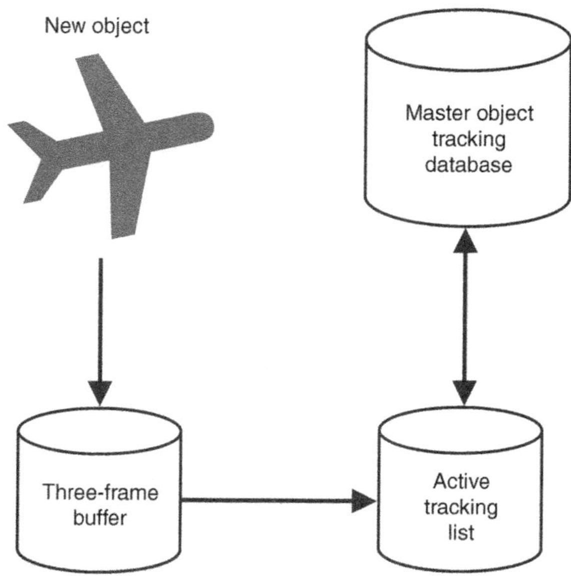

FIGURE 22.13 Keeping track of existing and candidate objects.

within the bounding box of object *A* as a possible candidate. If only one object meets this criteria, then we consider it a match, run an additional test to ensure that object cannot be considered a possible candidate for any other objects (since we now know what that object is), and update the tracking data for object *A*. This is illustrated in Figure 22.14. If there are more than one possible candidate for the new position of object *A*, we then compare object sizes, and that with the closest size to object *A* is considered a match. Finally, if at any given frame an object cannot be matched up with an object in the subsequent frame, that object is removed from the "active

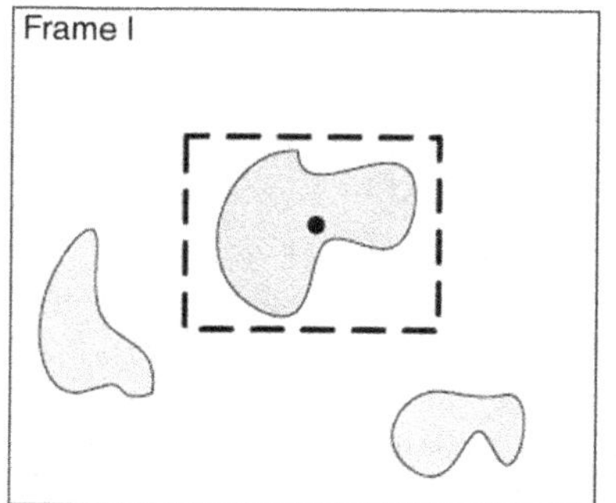

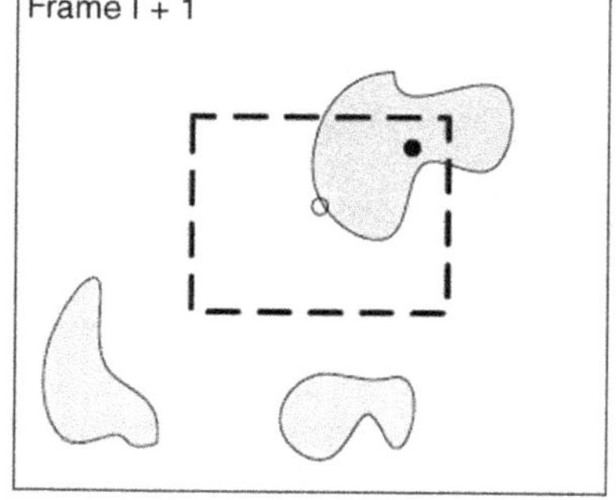

FIGURE 22.14 Updating the coordinates of an existing object.

FIGURE 22.15 Screenshot of the object detection and tracking system, showing the bounding box of an object being tracked, its trajectory since it started being tracked, and its properties (on a separate window).

tracking list," although its history remains in the "master object tracking database" (see Figure 22.12).

For each object, the system keeps track of its size, position, velocity, and color. The size information can be used to separate objects into application- and viewpoint-dependent classes, for example, large, medium, and small. Similarly, brightness and color information can be used to classify objects into categories such as "bright" or "red." Position is reported as the instantaneous pixel position of the centroid of the object within the current frame. Similarly, velocity is reported as an instantaneous change in position between consecutive frames, which can be easily calculated since we know the frame rate of the sequence.

The MATLAB implementation uses a modified version of the `implay` function, which adds the possibility of pausing the sequence and clicking within the bounding box of an object within the scene to open an external window, which displays the tracking information for that object and reports how many objects are being tracked. A screen shot of the movie player—with objects' bounding boxes and trajectory overlapped on the original frame—along with the external window is shown in Figure 22.15.

22.6 TUTORIAL 22.1: BLOCK-BASED MOTION ESTIMATION

Goal

The goal of this tutorial is to explore motion estimation algorithms in MATLAB.

Objectives

- Explore EBMA using both integer-pixel and half-pixel accuracy.
- Explore HBMA using both integer-pixel and half-pixel accuracy.
- Explore the phase correlation method.

What You Will Need

- *MATLAB Functions*: `ebma.m`, `hbma.m`, and `PhaseCorrelation.m`.[3]
- Test files `foreman69.Y` and `foreman72.Y`.

Procedure

Integer-Pixel EBMA

Before we perform the exhaustive block matching algorithm, let us first define a few necessary variables. Here, we will define the anchor frame, target frame, frame dimensions, block size, search field range (p), and finally accuracy (where 1 is for integer-pixel and 2 is for half-pixel).

1. Define initial variables.

```
anchorName = 'foreman69.Y';
targetName = 'foreman72.Y';
frameHeight = 352;
frameWidth = 288;
blockSize = [16,16];
p = 16;
accuracy = 1;
```

Next, we will read in the frame data.

2. Read the frame data.

```
fid = fopen(anchorName,'r');
anchorFrame= fread(fid,[frameHeight,frameWidth]);
anchorFrame = anchorFrame';
fclose(fid);
fid = fopen(targetName,'r');
targetFrame = fread(fid,[frameHeight,frameWidth]);
targetFrame = targetFrame';
fclose(fid);
```

[3]These functions were developed by Jeremy Jacob and can be downloaded from the book web site.

3. Run the function ebma and record the time to process the frames in variable time_full.

```
tic
[predictedFrame_Full, mv_d, mv_o] = ...
    ebma(targetFrame, anchorFrame, blockSize, p, accuracy);
time_full = toc
```

4. Display the anchor and target frames.

```
figure
subplot(1,2,1), imshow(uint8(anchorFrame)), title('Anchor Frame');
subplot(1,2,2), imshow(uint8(targetFrame)), title('Target Frame');
```

5. Display the motion vectors overlaid on the anchor frame.

```
figure, imshow(uint8(anchorFrame))
hold on
quiver(mv_o(1,:),mv_o(2,:),mv_d(1,:),mv_d(2,:)), ...
    title('Motion vectors: EBMA - Integer-pixel');
hold off
```

6. Display the predicted frame.

```
figure, imshow(uint8(predictedFrame_Full)), ...
    title('Predicted Frame Full-pixel');
```

7. Calculate the error frame by subtracting the predicted frame from the anchor frame.

```
errorFrame = imabsdiff(anchorFrame, predictedFrame_Full);
```

8. Calculate the PSNR by performing the following calculation.

```
PSNR_Full = 10*log10(255*255/mean(mean((errorFrame.^2))))
```

Half-Pixel EBMA

To perform half-pixel EBMA, we must first up-sample the frames.

9. Up-sample the target frame.

```
targetFrame2 = imresize(targetFrame,2,'bilinear');
```

10. Set the accuracy variable appropriately for half-pixel accuracy.

```
accuracy = 2;
```

11. Run the script to perform EBMA with half-pixel accuracy and compute the execution time.

```
tic
[predictedFrame_Half, mv_d, mv_o] = ...
    ebma(targetFrame2, anchorFrame, blockSize, p, accuracy);
time_half = toc
```

12. Compare the time of execution between integer-pixel and half-pixel accuracy.

```
time_half
time_full
```

Question 1 How does the difference in time compare with the additional number of pixels present in the up-sampled frame?

13. Display the motion vectors over the anchor frame.

```
figure, imshow(uint8(anchorFrame)), hold on
quiver(mv_o(1,:),mv_o(2,:),mv_d(1,:),mv_d(2,:)), ...
    title('Motion vectors: EBMA - Half-pixel');
hold off
```

14. Display the predicted frame.

```
figure, imshow(uint8(predictedFrame_Half)), ...
    title('Predicted Frame Half-pixel');
```

15. Calculate the error frame.

```
errorFrame = imabsdiff(anchorFrame, predictedFrame_Half);
```

16. Calculate the PSNR.

```
PSNR_Half = 10*log10(255*255/mean(mean((errorFrame.^2))))
```

Question 2 Compare the PSNR values between integer and half-pixel accuracy. Are the results what you expected?

Question 3 Visually, how do the results compare between integer and half-pixel accuracy (for the predicted frames)?

Multiresolution Motion Estimation: HBMA

To begin implementing HBMA motion estimation in MATLAB, we must first define several variables. Here, we are defining the names of the frames that will be read, the frames' dimensions, block size, start and end ranges, accuracy, and the number of levels to process (in our case, three).

17. Define variables needed to run HBMA.

```
anchorName = 'foreman69.Y';
targetName = 'foreman72.Y';
frameHeight = 352;
frameWidth = 288;
blockSize = [16,16];
rangs = [-32,-32];
range = [32,32];
accuracy = 1;
L = 3;
```

18. Read the frame data.

```
fid = fopen(anchorName,'r');
anchorFrame= fread(fid,[frameHeight,frameWidth]);
fclose(fid);
fid = fopen(targetName,'r');
targetFrame = fread(fid,[frameHeight,frameWidth]);
fclose(fid);
```

We can now run the hbma function by calling it with the parameters defined earlier. The function returns the predicted frame, direction of motion vectors, and the respective positions of the motion vectors (i.e., their orientation).

19. Run the hbma script and record the time needed to process it.

```
tic
[predict,mv_d,mv_o] = hbma(targetFrame, anchorFrame, blockSize, ...
    rangs, range, accuracy, L);
toc
```

20. Display the anchor and target frames to be able to compare with the predicted frame.

```
figure
subplot(1,2,1), imshow(uint8(anchorFrame')), title('Anchor Frame');
subplot(1,2,2), imshow(uint8(targetFrame')), title('Target Frame');
```

21. Display the motion vectors over the anchor frame.

```
figure, imshow(uint8(anchorFrame'))
hold on
quiver(mv_o(2,:),mv_o(1,:),mv_d(2,:),mv_d(1,:)), ...
    title('Motion vectors');
hold off
```

22. Show the predicted frame.

```
figure, imshow(uint8(predict')), ...
    title('Predicted Frame');
```

23. Calculate the error frame and PSNR.

```
errorFrame = imabsdiff(anchorFrame, predict);
PSNR=10*log10(255*255/mean(mean((errorFrame.^2))))
```

Question 4 How do the motion vectors compare between this and the EBMA technique?

Question 5 How does the quality of the predicted frame obtained with the HBMA algorithm compare with the one generated by the EBMA technique?

Phase Correlation Method

To perform phase correlation, we must first define and read two frames.

24. Read two frames and create the frames array to be used by the script.

```
anchorName = 'foreman69.Y';
targetName = 'foreman72.Y';
frameHeight = 352; frameWidth = 288;
fid = fopen(anchorName,'r');
anchorFrame= fread(fid,[frameHeight,frameWidth]);
anchorFrame = anchorFrame';
fclose(fid);
fid = fopen(targetName,'r');
targetFrame = fread(fid,[frameHeight,frameWidth]);
targetFrame = targetFrame';
fclose(fid);
frame(:,:,1) = anchorFrame;
frame(:,:,2) = targetFrame;
```

25. Run the `PhaseCorrelation.m` script to generate the predicted frame, a phase correlation plot, and a motion vector plot.

Question 6 How does the quality of this technique compare with the previous techniques (EBMA and HBMA)?

22.7 TUTORIAL 22.2: INTRAFRAME AND INTERFRAME FILTERING TECHNIQUES

Goal

The goal of this tutorial is to learn how to perform intraframe and interframe filtering in MATLAB.

Objectives

- Explore the averaging filter, median filter, and Wiener filter applied to video sequences.
- Learn how to use the `noisefilter` and `noisefilter2` functions to apply noise to selected frames in a video sequence and then filter them.

What You Will Need

- *MATLAB Functions*: `readYUV.m`, `noisefilter.m`, and `noisefilter2.m`.[4]
- Test file `football_cif_ori90.yuv`.

Procedure

Intraframe Filtering

To perform intraframe filtering in MATLAB, we will use the `readYUV` function to extract one frame from a YUV video sequence. We will perform filtering in the RGB color space. When using the `readYUV` function, recall that it returns two data structures; one that holds the MATLAB movie structure of the video (`mov`), and another that contains the YUV data for that sequence (`yuv`). The `CDATA` portion of the movie structure is in fact the RGB values for that particular frame—this is where we get the RGB values from.

1. Read in the first frame of the `football_cif_ori90` YUV sequence and extract the RGB data.

[4]These functions were developed by Jeremy Jacob and can be downloaded from the book web site.

```
[mov, yuv] = readYUV('football_cif_ori90.yuv',1,'CIF_PAL');
img = mov(1).cdata;
```

Before filtering, let us add artificial noise; this will give us something to filter out.

2. Create two noisy images: one with salt and pepper noise and another with Gaussian noise.

```
noisySP = imnoise(img, 'salt & pepper');
noisyGauss = imnoise(img,'gaussian');
```

3. Display the frames for later comparisons.

```
figure
subplot(1,3,1), imshow(img), title('Original frame');
subplot(1,3,2), imshow(noisySP), title('Salt & Pepper noise');
subplot(1,3,3), imshow(noisyGauss), title('Gaussian noise');
```

Our first filter is an averaging filter. Here, the pixel in question is simply taken as the average of the pixels in the given window size surrounding that pixel.

4. Apply an averaging filter to each noisy image.

```
filt = fspecial('average');
avgFilterFrame1 = imfilter(noisyGauss,filt);
avgFilterFrame2 = imfilter(noisySP,filt);
```

5. Display the results.

```
figure
subplot(1,3,1), imshow(img), title('Original frame');
subplot(1,3,2), imshow(avgFilterFrame2), ...
    title('Salt & Pepper averaged');
subplot(1,3,3), imshow(avgFilterFrame1), ...
    title('Gaussian averaged');
```

We will now filter a sequence of movie frames using the `noisefilter` function.

6. Clear all variables and close all figures.
7. Load 30 frames of the `football_cif_ori90.yuv` sequence.

```
[mov, yuv] = readYUV('football_cif_ori90.yuv',30,'CIF_PAL');
```

8. Add noise to all frames in the sequence and filter.

```
mov2 = noisefilter(mov,'salt & pepper','average',[]);
```

9. Play the resulting movie.

```
implay(mov2)
```

10. Add noise to only a select number of frames.

```
mov3 = noisefilter(mov,'salt & pepper','average',[5 10 20]);
```

11. Play the resulting movie.

```
implay(mov3)
```

Question 1 Was the noise noticeable even though the frames were filtered (for both `mov2` and `mov3`)?

Question 2 Filter the same sequences using the median filter. Is the noise noticeable?

Interframe Filtering Techniques

To perform interframe filtering in MATLAB, we will use the function `noisefilter2`. This is similar to the function `noisefilter` used earlier in this tutorial in that it will apply noise to designated frames (defined in the function call). It differs in the way the frames are filtered; frame data will be averaged over the temporal domain.

Let us perform interframe filtering on a video sequence to see its effect.

12. Load the video sequence.

```
[mov, yuv] = readYUV('football_cif_ori90.yuv',50,'CIF_PAL');
```

13. Apply noise to all frames, and average over three frames.

```
mov2 = noisefilter2(mov,'salt & pepper',[],3);
```

14. Play the resulting movie.

```
implay(mov2)
```

Question 3 What is the effect of the noise once the interframe filtering is applied (after the third frame)?

Question 4 Filter the video sequence again using a higher number of frames to be used in the filtering process. How does this affect noise? How does it affect image quality?

WHAT HAVE WE LEARNED?

- Motion estimation is the process of determining how much relative motion (between camera and scene) has occurred between two video frames. This information is usually encoded using motion vectors. Motion estimation is a required step for many video processing algorithms, from compression to spatiotemporal filtering.
- There are numerous motion estimation techniques and algorithms in the literature. They differ in the way they represent motion, the level of granularity used (from pixels to blocks to entire frames), the criteria used to determine whether motion was present, and the mathematical function that should be optimized as a result.
- Motion compensation is the process of using the results of the motion estimation stage (usually encoded in the form of motion vectors) to perform another video processing operation (e.g., spatiotemporal filtering) in such a way that takes into account the motion of relevant portions of the frame over time.
- Video filtering techniques usually fall into two categories: intraframe and interframe. The former treats one frame at a time and has strong resemblance to comparable image filtering techniques. The latter processes several (e.g., three) consecutive frames at a time and can benefit from results generated by a motion estimation step.

LEARN MORE ABOUT IT

- The topic of motion estimation has received book-length coverage (e.g., in [FGW96]), and appeared in many surveys (e.g., [SK99,HCT+06]).
- Chapters 6 and 7 of [WOZ02] discuss 2D and 3D motion estimation in great detail.
- Chapter 3.10 of [Bov00a] is entirely devoted to motion detection and estimation.
- Motion compensation techniques, especially the phase correlation method, are described in [Wat94a].
- Chapter 10 of [Woo06] covers intraframe and interframe filtering techniques, including many examples of motion-compensated video filtering.
- The topic of motion-compensated filtering—theory and algorithms—is covered in great depth in Chapters 13 and 14 of [Tek95].

- Section 16.2 of [SHB08] provides a good introduction to optical flow and its application in motion analysis.
- Chapter 5 of [Tek95] presents motion estimation using optical flow in great detail.
- Sections 16.5 and 16.6 of [SHB08] discuss contemporary methods used in video tracking solutions.
- Chapter 3.11 of [Bov00a] presents a detailed discussion on video enhancement and restoration techniques, including blotch detection and removal and intensity flicker correction.

22.8 PROBLEMS

22.1 Assume a very simple motion detection algorithm, in which to estimate the motion in a scene you examine the intensity of two captured video frames, $f(n_1, n_2, t_1)$ and $f(n_1, n_2, t_2)$ captured at t_1 and t_2 respectively, and you compute the absolute difference between them: $d(n_1, n_2) = |f(n_1, n_2, t_2) - f(n_1, n_2, t_1)|$.

(a) If the frame difference is nonzero, can you say that motion has occurred in the scene? Explain.

(b) If the frame difference is exactly zero, can you say that no motion has occurred in the scene? Explain.

22.2 Consider the phase correlation method for motion estimation and answer the following questions:

(a) What is this method's main strength?

(b) What is its principal weakness and how can it be overcome?

22.3 Answer the following questions (in your own words) about motion estimation.

(a) Why is motion estimation in video sequences such an *important* problem? (Select two–three specific video processing operations that benefit from or rely on motion estimation and describe them.)

(b) Why is motion estimation in video sequences such a *difficult* problem? (Select two–three specific problems and describe them.)

(c) What is the chief disadvantage of the exhaustive block matching algorithm?

22.4 Write a MATLAB script or function that will perform the three-step motion estimation algorithm and test it using frames 69 and 72 of the foreman sequence.

22.5 Write a MATLAB script or function that will perform the 2D log motion estimation algorithm and test it using frames 69 and 72 of the foreman sequence.

APPENDIX A

HUMAN VISUAL PERCEPTION

A.1 INTRODUCTION

The human visual system (HVS) is the final link in the perception of images and video sequences. A clear understanding of its capabilities and limitations can lead to better image and video processing solutions. In applications whose ultimate goal is to improve the image quality for human consumption, this knowledge allows designers to establish objective performance criteria and quality measures. In machine vision systems (MVS) whose goal is to emulate—and ultimately outperform—their human counterpart, it is absolutely necessary that we know how the human visual system works, which performance limits it imposes, and how this knowledge can be factored into the design of MVS.

In this appendix, we will provide a very brief overview of the human visual system with emphasis on aspects that are relevant—some may say essential—to the researcher and practitioner in the field. This is a long, deep, and fascinating topic for which there are almost as many open questions as there are answers. The interested reader may refer to the "Learn More About It" section at the end of the appendix for suggestions on books and other references that will broaden and deepen their understanding of the field of human vision science.

Practical Image and Video Processing Using MATLAB®. By Oge Marques.

A.2 THE HUMAN EYE

The HVS has two main components, the eye (input sensor) and the brain (information processing unit), connected by the optic nerve (transmission path) (Figure A.1). Image perception consists of capturing the image with the eye, then recognizing it, and finally interpreting its contents in the brain. First, light energy is focused by the lens of the eye onto the sensors on the retina, and then those sensors respond to light energy by an electrochemical reaction that sends an electrical signal down the optic nerve to the brain. The brain uses these nerve signals to create neurological patterns that we perceive as images. In this section, we look at selected anatomical and physiological aspects of the human eye.

Figure A.2 shows a simplified cross section of the human eye. The following are some of the anatomical properties of the eye that are of interest for our discussion:

- The eye contains a *lens* responsible for focusing an image. The movements of the lens are controlled by specialized *ciliary muscles*.
- The anterior portion of the lens contains an *iris diaphragm* responsible for controlling the amount of light that enters the eye. The central opening of the iris is called *pupil* and varies its diameter—approximately from 2 to 8 mm—in a way that is inversely proportional to the amount of incoming light.
- The innermost membrane of the eye is the *retina*, which is coated with photosensitive receptors called *cones* and *rods*. It is on the surface of the retina that an upside-down image of the scene is formed.

The combination of lens, diaphragm, and a back-projection surface is also present in a rudimentary camera, leading to a very popular analogy known as "the eye-camera analogy" (Figure A.3).

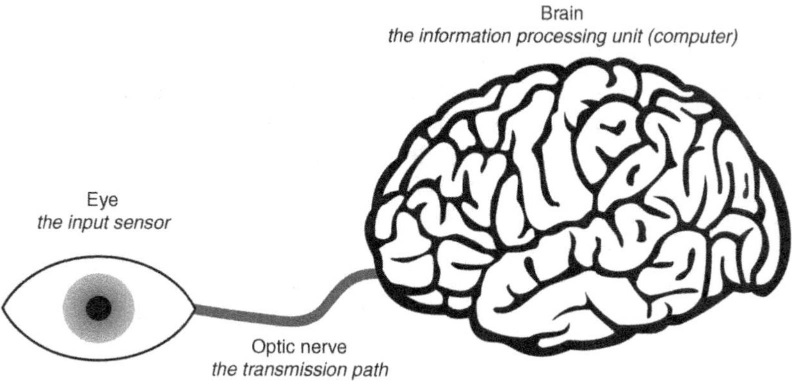

FIGURE A.1 Simplified view of the connection from the eye to the brain via the optic nerve. Adapted and redrawn from [Umb05].

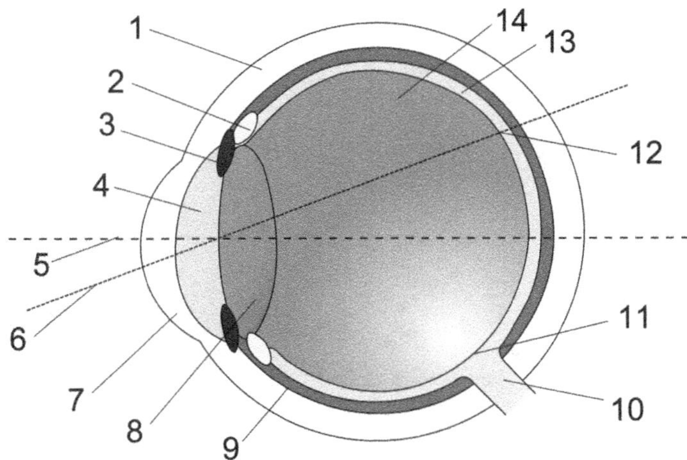

FIGURE A.2 The eye: a cross-sectional view. 1, sclera; 2, ciliary body; 3, iris; 4, pupil and anterior chamber filled with aqueous humor; 5, optical axis; 6, line of sight; 7, cornea; 8, crystalline lens; 9, choroid; 10, optic nerve; 11, optic disk; 12, fovea; 13, retina; 14, vitreous humor. Courtesy of Wikimedia Commons.

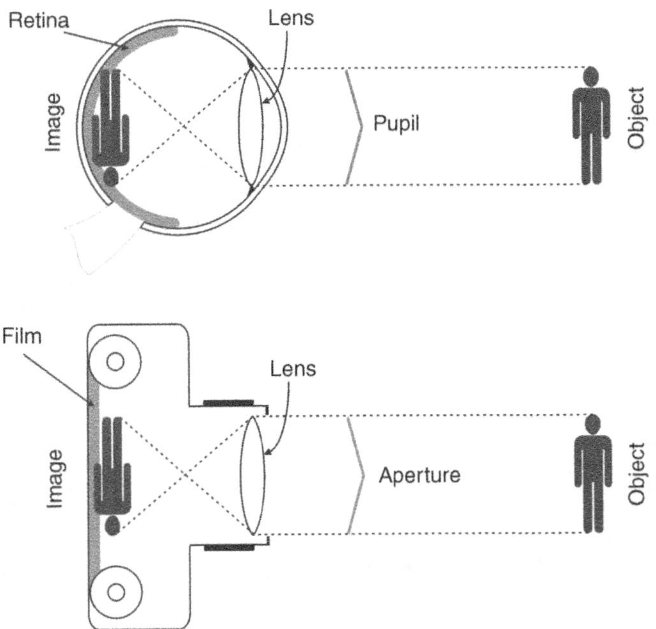

FIGURE A.3 The eye-camera analogy. Adapted and redrawn from [Pal99].

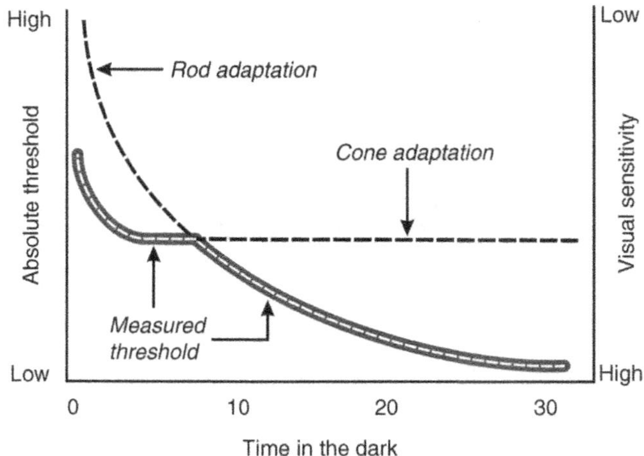

FIGURE A.4 Dark adaptation. Adapted and redrawn from [Pal99].

The surface of the human retina is coated with discrete photoreceptors, capable of converting light into electrochemical reactions that will eventually be transmitted to the brain. There are two types of photoreceptors, *cones* and *rods*, whose names were given based on their overall shape.

Cones (typically 6–8 million in total) are primarily concentrated in the *fovea*—the central part of the retina, aligned with the main visual axis—and are highly sensitive to color. Cones work well only in bright-light (*photopic*) vision scenarios, though. Under low lighting levels, they are not active and our ability to discriminate colors decreases dramatically.[1] Cones come in three varieties—*S*, *M*, and *L*, as in *short*, *medium*, and *long* (wavelengths), roughly meaning light in the red, green, and blue portions of the visible spectrum—each of which is primarily sensitive to certain wavelengths (and the colors associated with them).

The existence of three types of cones provides a physiological basis for the *trichromatic theory of vision*, postulated by Thomas Young in 1802, more than 150 years before it became possible to obtain physiological evidence of the existence of the three types of cones.

Rods outnumber cones (there are 75–150 million rods) and are primarily not concentrated in the fovea; instead, they are distributed over the entire retinal surface (except for the *optic disk*, a region of the retina that corresponds to the perceptual *blind spot*). Rods are not sensitive to color, but are sensitive to low levels of illumination, and therefore responsible for dim-light (scotopic) vision.

Figure A.4 shows how the sensitivity of the retina increases in response to decreasing incoming light in a process known as *brightness adaptation* or *dark adaptation*. The left portion of the curve corresponds to the adaptation experienced by the cones

[1]Anyone who has forgotten where he parked their car and tries to look for it after dark on a poorly lit parking lot knows firsthand how true this is.

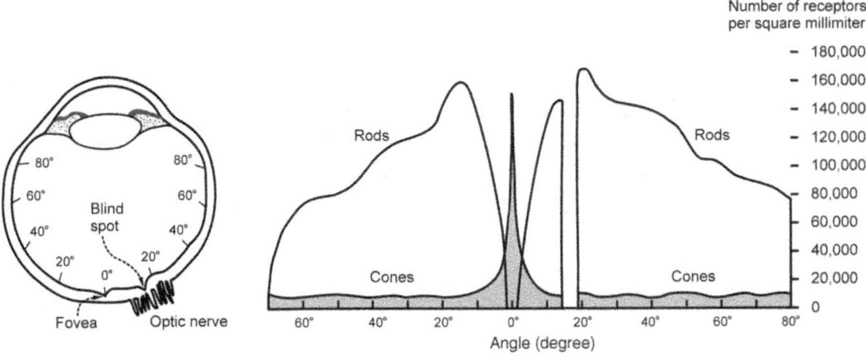

FIGURE A.5 Distribution of rods and cones in the human retina for the right eye (seen from the bottom). Adapted and redrawn from [Ost35].

(photopic vision). The right part indicates the range of time beyond which scotopic vision (primarily peripheral rod vision) becomes prevalent.

Figure A.5 shows the distribution of rods and cones in the retina. The pronounced peak at the center of the fovea (0° in relation to the visual axis) is indicative of the concentration of cones in that region. Note also that there are no cones or rods at a small region about 20° from the optical axis (toward the nasal side of each eye) known as the *blind spot*. This is a constructive limitation, after all the optic nerve must be attached to the retina at some point. It is truly remarkable that the brain "fills in" and allows us to see entire scenes even though some of the light reflected by objects in those scenes falls onto our blind spots (one for each eye).

A significant amount of visual processing takes place in the retina, thanks to a series of specialized (horizontal, bipolar, anacrine, and ganglion) cells. A detailed explanation of these cells and their behavior is beyond the scope of this appendix. It is important to note that although there are more than 100 million light receptors in the retina, the optic nerve contains only a million fibers, which suggests that a significant amount of data processing occurs before the electric impulses ever reach the brain. An additional curiosity is the fact that, since the human eye has evolved as an outgrowth of the brain, the retina appears to have been designed from the inside out: the photoreceptors are not the innermost cells; in fact, surprisingly enough, they point away from the incoming light.

After a retinal image is formed, it is converted into electric signals that traverse the optic nerve, cross the optic chiasm (where the left and right halves of the visual field of each eye cross), and reach the lateral geniculate nucleus (LGN) on its way to a region of the occipital lobe of the brain involved in visual perception: the *visual cortex*. The exact nature of the processing that occurs in the visual cortex, which cells and regions are in charge of what, and the number of possible visual pathways from photoreceptors in the retina to higher order regions of the cortex are among the many aspects over which there has been an enormous amount of research but not much agreement among researchers. Refer to the "Learn More About It" section at the end of the appendix for useful references.

A.3 CHARACTERISTICS OF HUMAN VISION

In this section, we look at a selected subset of characteristics of human vision that are of interest to the image and video processing researcher and system designer. Our goal is to highlight widely accepted facts about the way we perceive properties of scenes, such as brightness, contrast, sharpness (fine detail), color, motion, and flicker. We do not attempt to explain *why this is so* (we leave that to human vision researchers), but instead take an engineering approach and provide qualitative and—whenever possible—quantitative information that can be used in the design of imaging systems.

A.3.1 Resolution, Viewing Distance, and Viewing Angle

Resolution can be defined as the ability to separate two adjacent pixels, that is, resolve the details, in a test grating (such as the EIA test pattern shown in Figure A.6) or any other image. This ability depends on several factors, such as the picture (monitor) height (h) and the viewer's distance from the monitor (d), and the subtended viewing angle (θ) (Figure A.7).

The measure of the number of changes in image intensity for a certain test grating is referred to as its *spatial frequency*. The spatial frequency can be completely characterized by the variation frequencies in two orthogonal directions (e.g., horizontal and vertical). If we call f_x the horizontal frequency (expressed in cycles/horizontal

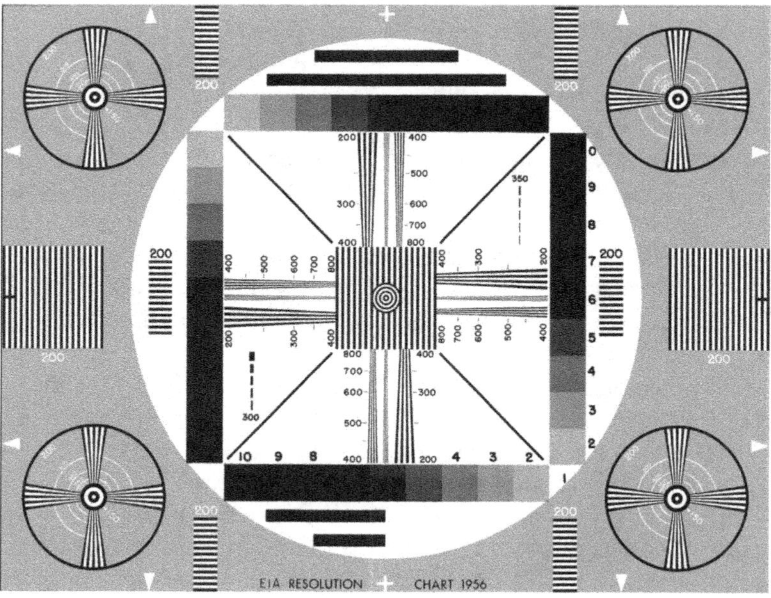

FIGURE A.6 EIA 1956 standard test pattern. Courtesy of http://www.bealecorner.com/trv900/respat/.

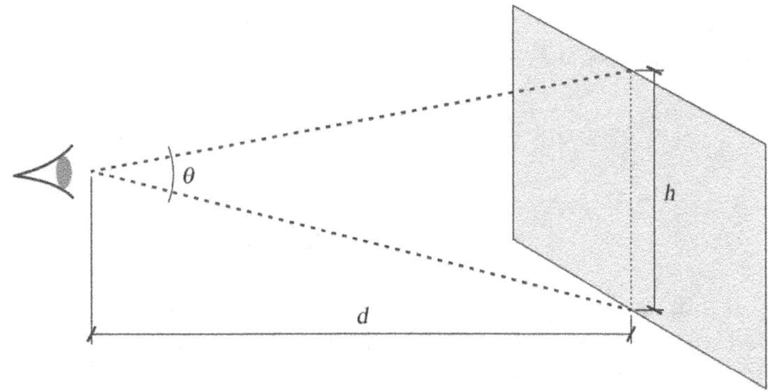

FIGURE A.7 Angular frequency concept.

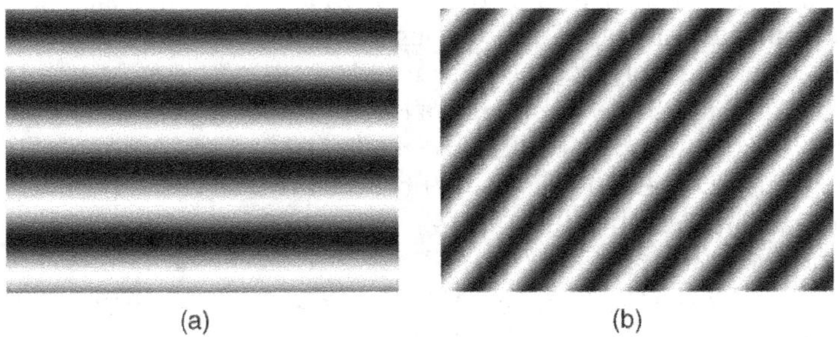

FIGURE A.8 Sinusoidal gratings commonly used for measures of resolution—based on MATLAB code by Alex Petrov: http://alexpetrov.com/softw/utils/.

unit distance) and f_y its vertical counterpart, the pair (f_x, f_y) characterizes the spatial frequency of a 2D image. For example, the test gratings on Figure A.8 have $(f_x, f_y) = (0, 4)$ (Figure A.8a) and $(f_x, f_y) = (7, 4.5)$ (Figure A.8b). These two values can be combined and expressed in terms of magnitude (f_{m}) and angle (θ):

$$f_{\mathrm{m}} = \sqrt{\left(f_x^2 + f_y^2\right)} \tag{A.1}$$

$$\theta = \arctan\left(\frac{f_y}{f_x}\right) \tag{A.2}$$

The examples in Figure A.8 have $(f_{\mathrm{m}}, \theta) = (4, 90^\circ)$ and $(f_{\mathrm{m}}, \theta) = (8.32, 32.7^\circ)$.

The spatial frequency of a test signal (as defined above) is not very useful in determining the user's perception of those signals because it does not account for the viewing distance. A more useful measure, which not only is a characterization of the signal, but also takes into account the viewing distance and the associated viewing angle, is the angular frequency[2] (f_θ), expressed in cycles per degree (cpd) of viewing angle and defined as follows (Figure A.7):

$$\theta = 2\arctan\left(\frac{h}{2d}\right) \approx \frac{h}{2d}\text{(radian)} = \frac{180h}{\pi d}\ \text{(degrees)} \tag{A.3}$$

$$f_\theta = \frac{f_s}{\theta} = \frac{\pi d}{180h} f_s\ \text{(cpd)} \tag{A.4}$$

A careful look at equation (A.4) reveals that, for the same picture (e.g., grating pattern) and a fixed picture height (PH), the angular frequency increases with the viewing distance; conversely, for a fixed viewing distance, larger display sizes lead to lower angular frequencies. This is consistent with our experience: the same test pattern appears to change more frequently if viewed from farther away and more slowly if displayed on a larger screen.

A common practical application of the concepts of resolution and viewing angle is the determination of the optimal viewing distance for a certain display. Although accurate calculations must take into account other parameters such as display brightness, ambient lighting, and visual acuity of the observer, a typical back of the envelope estimate of the optimal viewing distance can be obtained by assuming a subtended angle of 1 min of arc (1/60 of a degree) between two adjacent TV lines. For conventional standard definition TV (SDTV) displays (480 scan lines and a 4:3 aspect ratio), a viewing distance of about 7 × PH is usually recommended, whereas typical high-definition TV (HDTV) displays of the same height (with 1080 scan lines and a 16:9 aspect ratio) should be viewed at slightly less than half the distance (3.1 × PH) to fully appreciate the additional amount of detail available and maintain the same level of spatial frequency eye discrimination (also known as *visual acuity*) (Figure A.9). The horizontal picture angles at those viewing distances are 11° and 33°, respectively (Figure A.10).

A.3.2 Detail and Sharpness Perception

The ability to perceive fine details in an image or video sequence is one of the most important guiding factors in the design of image and video systems since it impacts parameters such as image definition, signal to noise ratio (SNR), and bandwidth. Perception of detail is intimately associated with the concept of *visual acuity*, which can be described as "the smallest angular separation at which individual lines in a grating pattern can be distinguished." The most familiar experience with visual acuity measures you may have had is a visit to the optometrist. In such visit, you are asked to read numbers or letters on a Snellen chart at a standardized test distance of 20 ft (6.1 m). At that distance, the width of the strokes of the letters in the 20/20 row subtends an angle of 1 min of arc. Being able to read that row is considered the standard for normal (20/20) vision. Visual acuity varies widely, from 0.5′ to 5′ (minutes of arc),

[2]Unfortunately, the expression *angular frequency* is not widely adopted in the literature; readers will often find the same concept expressed as *spatial frequency* instead, which is potentially confusing.

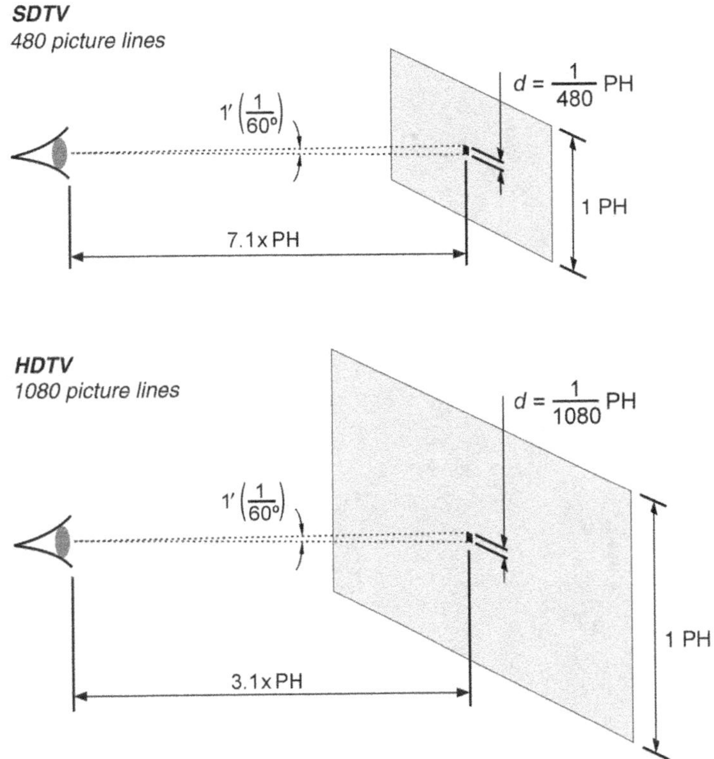

FIGURE A.9 Viewing distance for SDTV and HDTV displays. Adapted and redrawn from [Poy03].

depending on the contrast ratio and the quality of vision of each individual. An acuity of 1.7′ is usually assumed in the design of image and video systems [LI99].

A.3.3 Optical Transfer Function and Modulation Transfer Function

Most lenses, including the human eye's lens, are not perfect optical systems. Consequently, when visual stimuli with a certain amount of detail are passed through them, they may show a certain degree of degradation. The optical transfer function (OTF) is a way to measure how well spatially varying patterns are observed by an optical system, that is, a way to evaluate the extent of degradation. The OTF is usually expressed as a series of complex numbers—one for each spatial frequency—whose amplitude represents the reduction in signal strength and phase represents the corresponding phase shift. For the sake of simplicity, we will focus primarily on the amplitude component, which is known as *modulation transfer function* (MTF). The MTF is the spatial equivalent of frequency response in electronic circuits.

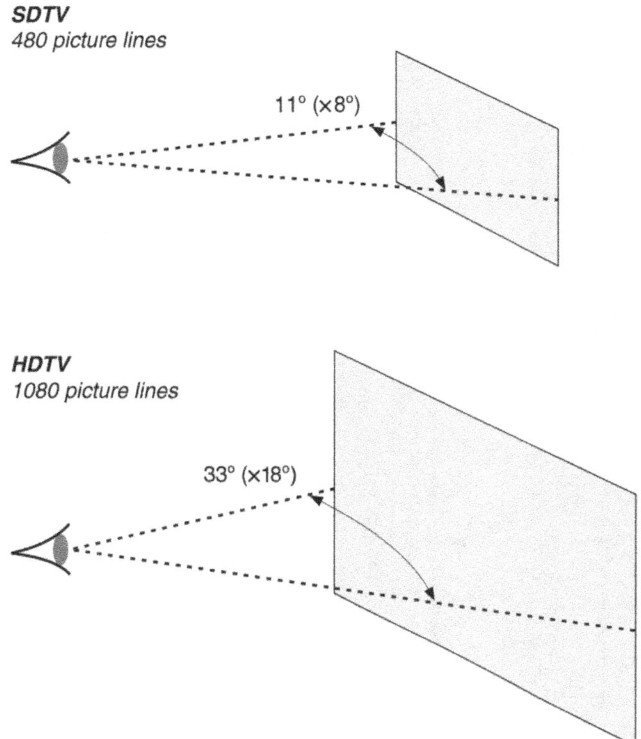

FIGURE A.10 Picture (viewing) angles for SDTV and HDTV displays. Adapted and redrawn from [Poy03].

Figure A.11 illustrates the MTF concept. Part (a) introduces the contrast index (CI), a measure of the amplitude differences between the darkest and brightest portions of the test image. Part (b) shows that a nonideal optical system tested with an input test image with constant CI will exhibit a CI that falls for higher spatial frequencies. Part (c) displays the MTF, which is the ratio between the output CI and the input CI. It is worth noting that while the MTF resolution test illustrated in Figure A.11 provides an objective evaluation of the possible optical degradation experienced by a test signal, the human perception of sharpness is subjective and it is also affected by contrast. As a result, an increase in the image contrast will cause an increased sensation of sharpness even though the MTF is unchanged, an aspect that has been exploited in the design of image display devices.

A.3.4 Brightness Perception

Brightness can be defined as "the attribute of a visual sensation according to which an area appears to emit more or less light" [Poy03].

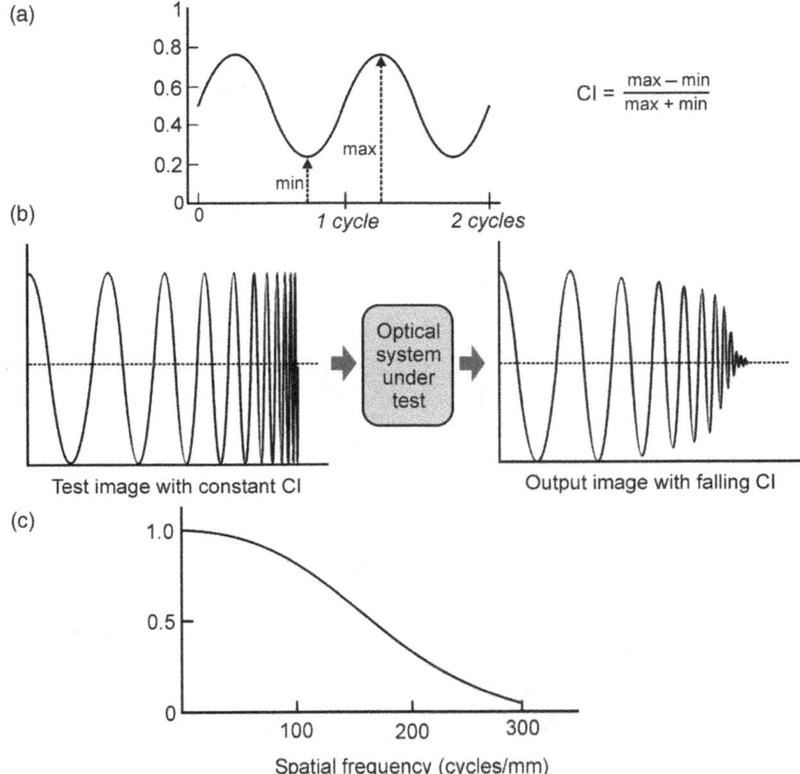

FIGURE A.11 (a) The definition of contrast index; (b) A test image with constant CI results in an output image with falling CI; (c) modulation transfer function: the ratio of output and input CIs. *Note*: When LF response is unity, CI and MTF are interchangeable. Redrawn from [Wat00].

The process of subjective brightness perception in humans is such that the perceived subjective brightness is proportional to the logarithm of the luminous intensity incident on the eye, as shown in Figure A.12. The long solid curve represents the (remarkably high, about eight orders of magnitude) range of luminous intensities that the HVS can adapt to. The plot also shows the transition between scotopic and photopic vision that takes place at low intensity levels. The region where there is overlap between cone-based and rod-based vision is called *mesopic vision*. More important, the small segment at the middle of the curve illustrates the phenomenon of *brightness adaptation*. In this case, after having adapted to a certain brightness level (B_a), the eye is capable of responding to stimuli around that value, provided that they are above another level (B_b in the figure). Any intensities below that will not be perceived. Should the average ambient intensity increase (or decrease), the eyes will adapt to another point in the main solid curve.

The perceived brightness of an area (or object) within an image also depends on the contrast between the object and its surroundings, in what is known as *simultaneous*

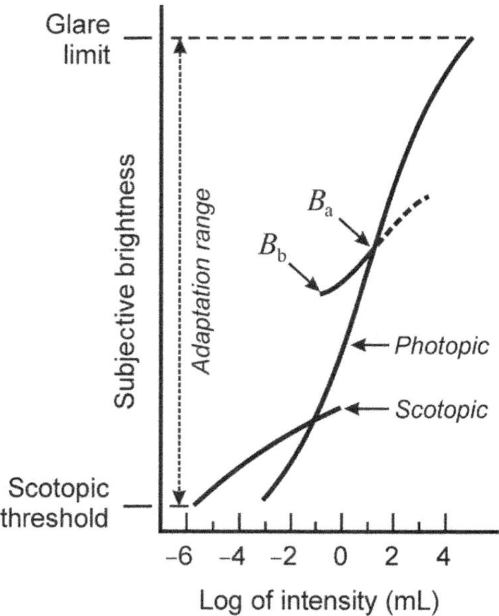

FIGURE A.12 Range of subjective brightness sensations showing a particular adaptation level. Redrawn from [GW08].

contrast (Figure A.13). In other words, we do not perceive gray levels as they are, but in terms of how they differ from their surroundings.

Another well-known way to show that perceived brightness is not a simple function of luminous intensity are the Mach bands (Figure A.14), named after Ernst Mach, who first described the phenomenon in 1865. These bands show that our visual system tends to undershoot or overshoot at the boundaries of regions with different intensities. This is due to the fact that the eye possesses a lower sensitivity to high- and low-spatial frequencies than to intermediate frequencies. It explains our ability to distinguish and separate objects, even in dimly lit scenes, thanks to the accentuated response around the actual edges. A possible implication of this property for designers of imaging

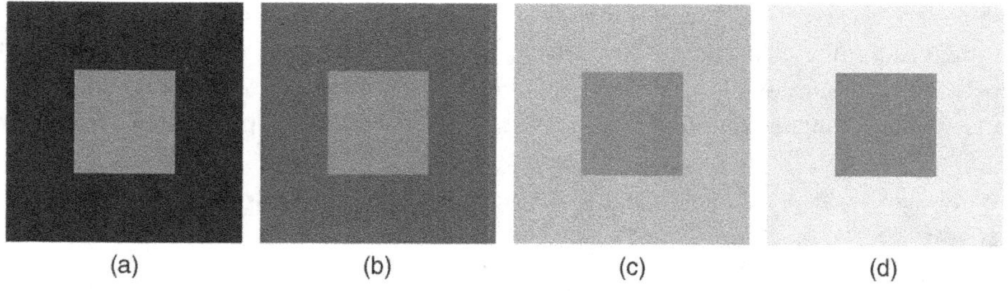

FIGURE A.13 Simultaneous contrast: the center square is perceived as progressively darker as the background becomes brighter (from (a) to (d)) even though it is identical in all four cases.

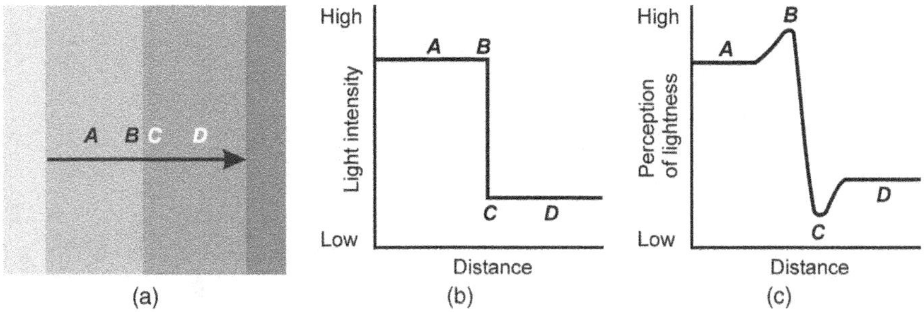

FIGURE A.14 Mach bands.

systems is that the reproduction of perfect edges is not a critical requirement, owing to the eye's imperfect response to high-frequency brightness transitions.

A.3.5 Contrast Ratio and Contrast Sensitivity Function

Contrast ratio is "the ratio of luminances of the lightest and darkest elements of a scene" [Poy03]. Typical contrast ratios are 80:1 (movie theater), 20:1 (TV in a living room), and 5:1 (computer monitor in an office).

The contrast sensitivity of the eye is defined as the smallest brightness difference that can be detected by an observer. Contrast sensitivity is usually measured as the ratio of luminances between two adjacent patches combined in a test pattern that is presented to a human subject (Figure A.15). The observer's field of vision is filled mostly by the surround luminance (Y_0). In the central area, the left portion of the circle has a test luminance value (Y), whereas the right half shows a slightly increased value ($Y + \Delta Y$). Subjects are asked to inform at which point the difference between the two

FIGURE A.15 Contrast sensitivity test pattern.

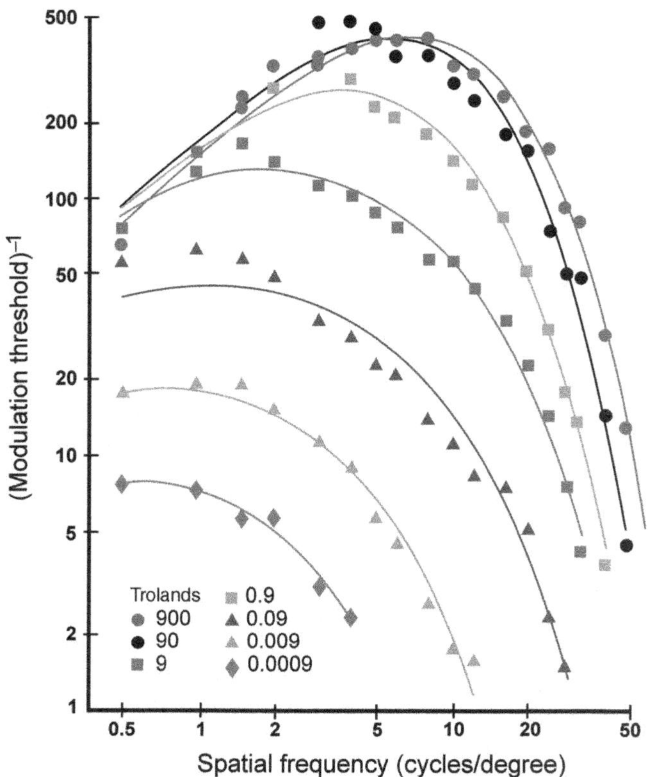

FIGURE A.16 Contrast sensitivity function for various retinal illuminance values (expressed in Td). Redrawn from [VNB67].

halves become noticeable[3] and the corresponding value of Y and ΔY are recorded. The process is repeated for a wide range of luminance values.

Experiments of this type have concluded that over a range of intensities of about 300:1, the discrimination threshold of vision is approximately a constant ratio of luminance. If one plots $\log(\Delta Y/Y)$ as a function of Y, it will show an interval of more than two decades of luminance over which the discrimination capability of vision is about 1% of the test luminance level. In other words, within that range, human vision cannot distinguish two luminance levels if the ratio between them is less than approximately 1.01.

In vision science, contrast sensitivity is also measured using a spatial grating test pattern. The resulting plot is called *contrast sensitivity function* (CSF) and it represents the contrast sensitivity as a function of the spatial frequency (in cycles/degree). Figure A.16 shows a family of curves, representing different adaptation levels, from very dark (0.0009 Td) to very bright (900 Td), where 9 Td is a representative value for

[3]This concept of *just noticeable difference* (JND) is also used in many other psychophysics experiments.

electronic displays.[4] The 9 Td curve peaks at about 4 cycles/degree. Below that spatial frequency, the eye acts as a differentiator; for higher spatial frequencies, the eye act as an integrator. Three important observations can be derived from this graph [Poy03]:

- Beyond a certain spatial frequency (around 50 cycles/degree for the 9 Td curve), the contrast sensitivity falls to very low values (less than 1% of the maximum value), which means our vision cannot perceive spatial frequencies greater than that. The implication for image and video systems designers is that there is no need to provide bandwidth or display resolution for those higher frequency contents, since, ultimately, they will not be noticed.
- Each curve peaks at a contrast sensitivity value that can be used to calculate the number of bits per pixel that must be used to quantize the image. Using more bits per pixel will allow the representation of luminance differences too subtle to be perceived by the human eye.
- The curve falls off for spatial frequencies lower than 1 cycle/degree, which suggests that luminance can be lower (within reasonable limits) in areas closer to the edges of the image without the viewer noticing it.

A.3.6 Perception of Motion

The electrochemical processes associated with the processing of incoming light in the human eye require several milliseconds to be performed and therefore act as a smoothing filter in the temporal domain: what the brain reconstructs is a time-averaged version of the actual input. As a result of this temporal smoothing, there is a *critical flicker*[5] *frequency* (CFF) below which we perceive the individual flashes of blinking light and above which those flashes merge into a continuous, smooth moving image sequence. This is a fundamental property in the design of movie, TV, and video systems. The CFF is directly proportional to the picture luminance and screen size and inversely proportional to the viewing distance.

The temporal frequency response of the HVS depends on several factors, such as viewing distance, display brightness, and ambient lighting. Figure A.17 shows the result of an experiment in which subjects where presented a flat screen whose brightness was modulated by a sinusoidal signal and were instructed to report the lowest modulation level at which the flicker became just noticeable. The reciprocal of that modulation level is referred to as contrast sensitivity and plotted versus the frequency of the modulating signal. Several conclusions can be derived from Figure A.17:

- The temporal response of the HVS is similar to a bandpass filter that peaks at intermediate frequencies and falls off quickly afterward.

[4] A *troland* (Td) is a unit of retinal illuminance equal to object luminance (in cd/m^2) times pupillary aperture area (in mm^2).

[5] The term *flicker* is also used to indicate an image defect usually caused by inadequate frame repetition rate, lower than the eye's temporal cutoff frequency.

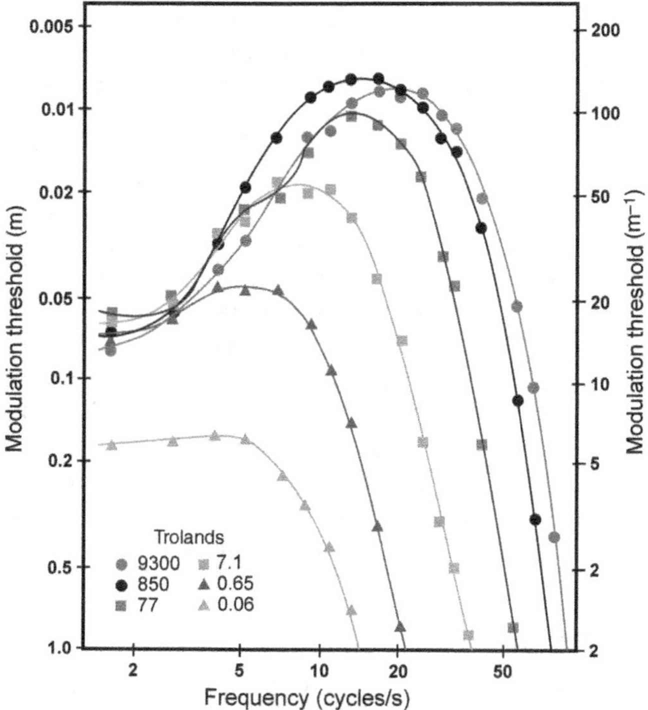

FIGURE A.17 Temporal frequency response of the HVS. Redrawn from [Kel61].

- The peak increases with the mean brightness of the image.
- One reason why the eye has reduced sensitivity at high frequencies is because it can retain the sensation of an image for a short interval after the image has been removed, a phenomenon known as *persistence of vision.*
- The critical flicker frequency is directly proportional to the average brightness of the display. In Figure A.17, the critical flicker frequency varies between 20 and 80 Hz.

A.3.7 Spatiotemporal Resolution and Frequency Response

After having seen the spatial and temporal frequency responses separately, we turn our attention to their combined effect. Figure A.18 shows experimental results by Robson [Rob66]. Figure A.18a shows that at higher temporal frequencies, both the peak and cutoff frequencies in the spatial frequency response shift downward. They also help confirm our intuitive expectation that the eye is not capable of resolving high spatial frequency details when an image moves very fast, compared to its spatial resolution capabilities for static images (Figure A.18b). The key implication of this finding for the design of TV and video systems is that it is possible to trade off spatial

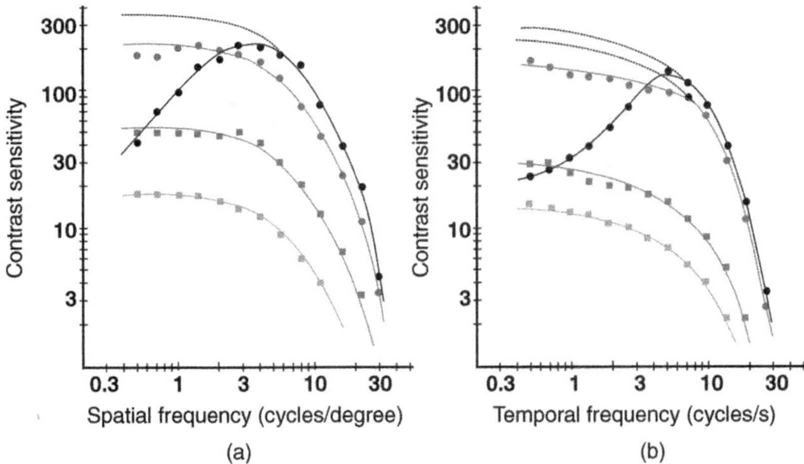

FIGURE A.18 Spatiotemporal frequency response of the HVS: (a) spatial frequency responses for different temporal frequencies (in cpd); (b) temporal frequency responses for different spatial (angular) frequencies (in Hz). Redrawn from [Rob66].

resolution with temporal resolution and vice versa. In Chapter 20, we discussed how this fact is exploited in the design of TV systems using interlaced scans.

The discussion thus far has assumed that the eye is not actually tracking any particular object within the image sequence, which is an obvious simplification that does not correspond to reality. Before we examine the different patterns that arise when tracking takes place, let us define a few important concepts regarding eye movements and their role in human visual perception.

The retina does not respond to the incoming light instantly; it requires between 0.15 and 0.30 s before the brain perceives an image. The early stages are often referred to as *preattentive* vision. After 0.15 s or so have elapsed, attentional mechanisms (some of which are triggered by the presence of salient objects in the scene in a *bottom-up* fashion, while other are dependent on the visual task at hand, the *top-down* component) are factored in and help guide the eyes toward regions of interest within the scene. The result of eye movements around a scene is called a *scanpath.*

Part of the eye movements registered in a scanpath results from involuntary unconscious vibrations known as *saccades*. The eye has a temporal filter mechanism that integrates spatial information from different positions of the retina covered by saccadic eye movements. This temporal filtering is responsible for the phenomenon of "persistence of vision" described earlier in this appendix.

Figure A.19 shows how the perceived temporal frequency changes as a function of eye movements. In both cases, the same object with the same amount of detail moves across the observer's field of view with the same constant speed. The temporal frequency is the product of the amount of detail in the object (usually expressed in lines/mm) and its speed. In Figure A.19a, the eye is fixed and the perceived temporal frequency is high, resulting in motion blur. In Figure A.19b, the eye tracks the moving

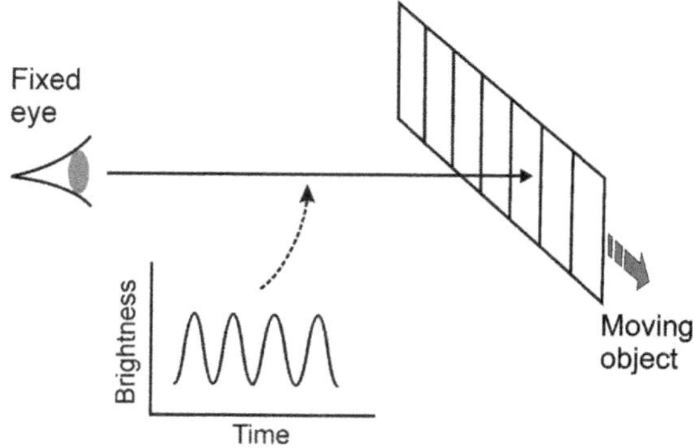

(a) *Temporal frequency = high*

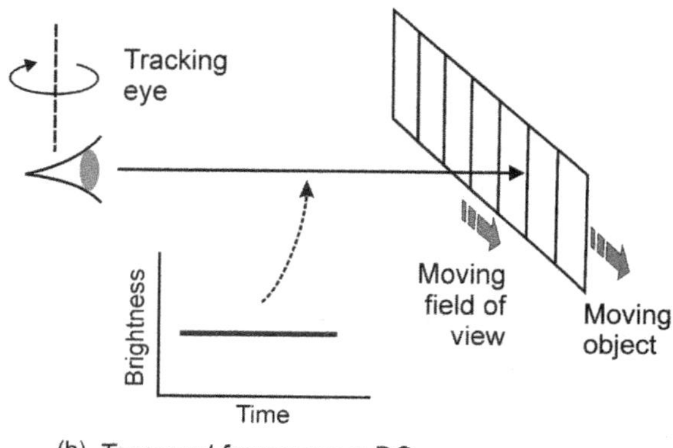

(b) *Temporal frequency = DC*

FIGURE A.19 Temporal frequency as a function of eye movements. Redrawn from [Wat00].

object (*smooth pursuit eye movement*), resulting in a temporal frequency of zero and improved ability to resolve spatial detail, besides the absence of motion blur. This is known as *dynamic resolution* and it is how humans judge the ability of reproducing detail in real moving pictures.

A.3.8 Masking

Masking is the reduction in the visibility of one image component (the target) due to the presence of another (the masker). The HVS is subject to several masking phenomena, such as [RP00]:

- *Texture Masking*: Errors in textured regions are usually harder to notice, whereas the HVS is very sensitive to errors in uniform areas.
- *Edge masking*: Errors near the edges are harder to notice.

- *Luminance Masking*: Visual thresholds increase with background luminance, as a result of brightness adaptation. Moreover, higher luminance levels increase the flicker effect.
- *Contrast Masking*: Errors (and noise) in light regions are harder to perceive as a result of the property of the HVS by which the visibility of an image detail is reduced by the presence of another.

A.4 IMPLICATIONS AND APPLICATIONS OF KNOWLEDGE ABOUT THE HUMAN VISUAL SYSTEM

In this section, we summarize some of the most relevant properties of the HVS that have implications for designers of image and video processing systems. They are listed as follows [Ric02]:

- The HVS is more sensitive to high contrast than low contrast regions within an image, which means that regions with large luminance variations (such as edges) are perceived as particularly important and should therefore be detected, preserved, and enhanced.
- The HVS is more sensitive to low spatial frequencies (i.e., luminance changes over a large area) than high spatial frequencies (i.e., rapid changes within small areas), which is an often exploited aspect of most image and video compression techniques. Discarding redundant *low* spatial frequency contents (while preserving edges) leads to computational savings.
- The HVS is more sensitive to image features that persist for a long duration, which means that it is important to employ techniques that minimize temporally persistent disturbances or artifacts in an image.
- HVS responses vary from one individual to the next, which means that subjective evaluations of image and video systems must be conducted with a large number of subjects. This aspect also reinforces the need to find quantitative measures of image and video quality that can be automatically calculated from the pixel data and yet reflect the subjective notion of perceived quality.

LEARN MORE ABOUT IT

The following are some of the books on human visual perception and related fields that may be of interest:

- Goldstein, E. B., *Sensation & Perception*, 7th ed., Belmont, CA: Thomson Wadsworth, 2007.
- Bruce, V., Green, P. R., and Georgeson, M.A., *Visual Perception: Physiology, Psychology and Ecology*, Philadelphia, PA: Psychology Press, 2003.

- Purves, D. and Lotto, R.B., *Why We See What We Do*, Sunderland, MA: Sinauer Associates, 2003.
- Yantis, S. (Ed.), *Visual Perception: Essential Readings*, Philadelphia, PA: Psychology Press, 2000.
- Palmer, S. E., *Vision Science: Photons to Phenomenology*, Cambridge, MA: Bradford Books/MIT Press, 1999.
- Gregory, R. L., *Eye and Brain: The Psychology of Seeing*, Princeton, NJ: Princeton University Press, 1998.
- Rodieck, R. W., *The First Steps in Seeing*, Sunderland, MA: Sinauer Associates, 1998.
- Wandell, B. A., *Foundations of Vision*, Sunderland, MA: Sinauer Associates, 1995.

The following are some of the scientific journals that publish research results in vision science and related areas (in alphabetical order): *Cognition*, *Journal of Vision*, *Nature*, *Nature Neuroscience*, *Perception*, *Perception & Psychophysics*, *Science*, *Spatial Vision*, *Vision Research*, *Visual Cognition*, and *Visual Neuroscience*.

ON THE WEB

- *Vision Science Portal*: an Internet resource for research in human and animal vision, with links to relevant conferences, journals, research groups, software, and much more.
  http://visionscience.com/
- *MATLAB Psychophysics Toolbox*: a widely used (and well-documented) collection of MATLAB functions for psychophysics experiments.
  http://psychtoolbox.org/
- *Project LITE (Boston University)*: a great collection of interactive visual illusions classified by category.
  http://lite.bu.edu/
- *Michael Bach: Optical Illusions & Visual Phenomena*: Another excellent collection of illusions. Many QuickTime movies.
  http://www.michaelbach.de/ot/
- *The Joy of Visual Perception*: a "web book" by Peter K. Kaiser, York University.
  http://www.yorku.ca/eye/

APPENDIX B

GUI DEVELOPMENT

B.1 INTRODUCTION

In this appendix, we provide a practical guide to develop graphical user interfaces, or GUIs, for MATLAB applications. This walk-through will give you the general idea of how a user interface works in MATLAB, from laying out the window in code to passing data around your application. We will refer to an example GUI throughout the appendix so that you can see the concepts in action.

First, we will look at the basic structure of the code that makes up a GUI. This is important because the structure of the code is related to its function, as we will see later. We will also take a look at how the control of the system is passed between MATLAB and the functions that make up a GUI. This will allow you to implement a fancy dynamic application, even though MATLAB executes code sequentially. Next, we will inspect the method by which data are saved in the GUI and how those data are passed around so that all necessary functions have access to them. Finally, we will dissect a working GUI demo that will wrap up all the concepts covered as well as solidify them.

B.2 GUI FILE STRUCTURE

The file that makes up a GUI interface is nothing more than an M-file. The typical uses of an M-file you have most likely seen so far are to save code in a script or create

Practical Image and Video Processing Using MATLAB®. By Oge Marques.

a stand-alone file, or a function, that performs a particular task. A GUI M-file is a lot like a function in that it performs one task: create and execute your GUI. It differs, however, from a function in the structure of the file. When writing a function, you have the option of writing subfunctions to take care of operations accessible only to your function. In a GUI M-file, there will always be subfunctions present, sometimes many. The basic structure of a GUI M-file is shown in the pseudocode that follows:

```
function mygui(parameter1, parameter2, ...)

if no parameters were passed, then
    initialize();
else
    switch over the parameters
        case 1:
            subfunction1();
        case 2:
            subfunction2();
        ...
    end
end

function initialize()
(code to initialize goes here)

function subfunction1()
(code for function1 goes here)

function subfunction2()
(code for function2 goes here)

...
```

The first line contains the function definition as usual. We will take a look at the uses of the parameters later.

Let us analyze the subfunctions. Three subfunctions are shown in the model, but more can exist (and usually will). The first subfunction is uniquely called `initialize()`. This should always be the first subfunction. It will initialize the GUI, as the name suggests. More specifically, this subfunction will create the GUI figure and generate all the buttons, axes, sliders, and any other graphical items necessary. It will also load any default parameters your application may contain. For example, if your application allows the user to blur an image dynamically by dragging a slider bar, you may want to start out with an initial amount of blur. Even if you wanted no blur initially, this must be defined, and you would do it in the `initialize()` subfunction. The initialization subfunction is executed only once—when the GUI runs for the first time. After that, it is up to all the other subfunctions to make sure all goes well.

The name of the initialization subfunction is not required to be `initialize()`, but this name is easy to understand and is self-explanatory. Ultimately, it is up to you how you wish to name it.

The other subfunctions, depicted as `subfunction1()` and `subfunction2()` in the model, represent all the other functionality that your GUI application will contain. Any action that could take place within the GUI will be coded in its own separate subfunction. So, when a button is pressed, there is a subfunction that handles that action. Same goes for sliders, pull-down menus, or any other interactive elements. Subfunctions can also be used for repetitive tasks. For instance, if there are many interactive elements in your GUI having a common component in their code, it is wise to factor that code out and place it in its own subfunction, thereby reducing the complexity of the code and making your code easier to maintain.

So far we have looked at the subfunctions of a GUI M-file, which have three basic uses: initialization of the GUI, encapsulating the functionality of interactive components in the user interface, and factoring out reusable code. It seems as if we have put all of the code into subfunctions, so you may ask if there is a use for the main function itself (in the case of the model, `mygui()`). The main function has a small but very crucial role in the execution of a user interface. Let us first look at what happens when this code is executed and later we will see why it is all necessary.

The only aspect to the main function is an `if-else` statement. Basically, the functionality is as follows: if no parameters are passed when the function is called, then its only job is to initialize the GUI. If parameters are passed, then it performs a switch in such a way that different combinations of parameters will give access to the subfunctions. That is it! That is the concept of the main function. You can think of it as a hub—from the main function, we can access all other subfunctions. As you will see later, this is what makes dynamic execution possible in MATLAB.

We will come back to our model later to see exactly how its structure meets its function. To continue, we must look at how control is passed in MATLAB.

B.3 PASSING SYSTEM CONTROL

One of the main challenges in developing a GUI in MATLAB is to give the user the feeling of complete control over the application, just as in any other program. By this we mean that the user can interact with the interface in a non-predetermined way, and the application will still execute as expected. This is a small problem in MATLAB because code is executed in a top-down fashion, meaning when a script or function is called, execution begins at the top of the file and works its way down and finally exits. If we look at the system control diagram depicted in Figure B.1, we can see this is consistent. When users run a script, they do so by calling the file name or a function name. Control is given to that function and when the function is complete, control is given back to MATLAB. You may have noticed that when running functions or scripts that take some time to execute, MATLAB is not responsive until that function is complete. This is because the function has control, not the main interface of MATLAB. Coming back to our problem, we can see that this type of execution is not consistent with a dynamic environment where the user may click on components in random order. By understanding how control is passed in MATLAB, we can understand the clever design of a GUI M-file that allows a dynamic environment.

When analyzing the passing of system control, keep in mind that the structure of the GUI M-file, as you will see, is consistent with the concepts depicted in Figure B.1. When MATLAB is first started, the user is presented with a command prompt. To start a GUI application, the user types the function call for that GUI, and it is loaded. From here, MATLAB passes control from the command prompt over to the main function (shown as step 1). Let us assume that the user did not pass any parameters to be consistent with our M-file model. Because there were no parameters, the main function sends control to the initialization subfunction (step 2). As stated previously, this subfunction creates the GUI window and all other components (step 3). Once the initialization subfunction is complete, control is returned back to the main function (4), but there is no more code to execute here (remember the main function was only a hub). So from here, the main function ends and control is returned to MATLAB (exit). Although control is returned to the command prompt, we now have another window open in MATLAB, our GUI. MATLAB is aware of this, and if we interact with any components, MATLAB will know.

It is probably a good time to introduce the concept of a callback. A callback is simply a line of code that executes when someone interacts with a component of your GUI. Every component has its own callback, and it is defined when the GUI is initialized. Remember that this code is small, so we cannot put the entire functionality of the component within the callback, but just one line of code that will call a function containing the real code that implements the component. So, to summarize the concept of a callback, the following events would take place: when the user interacts with a particular component (i.e., a button), MATLAB grabs the value of the callback for that component and then executes it as if it were typed into the command prompt. Now, because this code must be small, we want to design our callback in such a way that we will ultimately execute the code that gives our component its functionality.

If we observe Figure B.1, we see that when a user interacts with a component, MATLAB will execute the callback for that component (which is just one line of code). But, if you remember, the implementation for that component is one of the subfunctions of the M-file. This is a slight problem because subfunctions are only accessible from within the M-file. Remember, we are no longer executing code in the M-file, so currently we do not have access to that subfunction (i.e., MATLAB currently has control). To overcome this issue, we will define the callback such that it will again call the main function of our GUI M-file, but instead of calling with no parameters, we will specify the specific parameters that will gain us access to the subfunction we need. This is the heart of GUI functionality. Keep also in mind that each component has its own unique callback code, and therefore we can define unique ways of calling the M-file such that all the subfunctions are accessible. Once that callback code is executed (shown as step a), we are again inside the M-file. The

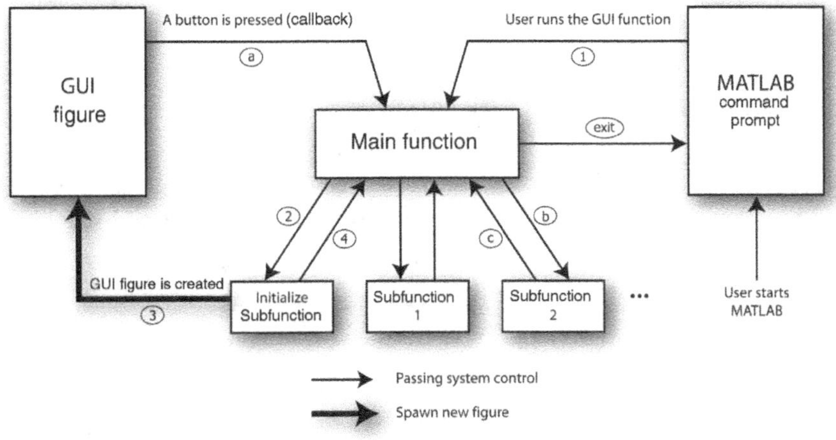

FIGURE B.1 System control diagram.

first chunk of code that is encountered is that hub-like `if-else` code. Here, we are forwarded to the appropriate subfunction (step b), where it does its business and then returns control to the main function (c). From here, we again see that there is nothing else to execute in the main function, so the function ends and control is returned to MATLAB (exit).

Once the GUI figure is initialized, the process of executing callbacks is repeated every time the user interacts with the GUI (following the sequence a–b–c–exit). The only time this stops is when the figure is closed. At that point, the only way to get back to the GUI is to start it up again by calling the main function.

So now we can see how the structure of the GUI M-file meets its function. The M-file must be structured so that the entire appropriate code is accessible by the GUI figure. Although it is nice to understand how this all works, you will not get far with a GUI unless you have a clear picture of how data can be saved, accessed later, and manipulated. Thus, we come to the `UserData` object.

B.4 THE USERDATA OBJECT

As you will see, the concept of the UserData object is actually quite simple. Before we have a look at it, let us first take note to a small feature of MATLAB. The variables available for use are organized in a stack, as illustrated in Figure B.2. This means that when MATLAB passes control from the command prompt over to the main function, any variables that were available are no longer accessible. They still exist, but they are accessible only from the command prompt. Similarly, when control is passed to a subfunction, any variables that might have been present in the main function are no longer available. Also, if variables are created within the subfunction and once we exit the subfunction and control is passed back to the main function, we would not

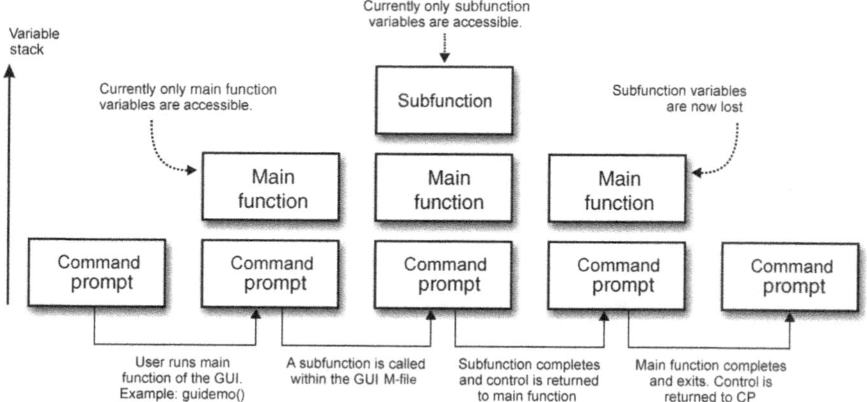

FIGURE B.2 Variable stack.

have access to those variables that were once accessible from the subfunction—those variables are in fact lost forever!

What this means for us in terms of a GUI is, at this point, we have no way of saving any data. If you notice, from the system control diagram, the main function is constantly called, but every time it exits, all variables are lost—the function is brought to life and killed constantly and right along with it follows its variables. In this entire GUI environment, there is in fact only one item that is always in existence other than MATLAB itself and that is the figure of the GUI. Because of this fact, we will store any persistent data within the GUI figure. The data structure that houses this information is traditionally known as the `UserData` object.

The `UserData` object is a child of the GUI figure and usually goes by the name `ud`, although technically the object could be named anything you want. The `UserData` object is actually just a variable, and as all other variables in MATLAB, it can take on any form that a variable may express. If the only persistent data in your application is an image, then `ud` can be equal to the image data. If your application holds different pieces of unrelated data, then `ud` can take on the form of a structure that is the most common. This is by far the most flexible because future changes may require an addition of information to the persistent data in your GUI and by initially taking on the form of a structure, this information can easily be added to the structure (remember a structure can contain elements of different types or classes).

With regard to the `UserData` object, the only thing left to look at is exactly how we initialize it, grab its contents when needed, and how to save back to it. This will be covered when we inspect the demo GUI in the next section.

B.5 A WORKING GUI DEMO

You should notice that the demo GUI M-file (available at book's web site) looks similar to our model pseudocode from earlier. There is one parameter that will be used by the GUI during run-time, and based on the switch statement, it can take on a value of 'average' or 'complement'. The initialization subfunction accounts for the majority of the code for this demo, so let us take a look at its code first. Remember that this is a simplified demo, and there are many other creative things that can be done in a GUI file; these are just the basics.

If the user runs the GUI function (simply by typing `guidemo` at the MATLAB command prompt), the `if-else` statement will direct system control over to the initialization subfunction. The first block of code will determine whether the GUI is already running. This is achieved by first searching MATLAB's list of objects for the name of the GUI. If it is present, we will close it. Of course, in your own application you can do whatever you want. Your functionality may allow multiple instances of the same GUI application running simultaneously. It is up to you and the design of your application.

The next block of code creates the GUI figure and sets its properties. After this, the figure is resized and relocated to the center of the screen. Next, all the components on the figure are created. You will notice, for example, that when the 'average' button is created, we are also saving its handle in the `UserData` object. A handle is simply MATLAB's language for a pointer to that component. We do this so that we may have access to the button and its properties if needed later in the application. Also, notice that the callback property of the 'average' button is set to `guidemo(''average'')`. Here, we are telling MATLAB to execute the `guidemo()` function with the parameter 'average' every time this button is pressed. It is important to know that the quotes around 'average' in this statement are two single quotes, *not* double quotes!

The last two components added to the figure are axes. Remember that an image is rendered on an axes—this is what these two axes are for. In the final steps of the initialization subfunction, we load the default image and save it to the `UserData` object. We then display the original loaded image in the `org_axes` axes. We could just stop there and let the user decide which filter to use by pressing one of the buttons, but at this point there is nothing rendered on the second axes, and so it will display as an axes. This might not be the best way to present the application when first loaded, so we execute the code to filter with an averaging filter by calling the appropriate subfunction. Notice that this is indeed the same function that is executed when the user presses the 'average' button and this is just fine.

The last two subfunctions implement the two buttons. Both for the most part are identical, except for the filtering part, which is not of importance here, so we will look at only one of them. In the `filter_average` subfunction, the first thing we do is get the `UserData` object from the figure by using the function `get()` and save it in our own variable `ud`. Remember that this variable will be gone once we leave this subfunction, so we should remember to save any changes back to the `UserData` object in the figure before exiting. The next, block of code simply filters the original image and saves this new image in an appropriate variable inside our copy of the `UserData` object. Next, we display the filtered image. Notice that we must first select the appropriate axes by using the axes function. If this code is not placed before the `imshow` function call, MATLAB will display the image in the first axes it finds, which in this case would most likely be the original image axes. Finally, because we have altered our version of the `UserData` object, we save it back to the figure by using the set function `set()`.

B.6 CONCLUDING REMARKS

Remember that this M-file was just for demonstration purposes and most likely your GUI M-files will be more complex to achieve your desired functionality. In this demo, we did not allow the user to pass any parameters to the function when calling it, but this could be done by modifying the `if-else` statement to allow it. Also, each component that can be used in a GUI has many features that allow you to customize the GUI, so it is a good idea to study the different components and their capabilities before designing your GUI. With your understanding of what goes on behind the scenes when a GUI is in execution, you should be able to design and implement a GUI that will fit the needs of your next MATLAB prototype.

REFERENCES

Ack01 S. Ackerman. Film sequence detection and removal in DTV format and standards conversion. Technical report, Teranex, Inc., 2001.

Ack02 S. Ackerman. Issues faced in DTV up-conversion. Technical report, Teranex, Inc., 2002.

AS07 S. Avidan and A. Shamir. Seam carving for content-aware image resizing. In *International Conference on Computer Graphics and Interactive Techniques*, ACM Press, New York, NY, 2007.

AT04 T. Acharya and P.-S. Tsai. *JPEG2000 Standard for Image Compression: Concepts, Algorithms and VLSI Architectures*, Wiley–Interscience, 2004.

Bax94 G. A. Baxes. *Digital Image Processing: Principles and Applications*, Wiley, New York, 1994.

BB08 W. Burger and M. J. Burge. *Digital Image Processing: An algorithmic Introduction Using Java*, Springer, New York, 2008.

BD00 A. C. Bovik and M. D. Desai. *Basic Binary Image Processing*, Chapter 2.2, Academic Press, San Diego, 2000, pp. 37–52.

BN92 T. Beier and S. Neely. Feature-based image metamorphosis. *ACM SIGGRAPH Computer Graphics*, 26(2):35–42, 1992.

Bov00a A. Bovik, editor. *Handbook of Image and Video Processing*, Academic Press, San Diego, 2000.

Bov00b A. C. Bovik. *Basic Gray-Level Image Processing*, Chapter 1.1, Academic Press, San Diego, 2000, pp. 3–17.

Bov00c A. C. Bovik. *Introduction to Digital Image and Video Processing*, Chapter 1.1, Morgan Kaufmann, 2000, pp. 3–17.

Practical Image and Video Processing Using MATLAB®. By Oge Marques.

Bra95 R. N. Bracewell. *Two-Dimensional Imaging*, Prentice Hall, 1995.

Bri74 E. O Brigham. *The Fast Fourier Transform*, Prentice Hall, 1974.

Can86 J. Canny. A computational approach to edge detection. *IEEE Transactions on Pattern Analysis and Machine Intelligence*, 8(6):679–698, 1986.

Cas96 K. R. Castleman. *Digital Image Processing*, Prentice Hall, Upper Saddle River, 1996.

CJSW01 H. D Cheng, X. H. Jiang, Y. Sun, and J. Wang. Color image segmentation: advances and prospects. *Pattern Recognition*, 34(12):2259–2281, 2001.

CMS+06 D. Ćulibrk, O. Marques, D. Socek, H. Kalva, and B. Furht. A neural network approach to Bayesian background modeling for video object segmentation. In *International Conference on Computer Vision Theory and Applications (VISAPP 2006)*, Setúbal, Portugal, February 2006, pp. 474–479.

CMS+07 D. Ćulibrk, O. Marques, D. Socek, H. Kalva, and B. Furht. Neural network approach to background modeling for video object segmentation. *IEEE Transactions on Neural Networks*, 18(6):1614–1627, November 2007.

Csi90 M. Csikszentmihalyi. *Flow: The Psychology of Optimal Experience*, Harper Collins, 1990.

Dav04 E. R. Davies. *Machine Vision: Theory, Algorithms, Practicalities*, Morgan Kaufmann Publishers Inc., San Francisco, CA, 2004.

Dea01 J. Deame. DTV format conversion: a buyer's guide. Technical report, Teranex, Inc., 2001.

DG87 E. R. Dougherty and C. R. Giardina. *Matrix Structured Image Processing*, Prentice Hall, 1987.

DHS01 R. O. Duda, P. E. Hart, and D. G. Stork. *Pattern classification*, 2nd edition, Wiley, 2001.

Dou92 E. R. Dougherty. *An Introduction to Morphological Image Processing*, SPIE Press, 1992.

Dou94 E. R. Dougherty. *Digital Image Processing Methods*, CRC Press, 1994.

DR78 L. S. Davis and A. Rosenfeld. Noise cleaning by iterated local averaging. *IEEE Transactions on Systems, Man, and Cybernetics*, 8:705–710, 1978.

DZ86 S. Di Zenzo. A note on the gradient of a multi-image. *Computer Vision, Graphics, and Image Processing*, 33(1):116–125, 1986.

Eff00 N. Efford. *Digital Image Processing: A Practical Introduction Using Java*, Addison-Wesley, 2000. Includes CD-ROM.

FB60 G. L. Fredendall and W. L. Behrend. Picture quality: procedures for evaluating subjective effects of interference. *Proceedings of the IRE*, 48(6):1030–1034, 1960.

FDHF+05 R. B. Fisher, K. Dawson-Howe, A. Fitzgibbon, C. Robertson, and E. Trucco. *Dictionary of Computer Vision and Image Processing*, Wiley, 2005.

FGW96 B. Furht, J. Greenberg, and R. Westwater. *Motion Estimation Algorithms for Video Compression*, Kluwer Academic Publishers, Norwell, MA, 1996.

FH04 P. F. Felzenszwalb and D. P. Huttenlocher. Efficient graph-based image segmentation. *International Journal of Computer Vision*, 59(2):167–181, 2004.

FM81 K. S. Fu and J. K. Mui. A survey on image segmentation. *Pattern Recognition*, 13(1):3–16, 1981.

FMR+02 J. Freixenet, X. Muñoz, D. Raba, J. Martí, and X. Cufí. Yet another survey on image segmentation: region and boundary information integration: In *Proceedings of the 7th European Conference on Computer Vision: Part III*, Springer, London, 2002, pp. 408–422.

FP03 D. A. Forsyth and J. Ponce. *Computer Vision: A Modern Approach*, Prentice Hall, 2003.

Fre77 W. Frei. Image enhancement by histogram hyperbolization. *Computer Graphics and Image Processing*, 6(3):286–294, 1977.

FS76 R. W. Floyd and L. Steinberg. An adaptive algorithm for spatial grey scale. *Proceedings of the Society of Information Display*, 17:7577, 1976.

GD88 C. R. Giardina and E. R. Dougherty. *Morphological Methods in Image and Signal Processing*, Prentice Hall, Englewood Cliffs, NJ, 1988.

GDCV99 J. Gomes, L. Darsa, B. Costa, and L. Velho. *Warping and Morphing of Graphical Objects*, Morgan Kaufmann, 1999.

GH99 B. Grob and C. E. Herndon. *Basic Television and Video Systems*, 6th edition, McGraw-Hill, New York, 1999.

Gil05 G. F. Gilder. *The Silicon Eye: How a Silicon Valley Company Aims to Make All Current Computers, Cameras and Cell Phones Obsolete*, W. W. Norton & Company, 2005.

Gir93 B. Girod. Motion compensation: visual aspects, accuracy, and fundamental limits. *Motion Analysis and Image Sequence Processing*, Kluwer Academic Publishers, 1993, pp. 126–152.

Gol07 E. B. Goldstein. *Sensation and Perception*, 7th edition, Thomson Wadsworth, Belmont, CA, 2007.

Gos05 A. A. Goshtasby. *2-D and 3-D Image Registration*, Wiley, 2005.

GW08 R. C. Gonzalez and R. E. Woods. *Digital Image Processing*, 3rd edition, Prentice Hall, Upper Saddle River, NJ, 2008.

GWE04 R. C. Gonzalez, R. E. Woods, and S. L. Eddins. *Digital Image Processing Using MATLAB*, Pearson Prentice Hall, 2004.

HCT+06 Y.-W. Huang, C.-Y. Chen, C.-H. Tsai, C.-F. Shen, and L.-G. Chen. Survey on block matching motion estimation algorithms and architectures with new results. *Journal of VLSI Signal Processing Systems*, 42(3):297–320, 2006.

HL05 D. Hanselman and B. Littlefield. *Mastering MATLAB 7*, Pearson Prentice Hall, Upper Saddle River, NJ, 2005.

Hou P. V. C. Hough. Method and means of recognizing complex patterns. US patent 306,965,418. December 1962.

HPN97 B. G. Haskell, A. Puri, and A. N. Netravali. *Digital Video: An Introduction to MPEG-2*, Kluwer Academic Publishers, Norwell, MA, 1997.

HS85 R. M. Haralick and L.G. Shapiro. Image segmentation techniques. *Applications of Artificial Intelligence II*, 548:2–9, 1985.

Hu62 M.-K. Hu. Visual pattern recognition by moment invariants. *IRE Transactions on Information Theory*, 8:179–187, 1962.

Hum75 R. Hummel. Histogram modification techniques. *Computer Graphics and Image Processing*, 4:209–224, 1975.

HYT79 T. Huang, G. Yang, and G. Tang. A fast two-dimensional median filtering algorithm. *IEEE Transactions on Acoustics, Speech, and Signal Processing*, 27(1):13–18, 1979.

ITU00 Recommendation 500-10. Methodology for the subjective assessment of the quality of television pictures. ITU-R BT.500, 2000.

Jac01 K. Jack. *Video Demystified: A Handbook for the Digital Engineer*, 3rd edition, LLH Technology Publishing, Eagle Rock, VA, 2001.

Jah05 B. Jahne. *Digital Image Processing: Concepts, Algorithms, and Scientific Applications*, 6th edition, Springer-Verlag New York Secaucus, NJ, 2005.

Jai89 A. K. Jain. *Fundamentals of Digital Image Processing*, Prentice Hall, Englewood Cliffs, NJ, 1989.

JC90 B. K. Jang and R. T. Chin. Analysis of thinning algorithms using mathematical morphology. *IEEE Transactions on Pattern Analysis and Machine Intelligence*, 12(6):541–551, 1990.

JKS95 R. Jain, R. Kasturi, and B. G. Schunck. *Machine Vision*, McGraw-Hill, 1995.

JW75 D. Judd and G. Wyszecki. *Color in Business, Science, and Industry*, Wiley, 1975.

Kel61 D. H. Kelly. Visual responses to time-dependent stimuli. I. Amplitude sensitivity measurements. *Journal of the Optical Society of America*, 51:422–429, 1961.

Kir71 R. Kirsch. Computer determination of the constituent structure of biological images. *Computers and Biomedical Research*, 4:315–328, 1971.

LA01 D. L. Lau and G. R. Arce. *Modern Digital Halftoning*, Marcel Dekker, 2001.

LD04 Z.-N. Li and M. Drew. *Fundamentals of Multimedia*, Pearson Prentice Hall, Upper Saddle River, NJ, 2004.

LI99 A. C. Luther and A. F. Inglis. *Video Engineering*, 3rd edition, McGraw-Hill, New York, 1999.

Lim90 J. S. Lim. *Two-Dimensional Signal and Image Processing*, Prentice Hall, Englewood Cliffs, NJ, 1990.

LP06 R. Lukac and K. N. Plataniotis, editors. *Color Image Processing: Methods and Applications,* CRC Press / Taylor & Francis, Boca Raton, FL, 2006.

Luk07 R. Lukac. Guest editorial: special issue on applied color image processing: editorials. *International Journal of Imaging Systems and Technology*, 17(3):103–104, 2007.

Mar82 D. Marr. *Vision: A Computational Investigation into the Human Representation and Processing of Visual Information*, W. H. Freeman, September 1982.

McA04 A. McAndrew. *An Introduction to Digital Image Processing with MATLAB*, Brooks/Cole, 2004.

MH80 D. Marr and E. Hildreth. Theory of edge detection. *Proceedings of the Royal Society of London B*, 207(1167):187–217, 1980.

Mia99 J. Miano. *Compressed Image File Formats*, Addison-Wesley Professional, 1999.

MPG85 H. Mussmann, P. Pirsch, and H. Garllet. Advances in picture coding. *Proceedings of the IEEE*, 73:523–548, April 1985.

MW93 H. R. Myler and A. R. Weeks. *Computer Imaging Recipes in C*, Prentice Hall, Upper Saddle River, NJ, 1993.

MXS08 M. Mirmehdi, X. H. Xie, and J. Suri. *Handbook of Texture Analysis*, World Scientific, 2008.

NG95 M. Nelson and J. L. Gaily. *The Data Compression Book*, 2nd edition, M&T Books, 1995.

Ost35 G. Osterberg. *Topography of the Layer of Rods and Cones in the Human Retina*, Levin & Munksgaard, 1935.

Ots79 N. Otsu. A threshold selection method from gray-level histograms. *IEEE Transactions on Systems, Man, and Cybernetics*, 9(1):62–66, 1979.

OWY83 A. V. Oppenheim, A. S. Willsky, and I. T. Young. *Signals and Systems*, Prentice Hall, 1983.

Pal99 S. E. Palmer. *Vision Science: Photons to Phenomenology*, MIT Press, 1999.

Pap62 A. Papoulis. *The Fourier Integral and Its Applications*, McGraw-Hill, 1962.

Par96 J. R. Parker. *Algorithms for Image Processing and Computer Vision*, Wiley, 1996.

Pav82 T. Pavlidis. *Algorithms for Graphics and Image Processing*, Springer, 1982.

PCK85 W. K. Pratt, T. J. Cooper, and I. Kabir. Pseudomedian filter. *Architectures and Algorithms for Digital Image Processing II, SPIE Proceedings*, 1985.

Pet06 M. Petrou. *Image Processing: Dealing with Texture*, Wiley, 2006.

PM92 W. B. Pennebaker and J. L. Mitchell. *JPEG Still Image Data Compression Standard*, Kluwer Academic Publishers, Norwell, MA, 1992.

Poy C. Poynton. Color FAQ: frequently asked questions about color. Web.

Poy96 C. Poynton. *A Technical Introduction to Digital Video*, Wiley, New York, 1996.

Poy03 C. Poynton. *Digital Video and HDTV Algorithms and Interfaces*, Morgan Kaufmann Publishers, San Francisco, 2003.

PP93 N. R. Pal and S. K. Pal. A review on image segmentation techniques. *Pattern Recognition*, 26(9):1277–1294, 1993.

Pra02 R. Pratap. *Getting Started with MATLAB*, Oxford University Press, New York, 2002.

Pra07 W. K. Pratt. *Digital Image Processing*, 4th edition, Wiley, New York, 2007.

Pre70 J. M. Prewitt. Object enhancement and extraction. *Picture Processing and Psychopictorics*, Academic Press, 1970.

PV90 I. Pitas and A. N. Venetsanopoulos. *Nonlinear Digital Filters: Principles and Applications*, Kluwer Academic Publishers, 1990.

Ric02 I. E. G. Richardson. *Video Codec Design*, Wiley, 2002.

Rob66 J. G. Robson. Spatial and temporal contrast sensitivity functions of the visual systems. *Journal of the Optical Society of America*, 56:1141–1142, 1966.

Rob77 G. S. Robinson. Edge detection by compass gradient masks. *Computer Graphics and Image Processing*, 6:492–501, 1977.

ROC97 R. A. Rensink, J. K. O'Regan, and J. J. Clark. The need for attention to perceive changes in scenes. *Psychological Science*, 8(5):368–373, 1997.

RP00 M. Robin and M. Poulin. *Digital Television Fundamentals: Design and Installation of Video and Audio Systems*, 2nd edition, McGraw-Hill, New York, 2000.

Sal06 D. Salomon. *Data Compression: The Complete Reference*, 4th edition, Springer, 2006.

Say05 K. Sayood. *Introduction to Data Compression*, 3rd edition, Morgan Kaufmann, 2005.

SB91 M. J. Swain and D. H. Ballard. Color indexing. *International Journal of Computer Vision*, 7(1):11–32, 1991.

SC82 J. Serra and N. A. C. Cressie. *Image Analysis and Mathematical Morphology*, Vol. 1, Academic Press, 1982.

Sch89 R. J. Schalkoff. *Digital Image Processing and Computer Vision*, Wiley, 1989.

SCM+05 D. Socek, D. Ćulibrk, O. Marques, H. Kalva, and B. Furht. A hybrid color-based foreground object detection method for automated marine surveillance. In *Advanced Concepts for Intelligent Vision Systems (ACIVS 2005)*, Antwerp, Belgium, September 2005, pp. 340–347.

Ser82 J. Serra. *Image Analysis and Mathematical Morphology*, Academic Press, London, 1982.

Ser88 J. Serra. *Image Analysis and Mathematical Morphology*, Vol. 2, Academic Press, London, 1988.

SHB08 M. Sonka, V. Hlavac, and R. Boyle. *Image Processing, Analysis, and Computer Vision*, 3rd edition, Thomson, Ontario, Canada, 2008.

SK99 C. Stiller and J. Konrad. Estimating motion in image sequences. *IEEE Signal Processing Magazine*, 16(4):70–91, 1999.

SKH08 T. Svoboda, J. Kybic, and V. Hlavac. *Image Processing, Analysis, and Machine Vision: A MATLAB Companion*, Thomson Learning, Toronto, Ontario, 2008.

SM00 J. Shi and J. Malik. Normalized cuts and image segmentation. *IEEE Transactions on Pattern Analysis and Machine Intelligence*, 22(8):888–905, 2000.

SOS00 M. Seul, L. O'Gorman, and M. J. Sammon. *Practical Algorithms for Image Analysis*, Cambridge University Press, Cambridge, UK, 2000.

SS96 J. P. Serra and P. Soille. *Mathematical Morphology and Its Applications to Image Processing*, Kluwer Academic Publishers, 1996.

SS01 G. Stockman and L. G. Shapiro. *Computer Vision*, Prentice Hall, Upper Saddle River, NJ, 2001.

SSVPB02 G. Sandini, J. Santos-Victor, T. Pajdla, and F. Berton. Omniviews: direct omnidirectional imaging based on a retina-like sensor. *Sensors, 2002. Proceedings of IEEE*, 1:27–30, 2002.

Sta00 J. Stark. Adaptive image contrast enhancement using generalizations of histogram equalization. *IEEE Transactions on Image Processing*, 9(5):889–896, 2000.

Tek95 A. M. Tekalp. *Digital Video Processing*, Prentice Hall, Upper Saddle River, NJ, 1995.

TM01 D. Taubman and M. Marcellin, editors. *JPEG2000: Image Compression Fundamentals, Standards and Practice*, Springer, 2001.

TSV05 H. J. Trussell, E. Saber, and M. Vrhel. Color image processing [basics and special issue overview]. *IEEE Signal Processing Magazine*, 22(1):14–22, 2005.

TT90 F. Tomita and S. Tsuji. *Computer Analysis of Visual Textures*, Kluwer, 1990.

TTP08 A. Trémeau, S. Tominaga, and K.N. Plataniotis. Color in image and video processing: Most recent trends and future research directions. *EURASIP Journal on Image and Video Processing*, vol. 2008 Article ID 581371, 26 pages, 2008. doi:10.1155/2008/581371

Uli87 R. Ulichney. *Digital Halftoning*, MIT Press, Cambridge, MA, 1987.

Umb05 S. E. Umbaugh. *Computer Imaging: Digital Image Analysis and Processing*, CRC Press, Boca Raton, FL, 2005.

UPH05 R. Unnikrishnan, C. Pantofaru, and M. Hebert. A measure for objective evaluation of image segmentation algorithms. In *Proceedings of the 2005 IEEE Conference on Computer Vision and Pattern Recognition (CVPRŠ05), Workshop on Empirical Evaluation Methods in Computer Vision*, 2005.

UPH07 R. Unnikrishnan, C. Pantofaru, and M. Hebert. Toward objective evaluation of image segmentation algorithms. *IEEE Transactions on Pattern Analysis and Machine Intelligence*, 29(6):929–944, 2007.

vdEV89 A. W. M. van den Enden and N.A.M. Verhoeckx. *Discrete-Time Signal Processing: An Introduction*, Prentice Hall, 1989.

vdHDdRT04 F. van der Heijden, R. P. W. Duin, D. de Ridder, and D. M. J. Tax. *Classification, Parameter Estimation and State Estimation: An Engineering Approach Using MATLAB*, Wiley, 2004.

VF08 A. Vedaldi and B. Fulkerson. VLFeat: an open and portable library of computer vision algorithms. http://www.vlfeat.org/, 2008.

Vin93 L. Vincent. Morphological grayscale reconstruction in image analysis: applications and efficient algorithms. *IEEE Transactions on Image Processing*, 2(2):176–201, 1993.

VNB67 F. L. Van Ness and M. A. Bouman. Spatiotemporal transfer in the human eye. *Journal of the Optical Society of America*, 57(9):1082–1088, 1967.

Wan08 J. Wang. *Graph Based Image Segmentation: A Modern Approach*, VDM Verlag Dr. Müller, 2008.

Wat94a J. Watkinson. The engineer's guide to motion compensation. Technical report, Snell & Wilcox, 1994.

Wat94b J. Watkinson. The engineer's guide to standards conversion. Technical report, Snell & Wilcox, 1994.

Wat00 J. Watkinson. *The Art of Digital Video*, 3rd edition, Focal Press, Oxford, 2000.

WB00 J. C. Whitaker and K. B. Benson, editors. *Standard Handbook of Video and Television Engineering*, 3rd edition, McGraw-Hill, New York, 2000.

Wol90 G. Wolberg. *Digital Image Warping*, Wiley-IEEE Computer Society Press, 1990.

Won00 P. W. Wong. *Image Quantization, Halftoning, and Printing*, Chapter 1.1, Morgan Kaufmann, 2000, pp. 657–667.

Woo06 J. W. Woods. *Multidimensional Signal, Image, and Video Processing and Coding*, Academic Press, San Diego, 2006.

WOZ02 Y. Wang, J. Ostermann, and Y.-Q. Zhang. *Video Processing and Communications*, Prentice Hall, Upper Saddle River, NJ, 2002.

WR04 S. Westland and C. Ripamonti. *Computational Colour Science Using MATLAB*, Wiley, 2004.

WS82 G. Wyszecki and W. S. Styles. *Color Science: Concepts and Methods*, Wiley, New York, 1982.

Yar59 A. L. Yarbus. *Eye Movements and Vision*, Plenum Press, New York, 1959.

YGvV I. T. Young, J. J. Gerbrands, and L. J. van Vliet. Fundamentals of image processing. Delft University of Technology 1998, 113 Pages, ISBN:9075691017.

ZF03 B. Zitová and J. Flusser. Image registration methods: a survey. *Image and Vision Computing*, 21(11):977–1000, 2003.

Zha96 Y. J. Zhang. A survey on evaluation methods for image segmentation. *Pattern Recognition*, 29(8):1335–1346, 1996.

Zha01 Y. J. Zhang. A review of recent evaluation methods for image segmentation. In *Sixth International Symposium on Signal Processing and Its Applications*, Vol. 1, 2001, pp. 148–151.

Zha06a Y. J. Zhang. *Advances in Image and Video Segmentation*, IRM Press, 2006.

Zha06b Y. J. Zhang. An overview of image and video segmentation in the last 40 years. In *Advances in Image and Video Segmentation*, Y. J. Zhang (Ed.), pp. 1–16, 2006.

INDEX

Practical Image and Video Processing Using MATLAB®. By Oge Marques.

Printed in the USA
J075384SCI091714 01S29053000000000118